Gulf Studies

Volume 15

This book series, indexed in Scopus, serves the growing scholarly interest in the dynamic, complex, and strategically important Arab Gulf region by offering an academic publication platform to scholars in the region and beyond. The series takes an interdisciplinary approach to documenting the changes occurring in Gulf societies. It aims to advance the understanding of Gulf societies and their interactions with the rest of the world from a non-Western perspective. The series explicitly gives scholars from the Arab Gulf region more voice and visibility in academic publications. It includes studies of the member states of the Gulf Cooperation Council (GCC), namely: Bahrain, Kuwait, Oman, Qatar, Saudi Arabia, and the United Arab Emirates, as well as Iran, Iraq, and Yemen. It accepts proposals for research monographs, edited volumes, and handbooks that explain how the Gulf societies' social, economic, and political landscapes have evolved over the past few decades. The series covers a wide range of topics within four broad themes, as follows:

(1) Social and Cultural Issues in the Gulf: Demography, migration, citizenship, gender, culture and identity, urbanization, education, new media, unemployment, youth, family, women empowerment, leadership, aging, human rights, sports, Islamic ethics, and Islam and society

(2) Politics and Security in the Gulf: Gulf defence and security, the GCC integration, Gulf politics and political systems, and the GCC's foreign policy

(3) Energy and Economics in the Gulf: The political economy of the Gulf, economic diversification, international investments, international trade, knowledge economy, energy security and geopolitics of energy, renewable energy development, environmental regulations, urban planning, and Islamic finance

(4) The GCC States and the World: The GCC and Iran, the GCC and Africa, the GCC and South Asia, the GCC and China, the GCC and East Asia, the GCC and Southeast Asia, and the GCC and Western powers.

Submission and Peer Review

All proposals will undergo peer review by the editorial board members. If accepted, the final manuscript will be peer reviewed internally by the editorial board as well as externally (single/double blind) by Springer ahead of acceptance and publication. Please submit your book proposal to the Series Editor, Md Mizanur Rahman: mizan@qu.edu.qa, and to Springer Publishing Editor, Alex Westcott Campbell: alexandra.campbell@springernature.com

Marwa Elnahass • Amr ElAlfy
Toka S. Mohamed
Editors

A Greenprint for Net Zero

The Future of Sustainability in the Middle East and North Africa Region (MENA)

Springer

Editors
Marwa Elnahass
King Fahd University of Petroleum and Minerals (KFUPM)
Dhahran, Saudi Arabia

Amr ElAlfy
University of Waterloo
Waterloo, ON, Canada

Toka S. Mohamed
Qatar University
Doha, Qatar

ISSN 2662-4494 ISSN 2662-4508 (electronic)
Gulf Studies
ISBN 978-981-92-1761-8 ISBN 978-981-92-1762-5 (eBook)
https://doi.org/10.1007/978-981-92-1762-5

This Springer imprint is published by the registered company Springer Nature Singapore Pte Ltd.
The registered company address is: 152 Beach Road, #21-01/04 Gateway East, Singapore 189721, Singapore

This series is dedicated to serving growing academic interests in the dynamic, complex, and strategically important Gulf region, offering a publication platform to scholars in the region and globally. The series takes an interdisciplinary approach to documenting the changes taking place in the Gulf societies, and examines the evolving relationship between the Gulf and other regions. The series advances non-Western perspectives in studying the Gulf societies, and their interactions with the rest of the world. The series includes work being done on the member states of the Gulf Cooperation Council (GCC), namely: Saudi Arabia, Oman, the United Arab Emirates, Qatar, Bahrain, and Kuwait, in addition to Iran, Iraq, and Yemen.

The series accepts proposals for monographs, edited volumes, and handbooks that provide an understanding of the Gulf societies' contemporary social, economic, and political landscapes. The series covers a wide range of topics within four broad themes, as follows:

Social and Cultural Issues in the Gulf:
Demography, migration, citizenship, gender, culture and identity, urbanization, education, new media, unemployment, youth, family, women empowerment, leadership, ageing, human rights, sports, Islamic ethics, and Islam and society

Politics and Security in the Gulf:
Gulf defence and security, the GCC integration, democracy and political reform, Gulf politics and political systems, the GCC's foreign policy

Energy and Economics in the Gulf:
The political economy of the Gulf, economic diversification, international investments, international trade, knowledge economy, energy security and geopolitics of energy, renewable energy development, environmental regulations, infrastructure and urban planning, Islamic finance

The GCC and the World:
The GCC and Iran, the GCC and Africa, the GCC and South Asia, the GCC and East Asia, the GCC and Southeast Asia, and the GCC and Western Powers

Contents

Editors and Contributors

About the Editors

Amr ElAlfy is an Assistant Professor at the School of Environment, Enterprise and Development (SEED), University of Waterloo, Canada. Dr. ElAlfy is a leading scholar in sustainability disclosure and ESG (Environmental, Social, and Governance) issues, with a particular focus on non-financial reporting practices. His research explores the significance of sustainability risks and opportunities through a double materiality lens, connecting financial and non-financial metrics to aid decision-makers in promoting sustainable growth and responsible investment strategies. Dr. ElAlfy is a Co-investigator for the municipal net-zero action research partnership (N-ZAP) project, reflecting his commitment to advancing sustainability at the local level. Additionally, he serves as the Associate Director of the Waterloo Institute for Sustainable Aeronautics (WISA), where he focuses on enhancing the ESG performance within the aeronautics sector.

Dr. ElAlfy brings extensive experience from his tenure in business development roles within multinational conglomerates such as Nestlé and 3M. During his time in the Middle East and Africa (MEA) region, he led marketing and Corporate Social Responsibility (CSR) initiatives, showcasing his ability to drive responsible business practices in diverse and complex markets. Furthermore, Dr. ElAlfy has been instrumental in the development and translation of the Global Reporting Initiative (GRI) Sustainability Universal and Sector Standards, reinforcing his status as a key contributor to the field of sustainability disclosure.

Marwa Elnahass is a Professor of Accounting and Finance at **King Fahd University of Petroleum and Minerals (KFUPM)**. She holds a **PhD from Lancaster University (UK)** and is an internationally recognised scholar in **sustainability and financial reporting, green finance, FinTech, and banking (conventional and Islamic)**. She has published **over 60 peer-reviewed articles** in leading AJG 3 and 4 ranked journals, alongside **edited books, book reviews, policy reports, technical papers, industry reports, and media articles**. Her research addresses critical global challenges in **net-zero transitions, climate governance,**

sustainable finance, and financial sector resilience, and has secured **over £7 million in external and internal research funding** from the UK, EU, Middle Eastern, and international funding bodies. Professor Elnahass serves as **Associate Editor** for several leading academic journals and is currently **Guest Editor of multiple special issues** in top-tier journals on **Net-Zero, Climate Risk, Sustainability Reporting, Cybersecurity and AI**. She also contributes as a **review panel member and referee** for major research councils, including **UKRI (AHRC, ESRC)** and **QDRI**. Her work has strong policy and impact orientation. She has contributed to **UK Government–led climate policy initiatives**, including research featured on **GOV.UK** (COP26_GOV.UK **and** COP27), and regularly participates in **policy roundtables and consultation reviews** with government authorities and international institutions. She collaborates extensively with **industry partners and stakeholders** across **Qatar, Saudi Arabia, Canada, the UK, and Southeast Asia** on funded impact projects related to **carbon accounting, biodiversity, water management, and sustainability governance** management, see, e.g., Publication and the THE_Conversation. Her contributions have been recognised by prominent organisations, including the **World Health Organization**, and she is a frequent **keynote speaker** at international policy forums, central banks, ministries, public authorities, professional bodies, and ranked international universities. Recent invitations include the **Qatar Central Bank**, **Qatar National Dialogue on Climate Change (2025)**, **Bank Indonesia Institute**, **Turkey Chamber of Commerce**, **European Accounting Association**, **Qatar Association for Certified Public Accountants**, and multiple keynote speeches at universities across Europe, the Middle East, and Asia. Professor Elnahass has been featured continuously since 2020 in the **Top 200 Most Influential Women in Islamic Business and Finance** (Cambridge International Financial Advisory). She serves on advisory boards for international networks including the UK innovation, **UK Universities Climate Network**, **FinTech UK**, and acts as a **visiting fellow and external examiner** for several **European and International universities**.

Toka S. Mohamed is a Postdoctoral Researcher at the College of Business and Economics at Qatar University. Her research focuses on financial inclusion, Islamic finance, and governance. She has published multiple articles in top-tier peer-reviewed academic journals and has been involved in several externally funded projects. Complementing her academic work is over a decade of industry experience in dynamic leadership roles where she has been instrumental in advancing sustainability initiatives within her organization and across its supply chain partners, driving the adoption of eco-friendly practices. Her work aims to bridge the gap between academia and industry, exploring innovative solutions that align with developmental and sustainability goals.

Contributors

Lucia Alessi European Commission, Joint Research Centre, Ispra, Italy

Silvio Andrae Independent Researcher, Berlin, Germany

Halima Baghad Department of Finance Research (LAREF), The Higher Institute of Commerce and Business Administration (Groupe ISCAE), Casablanca, Morocco

Annette Becker Deutsche Bundesbank, Frankfurt, Germany

Mohamed Anas Belidan Department of Finance Research (LAREF), The Higher Institute of Commerce and Business Administration (Groupe ISCAE), Casablanca, Morocco

Hrishikesh Desai Department of Accounting, Neil Griffin College of Business, Arkansas State University, Jonesboro, AR, USA

Tia Rebecca Driver School of Environment, Enterprise and Development, University of Waterloo, Waterloo, ON, Canada

Bahareh Esmaeili Kingston University, London, UK

Shakila Halim Finance and Economics Department, Business and Law Faculty, Manchester Metropolitan University, Manchester, UK

Md Tanvir Hamim Department of Banking and Insurance, University of Dhaka, Dhaka, Bangladesh

Tinotenda Douglas Hwara School of Finance and Professional Studies, Botswana Accountancy College, Gaborone, Botswana

Karim Iddouch Université Internationale de Casablanca, Casablanca, Morocco

Imen Khanchel University of Manouba Higher School of Commerce of Tunis, Manouba, Tunisia

LARIME LR11ES02, University of Tunis, Tunis, Tunisia

Naima Lassoued University of Manouba Higher School of Commerce of Tunis, Manouba, Tunisia

QuAnLab, LR24ES21, ESCT, University of Manouba, Manouba, Tunisia

Mélina London European Commission Joint Research Centre, Ispra, Italy

Ramshah Rashid Lone Faculty of Business and Law, Anglia Ruskin University, Chelmsford, UK

Irma Malafronte Roehampton University, London, UK

Anas Alaoui Mdaghri National School of Business and Management (ENCG), Ibnou Zohr University, Agadir, Morocco

Alain Neher School of Business, Charles Sturt University, Bathurst, NSW, Australia

Mohamed Nurullah Kingston University, London, UK

Ayodeji Michael Obadire School of Finance and Professional Studies, Botswana Accountancy College, Gaborone, Botswana

John Pereira Kingston University, London, UK

Gamini Premaratne UBD School of Business and Economics, Bandar Seri Begawan, Brunei

Albana Berisha Qehaja Faculty of Economics, University of Prishtina, Prishtina, Republic of Kosovo

Abdessamad Raghibi Ibnou Zohr University, Agadir, Morocco

Caterina Rho European Commission, Joint Research Centre, Ispra, Italy

Sana Saleem UBD School of Business and Economics, Bandar Seri Begawan, Brunei

Rasim Simsek Faculty of Economics and Administrative Sciences, Bilecik Seyh Edebali University, Bilecik, Turkey

Zahra Souguir University of Manouba Higher Institute of Accountancy and Business Administration, Manouba, Tunisia

QuAnLab, LR24ES21, ESCT, Manouba University, Manouba, Tunisia

Alfred Wong School of Business, Charles Sturt University, Bathurst, NSW, Australia

Chapter 1
Navigating Net-Zero Transitions in the Middle East and North Africa Region: Climate Imperatives, Economic Risks, and Strategic Opportunities

Marwa Elnahass, Amr ElAlfy, and Toka S. Mohamed

Abstract Global efforts to achieve net-zero greenhouse gas emissions are gaining momentum, yet progress varies widely across regions. While advanced economies have achieved notable strides in renewable deployment, sustainable finance, and climate governance, the Middle East and North Africa (MENA) region continues to face structural barriers that complicate its decarbonization trajectory. This edited book integrates global best practices and recent empirical evidence with region-specific insights to critically assess the conditions under which sustainability initiatives can yield transformative outcomes in the MENA region. Drawing on evidence from foreign direct investment, particularly investments associated with the Belt and Road Initiative spearheaded by China, some analyses highlight both opportunities and risks associated with external capital flows. Empirical chapters further examine financial market mechanisms, emphasizing how rising environmental costs, greenflation, and climate-related credit risks influence financial stability and investment dynamics. At the institutional level, board-level incentives are considered as emerging governance tools to embed sustainability into corporate decision-making, albeit with ethical and transparency concerns. The edited book further assesses the role of sustainable finance instruments, such as green bonds and blended finance, in mobilizing capital, emphasizing the need for robust environmental, social, and governance (ESG) disclosure and institutional capacity. Finally, it critiques the fragmented policy frameworks across MENA, contrasting them with binding climate legislation in advanced economies, and argues for integrated

M. Elnahass (✉)
King Fahd University of Petroleum and Minerals (KFUPM), Dhahran, Saudi Arabia
e-mail: Marwa.Elnahass@kfupm.edu.sa

A. ElAlfy
University of Waterloo, Waterloo, ON, Canada
e-mail: alelalfy@uwaterloo.ca

T. S. Mohamed
College of Business and Economics, Qatar University, Doha, Qatar
e-mail: tmohamed@qu.edu.qa

M. Elnahass et al. (eds.), *A Greenprint for Net Zero*, Gulf Studies 15,
https://doi.org/10.1007/978-981-92-1762-5_1

governance models to overcome siloed approaches. Collectively, the findings position MENA at a critical crossroads: the region must bridge governance gaps, align financial and policy frameworks, and adopt inclusive strategies to transform its renewable potential into credible net-zero pathways.

Keywords Net-zero transitions · Sustainability practices · Middle East and North Africa (MENA) · Climate governance · Sustainable finance

1.1 Sustainability Practices and Transitions to Net-Zero (Global Focus)

Global efforts to achieve net-zero greenhouse gas (GHG) emissions by mid-century have catalyzed ambitious policy reforms, technological advancements, and sustainability practices across sectors (IPCC, 2022). However, progress remains uneven, particularly in the Middle East and North Africa (MENA) region, where acute environmental challenges intersect with governance deficiencies, limited institutional capacity, and complex geopolitical considerations. This introduction draws on the best global practices and recent MENA-focused research to examine the conditions under which sustainability initiatives yield transformative outcomes. It explores several key dimensions of the net-zero transition in the MENA: energy systems and foreign investment, financial market dynamics, corporate governance incentives, sustainable finance and social factors, policy frameworks, and the need for an integrated regional strategy.

1.1.1 Energy Systems and Foreign Investment: Lessons from the Belt and Road Initiative

The transition to low-carbon energy systems remains foundational to achieving net-zero targets. Globally, investment in renewables has surged, exceeding $2 trillion in 2024 (BNEF, 2025). Yet in the MENA, foreign-led investment, particularly under China's Belt and Road Initiative (BRI), plays an important role in financing clean energy infrastructure. The experience of North African countries such as Egypt, Morocco, and Algeria demonstrates that BRI investments have supported significant expansion of solar and wind capacity. At the same time, these projects highlight risks related to technological dependence on external providers, limited local capacity-building, and weak regulatory oversight (Zhang et al., 2025; Christy et al., 2024). Applying a foreign direct investment (FDI)-Supported Sustainable Transition (FDI-SST) model suggests that political stability, clear legal frameworks, and harmonized sustainability standards are critical preconditions for aligning foreign capital with long-term climate objectives in the region.

1.1.2 Financial Market Mechanisms: Greenflation and Credit Risk Dynamics

Another underexplored aspect of the net-zero transition is the interplay between environmental costs and financial stability. Moreover, the COVID-19 outbreak has severely affected the financial stability of financial institutions located in MENA countries, as evidenced by declines in multiple financial performance indicators, both accounting-based and market-based measures, as well as heightened financial stability risks, including default, liquidity, and asset risks. Similar patterns were consistently observed across other regions, such as Europe and Asia (Elnahass et al., 2021). Rising environmental costs have been found to significantly elevate corporate default risk across 11 MENA countries, particularly for firms in resource-intensive sectors (ElAlfy et al., 2025). Similarly, greenflation, which refers to the inflation driven by supply-side constraints in critical minerals needed for clean energy, has paradoxically improved stock market liquidity in parts of the Gulf Cooperation Council (GCC) due to investors reallocating capital toward sustainable assets (Hartzmark & Shue, 2023). These financial dynamics highlight the need for policy mechanisms that manage climate-related risks without destabilizing capital markets. In particular, climate-risk disclosure frameworks must be both material (focused on relevant, substantive exposures) and forward-looking in order to guide investment decisions effectively (Farooq et al., 2021; Marie et al., 2024). In this regard, emerging global baselines such as the International Financial Reporting Sustainability Disclosure Standards (IFRS S1 and IFRS S2), which overtake the earlier recommendations of the Task Force on Climate-related Financial Disclosures (TCFD), also the supervisory principles developed by the Network for Greening the Financial System (NGFS) are increasingly shaping regulatory expectations, with several MENA central banks and securities regulators beginning to align listing rules, scenario-analysis practices, and reporting templates with these frameworks (ElAlfy et al., 2025).

1.1.3 Institutional Incentives and Board-Level Sustainability Integration

The role of corporate boards in advancing sustainability objectives is gaining empirical support. In the MENA context, board compensation has emerged as a novel mechanism for driving environmental performance. Evidence from comparative studies of "green" versus "brown" firms suggests that higher board remuneration in sustainability-focused companies correlates with reductions in energy use, highlighting how performance-linked incentives can help institutionalize environmental stewardship (Elnahass et al., 2022). However, this approach also raises ethical concerns regarding equity, transparency, and alignment of governance practices, where boards must balance incentivizing sustainability with ensuring high financial reporting and accountable decision-making (Krane, 2022; Elnahass et al., 2024).

1.1.4 Scaling Sustainable Finance and Addressing Social Dimensions

Efforts to scale up sustainable finance, through instruments such as green bonds, catastrophe bonds, and blended finance, are crucial for de-risking climate projects in MENA's diverse political and economic contexts (World Bank, 2021). Yet disparities in ESG data quality, weak regulatory enforcement, and nascent financial ecosystems continue to hinder the full deployment of these financing tools. An illustrative example outside the region, the case of Botswana, offers comparative insight: in an upper-middle-income country with stable governance, robust disclosure standards and financial innovation have positioned it as a regional exemplar in ESG reporting. This suggests that strengthening institutional quality and transparency in MENA will be key to attracting sustainable investment and ensuring that financial initiatives also address social inclusion and development needs. At the same time, recent work by the International Monetary Fund and the World Bank on climate finance mobilization in the MENA, together with Organisation for Economic Co-Operation and Development and United Nations Environment Programme Finance Initiative analyses of green finance in emerging markets, underlines that scaling sustainable finance in MENA will require not only new instruments but also coherent standards, robust project pipelines, and carefully designed public–private risk-sharing arrangements. By situating MENA's sustainable finance trajectory in this wider global debate, the chapters in this volume highlight where regional experiences converge with, or diverge from, broader emerging-economy patterns and what this implies for future policy design.

1.1.5 Policy Frameworks and Multi-level Governance

Effective net-zero transitions depend on robust, coherent policy frameworks. Lessons from countries with legally binding climate legislation (e.g., the UK's Climate Change Act) indicate that enshrining emission targets in law and establishing independent oversight bodies enhances the credibility and continuity of climate policy (Meadowcroft & Rosenbloom, 2023). By contrast, in the MENA region, most climate policies remain aspirational, often lacking enforceable implementation plans or cross-government coordination. For example, while Morocco, Qatar, and Saudi Arabia have each launched "Vision 2030" national strategies aligned with the Sustainable Development Goals (SDGs) (e.g., NDS3, 2024), and national climate change action plans have been introduced (MME, 2021), actual climate outcomes will depend on the depth of regulatory design and the extent of multi-stakeholder engagement embedded in these initiatives (Palaschuk et al., 2022; ElAlfy et al., 2021).

The challenge is compounded by governance fragmentation, as ministries of energy, environment, and finance often operate in silos that undermine policy coherence. One approach to remedy this is to establish national climate councils or similar multi-agency coordination bodies to unite efforts under a single net-zero framework. Such unified governance structures could align fiscal policy, industrial strategy, and regulatory oversight, thereby significantly enhancing policy effectiveness and mobilizing investment. In a MENA context, a functional interministerial or cross-governmental body would typically take the form of a high-level national climate and sustainable finance council, chaired by the head of government or minister of finance and bringing together ministers responsible for energy, environment, planning, industry, and social affairs, alongside the central bank governor and financial-market regulators. Its mandate would be to translate long-term net-zero pledges and Nationally Determined Contributions into medium-term expenditure frameworks, align national budgets, sovereign wealth fund strategies, and state-owned enterprise investment plans with climate objectives, and resolve cross-sector trade-offs (e.g., between energy pricing, social protection, and industrial policy). To be effective, such a council requires a clear legal basis, whether through a framework climate law, a cabinet decree, or amendments to public financial management and budget laws, which specifies its coordinating authority, reporting obligations to parliament and the public, and the mechanisms through which it can require line ministries, subnational authorities, and state-owned enterprises to align their policies and investment decisions with national climate goals. While some MENA countries have begun to experiment with supreme councils for climate change or sustainable development, these bodies often remain advisory; strengthening their analytical capacity and regulatory authority is therefore a critical step toward operationalizing ambitious national visions and net-zero strategies.

1.1.6 Toward an Integrated Regional Strategy

In summary, the MENA region stands at a critical crossroads. It possesses substantial renewable energy potential and is attracting growing investor interest in sustainable projects, yet systemic gaps in governance, disclosure practices, and institutional capacity continue to impede the effective implementation of net-zero strategies. The chapters reviewed herein collectively affirm that successful transitions require not only technological and financial inputs but also deeper structural reforms, widespread behavioral change, and inclusive governance. Integrating global best practices with region-specific pathways, underpinned by rigorous academic analysis, will be essential for closing the emissions gap and ensuring a just, sustainable future for the MENA region. Developing an integrated regional strategy, therefore, entails going beyond shared declarations of intent to identify concrete institutional, legal, and monitoring mechanisms that can be implemented at the regional level. One promising avenue would be the establishment of a common disclosure and ESG platform for MENA stock markets, potentially hosted by a consortium of regional exchanges or development banks, that harmonizes taxonomies, minimum disclosure requirements, and climate-risk metrics in line with IFRS S1/S2, the legacy TCFD recommendations, and NGFS guidance, while allowing for national specificities. A second example is the creation of a GCC- or MENA-wide climate finance working group, convened by institutions such as the Arab Monetary Fund, the Islamic Development Bank, or regional development banks, tasked with coordinating green project pipelines, standardizing blended-finance structures, and sharing good practices on de-risking tools for cross-border sustainable infrastructure. A third, complementary, mechanism could be a regional peer-review process under the auspices of the League of Arab States or a dedicated MENA climate forum, through which countries periodically benchmark their climate legislation, fossil-fuel subsidy reforms, and just-transition policies, and issue joint recommendations on enforcement and capacity-building needs.

1.2 MENA at Crossroads: Climate Action and the Pursuit of Net-Zero Emissions

The global urgency to address climate change and achieve net-zero emissions by mid-century continues to shape the core of contemporary climate policy and research. Realizing net-zero CO_2 emissions by 2050, aligned with limiting global temperature rise to 1.5 °C, necessitates deep systemic transformations in energy, industry, agriculture, transport, and finance (Sachs et al., 2019; IPCC, 2022). While over 145 countries have pledged net-zero targets, collectively accounting for more than 70% of global emissions, the pathway to operationalizing these ambitions is fraught with technological, institutional, and economic complexities (Köhler et al., 2019; Markard, 2018). These challenges and opportunities are most apparent in the MENA region, a geography marked by both acute climate vulnerabilities and abundant clean energy potential.

1.2.1 Structural Vulnerabilities and Climate Risk

Climate change in the MENA region is no longer an abstract environmental issue; it constitutes an existential development challenge. The region is experiencing warming at nearly twice the global average, with extreme temperatures, chronic water scarcity, and accelerating desertification threatening food security, health outcomes, and economic productivity (Al-Saidi & Elagib, 2017; Logan & Al-Hababi, 2023). Per capita CO_2 emissions also exceed the world average (see Fig. 1.1). According to

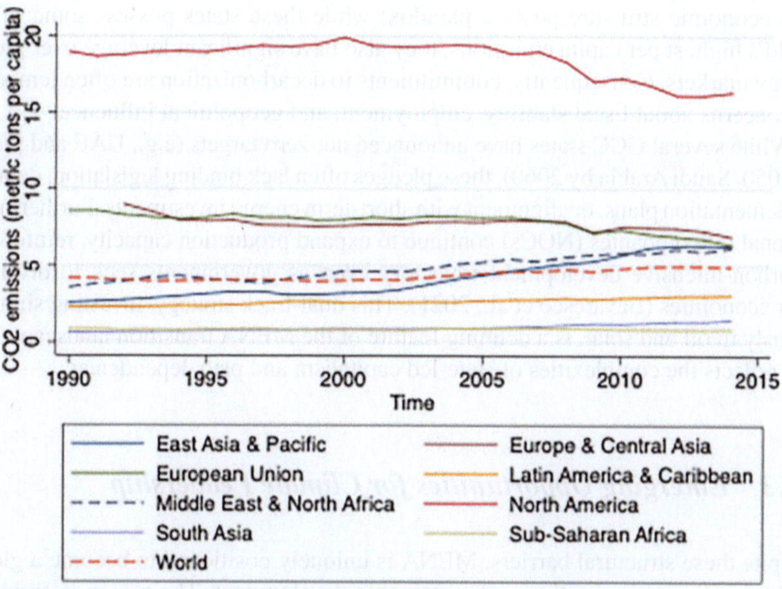

Fig. 1.1 CO_2 emissions per geographic region (Climate Watch, 2023)

the World Bank, if no adaptive measures are taken, climate-induced disruptions could reduce regional gross domestic product by up to 14% by 2050 (Hallegatte et al., 2016). This projection encompasses reduced agricultural yields, stress on water and energy systems, increased frequency of heat-related illnesses, and heightened risks of socio-political instability. In fragile and conflict-affected states within MENA, climate change also acts as a threat multiplier, intensifying displacement, resource conflicts, and governance deficits (Béné et al., 2016; Buhaug & Urdal, 2013; McGuirk & Nunn, 2025).

Moreover, urbanization across MENA is rapidly expanding, often in climate-vulnerable coastal zones and arid hinterlands. With over 60% of the region's population residing in cities, the exposure of urban infrastructure to climate extremes places additional strain on public services and energy systems (UN-Habitat, 2020). The increasing demand for air conditioning and desalinated water, driven by extreme heat, paradoxically raises emissions unless these systems are powered by renewable sources, highlighting the interdependence of mitigation and adaptation strategies.

1.2.2 The Fossil Fuel Paradox and Economic Dependencies

A central obstacle to climate action in MENA is the region's entrenched dependence on fossil fuel exports, which underpin both national economies and political institutions. Hydrocarbon revenues account for more than 70% of government income in countries such as Saudi Arabia, Kuwait, and Iraq (Atalay et al., 2016).

This economic structure poses a paradox: while these states possess some of the world's highest per capita emissions, they also have significant leverage over global energy markets. Consequently, commitments to decarbonization are often tempered by concerns about fiscal stability, employment, and geopolitical influence.

While several GCC states have announced net-zero targets (e.g., UAE and Oman by 2050, Saudi Arabia by 2060), these pledges often lack binding legislation, detailed implementation plans, or alignment with short-term energy investments. Furthermore, national oil companies (NOCs) continue to expand production capacity, reinforcing a carbon-intensive development trajectory even as governments seek to diversify their economies (Bavaresco et al., 2021). This dual-track strategy, investing simultaneously in oil and solar, is a defining feature of the MENA transition landscape, one that reflects the complexities of state-led capitalism and path dependency.

1.2.3 Emerging Opportunities for Climate Leadership

Despite these structural barriers, MENA is uniquely positioned to become a global leader in climate innovation and sustainable development. The region is endowed with vast solar irradiance, wind potential, and ample desert land, factors that make it ideal for large-scale renewable energy deployment (IRENA, 2020). Countries such as Morocco, Jordan, Egypt, and the UAE have already initiated ambitious renewable energy projects, with solar parks such as Noor Ouarzazate and Mohammed bin Rashid Al Maktoum Solar Park serving as global benchmarks.

Green hydrogen is also emerging as a strategic opportunity. Projects in Saudi Arabia (NEOM) and Oman aim to leverage renewable electricity to produce hydrogen for domestic use and export, positioning the region as a potential hub for low-carbon fuels (Logan & Al-Hababi, 2023). These ventures are bolstered by sovereign wealth funds and public–private partnerships that have begun to channel investment toward green infrastructure, circular economy initiatives, and clean technology incubation.

At the urban scale, cities such as Doha, Dubai, and Abu Dhabi are experimenting with smart-city models, integrating energy-efficient building codes, public transport modernization, and digital monitoring of resource use. These initiatives not only reduce emissions but also signal a broader reimagining of urban resilience in response to climate stressors.

1.2.4 Governance, Institutional Reform, and Global Engagement

Effective climate action in MENA, however, will depend on more than just technological deployment; it requires robust institutional reform and coherent governance structures. Many MENA countries face capacity constraints, fragmented regulatory environments, and weak enforcement mechanisms, particularly in water and energy management (World Bank, 2021). A transition to net-zero necessitates integrated national strategies that coordinate across ministries, align public and private incentives, and embed climate objectives into fiscal and trade policy.

Notably, the region's engagement in global climate diplomacy has deepened. Hosting COP27 and COP28 reflects a shift in regional status from passive participant to active agenda-setter. These forums offer MENA states the platform to shape narratives around climate finance, loss and damage, and just transition frameworks, issues of paramount importance to both oil producers and climate-vulnerable countries in the region.

International cooperation and climate finance are also critical. Access to concessional finance, technology transfer, and capacity-building mechanisms under the Paris Agreement can help de-risk clean energy investments and support adaptation projects. Initiatives such as the African Adaptation Acceleration Program and Green Climate Fund have already begun targeting MENA priorities, but greater coordination is needed to ensure these that resources are integrated into national development plans.

1.2.5 *Pathways Forward: Inclusive, Adaptive, and Strategic Transitions*

A just and effective transition in MENA must navigate a delicate balancing act: reducing emissions while maintaining economic and political stability. This demands targeted investments in human capital, green jobs, and institutional transparency. Vocational training for youth in renewable energy sectors, reform of fossil fuel subsidies, and participatory governance frameworks can contribute to building public support and enhancing resilience (Geels et al., 2017). Cross-sectoral policy frameworks, such as integrated water-energy-food nexus approaches, can provide a systems-level view of sustainability planning. Moreover, national visions (e.g., Saudi Vision 2030, Qatar Vision 2030, and UAE Green Agenda) offer useful strategic platforms to mainstream climate objectives across sectors. These need to be strengthened with measurable indicators, inclusive stakeholder engagement, and independent accountability mechanisms.

Ultimately, the MENA region stands at a critical crossroads. It can either remain locked in a high-emissions trajectory shaped by fossil fuel dependence or leverage its unique advantages to pioneer net-zero transitions that are economically viable, socially inclusive, and environmentally sustainable. The urgency of the climate crisis demands that the latter path be pursued with ambition, coordination, and resolve.

1.3 Book Contribution and Significance

This edited book assesses how these global transformations manifest in the MENA region, which is characterized by both acute climate vulnerabilities and strategic economic leverage. Climate change impacts in MENA are pronounced: the region is warming at approximately twice the global average and faces escalating risks from water scarcity, desertification, and extreme heat (Al-Saidi & Elagib, 2017; Logan & Al-Hababi, 2023). These environmental pressures intersect with socio-economic fragility, intensifying inequalities and threatening long-term development outcomes.

This book makes a timely and significant contribution to the evolving discourse on climate change mitigation by addressing a persistent gap in our understanding of net-zero strategies and sustainability practices, particularly in the context of the MENA region. Despite global momentum following the Paris Climate Accords, there remains limited integrative knowledge about how different sectors, across both emerging and advanced economies, are operationalizing net-zero ambitions in relation to financial reporting, governance, socio-economic development, and environmental resilience.

The book adopts a comprehensive, interdisciplinary approach to explore the complexity of these transitions, presenting 13 novel chapters authored by leading international scholars and practitioners. The contributors represent a diverse array of academic, policy, and industry backgrounds, bringing expertise across climate change policy, sustainability science, green finance, innovation, risk governance, and accountability. This rich intellectual diversity ensures that the book captures the multi-scalar dynamics and region-specific challenges of implementing net-zero strategies, with particular emphasis on under-researched geographies such as the MENA region.

A key strength of this book lies in its exploration of cross-cutting themes that have received limited attention in the existing literature. It investigates how emerging data-driven innovations, including artificial intelligence, machine learning, and advanced analytics, can facilitate the achievement of net-zero targets by optimizing resource allocation, detecting emissions hotspots, and enhancing organizational efficiency. The book also critically engages with the escalating risks posed by climate change to businesses and economies, highlighting the importance of proactive climate risk management to ensure resilience across global value chains.

In parallel, the edited book sheds light on the role of climate governance frameworks in aligning institutional behavior with sustainability goals. It highlights the necessity of embedding climate considerations into corporate and public-sector decision-making processes and of ensuring that governance mechanisms foster transparency, accountability, and strategic coherence. Importantly, the book addresses the increasingly pressing concern of greenwashing, advocating for transparent disclosures, verifiable sustainability claims, and stronger assurance systems to uphold the credibility of environmental performance reporting.

Green finance is positioned as a central enabler of climate action. By mobilizing capital for low-carbon technologies, renewable energy projects, and sustainable infrastructure, green finance can catalyze transformative change, especially in resource-abundant regions like the MENA, where targeted investments and policy alignment are essential for unlocking latent potential. The book highlights the need for integrated financial-environmental approaches that consider carbon and biodiversity accounting alongside economic indicators of resilience and performance.

The novelty of this book lies in its interdisciplinary breadth and its regionally grounded analysis of net-zero trajectories. While there is a growing body of work focused on climate transitions in developed contexts, there is comparatively less literature examining the institutional, economic, and ecological complexities in emerging economies. This book helps fill that gap by offering a pioneering exploration of the synergies and trade-offs between economic development, environmental sustainability, and social equity in transitioning to a low-carbon future, with the MENA region as a focal lens.

By providing new research insights, practical frameworks, and state-of-the-art examples, this book offers an essential reference for academics, practitioners, students, and policymakers. It aims to support the alignment of financial and non-financial systems with the global climate action agenda and the SDGs. Its accessible yet rigorous content ensures that the book serves both experts seeking advanced analysis and newcomers requiring a foundational understanding of net-zero transitions across both emerging and developed economies.

1.4 Contributed Chapters

The closing section of this introduction offers a chapter-by-chapter overview of the book, drawing attention to the key elements that, as editors, we found particularly noteworthy.

In Chap. 2, Bahareh Esmaeili, John Pereira, Irma Malafronte, and Mohamed Nurullah investigate the relationship between carbon performance, climate change commitments, and corporate credit risk in the MENA region, analyzing data from publicly listed firms across 14 countries between 2013 and 2023. The authors adopt a mixed-method approach, using market-based measures such as Credit Default Swap spreads and distance-to-default (Merton, 1974), along with accounting-based indicators like the Z-score, to assess credit risk. Carbon performance is measured through firm-level Scope 1, 2, and 3 emissions, while climate commitments are evaluated based on disclosures of environmental governance, targets, and actions, in line with the Carbon Disclosure Project (CDP) (2023) and GHG Protocol standards. The study finds significant variation in emissions, credit risk, and climate engagement across sectors and countries. Despite rising disclosure, many firms, especially in high-emission economies, remain reliant on fossil fuels, exposing a gap between policy ambition and actual transition. The analysis draws on stakeholder, legitimacy, and resource-based theories to explain how credible climate strategies enhance firm reputation and creditworthiness. This chapter makes a significant contribution by offering a region-specific analysis of climate-finance linkages in MENA and emphasizing the need for stronger disclosure standards and enforceable, science-aligned policies to support a sustainable and financially resilient corporate sector.

Chapter 3, by **Rasim Simsek, Ramshah Rashid Lone, Tanvir Hamim, and Shakila Halim,** provides novel empirical evidence on the relationship between climate-related natural disasters and credit creation, with a focus on MENA countries. Utilizing a fixed-effects panel regression model and data from 222 countries (including 22 MENA nations) spanning 1974–2022, the study finds that natural disasters significantly stimulate credit creation in MENA countries, in contrast to statistically insignificant effects in non-MENA regions. Further disaggregation reveals that high-income MENA countries experience a stronger positive relationship, while low-income non-MENA countries face credit contraction post-disaster. This chapter makes a significant scholarly contribution by being among the first to examine the impact of natural disasters on credit creation, specifically within the MENA context, an area highly vulnerable to climate change but largely under-researched. The findings challenge conventional assumptions, which suggest that disasters typically impede credit activity, by demonstrating a unique regional dynamic in which disasters trigger increased credit demand and supply. Robustness is ensured through the use of the generalized method of moments and propensity score matching techniques. The results hold important policy implications for financial regulators and institutions in the MENA region, emphasizing the need for disaster-resilient financial systems and targeted risk management strategies to sustain credit flow in the face of escalating climate risks.

Chapter 4, by **Lucia Alessi, Annette Becker, Mélina London, and Caterina Rho,** offers a critical and timely examination of sustainable finance in the MENA region, emphasizing its importance in addressing the region's acute vulnerability to climate change and persistent socio-economic and governance challenges. It assesses the current development of sustainable finance across MENA, situating it within the broader macro-financial context and benchmarking it against other global regions. The chapter highlights key environmental risks, such as extreme heat, water scarcity, and rising sea levels, and their implications for health, food security, and displacement. It also explores how sustainable finance can address pressing social issues, including poverty, youth unemployment, and gender inequality, while identifying the obstacles posed by weak governance and institutional inefficiencies. The chapter's importance lies in its dual contribution to scholarship and policy. First, it provides one of the most comprehensive region-wide, data-informed assessments of sustainable finance development in MENA, filling a notable gap in a literature often limited by fragmented case studies and poor ESG data coverage. Second, it offers a forward-looking analysis by evaluating global financing tools (e.g., blended finance and catastrophe bonds) and adapting them to the MENA context. By connecting global best practices to regional realities, this chapter delivers a strategic roadmap for scaling sustainable finance and advancing inclusive, climate-resilient development in MENA.

In Chap. 5, Sana Saleem and Gamini Premaratne examine the dynamic relationship between carbon emissions, natural resource dependence, and green finance across 23 MENA countries from 2000 to 2022. Using robust econometric techniques, the study demonstrates that green finance plays a significant role in reducing carbon emissions, particularly in urbanized contexts where green investments are more efficiently absorbed. In contrast, over-reliance on natural resource rents exacerbates environmental degradation, highlighting a structural challenge for resource-dependent economies in the region. The importance of this chapter lies in its multifaceted contribution. Theoretically, it extends the Environmental Kuznets Curve framework by incorporating financial development and natural resource rents, providing a nuanced understanding of how economic and financial structures impact environmental outcomes. Empirically, the study leverages updated datasets from IRENA, the World Bank, and the Global Carbon Atlas to deliver region-specific insights on investment patterns and renewable energy adoption. This chapter is among the first to empirically link green finance effectiveness to levels of urbanization in MENA, offering practical policy recommendations for integrating green finance into national climate strategies. It underscores the urgent need for MENA countries to transition away from fossil fuels and adopt sustainable development models supported by targeted green investment and financial innovation, positioning the chapter as a vital resource for climate and economic policymakers.

Chapter 6, by **Abdessamad Raghibi, Anas Alaoui Mdaghri, and Karim Iddouch,** examines the impact of *greenflation*, which refers to the inflationary pressures resulting from environmental policies and the green transition, on stock market liquidity across 12 MENA countries between November 2019 and December 2023. Using a high-frequency panel dataset of 823 publicly listed firms (over 325,000 firm-day observations), the study measures liquidity through market tightness (transaction costs) and depth (trading volume). Findings reveal that greenflation enhances market liquidity, driven by increased investor demand for sustainable assets and reallocation behavior influenced by rising prices of critical minerals. However, the effect is uneven across the region: Middle Eastern markets benefit significantly, while North African markets show a more muted response, reflecting structural and regulatory differences. The contribution of this chapter is both empirical and conceptual. It fills a gap in the literature by shifting the greenflation discourse from macroeconomic price effects to financial market dynamics, particularly in emerging economies. By focusing on the supply-side dimensions of the energy transition, especially the role of critical minerals, this chapter offers a novel angle on green investment behavior. Its findings emphasize the importance of resilient and inclusive capital markets in facilitating sustainable transitions. For policymakers, it offers clear guidance on enhancing financial resilience, incentivizing private green investment, and fostering regional coordination to bridge regulatory and developmental gaps across MENA.

Chapter 7, authored by **Ayodeji Michael Obadire and Tinotenda Douglas Hwara,** explores the adoption of sustainability practices among listed firms in Botswana during 2022 and 2023, addressing a critical research gap in the study of sustainability in emerging African economies. Through a quantitative content analysis of ESG disclosures, the study assesses how various sectors engage with sustainability frameworks. Results reveal a dominance of social sustainability reporting, accounting for over 50% of disclosures, followed by governance, while environmental practices remain significantly underrepresented. Sector-wise, the mining industry leads in sustainability disclosures, with agriculture also demonstrating growing engagement. This chapter offers rare empirical insights into corporate sustainability reporting in Botswana, a country with socio-economic parallels to North African nations such as Libya and Morocco. This makes it a relevant case for comparison and regional learning. Second, it offers practical and policy-level implications: firms can use these findings to enhance their environmental and governance practices, while policymakers can leverage evidence to design targeted sustainability regulations. The study's methodological approach and focus on a stable, data-accessible context like Botswana make it a valuable model for other upper-middle-income MENA countries seeking to advance sustainability integration and align corporate behavior with national and regional development goals.

In Chap. 8, Albana Berisha Qehaja explores the strategic integration of sustainability practices by businesses in the MENA region, emphasizing how sustainability can drive long-term performance and competitive advantage. It offers a comprehensive analysis of current sustainability efforts, strategic implementation approaches, and their practical impacts on business outcomes. Drawing on real-world case studies, including Masdar City (UAE), Saudi Aramco (Saudi Arabia), and the Office Chérifien des Phosphates Group (Morocco), the chapter illustrates how firms are embedding sustainability into their core operations, enhancing innovation, operational efficiency, and stakeholder engagement. The chapter's contribution lies in its combination of theoretical insight and practical relevance tailored specifically to the MENA region. It fills a critical gap in the literature by providing an integrated framework for understanding how leadership, organizational culture, and stakeholder involvement influence sustainability strategy execution. It also addresses pressing questions related to challenges in implementation, offering actionable solutions for firms aiming to align with global sustainability goals such as net zero emissions. Importantly, the chapter underscores the transformative role of sustainability in enhancing corporate competitiveness and advancing regional development goals. By bridging academic theory and business practice, it offers a roadmap for firms and policymakers seeking to mainstream sustainability as a strategic priority for resilience, growth, and regional prosperity in the MENA context.

Chapter 9, by **Mohamed Anas Belidan and Halima Baghad,** offers a comprehensive empirical assessment of the evolution and determinants of sustainability disclosure in the MENA region, with a focused case study on Morocco. Using a Sustainability Disclosure Index (SDI) constructed from 118 indicators, the study tracks trends across all industries from 2014 to 2023. Employing both descriptive analysis and econometric methods, specifically random-effects and feasible generalized least squares panel regressions, the research identifies key factors influencing sustainability reporting. The findings reveal a gradual improvement in SDI scores over time, with firm size, age, and listing on the MASI ESG index positively associated with greater disclosure. However, despite Morocco's mandatory ESG reporting since 2019, overall disclosure quality remains moderate, underscoring the need for enhanced reporting frameworks and practices. The chapter's significance lies in its originality and contextual relevance. It represents the first large-scale, industry-wide study in Morocco to empirically evaluate the drivers of sustainability disclosure, filling a notable gap in the regional literature. By analyzing reporting behavior in an environment where disclosure is legally mandated but poorly standardized, the study highlights the institutional and regulatory complexities firms face. The insights are crucial for academics, policymakers, and practitioners aiming to promote transparency, align business practices with the SDGs, and advance sustainability reporting in emerging MENA economies.

In Chap. 10, Hrishikesh Desai presents a detailed analysis of Saudi Aramco's sustainability accounting practices, offering critical insights into how the world's largest integrated oil and gas company addresses the complexities of environmental responsibility amid the global energy transition. Using a mixed-method approach—combining qualitative corporate report analysis and quantitative Natural Language Processing techniques—the study evaluates Aramco's carbon accounting frameworks, ESG governance, and the adoption of advanced technologies such as artificial intelligence and blockchain in sustainability reporting. Comparative insights are drawn from benchmarked assessments of the Abu Dhabi National Oil Company (ADNOC) and ExxonMobil, and alignment with global reporting standards (Global Reporting Initiative (GRI), TCFD, and Sustainability Accounting Standards Board (SASB) is also examined. This chapter's key contribution lies in bridging the gap between sustainability accounting theory and practice within a high-impact industry. It provides one of the first data-driven, technology-enhanced analyses of an NOC's ESG strategies in the MENA region, emphasizing the integration of digital innovation in carbon and ESG reporting. The findings reveal Aramco's progressive governance structure and technological investments but also highlight persistent gaps, notably in Scope 3 emissions disclosure. The chapter is significant for policymakers shaping regulatory frameworks, investors evaluating energy firms' long-term sustainability, and researchers exploring ESG accountability in fossil-fuel-dependent economies. It advances the understanding of how large oil corporations can evolve toward credible, transparent sustainability leadership.

Chapter 11 by **Tia Rebecca Driver** offers a qualitative investigation into how firms in the GCC region disclose climate change risks, focusing on the adoption and implementation of ESG taxonomies. While global research on climate risk disclosure, particularly within the European context, has centered on quantitative analyses and financial implications, this study shifts the focus to the depth and quality of disclosures in the MENA region. Utilizing directed qualitative content analysis and thematic analysis of public corporate reports, the study assesses how GCC firms report on physical and transition risks related to climate change. The findings reveal that climate risk disclosures in the GCC are largely superficial and fail to meaningfully address the material nature of these risks. Although ESG frameworks are being adopted, most firms do not fully integrate climate risks into their strategic communications, exposing a significant transparency gap. Applying legitimacy theory, the chapter argues that current disclosure practices appear more symbolic than substantive, serving to maintain external legitimacy rather than reflect internal climate risk preparedness. This chapter makes an important contribution by addressing a critical gap in MENA-focused climate disclosure literature. It offers practical insights for policymakers, regulators, and corporate leaders to enhance ESG reporting quality, aligning MENA firms with emerging global standards and better preparing them for climate-related financial risks.

Chapter 12 by **Naima Lassoued, Imen Khanchel, and Zahra Souguir** examines the impact of environmental costs on corporate default risk across 326 firms in 11 MENA countries over the period 2010–2022. Employing panel data analysis, the study finds that high environmental costs, particularly those related to natural resources, air pollution, GHG emissions, and water consumption, are significantly associated with increased default risk, indicating both financial strain and reputational damage for firms. Waste management costs, however, do not show a significant effect. These results are interpreted through the lenses of institutional theory and the resource-based view, suggesting that firms face challenges in balancing legitimacy with resource sufficiency in a region undergoing growing environmental and regulatory pressures. This chapter contributes to the literature in three important ways. First, it narrows the focus to the environmental pillar of CSR, offering a more targeted and quantifiable analysis using the dollar value of environmental costs rather than broad CSR indices. Second, it extends existing research by directly linking environmental expenditure to corporate default risk, a less explored but highly relevant financial outcome. Third, it addresses a critical gap in the MENA region, where empirical studies on the financial consequences of environmental responsibility remain scarce. The findings highlight the importance of integrating environmental accountability into corporate risk management strategies across the MENA region.

Chapter 13 by **Silvio Andrae** provides a comprehensive examination of the role of China's BRI in shaping sustainability outcomes in the MENA region, with a focus on five North African countries, namely Morocco, Algeria, Tunisia, Libya, and Egypt. While the BRI initially prioritized fossil fuel investments, it has increasingly shifted toward funding clean energy projects, facilitating the integration of renewable energy into local energy mixes. The study investigates how BRI investments promote or hinder sustainable development, employing a mixed-method approach that combines the FDI-SST theoretical model with empirical analysis of secondary data and comparative case studies. The chapter identifies several critical factors influencing the sustainability of BRI projects: the political and regulatory context of recipient countries, the economic interests at stake, and the structure of Chinese investments. While the BRI offers economic opportunities and accelerates renewable energy deployment, challenges remain, such as dependency on Chinese technology and financing, limited adherence to international environmental standards, and socio-economic mismatches. This chapter's contribution is significant. It fills a gap in the literature on foreign direct investment and sustainability in the MENA region, offering nuanced insights into how global geopolitical and financial strategies intersect local development goals. It provides policymakers and stakeholders with grounded recommendations for fostering more inclusive, environmentally aligned BRI investment frameworks in the region.

Finally, in Chap. 14, Alfred Wong and Alain Neher present a rigorous empirical investigation that explores the role of Board of Directors (BoD) compensation in influencing the environmental performance of firms in the MENA region. This study differentiates between "Green" and "Brown" firms based on their carbon emissions and examines how BoD compensation correlates with environmental outcomes. The study finds that Green firms, which offer higher BoD compensation, tend to exhibit lower environmental damages; specifically, a 10% increase in BoD pay correlates with a 3.3–6.3% reduction in energy use relative to Brown firms. These results underscore the strategic use of compensation incentives to promote environmental responsibility, suggesting that sustainable development efforts are more embedded in the governance structures of Green firms. This chapter makes a significant contribution to the literature on corporate governance and sustainability in emerging economies. It extends prior work by providing empirical evidence from the MENA region, a context often underrepresented in global ESG studies. Importantly, the research highlights how compensation structures can serve as mechanisms for environmental accountability, offering practical insights for policymakers, regulators, and corporate leaders. It also calls for ethical considerations and collaborative innovation among public and private stakeholders to support broader climate goals, including the achievement of the SDGs.

References

Al-Saidi, M., & Elagib, N. A. (2017). Towards understanding the integrative approach of the water–energy–food nexus. *Science of the Total Environment, 574*, 1131–1139. https://doi.org/10.1016/j.scitotenv.2016.09.046

Atalay, Y., Biermann, F., & Kalfagianni, A. (2016). Adoption of renewable energy technologies in oil-rich countries: Explaining policy variation in the Gulf Cooperation Council states. *Renewable Energy, 85*, 206–214.

Bavaresco, M. V., Ghisi, E., D'Oca, S., & Pisello, A. L. (2021). Triggering occupant behaviour for energy sustainability: Exploring subjective and comfort-related drivers in Brazilian offices. *Energy Research & Social Science, 74*, 101959.

Béné, C., Headey, D., Haddad, L., & von Grebmer, K. (2016). Is resilience a useful concept in the context of food security and nutrition programmes? *Food Security, 8*, 123–138.

BNEF. (2025). *Energy Transition Investment Trends: Tracking global investment in the low-carbon transition.*

Buhaug, H., & Urdal, H. (2013). An urbanization bomb? Population growth and social disorder in cities. *Global Environmental Change, 23*(1), 1–10.

CDP. (2023). *Scoring Introduction 2023*. https://cdn.cdp.net/cdp-production/cms/guidance_docs/pdfs/000/000/233/original/Scoring-Introduction.pdf?1639144388

Christy, A., Elnahass, M., Amezaga, J., Browne, A., & Heidrich, O. (2024). A dynamic framework to align company climate reporting and action with global climate targets. *Business Strategy and the Environment, 33*(4), 3103–3128.

Climate Watch. (2023). *Historical GHG Emissions*. World Resources Institute. https://www.climatewatchdata.org/ghg-emissions

ElAlfy, A., Elgharbawy, A., Driver, T. R., & Ibrahim, A. J. (2025). Sustainability disclosure in the Gulf Cooperation Council (GCC) countries: opportunities and challenges. *Green Finance, 7*(1), 40–82.

Elalfy, A., Weber, O., & Geobey, S. (2021). The Sustainable Development Goals (SDGs): A rising tide lifts all boats? Global reporting implications in a post SDGs world. *Journal of Applied Accounting Research, 22*(3), 557–575.

Elnahass, M., Marie, M., & Elgammal, M. (2022). Terrorist attacks and bank financial stability: Evidence from MENA economies. *Review of quantitative finance and accounting, 59*(1), 383–427.

Elnahass, M., Tahir, M., Abdul Rahman Ahmed, N., & Salama, A. (2024). Internal governance mechanisms and information value of banks' earnings. *Journal of Accounting in Emerging Economies, 14*(2), 450–488.

Elnahass, M., Trinh, V. Q., & Li, T. (2021). Global banking stability in the shadow of Covid-19 outbreak. *Journal of International Financial Markets, Institutions and Money, 72*, 101322.

Farooq, F., Ahmed, W., Akbar, A., Aslam, F., & Alyousef, R. (2021). Predictive modeling for sustainable high-performance concrete from industrial wastes: A comparison and optimization of models using ensemble learners. *Journal of Cleaner Production, 292*, 126032.

Geels, F. W., Sovacool, B. K., Schwanen, T., & Sorrell, S. (2017). Sociotechnical transitions for deep decarbonization. *Science, 357*(6357), 1242–1244.

Hallegatte, S., Bangalore, M., Bonzanigo, L., Fay, M., Kane, T., Narloch, U., & Vogt-Schilb, A. (2016). *Shock waves: Managing the impacts of climate change on poverty*. World Bank Publications.

Hartzmark, S. M., & Shue, K. (2023). *Counterproductive impact investing: The impact elasticity of brown and green firms*. Available at SSRN, 4359282.

IPCC. (2022). *Climate Change 2022: Mitigation of climate change*. Cambridge University Press.

IRENA. (2020). *Power sector planning in Arab countries: Incorporating variable renewables*. International Renewable Energy Agency.

Köhler, J., Geels, F. W., Kern, F., et al. (2019). An agenda for sustainability transitions research: State of the art and future directions. *Environmental Innovation and Societal Transitions, 31*, 1–32.

Krane, J. (2022). Climate ambition and the petrostates: Toward a political economy of net-zero. *Energy Research & Social Science, 91*, 102739.

Logan, N., & Al-Hababi, R. (2023). *Sustainable Qatar: Social, political and environmental perspectives*. Springer.

Marie, M., Qi, B., Elgammal, M., & Elnahass, M. (2024). A more sustainable future: Can politically connected CEOs spur the nexus between ESG performance and firm financial performance? *Journal of International Financial Markets, Institutions and Money, 96*, 102056.

Markard, J. (2018). The next phase of the energy transition and its implications for research and policy. *Nature Energy, 3*, 628–633.

McGuirk, E. F., & Nunn, N. (2025). Transhumant pastoralism, climate change, and conflict in Africa. *Review of Economic Studies, 92*(1), 404–441.

Meadowcroft, J., & Rosenbloom, D. (2023). Governing the net-zero transition: Strategy, policy, and politics. *Proceedings of the National Academy of Sciences, 120*(47), e2207727120.

Merton, R. C. (1974). On the pricing of corporate debt: The risk structure of interest rates. *The Journal of Finance, 29*(2), 449–470.

MME (Ministry of Municipality and Environment). (2021). *Qatar National Climate Change Action Plan 2030*.

NDS3. (2024). *Third Qatar National Development Strategy 2024-2030*. https://www.npc.qa/en/planning/nds3/Documents/QNDS3_EN.pdf

Palaschuk, N., El Alfy, A., & Wilson, J. (2022). Positioning corporations as sustainability leaders: A roadmap to align strategic corporate social responsibility with the SDGs. In *Future advancements for CSR and the Sustainable Development Goals in a post-COVID-19 world* (pp. 24–49). IGI Global.

Sachs, J. D., Woo, W. T., Yoshino, N., & Taghizadeh-Hesary, F. (Eds.). (2019). *Handbook of green finance: Energy security and sustainable development*. Springer Singapore.

UN-Habitat. (2020). *World Cities Report 2020: The value of sustainable urbanization*. UN-Habitat.

World Bank. (2021). *Water in the Arab world: Management perspectives and innovations*. World Bank.

Zhang, L., Fu, D., Wu, C., Liu, Y., & Xi, Y. (2025). How does the belt and road initiative improve China's product export potential: A global value chain perspective. *China & World Economy, 33*(1), 195–222.

Chapter 2
Carbon Performance and Climate Change Commitment: Implications for Firm Credit Risk in the MENA Region

Bahareh Esmaeili, John Pereira, Irma Malafronte, and Mohamed Nurullah

Abstract This chapter investigates how carbon performance and climate change commitments affect corporate credit risk in the Middle East and North Africa (MENA) region, using a sample of publicly listed firms across 14 MENA countries over the period 2013–2023. We use market-based and accounting-based measures to assess credit risk; we measure carbon performance using firm-level Scope 1, 2, and 3 emissions data, adjusted for firm size and industry norms; we estimate climate commitment using firm disclosure data on targets, governance, and sustainability actions. The findings reveal wide variations in emissions, carbon performance, and credit risk across sectors and countries. This chapter contributes to the literature by offering a novel empirical analysis specific to the under-researched MENA region; we integrate stakeholder, legitimacy, and resource-based theories to explain how firms' climate actions influence creditworthiness. These findings underscore the importance of improving disclosure practices and implementing sector-specific policies to support a more sustainable financial ecosystem.

Keywords Credit risk · Carbon emission · Carbon performance · Climate commitment · MENA region

---

B. Esmaeili (✉) · J. Pereira · M. Nurullah
Kingston University, London, UK
e-mail: Bahareh.Esmaeili@kingston.ac.uk; J.Pereira@kingston.ac.uk; M.Nurullah@kingston.ac.uk

I. Malafronte
Roehampton University, London, UK
e-mail: Irma.Malafronte@roehampton.ac.uk

M. Elnahass et al. (eds.), *A Greenprint for Net Zero*, Gulf Studies 15,
https://doi.org/10.1007/978-981-92-1762-5_2

2.1 Introduction

The Middle East and North Africa (MENA) region is witnessing a dynamic shift in how it approaches climate change, with regulations on greenhouse gas (GHG) emissions and climate pledges becoming more robust. Saudi Arabia's launch of the Saudi Green Initiative is a testament to the region's increasing commitment to sustainability (United Nations, 2023). Furthermore, the growth of the voluntary carbon market is gaining momentum, empowering private companies to pursue net-zero targets and potentially shaping future regulatory frameworks (Middle East Institute, 2023). Countries in the Gulf Cooperation Council, such as Saudi Arabia, the United Arab Emirates (UAE), Kuwait, and Oman, are making significant strides toward a more climate-ambitious future, pledging to reach net-zero emissions and channeling substantial investments into renewable energy. For example, the UAE's goal to triple its renewable energy capacity by 2030 and its development of the world's largest single-site solar park reflect a deep commitment to transitioning to a low-carbon economy (International Renewable Energy Agency, 2023). MENA's active participation in international climate initiatives, such as the UN Climate Change Conferences of Parties (COP27 and COP28) and MENA Climate Week, highlights a concerted regional effort to enhance climate resilience, reduce emissions, and achieve the Paris Agreement's (COP21) (United Nations, 2015) goals. These initiatives foster collaboration among stakeholders and aim to bolster both regional and global climate action (United Nations Climate Change, 2022).

However, despite these promising developments, MENA countries still grapple with challenges such as limited funding, knowledge gaps, and weak regulatory frameworks that impede the decarbonization of small- and medium-sized enterprises (SMEs). Yet, opportunities abound, especially in harnessing the region's rich renewable energy resources and positioning itself as a green hydrogen hub, leveraging its strategic location between Europe and Asia (World Economic Forum, 2024). Although the MENA region has made significant strides in climate change initiatives, there remains a critical gap in understanding how firms in this area are responding to these challenges.

This chapter seeks to fill this gap by examining the evolution of carbon performance, climate change commitment, and credit risk among publicly listed firms in 14 MENA countries over the last decade (2013–2023). We aim to provide insights into the evolution of carbon management practices, corporate climate commitment, and credit risk over time, across sectors, and countries. Despite growing disclosure and commitment efforts, evidence suggests that firms, particularly in high-emission economies, still rely heavily on non-renewable energy sources (Chen et al., 2023). This underscores the persistent gap between policy ambition and actual transitions, reinforcing the need for science-aligned, enforceable corporate strategies in carbon-intensive regions. This underscores the persistent gap between policy ambitions and actual energy transitions, reinforcing the need for robust, science-aligned corporate strategies to achieve meaningful carbon reductions.

This study employs a diverse sample of firms from Bahrain, Egypt, Iraq, Jordan, Kuwait, Lebanon, Morocco, Oman, Palestine, Qatar, Saudi Arabia, Syria, Tunisia, and the UAE. We employ a comprehensive analytical approach, combining the market-based Distance-to-Default (DTD) (Merton, 1974), Credit Default Swap (CDS) spreads, and the Z-score measures (Altman et al., 2017) to assess the likelihood of default and evolutions of corporate credit risk over the last decade. Following the Carbon Disclosure Project (CDP), we assess emissions in line with the GHG Protocol (World Resources Institute (WRI) and World Business Council for Sustainable Development (WBCSD), 2020), which encompasses direct (Scope 1), indirect from purchased energy (Scope 2), and all other indirect emissions (Scope 3) throughout the supply chain. Furthermore, we examine how firms demonstrate their commitment to climate action through efforts such as environmental stewardship, waste management, and energy-saving initiatives (Albitar et al., 2023; Erdoğan et al., 2020). Using the CDP's consolidated metric, we assess the extent of firms' commitment to mitigating and adapting to climate change (CDP, 2023).

The remainder of this chapter is structured as follows. Section 2.2 outlines the theoretical framework and reviews the relevant literature. Section 2.3 details the methodological approach and describes the variables employed in the analysis. Section 2.4 presents and discusses the empirical findings. Section 2.5 offers concluding remarks, and Sect. 2.6 highlights the policy and practical implications derived from the chapter.

2.2 Literature Review

2.2.1 Theoretical Framework

Disclosing carbon emissions helps firms manage perceptions and meet calls for transparency, as explained by stakeholder theory (Freeman, 1984) and legitimacy theory (Suchman, 1995). Both frameworks interpret disclosure as a response to external pressures (Hahn et al., 2015), though stakeholder theory focuses on the influence of specific interest groups (Bendig et al., 2024; Cotter & Najah, 2012) and their role in shaping firm strategies to ensure long-term success (Roberts, 1992), while legitimacy theory considers broader societal expectations (Cotter & Najah, 2012). Stakeholder theory sees disclosure as a response to climate-related demands, shaped by stakeholder power (Deegan, 2002), and contributing to the evolution of ethical reporting (Parmar et al., 2010). Legitimacy theory highlights the role of institutional and societal pressures, such as regulations, media, or public awareness, in driving environmental disclosures (Chelli et al., 2014; Liu et al., 2023; O'Donovan, 2002; Patten, 2002; Vershinina et al., 2020). The resource-based view (RBV) (Hart, 1995) suggests that financially strong firms are better positioned to absorb the costs associated with environmental strategies (Amit & Schoemaker, 1993; Qiu et al., 2016). Building on this perspective, it is also argued that proactively embedding climate change mitigation into core business policies can become a valuable internal capability. This, in turn, may enhance firm performance by enabling more effective carbon disclosure to stakeholders, thereby reinforcing the firm's competitive advantage (Alsaifi et al., 2020). Together, these theories suggest increasing pressure on firms to disclose carbon data and adopt low-carbon strategies, supported by policies like mandatory standards, green finance, and sector-specific measures. In the MENA region, where carbon disclosure remains limited, these frameworks offer a useful lens to understand how firms' responses to stakeholder expectations, institutional legitimacy, and resource capacity, through carbon emissions and climate change commitment, impact credit risk. Integrating these perspectives helps explain why some firms lead in climate action, while others lag, reflecting varying degrees of stakeholder engagement, legitimacy threats, and resource constraints.

2.2.2 Carbon Performance and Credit Risk

The risks associated with climate change have attracted increasing attention in both academic research and policymaking circles (Aldy & Gianfrate, 2019; Bento et al., 2020). The Paris Agreement, which was formalized at COP 21 in 2015, is widely acknowledged as a critical milestone in global climate policy, as it prompted nations to commit to reducing GHG emissions (Kinley, 2017), Since COP 21's adoption, numerous countries have developed strategies to limit emissions and incentivize private investment in sustainable technologies, particularly through the implementation of GHG emission pricing mechanisms (Bento et al., 2020). Consequently, assets with high carbon emissions face an increasing risk of regulatory sanctions, which could alter investor views on their future profitability, sustainability, and creditworthiness (Koten, 2018). Abrupt policy shifts may provoke a rapid sell-off of such assets, potentially resulting in significant financial instability (Battiston et al., 2017).

Carbon emissions refer to the GHG emissions linked to a company's operations or products (Harangozo & Szigeti, 2017; Wiedmann, 2009), providing a measure of a company's impact on global warming. Companies participating in the CDP disclose their emissions in accordance with the GHG Protocol, a framework developed by the WRI and WBCSD (2020). The GHG Protocol divides emissions into three categories or scopes: Scope 1 encompasses direct GHG emissions from sources owned or controlled by the company, such as emissions from power generation, industrial processes, or fuel consumption. Scope 2 includes indirect emissions resulting from the purchase of energy consumed by the company. Finally, Scope 3 covers all other indirect emissions across a company's value chain, including those generated by upstream and downstream activities, such as the transportation of materials or the processing of sold products.

In the MENA region, several studies have examined the key drivers of carbon emissions and highlighted the role of energy sources in shaping environmental outcomes. Ben Cheikh and Ben Zaied (2021) identify a threshold effect in the

CO_2–income relationship, showing that income's impact on emissions depends on energy-related variables and emphasizing the need for a shift to low-carbon energy. Albaker et al. (2023) find that renewable energy consistently helps reduce emissions, while higher levels of energy intensity, green innovation, and GDP are associated with increased emissions across different levels of carbon output. These results highlight the need for stronger, well-targeted policies to promote energy efficiency and sustainable innovation in line with Sustainable Development Goals (SDGs) 7 (clean energy), 9 (industry, innovation, and infrastructure), and 13 (climate action). Similarly, Omri and Saidi (2022) demonstrate that renewable energy enhances environmental quality, whereas non-renewable sources and industrial activities contribute to environmental degradation, reinforcing the need for targeted, sector-specific low-carbon policies to support a green transition.

Firms with higher carbon emissions are more likely to encounter increasingly stringent climate regulations compared to those with lower emissions (Hassan et al., 2022; Ji et al., 2023). Research indicates that lenders incorporate a company's carbon profile into their lending decisions, considering it a key factor (Eliwa et al., 2021). A significant negative correlation has been observed between carbon emission levels and the cost of bank loans; companies that disclose emissions and maintain lower carbon output tend to secure more favorable loan terms (Kleimeier & Viehs, 2018), whereas firms with substantial fossil fuel reserves may face penalties (de Greiff et al., 2018). Furthermore, studies show that companies not participating in the CDP survey are often subject to higher debt costs associated with carbon risks (Jung et al., 2018). However, as Christy et al. (2024) highlight, there is a gap between high-level climate guidance and its practical implementation at the company level. This result underscores the need for clearer, standardized carbon accounting frameworks, especially the GHG Protocol and Science-Based Targets initiative, to support effective climate action and investment decisions. Additionally, Jiang et al. (2019) highlight significant sectoral differences in carbon risk exposure, showing that sectors with high strength-out and closeness-out roles (e.g., downstream producers) contribute most to emissions, while sectors with high strength-in roles tend to have lower emissions. These insights suggest that tailored mitigation strategies must consider the structural position of sectors within the global carbon emission network to effectively reduce emissions.

Historically, firms were able to mitigate carbon risk by externalizing the costs associated with GHG emissions. However, the regulatory landscape has undergone a significant transformation with the implementation of carbon-related policies, including cap-and-trade systems and carbon allowance frameworks. These developments have compelled companies to internalize the costs of carbon emissions, thereby positioning carbon risk as a critical strategic issue (Clarkson et al., 2015). Recent empirical evidence by Owolabi et al. (2024) indicates that firms with elevated carbon risk, quantified through Scope 1 and 2 emissions relative to revenue, face higher borrowing costs. Their analysis, which focuses on G7 economies in the post-Paris Agreement era, underscores the growing financial implications of carbon exposure.

Recent research has employed various credit risk measures to assess the influence of carbon emissions on a company's creditworthiness. For instance, Capasso et al. (2020), Kabir et al. (2021), and Zhang and Zhao (2022) found an inverse correlation between the credit risk and both the volume and intensity of a firm's emissions. Also, Costola and Vozian (2025) indicate that higher Scope 1 GHG emissions are associated with increased credit risk across all CDS tenors, particularly following the Paris Agreement. In contrast, Nguyen et al. (2023) note that this effect is significant only for climate risk disclosures in annual filings, rather than for emissions themselves. Additionally, a significant reduction in default risk has been observed in U.S. firms with higher levels of green innovation; this effect appears to be more substantial among firms characterized by greater carbon intensity in their operations (Safiullah et al., 2024).

Building on stakeholder theory, legitimacy theory, and the RBV, we hypothesize that higher carbon emissions increase firm credit risk. From a stakeholder and legitimacy perspective, firms with greater emissions are subject to intensified scrutiny and pressure from stakeholders, regulatory bodies, and the public, which can damage their perceived legitimacy and financial standing. Additionally, consistent with the RBV, carbon-intensive firms may lack the strategic capabilities needed to navigate environmental regulation effectively. As highlighted by Nordhaus (2017), carbon emissions introduce volatility into future cash flows due to the uncertainty surrounding evolving climate policies—further exacerbating the perceived risk among creditors and investors.

2.2.3 Climate Change Commitment and Credit Risk

In today's business landscape, marked by limited natural resources and rising socio-environmental pressures, companies are increasingly compelled to embed more robust climate change initiatives within their strategic decision-making frameworks (Albitar et al., 2020; Cho & Patten, 2007). To demonstrate a stronger commitment to addressing climate change, organizations are implementing innovative strategies focused on preventing environmental harm, reducing negative impacts, and fostering sustainable practices such as waste recycling and energy efficiency. These initiatives underscore a deeper commitment to tackling environmental issues (Albitar et al., 2023; Erdoğan et al., 2020).

A strong commitment to climate action offers companies several tangible advantages, such as an enhanced corporate image, reduced capital costs, and greater competitiveness in the market (Bae et al., 2018; Sarumpaet et al., 2017). Financial markets are increasingly attentive to environmental considerations, rewarding companies that demonstrate environmental responsibility with lower after-tax weighted average costs of capital. This pattern is evident across industries, regardless of their emission levels, ranging from high-emission to low-emission sectors (Mariani et al., 2021).

Companies that actively engage in climate-related initiatives tend to allocate a larger portion of their debt issuance to long-term financial instruments, irrespective of their carbon risk exposure. However, there is still a lack of clear evidence establishing a direct link between a firm's carbon risk exposure and its debt financing decisions. Theoretical models suggest that firms dedicated to climate action benefit from reputational gains, improved creditworthiness, and lowered agency costs and information asymmetries, making access to long-term debt markets more attainable (Lemma et al., 2021). Furthermore, Birindelli et al. (2022) highlight that when banks place medium to high importance on climate issues, a firm's commitment to these matters can reduce the risks associated with bank loans.

Although previous studies document a positive relationship between initiatives to reduce GHG emissions and improved financial performance (e.g., Boiral et al., 2012), the empirical evidence regarding the impact of such commitments on a company's financial results and market valuation remains relatively scarce. Bui et al. (2020) reveal that climate governance enhances the alignment between reported and actual carbon performance while discouraging exaggerated disclosures. Kabir et al. (2021) identify a negative correlation between a firm's environmental commitments—such as public declarations and emissions reduction targets—and its credit risk. Liêu et al. (2024) show that stronger climate governance reduces carbon emissions, primarily through increased eco-innovation, with the effect being more pronounced in firms with strong ESG practices and robust across various firm- and country-level contexts.

Drawing on stakeholder theory, legitimacy theory, and the RBV, we hypothesize that a firm's strong commitment to climate change mitigation is associated with lower credit risk. From a stakeholder perspective, proactive climate strategies enhance transparency and trust, fulfilling stakeholder expectations and contributing to long-term value creation (Lemma et al., 2021). Legitimacy theory further posits that such actions help maintain or restore organizational legitimacy in the face of growing societal and regulatory pressures. Additionally, RBV suggests that firms investing in climate-resilient capabilities may achieve competitive advantages that improve operational stability and financial performance, thereby enhancing creditworthiness.

2.3 Method and Variables

2.3.1 Sample

This study uses a sample of 1519 listed firms from the 14 MENA countries, based on the availability of relevant data. Table 2.1 presents the distribution of the sample across Global Industry Classification Standards (GICS) in Panel A and Country in Panel B. Financials with 30.41%, Industrials with 13.89%, Real Estate with 11.19%, and Materials with 11.06% are sectors with the highest representation in the sample. As per country classification, 22.38% of firms are in Saudi Arabia, 15.80% in Egypt, and 11.32% in the UAE.

Table 2.1 Sample distribution

	N	%
Panel A. Industry		
Communication services	41	2.70
Consumer discretionary	157	10.34
Consumer staples	139	9.15
Energy	39	2.57
Financials	462	30.41
Health care	69	4.54
Industrials	211	13.89
Information technology	31	2.04
Materials	168	11.06
Real estate	170	11.19
Utilities	32	2.11
Total	1519	100.00
Panel B. Country		
Bahrain	33	2.17
Egypt	240	15.80
Iraq	70	4.61
Jordan	156	10.27
Kuwait	141	9.28
Lebanon	10	0.66
Morocco	73	4.81
Oman	109	7.18
Palestine	41	2.70
Qatar	56	3.69
Saudi Arabia	340	22.38
Sudan	2	0.13
Tunisia	76	5.00
UAE	172	11.32
Total	1519	100.00

Notes: This table reports descriptive statistics on the sample distribution. The sample is composed of 1519 companies belonging to 11 GICS industries (Panel A) and based in 14 countries (Panel B)

2.3.2 *Variables*

The following variables are adapted and constructed following Esmaeili et al. (2024).

2.3.2.1 Carbon Emissions and Measures of Carbon Performance

We follow previous literature (Misani & Pogutz, 2015) and measure carbon performance as in Eq. (2.1). As carbon emissions are highly dependent on the industry of operations, this value is divided by the firm's GICS sector median emission for the year. To avoid skewness, a natural logarithmic transformation is applied. Finally, the result is multiplied by (–1) and is rescaled as a positive scale of values, such that higher values of this measure represent better performance:

$$CP_{i,t} = \left[ln\left(\frac{Firm's\,GHG\,emission_{i,t} / Sale_{i,t}}{Firm's\,sector\;median\,GHG\,emission_{j,t}} \right) \right] * (-1) \tag{2.1}$$

where i represents the firm in year t, and j refers to the GICS for the year of observation.

Carbon performance (*CP*) is determined based on a company's total carbon emissions (CP_1), encompassing all three scopes. By including Scope 1 and 2, the firm's major operational changes, investments, and innovations that influence GHG emissions are captured. However, Scope 3 emissions represent corporate efforts to reduce carbon impacts in specific industries (such as automotive emissions in the automobile sector) and thus are essential in assessing the company's overall carbon footprint. This metric is further adjusted to account for the industry in which the company operates, recognizing that certain sectors, like energy, naturally generate significantly higher carbon emissions. In addition to the overall CP score, we also calculated carbon performance separately for Scope 1 (*CP_1*), Scope 2 (*CP_2*), and Scope 3 (*CP-3*).

2.3.2.2 Measures of Climate Change Commitment

Drawing from Albitar et al. (2023), a company's commitment to climate change is assessed through the development of an index, calculated as a composite score, which aggregates four key components of climate commitment. The corporate climate change commitment (*CC_RF*) index ranges from 0 to 4, with higher scores indicating stronger corporate commitment to climate action. It reflects whether a company supports SDG 13, whether it recognizes climate change as a source of commercial risk or opportunity, whether it discloses Scope 3 CO_2 emissions, and whether it has set an emissions reduction target. Each component is assigned a value of 1 if the criterion is met, and 0 otherwise, based on data sourced from Refinitiv.

As an alternative approach, and consistent with prior studies by Birindelli et al. (2022) and Lemma et al. (2021), we employ the CDP's consolidated metric to assess a company's commitment to climate change mitigation, adaptation, and transparency. This metric evaluates not only the quality of a firm's carbon disclosures but also broader aspects, including the company's consideration of the environmental and social impacts of its operations, tangible evidence of its commitment to strong environmental management, and its adherence to industry best practices in addressing climate change challenges (CDP, 2023). CDP provides disclosure scores to firms based on the quality and comprehensiveness of their climate-related reporting, ranging from low transparency to leadership in climate action. These scores reflect levels of disclosure, awareness, management, and strategic action on climate change. Bloomberg converts these letter grades into numerical values, which are used in this study as proxies for corporate commitment to carbon management (*CC_BB*).

2.3.2.3 Measures of Credit Risk

In line with prior research (Capasso et al., 2020; Kabir et al., 2021; Nguyen et al., 2023; Zhang & Zhao, 2022), we employ a combination of market-based measures, Merton's Distance-to-Default and CDS spreads, and accounting-based measure, the Z-score, to provide a comprehensive analysis and enhance the reliability of our findings. The selection of both market and accounting-based credit risk measures is supported by precedence in literature, as each captures a distinct dimension of creditworthiness. DTD, derived from Merton's option-theoretic model, offers a forward-looking, theoretically grounded indicator of default risk based on a firm's asset volatility and capital structure, and has been widely adopted in studies examining firm-level risk under uncertainty. CDS spreads reflect market perceptions of default probability in real time, making them highly responsive to external shocks such as regulatory changes or climate-related financial risks. On the other hand, the Z-score serves as a robust accounting-based measure that incorporates liquidity, profitability, and leverage indicators, making it particularly useful for evaluating firms with limited market data or high information asymmetry.

Distance-to-Default To evaluate a company's likelihood of default over time, we adopt the well-established DTD framework introduced by Merton (1974). This structural model conceptualizes a firm's equity as a call option on the company's asset value, with the strike price corresponding to the book value of its liabilities. While asset value and volatility are unobservable, they can be inferred from equity value, equity volatility, and capital structure. The *DTD* represents the number of standard deviations separating the firm's asset value from the default point and is derived using standard option-pricing principles. Specifically, the market value of assets and asset volatility are estimated through the Black-Scholes-Merton model, as outlined by Gilchrist and Zakrajsek (2012) and Vassalou and Xing (2004). Assuming a 1-year horizon (T=1), asset volatility is computed iteratively from daily equity data until convergence, and the resulting estimates are used to calculate DTD. This method provides a forward-looking, market-based measure of credit risk.

CDS Spreads Over the past two decades, CDS spreads have gained popularity as a widely used proxy for credit risk (Anand et al., 2023). A higher spread, expressed in basis points, reflects a greater cost of insuring the debt of the underlying company, implying increased credit risk. Since CDS contracts are typically more liquid than certain bonds, their spreads are more readily accessible and less expensive to obtain. As a result, CDS spreads are often favored over bond yields (Longstaff et al., 2011; Rodríguez et al., 2019). Additionally, while bond spreads can be affected by various factors, CDS spreads are widely regarded in the literature as a more accurate measure of default risk (Ang & Longstaff, 2013). In line with previous research (Pereira et al., 2018), we use 5-year constant maturity CDS spreads, as they represent the most liquid instruments available.

Accounting-Based Linear Discriminant Model—Altman Z-score The original Z-score, introduced by Altman (1968), provided an accounting-based approach to estimate a firm's credit risk. However, this measure was initially restricted to manufacturing companies. Therefore, we employ a modified version of the Z-score, as developed by Altman et al. (2017). This variation is applicable to both manufacturing and non-manufacturing firms, as well as to both private and public entities. As noted by MacKie-Mason (1990) and Agrawal and Matsa (2013), a lower Z-score indicates an increased probability that a company will face insolvency and be unable to meet its debt obligations, thereby signifying higher credit risk.

Table 2.2 summarizes the variables used in this study. Firm-specific variables are sourced from Bloomberg, and climate change commitment data are retrieved from the Refinitiv Eikon database.

Table 2.2 Definition of variables

Category	Variable	Description	Source
Carbon emissions	*CE*	Total GHG emissions using Scope 1, Scope 2, Scope 3, and total (Scope 1+ Scope 2+ Scope 3) emissions	Bloomberg
Carbon performance	*CP*	Using Scope 1, Scope 2, Scope 3, and total (Scope 1+ Scope 2 + Scope 3) emissions (Misani & Pogutz, 2015)	Authors' estimation of Eq. (2.1) using Bloomberg data
Per capita emission	*PCE*	Total GHG emissions per million population	Authors' estimation from Bloomberg and the World Bank
Climate change commitment	*CC_RF*	Total score of SDG 13 on climate action, climate change commercial risk opportunities, Scope 3 CO_2 emission, and emission reduction target (Albitar et al., 2023)	Authors' estimation from Refinitiv Eikon
	CC_BB	CDP's metric (Birindelli et al., 2022; Lemma et al., 2021)	Bloomberg
Credit risk	*DTD*	Market-based measure of firm credit risk DTD (Gilchrist & Zakrajsek, 2012; Vassalou & Xing, 2004)	Authors' estimation from Bloomberg
	CDS	Natural log of 5-year CDS spreads	Bloomberg
	ZSCORE	Accounting-based measure of firm credit risk (Altman et al., 2017)	Authors' estimation from Bloomberg

Notes: The table provides details on the variables employed in the descriptive analysis, including their definitions and sources. Firm-level variables are acquired and built using Bloomberg, while firm-specific climate change commitment is retrieved from both Bloomberg and Refinitiv databases

2.4 Results

2.4.1 *Carbon Emission and Carbon Performance*

Data on total carbon emissions is limited to 470 observations, as most of the MENA companies do not seem to publish their carbon emissions data. To make the context clear, Table 2.3 shows the annual disclosure of total carbon emissions per capita (thousands of metric tons of CO_2 equivalent (CO_2e) per million population) for each country. Tables 2.4 and 2.5 present firms' total carbon emissions (raw metrics) and corresponding carbon performance. Similarly, Tables 2.6 and 2.7 report Scope 1, Scope 2, and Scope 3 emissions and performance, respectively. Each set of tables includes both the absolute emission values and performance metrics and is structured across three panels: by industry (Panel A), by country (Panel B), and by year (Panel C). It should be noted that better carbon performance indicates lower total carbon emissions with respect to the activity and GICS sector for the year, by adjusting for the sales and industry median GHG emission of the firms. Thus, this measure does not allow comparison of carbon performance across different sectors. Notably, several countries, including Iraq, Palestine, Sudan, and Tunisia, lack data, which indicates potential gaps in reporting and emphasizes the need for more comprehensive emissions data across the region.[1]

An analysis of per capita emissions across MENA countries from 2013 to 2023 reveals significant disparities in emissions intensity. Saudi Arabia consistently reported the highest per capita emissions, peaking at 17,574 thousand metric tons of total carbon emissions per million population in 2021, followed by the UAE and Qatar with similar upward trends in recent years. These figures reflect the industrial concentration and fossil-fuel dependence of their economies. In contrast, countries such as Egypt, Jordan, and Lebanon recorded much lower per capita emissions throughout the period. Notably, several countries exhibited sharp spikes in reported emissions, such as Qatar in 2018 and Bahrain in 2019, likely reflecting changes in disclosure practices or improved data availability, rather than sudden increases in actual emissions. These differences underscore the heterogeneity in emissions profiles and reporting maturity across the region, highlighting the importance of context-specific policy interventions and the need for consistent, transparent emissions reporting.

Carbon emissions (*CE*) across MENA firms show considerable variation by industry. Unsurprisingly, utilities report the highest average emissions (71,408.13 thousand metric tons), followed by materials (28,951.01 thousand metric tons) and energy (12,380.26 thousand metric tons), reflecting their dependence on fossil fuels and heavy industrial processes. In contrast, sectors such as information technology

[1] The chapter does not use imputation or exclude any data. All available firm-reported carbon emissions are included to reflect actual disclosure practices and highlight gaps in voluntary reporting across the MENA region.

Table 2.3 Per capita total carbon emission (*PCE*) in MENA countries

	Bahrain	Egypt	Jordan	Kuwait	Lebanon	Morocco	Oman	Qatar	Saudi Arabia	UAE
2013	0.00	0.00	10.41	0.00	3.71	7.73	32.69	0.00	2027.22	115.48
2014	0.00	0.00	10.65	0.00	3.32	0.00	53.13	7.40	1978.15	116.71
2015	0.00	0.00	13.24	0.00	3.36	0.00	20.90	29.10	2020.15	120.99
2016	0.00	0.00	11.55	54.81	3.16	0.00	22.34	50.87	8438.91	84.50
2017	0.00	0.60	15.06	87.42	3.78	406.25	16.53	40.12	9219.38	215.73
2018	0.00	1.17	14.49	96.61	3.70	420.64	18.28	1857.15	10,965.14	386.25
2019	6656.68	3.08	57.18	107.87	0.00	410.04	54.42	2013.07	11,472.96	11,873.27
2020	8292.71	3.83	57.23	503.61	0.00	4003.64	50.72	4706.46	14,078.72	11,720.40
2021	8294.09	10.73	113.99	554.87	0.00	12.49	53.74	4547.45	17,574.00	12,933.03
2022	8492.47	31.64	127.29	880.90	0.00	12.39	2.35	4685.54	17,291.99	12,602.07
2023	4.14	28.27	129.76	1157.64	0.00	11.94	2.12	29.07	6943.37	11,719.77

Notes: This table reports per capita emissions (*PCE*) in thousands of metric tons of CO_2 equivalent (CO_2e) per million population in MENA countries. There is no carbon emission (*CE*) data available for the firms within Iraq, Palestine, Sudan, and Tunisia. See Table 2.2 for details on the variable

Table 2.4 Descriptive statistics—total carbon emission (*CE*) in MENA countries

	N	Mean	Median	Min	Max	Std dev.
Panel A. Industry						
Communication services	50	596.87	408.10	15.92	2653.75	639.68
Consumer discretionary	10	101.55	139.38	6.37	203.51	83.80
Consumer staples	19	1328.15	1001.70	147.15	2758.54	1120.20
Energy	38	12,380.26	418.18	23.02	72,600.00	24,936.28
Financials	173	37.53	14.36	0.01	861.72	100.82
Health care	13	299.00	124.81	44.81	1051.82	386.75
Industrials	63	1082.92	157.99	0.01	37,450.77	4736.52
Information technology	2	2.86	2.86	2.80	2.91	0.07
Materials	46	28,951.01	9588.16	369.98	170,966.10	46,356.75
Real estate	26	520.75	255.71	143.87	2008.26	517.32
Utilities	30	71,408.13	43,485.85	1.50	212,732.90	76,864.97
Total	470	8707.84	130.04	0.01	212,732.90	31,124.65
Panel B. Country						
Bahrain	11	4319.79	40.68	6.53	12,900.70	6002.97
Egypt	16	558.01	85.62	43.46	3238.21	960.04
Jordan	39	157.31	16.40	0.68	1051.82	275.52
Kuwait	38	415.18	101.59	2.18	3838.73	811.35
Lebanon	6	21.59	21.53	19.98	23.29	1.11
Morocco	12	16,035.73	459.75	6.77	146,000.00	41,407.75
Oman	27	52.80	21.03	0.21	173.18	56.40
Qatar	47	1010.35	15.14	2.46	6945.92	2108.49
Saudi Arabia	103	30,753.75	2653.75	2.80	212,732.90	57,173.96
UAE	171	3539.29	129.92	0.00	85,130.00	14,179.23
Total	470	8707.84	130.04	0.01	212,732.90	31,124.65
Panel C. Year						
2013	11	5216.32	118.41	1.76	56,000.00	16,843.76
2014	12	4773.66	42.29	1.50	56,000.00	16,133.19
2015	19	3240.34	37.30	1.86	56,000.00	12,793.51
2016	18	14,586.33	78.42	7.20	200,223.90	48,147.71
2017	30	10,092.07	111.98	2.09	206,259.40	38,479.23
2018	40	8884.60	143.46	2.14	191,872.50	32,468.38
2019	56	8728.21	165.84	1.44	192,177.70	29,853.79
2020	66	11,056.33	151.18	0.21	182,105.20	36,478.14
2021	74	9382.22	189.23	0.01	203,600.10	33,100.62
2022	84	8550.62	160.51	0.03	212,732.90	31,979.78
2023	60	6078.65	169.87	0.01	74,217.40	18,166.48
Total	470	8707.84	130.04	0.01	212,732.90	31,124.65

Notes: This table reports descriptive statistics on the total *CE* (reported in thousands of metric tons of carbon dioxide equivalent (CO_2) across GICS sectors (Panel A), countries (Panel B), and years (Panel C). There is no carbon emission (*CE*) data available for the firms within Iraq, Palestine, Sudan, and Tunisia. See Table 2.2 for details on the variable

Table 2.5 Descriptive statistics—total carbon performance (*CP*) in MENA countries

	N	Mean	Median	Min	Max	Std dev.
Panel A. Industry						
Communication services	46	8.60	8.52	6.37	10.30	0.96
Consumer discretionary	10	9.14	9.11	7.52	11.62	1.26
Consumer staples	19	7.69	7.92	4.97	10.74	1.61
Energy	38	7.19	7.30	4.27	10.87	1.47
Financials	173	8.15	7.86	4.17	15.27	1.96
Health care	13	7.63	7.57	5.37	9.62	1.04
Industrials	63	7.65	7.42	4.72	17.83	1.98
Information technology	2	7.87	7.87	7.76	7.99	0.16
Materials	43	8.44	8.23	3.69	10.83	1.23
Real estate	26	7.35	6.69	5.33	11.87	1.75
Utilities	30	8.03	8.60	3.29	10.79	1.92
Total	463	8.01	7.88	3.29	17.83	1.76
Panel B. Country						
Bahrain	11	8.21	7.69	6.77	9.88	1.14
Egypt	16	7.40	6.85	6.12	12.20	1.54
Jordan	39	8.25	7.85	5.37	14.08	1.84
Kuwait	38	7.51	7.41	5.55	14.04	1.40
Lebanon	6	8.83	7.94	7.61	13.52	2.31
Morocco	11	7.19	8.13	3.29	8.76	2.07
Oman	26	7.77	7.57	4.71	13.51	2.04
Qatar	47	8.08	8.09	4.27	13.93	1.63
Saudi Arabia	100	7.91	7.94	3.65	12.55	1.56
UAE	169	8.21	7.94	4.72	17.83	1.91
Total	463	8.01	7.88	3.29	17.83	1.76
Panel C. Year						
2013	11	7.73	7.80	4.71	10.83	1.70
2014	12	7.80	7.76	4.77	10.83	1.53
2015	19	7.65	7.94	5.14	9.93	1.29
2016	18	7.78	7.89	5.54	9.91	1.13
2017	30	7.87	8.00	5.55	10.30	1.12
2018	33	9.93	9.98	3.29	14.08	3.45
2019	56	7.92	7.95	4.72	10.87	1.17
2020	66	7.73	7.72	3.69	11.02	1.40
2021	74	7.83	7.75	4.20	15.27	1.59
2022	84	7.93	7.89	4.30	13.23	1.51
2023	60	7.99	7.93	4.17	17.83	1.86
Total	463	8.01	7.88	3.29	17.83	1.76

Notes: This table reports descriptive statistics on the total *CP* across GICS sectors (Panel A), countries (Panel B), and years (Panel C). See Table 2.2 for details on the variable. There is no carbon emission (*CE*) data available for the firms within Iraq, Palestine, Sudan, and Tunisia; consequently, no carbon performance (*CP*)

Table 2.6 Descriptive statistics—Scope 1, 2, and 3 carbon emission in MENA countries

	Scope 1 (*CE_1*)		Scope 2 (*CE_2*)		Scope 3 (*CE_3*)	
	N	Mean	N	Mean	N	Mean
Panel A. Industry						
Communication services	49	138.4054	49	338.07	24	270.68
Consumer discretionary	7	19.24629	10	88.08	0	–
Consumer staples	15	678.1641	14	500.33	11	732.53
Energy	30	11,931.65	31	3602.04	8	104.65
Financials	151	4.580444	168	17.69	97	29.16
Health care	13	30.12515	13	74.73	4	630.99
Industrials	46	895.1933	63	77.61	48	461.58
Information technology	2	0.275	2	2.58	0	–
Materials	46	14,487.94	45	5484.07	6	69,753.00
Real estate	23	186.5777	21	194.35	12	430.58
Utilities	27	56,820.76	23	21,340.78	8	14,655.64
Total	409	6411.243	439	2019.71	218	2678.12
Panel B. Country						
Bahrain	10	4681.643	11	63.74	3	0.06
Egypt	12	13.78825	16	53.77	16	493.90
Jordan	33	73.03524	39	30.79	7	360.62
Kuwait	35	75.38803	38	118.19	34	254.32
Lebanon	6	4.861833	6	15.39	6	1.34
Morocco	11	13,710.77	12	1050.17	9	3223.14
Oman	16	10.21994	24	46.92	11	12.36
Qatar	47	929.2942	42	90.61	12	0.34
Saudi Arabia	99	19,453.19	96	8809.78	19	20,843.76
UAE	140	3211.528	155	103.45	101	1381.87
Total	409	6411.243	439	2019.71	218	2678.12
Panel C. Year						
2013	9	4234.995	9	2079.20	7	78.81
2014	9	4235.06	10	1854.06	6	104.62
2015	16	2577.666	17	1163.17	6	91.64
2016	17	11,019.28	18	4144.07	4	158.21
2017	28	7175.248	30	3350.58	12	111.49
2018	36	6608.081	36	3222.63	17	86.98
2019	48	7187.528	51	2304.64	21	1249.60
2020	58	7539.737	60	2009.05	28	6138.22
2021	63	6572.523	70	1864.14	39	3839.11
2022	73	5823.556	80	1617.60	44	3721.01
2023	52	4944.322	58	698.69	34	1973.24
Total	409	6411.243	439	2019.71	218	2678.12

Notes: This table reports descriptive statistics on the *CE_1*, *CE_2*, and *CE_3* (reported in thousands of metric tons of carbon dioxide equivalent (CO_2) across GICS sectors (Panel A), countries (Panel B), and years (Panel C). There is no carbon emission (*CE*) data available for the firms within Iraq, Palestine, Sudan, and Tunisia. See Table 2.2 for details on the variable

Table 2.7 Descriptive statistics—Scope 1, 2, and 3 carbon performance in MENA countries

	Scope 1 *CP_1*		Scope 2 *CP_2*		Scope 3 *CP_3*	
	N	Mean	N	Mean	N	Mean
Panel A. Industry						
Communication services	45	8.84	49	9.15	24	9.76
Consumer discretionary	7	9.14	10	8.86	0	–
Consumer staples	15	8.30	14	8.30	11	7.66
Energy	30	7.99	31	6.83	8	6.21
Financials	151	8.68	168	7.81	97	7.88
Health care	13	7.04	13	7.55	4	7.82
Industrials	46	7.34	63	7.42	48	7.41
Information technology	2	10.80	2	7.87	0	–
Materials	46	7.60	45	8.94	6	9.71
Real estate	23	9.30	21	6.78	12	8.18
Utilities	27	7.93	23	7.86	8	7.41
Total	405	8.31	439	7.94	218	7.96
Panel B. Country						
Bahrain	10	9.56	11	9.08	3	8.96
Egypt	12	7.33	16	7.66	16	6.96
Jordan	33	7.46	39	8.00	7	8.72
Kuwait	34	9.02	38	7.49	34	7.80
Lebanon	6	6.42	6	8.05	6	8.42
Morocco	10	9.00	12	8.07	9	11.06
Oman	16	8.03	24	8.28	11	6.42
Qatar	47	8.65	42	7.99	12	10.03
Saudi Arabia	99	8.24	96	7.77	19	7.23
UAE	138	8.32	155	8.00	101	7.84
Total	405	8.31	439	7.94	218	7.96
Panel C. Year						
2013	9	7.88	9	7.93	7	7.03
2014	9	7.76	10	7.86	6	7.41
2015	16	7.75	17	7.69	6	7.45
2016	17	7.92	18	7.71	4	6.90
2017	28	7.70	30	8.10	12	7.75
2018	36	8.25	36	8.41	17	7.53
2019	48	7.97	51	7.89	21	8.14
2020	58	8.10	60	7.85	28	8.03
2021	63	7.91	70	7.81	39	7.97
2022	73	8.54	80	7.84	44	8.36
2023	48	9.96	58	8.14	34	8.04
Total	405	8.31	439	7.94	218	7.96

Notes: This table reports descriptive statistics on the Scope 1, 2, and 3 *CP* across GICS sectors (Panel A), countries (Panel B), and years (Panel C). See Table 2.2 for details on the variable. There is no carbon emission (*CE*) data available for the firms within Iraq, Palestine, Sudan, and Tunisia; consequently, no carbon performance (*CP*) and financials exhibit the lowest emissions, consistent with their low operational carbon intensity.

At the country level, Saudi Arabia contributes the highest *CE* (30,753.75 thousand metric tons), driven by its concentration of high-emitting industries such as energy and manufacturing. Morocco and the UAE also report relatively high averages. Conversely, countries such as Lebanon and Oman show much lower emissions, though this may partly reflect the absence of large industrial players or limited disclosure. Variations in national emissions are therefore shaped not only by economic structure but also by differing levels of transparency.

Emissions patterns over time reveal a peak in 2016, likely due to elevated industrial activity, followed by a gradual decline to 2023. This drop may reflect improved emissions management, shifts in production, or increasing pressure to disclose environmental data. Nonetheless, the year-on-year variability highlights both operational dynamics and inconsistent reporting trends across the sample.

Overall, while the mean total *CE* for the sample is 8707.84 tons, the wide standard deviation (31,124.65 tons) reflects both sectoral concentration and reporting gaps. Firms in high-emission industries dominate the regional footprint, yet firm-level comparisons must account for disclosure bias—especially given the significant number of companies in the region that do not report emissions at all.

These data highlight the diverse *CE* profiles of MENA firms, emphasizing the impact of industry-specific practices and regional differences in determining carbon outputs. The observed trends over time reflect evolving corporate strategies and regulatory pressures, highlighting the critical need for ongoing monitoring and action to further reduce carbon emissions and address climate change challenges in the region.

These data highlight the diverse carbon emissions profiles of MENA firms, emphasizing the impact of industry-specific practices and regional differences in determining carbon outputs. The observed trends over time reflect evolving corporate strategies and regulatory pressures, highlighting the critical need for ongoing monitoring and action to further reduce carbon emissions and address climate change challenges in the region.

Carbon performance (*CP*) should be interpreted within the context of each sector, as scores are adjusted to reflect how firms perform relative to their industry peers. This makes *CP* most useful for comparing companies within the same sector, not across sectors. For instance, the energy sector has a relatively low mean *CP* score of 7.19, reflecting the structural emissions intensity of fossil fuel operations. However, firms with higher *CP* scores within this sector are managing their emissions more efficiently than their peers, even if their absolute emissions remain high. In contrast, sectors with inherently lower emissions—such as consumer discretionary (9.14) and communication services (8.60)—tend to score higher on *CP* overall. Moderate scores in heavy-emitting sectors such as materials (8.44) and utilities (8.03) suggest mixed progress in carbon management under more challenging conditions.

At the country level, Lebanon records the highest mean *CP* score (8.83), followed by Jordan (8.25), Bahrain (8.21), and the UAE (8.21), suggesting that firms in these countries demonstrate relatively efficient emissions management within their sectors. However, these high scores often coincide with lower absolute emissions and may reflect limited industrial activity or selective reporting by less carbon-intensive firms. In contrast, Saudi Arabia, despite showing a moderate *CP* score (7.91), contributes the highest absolute emissions in the region. This likely reflects its concentration of large, carbon-intensive firms in sectors such as energy and materials. Notably, Saudi firms are also more consistent in disclosing carbon data, which may make their *CP* figures more reliable compared to countries with limited or selective disclosure practices. This distinction reinforces the importance of interpreting *CP* alongside disclosure coverage and raw emissions to gain a fuller picture of corporate carbon performance.

Over time, *CP* scores remain relatively stable, with modest fluctuations from 2013 to 2023. The mean score hovers around 7.7–7.9 in most years, with a sharp peak observed in 2018 (mean *CP* of 9.93), which may suggest improvements in emissions reporting, more proactive management, or changes in disclosure coverage. Post-2018, *CP* values appear to normalize, pointing to a more consistent regional approach to carbon management. While *CP* reflects how well firms are managing emissions relative to their sector peers, the raw data provides a clearer picture of the absolute emissions trends and highlights the evolving nature of carbon emissions in the MENA region.

Scope-specific analysis reveals stark differences in carbon emissions across industries and countries. Scope 1 emissions (*CE_1*), primarily from direct operations, are highest in utilities (56,820.76 thousand metric tons), materials (14,487.94 thousand metric tons), and energy (11,931.65 thousand metric tons), reflecting their fossil fuel dependence and energy-intensive processes. In contrast, sectors such as information technology and financials exhibit minimal Scope 1 emissions, consistent with their low operational footprints.

Scope 2 emissions (*CE_2*), arising from purchased electricity, show similar patterns. Utilities again lead (21,340.78 thousand metric tons), followed by materials (5484.07 thousand metric tons) and energy (3602.04 thousand metric tons), pointing to high external energy demand. Meanwhile, Scope 3 emissions (*CE_3*), which include value-chain activities, are heavily concentrated in materials (69,753.00 thousand metric tons) and utilities (14,655.64 thousand metric tons). These figures not only suggest a significant upstream and downstream carbon burden in certain industries but are also heavily affected by disclosure quality, with some firms underreporting or omitting Scope 3 data.

Geographically, Saudi Arabia consistently reports the highest emissions across all scopes, driven by its concentration of large-scale industrial and energy firms. The UAE and Morocco also rank high in *CE_2* and *CE_3*, while Lebanon, Qatar, and Bahrain report lower levels, though this may be more reflective of underreporting or lower disclosure rates than genuinely lower emissions. Saudi Arabia's more comprehensive disclosure coverage likely makes its figures more robust for comparison.

Temporally, *CE_1* and *CE_2* peaked between 2016 and 2018, with gradual declines observed thereafter, potentially due to improved efficiency, regulatory pressure, or reporting shifts. Scope 3 data, however, is more volatile and increases notably post-2019, likely tied to expanded reporting rather than an actual rise in emissions. These trends underscore both the emissions intensity of key sectors and the importance of improving transparency across the MENA region.

Carbon performance (*CP*) scores reflect how efficiently firms manage emissions relative to peers within the same sector, meaning that comparisons across sectors are not meaningful. However, these scores remain useful for identifying patterns at the country level and over time.

Across countries, higher *CP* scores are observed in Bahrain, Kuwait, and Qatar, especially in Scope 1 and Scope 3, indicating relatively stronger emissions management practices within firms. For example, Qatar records a *CP_3* of 10.03, while Bahrain scores 9.56 in *CP_1*. These results should be interpreted alongside disclosure limitations, as high *CP* may partially reflect selective reporting by low-emitting firms. Conversely, Saudi Arabia, despite having more moderate *CP* scores (e.g., *CP_1*: 8.24 and *CP_3*: 7.23), contributes the largest absolute emissions across all scopes. Given its extensive reporting, *CP* values for Saudi firms may offer a more reliable and transparent view of emissions management within a high-emission economy.

Over time, *CP* scores remain relatively stable, with a slight upward trend in recent years. *CP_1* rose from 7.88 in 2013 to 9.96 in 2023, while *CP_3* increased from 7.03 to 8.04 in the same period. These improvements may reflect gradual enhancements in firm-level practices, growing awareness, and evolving disclosure norms across the region.

Overall, the carbon performance data offers valuable insights into how firms manage emissions relative to sector norms, but it must be complemented by the raw emissions data to fully understand the scope and scale of carbon emissions across sectors and countries. While higher carbon performance scores indicate relative efficiency, they may obscure the substantial environmental impact of high-emitting sectors or economies. For instance, sectors such as energy and countries like Saudi Arabia continue to exhibit high absolute emissions despite moderate carbon performance, underscoring the structural carbon intensity of their activities. These findings highlight the importance of integrating both relative and absolute measures when assessing corporate emissions management. They also point to the urgent need for targeted policies and sector-specific strategies aimed at driving down total emissions, particularly in carbon-intensive industries, if the MENA region is to make meaningful progress toward climate goals.

2.4.1.1 Climate Change Commitment

Tables 2.8 and 2.9 provide descriptive statistics of climate change commitment, for *CC_RF* and *CC_BB*, respectively, based on the industry they operate in (Panel A), the country (Panel B), and the trend over the years (Panel C). The *CC_RF* index

Table 2.8 Descriptive statistics—climate change commitment (CC_RF) in MENA countries

	N	Mean	Median	Min	Max	Std dev.
Panel A. Industry						
Communication services	451	0.21	0.00	0.00	3.00	0.62
Consumer discretionary	1727	0.03	0.00	0.00	3.00	0.23
Consumer staples	1529	0.03	0.00	0.00	3.00	0.22
Energy	429	0.06	0.00	0.00	2.00	0.27
Financials	5082	0.06	0.00	0.00	3.00	0.33
Health care	759	0.04	0.00	0.00	3.00	0.28
Industrials	2321	0.05	0.00	0.00	4.00	0.33
Information technology	341	0.00	0.00	0.00	0.00	0.00
Materials	1848	0.04	0.00	0.00	3.00	0.28
Real estate	1870	0.03	0.00	0.00	3.00	0.24
Utilities	352	0.09	0.00	0.00	3.00	0.44
Total	16,709	0.05	0.00	0.00	4.00	0.30
Panel B. Country						
Bahrain	363	0.08	0.00	0.00	3.00	0.36
Egypt	2640	0.03	0.00	0.00	3.00	0.23
Iraq	770	0.00	0.00	0.00	0.00	0.00
Jordan	1716	0.02	0.00	0.00	3.00	0.21
Kuwait	1551	0.05	0.00	0.00	4.00	0.34
Lebanon	110	0.00	0.00	0.00	0.00	0.00
Morocco	803	0.07	0.00	0.00	2.00	0.29
Oman	1199	0.01	0.00	0.00	1.00	0.11
Palestine	451	0.00	0.00	0.00	0.00	0.00
Qatar	616	0.08	0.00	0.00	2.00	0.31
Saudi Arabia	3740	0.05	0.00	0.00	3.00	0.29
Sudan	22	0.00	0.00	0.00	0.00	0.00
Tunisia	836	0.00	0.00	0.00	0.00	0.00
UAE	1892	0.17	0.00	0.00	3.00	0.57
Total	16,709	0.05	0.00	0.00	4.00	0.30
Panel C. Year						
2013	1519	0.00	0.00	0.00	1.00	0.06
2014	1519	0.00	0.00	0.00	1.00	0.06
2015	1519	0.00	0.00	0.00	1.00	0.06
2016	1519	0.00	0.00	0.00	1.00	0.05
2017	1519	0.01	0.00	0.00	2.00	0.10
2018	1519	0.02	0.00	0.00	3.00	0.18
2019	1519	0.04	0.00	0.00	4.00	0.30
2020	1519	0.07	0.00	0.00	3.00	0.37
2021	1519	0.13	0.00	0.00	3.00	0.47
2022	1519	0.17	0.00	0.00	3.00	0.56
2023	1519	0.10	0.00	0.00	3.00	0.40
Total	16,709	0.05	0.00	0.00	4.00	0.30

Notes: This table reports descriptive statistics on *CC_RF* across GICS sectors (Panel A), countries (Panel B), and years (Panel C). See Table 2.2 for details on the variable

Table 2.9 Descriptive statistics—climate change commitment (*CC_BB*) in MENA countries

	N	Mean	Median	Min	Max	Std dev.
Panel A. Industry						
Communication services	451	0.15	0.00	0.00	7.00	0.89
Consumer discretionary	1727	0.02	0.00	0.00	5.00	0.27
Consumer staples	1529	0.00	0.00	0.00	2.00	0.05
Energy	429	0.08	0.00	0.00	7.00	0.69
Financials	5082	0.02	0.00	0.00	7.00	0.32
Health care	759	0.00	0.00	0.00	0.00	0.00
Industrials	2321	0.00	0.00	0.00	4.00	0.12
Information technology	341	0.00	0.00	0.00	0.00	0.00
Materials	1848	0.01	0.00	0.00	6.00	0.27
Real estate	1870	0.01	0.00	0.00	6.00	0.18
Utilities	352	0.02	0.00	0.00	6.00	0.32
Total	16,709	0.02	0.00	0.00	7.00	0.30
Panel B. Country						
Bahrain	363	0.00	0.00	0.00	0.00	0.00
Egypt	2640	0.00	0.00	0.00	0.00	0.00
Iraq	770	0.00	0.00	0.00	0.00	0.00
Jordan	1716	0.00	0.00	0.00	4.00	0.14
Kuwait	1551	0.03	0.00	0.00	7.00	0.38
Lebanon	110	0.00	0.00	0.00	0.00	0.00
Morocco	803	0.00	0.00	0.00	0.00	0.00
Oman	1199	0.00	0.00	0.00	0.00	0.00
Palestine	451	0.00	0.00	0.00	0.00	0.00
Qatar	616	0.05	0.00	0.00	6.00	0.41
Saudi Arabia	3740	0.03	0.00	0.00	7.00	0.41
Sudan	22	0.00	0.00	0.00	0.00	0.00
Tunisia	836	0.00	0.00	0.00	0.00	0.00
UAE	1892	0.05	0.00	0.00	7.00	0.50
Total	16,709	0.02	0.00	0.00	7.00	0.30
Panel C. Year						
2013	1519	0.00	0.00	0.00	0.00	0.00
2014	1519	0.00	0.00	0.00	0.00	0.00
2015	1519	0.01	0.00	0.00	6.00	0.22
2016	1519	0.01	0.00	0.00	7.00	0.21
2017	1519	0.01	0.00	0.00	6.00	0.16
2018	1519	0.01	0.00	0.00	6.00	0.23
2019	1519	0.03	0.00	0.00	7.00	0.37
2020	1519	0.03	0.00	0.00	7.00	0.41
2021	1519	0.04	0.00	0.00	7.00	0.43
2022	1519	0.06	0.00	0.00	7.00	0.55
2023	1519	0.00	0.00	0.00	0.00	0.00
Total	16,709	0.02	0.00	0.00	7.00	0.30

Notes: This table reports descriptive statistics on *CC_BB* across GICS sectors (Panel A), countries (Panel B), and years (Panel C). See Table 2.2 for details on the variable

ranges from 0 to 4, with higher scores reflecting greater corporate commitment to addressing climate change. The *CC_BB* index, derived from CDP, ranges from 0 to 8, with higher scores similarly indicating increased corporate commitment to climate action.

The mean corporate *CC_RF* scores across industries reveal varying levels of commitment to climate action. The communication services sector exhibits the highest mean score of 0.21, indicating relatively greater attention to climate change initiatives compared to other sectors. This is followed by the utilities sector with a mean of 0.09, suggesting very low engagement with climate commitment. In contrast, information technology has a mean score of 0.00, reflecting no recorded commitment to climate action.

Among countries, the UAE shows the highest average commitment level, with a mean score of 0.17. This suggests that companies based in the UAE are relatively more engaged in climate change initiatives. Bahrain and Qatar follow with mean scores of 0.08 each, indicating a moderate level of commitment. Conversely, several countries, such as Iraq, Lebanon, Palestine, Sudan, and Tunisia, have a mean score of 0.00, indicating no measurable climate commitment among companies in these countries.

The data indicates a temporal trend in climate commitment. From 2013 to 2016, the mean scores remain consistently low at 0.00, suggesting that very few companies focus on climate action during these years. However, there is a gradual increase starting from 2017 with a mean score of 0.01, reaching 0.17 in 2022 before slightly declining to 0.10 in 2023. This trend suggests that corporate commitment to climate action is starting to gain traction in recent years, particularly from 2018 onward, reflecting a potential increase in awareness or regulatory pressure resulting from various rounds of COPs.

The overall *CC_RF* mean score for the entire dataset is 0.05, indicating that, on average, companies demonstrate a low level of commitment to climate change action. The standard deviation of 0.30 shows moderate variability, suggesting differences in commitment levels across various sectors, countries, and years. The data reflects that while some companies and countries are beginning to prioritize climate commitments, many still have low or negligible engagement.

The data reveals that corporate climate change commitment is generally low, with some sectors like utilities and countries like the UAE showing relatively higher levels of engagement. Over time, there has been an upward trend in commitment, particularly from 2018 onward, suggesting a growing recognition of the importance of addressing climate change. However, many companies, especially in certain MENA countries and industries, still exhibit minimal to no commitment, indicating substantial room for improvement in corporate climate action initiatives.

The *CC_BB*, which ranges from 0 to 8 with higher scores indicating increased corporate commitment to addressing climate change, reveals varying levels of commitment across industries, countries, and over time. Among industries, the communication services sector shows the highest mean score of 0.15, suggesting a relatively greater focus on climate change initiatives compared to other sectors. This is followed by the energy sector, with a mean score of 0.08, indicating moderate engagement in climate actions. Sectors such as consumer discretionary and financials have low mean scores of 0.02, reflecting minimal commitment to climate action. Notably, the information technology and health care sectors have mean scores of 0.00, showing no recorded climate commitment.

At the country level, the data indicates that the UAE and Qatar exhibit the highest average commitment levels, with mean scores of 0.05 each. This suggests that companies in these countries are relatively more engaged in climate change initiatives compared to others. Saudi Arabia also shows some level of commitment, with a mean score of 0.03. In contrast, countries such as Bahrain, Egypt, Iraq, Lebanon, Morocco, Oman, Palestine, Sudan, and Tunisia all have mean scores of 0.00, indicating no measurable climate commitment among companies.

Over time, the temporal analysis shows that the mean *CC_BB* scores were consistently low from 2013 to 2014, remaining at 0.00, suggesting minimal focus on climate action during these years. However, a slight upward trend begins in 2015 with a mean score of 0.01, gradually increasing to 0.06 by 2022, before dropping back to 0.00 in 2023. This trend indicates that corporate commitment to climate action gained some traction from 2015 onward, peaking in 2022, possibly influenced by growing awareness, regulatory pressures, or initiatives such as the various rounds of COPs.

Overall, the *CC_BB* data reveals that corporate commitment to climate change action remains generally low, with an overall mean score of 0.02 and a standard deviation of 0.30, indicating moderate variability across sectors, countries, and years. Although there has been some improvement over time, particularly from 2018 onward, many companies and industries, especially in certain MENA countries, still exhibit minimal engagement in climate action, highlighting significant room for improvement in corporate climate commitments.

The *CC_RF* and *CC_BB* scores generally reflect similar outcome profiles, highlighting that firms across regions and industries exhibit limited progress in climate commitment despite rising awareness and regulatory pressures. While there has been some recent improvement, particularly post-2018, substantial gaps remain, suggesting significant room for enhanced climate engagement.

2.4.1.2 Credit Risk

Tables 2.10, 2.11 and 2.12 provide descriptive statistics of sampled MENA firms' credit risk measures, *DTD*, CDS spreads, and Altman Z-score (*ZSCORE*), respectively, based on the industry they operate in (Panel A), based on the country (Panel B), and the trend over the years (Panel C).

The descriptive statistics of *DTD* among sampled MENA countries illustrate significant variations across sectors, countries, and over time. Higher *DTD* translates into lower credit risk. Among sectors, the financials sector has the highest mean *DTD* score of 98.61, indicating lower credit risk, with a broad range reaching up to 26,334.91 and a high standard deviation of 1091.46, reflecting substantial

Table 2.10 Descriptive statistics—DTD in MENA countries

	N	Mean	Median	Min	Max	Std dev.
Panel A. Industry						
Communication services	332	10.96	8.18	0.40	162.29	13.34
Consumer discretionary	906	61.49	7.34	0.00	4664.91	319.64
Consumer staples	922	71.52	8.45	0.00	20,794.86	752.28
Energy	235	12.36	5.68	0.00	398.42	30.75
Financials	2738	98.61	6.10	0.00	26,334.91	1091.46
Health care	418	12.45	9.12	0.00	364.48	22.72
Industrials	1371	37.65	5.53	0.00	8940.06	345.22
Information technology	143	7.99	6.15	0.00	79.64	8.31
Materials	1337	23.01	7.07	0.00	3293.45	152.77
Real estate	1138	29.51	5.89	0.00	18,721.41	558.64
Utilities	214	50.47	5.84	0.00	6073.26	435.43
Total	9754	54.47	6.55	0.00	26,334.91	677.34
Panel B. Country						
Bahrain	200	10.87	8.25	0.00	131.35	12.47
Egypt	1693	9.89	4.94	0.00	1306.33	37.61
Iraq	10	10.42	8.50	5.07	22.91	5.92
Jordan	1089	13.33	7.73	0.00	378.31	27.18
Kuwait	1152	7.05	4.66	0.00	475.24	21.42
Lebanon	24	6.81	5.87	1.19	18.46	5.41
Morocco	689	9.92	7.27	0.00	72.29	9.76
Oman	874	201.64	8.64	0.00	20,794.86	932.54
Palestine	216	12.85	8.61	1.78	145.11	14.94
Qatar	415	8.28	6.33	0.05	38.26	6.14
Saudi Arabia	1711	11.16	8.22	0.19	881.82	24.61
Sudan	8	3.22	2.84	2.49	5.31	1.00
Tunisia	620	7.60	5.64	0.00	57.19	7.34
UAE	1053	262.47	6.24	0.00	26,334.91	1,857.69
Total	9754	54.47	6.55	0.00	26,334.91	677.34
Panel C. Year						
2013	709	14.06	6.68	0.00	650.52	44.06
2014	778	39.64	6.05	0.00	10,768.05	495.64
2015	799	62.94	6.35	0.00	20,794.86	799.53
2016	822	62.16	5.55	0.00	12,084.46	623.83
2017	837	40.64	6.61	0.00	6073.26	310.13
2018	858	29.93	6.38	0.00	2639.28	174.19
2019	949	82.68	7.13	0.00	17,671.34	809.77
2020	958	92.66	6.18	0.00	25,641.54	1137.55
2021	989	105.81	7.38	0.00	26,334.91	1129.40
2022	1033	20.27	6.21	0.00	8541.40	267.72
2023	1022	35.80	7.28	0.00	8476.87	375.08
Total	9754	54.47	6.55	0.00	26,334.91	677.34

Notes: This table reports descriptive statistics on *DTD* across GICS sectors (Panel A), countries (Panel B), and years (Panel C). See Table 2.2 for details on the variable

Table 2.11 Descriptive statistics—*CDS* spreads in MENA countries

	N	Mean	Median	Min	Max	Std dev.
Panel A. Industry						
Communication services	371	4.54	4.57	1.95	6.28	0.66
Consumer discretionary	1181	4.95	5.06	2.08	6.61	0.64
Consumer staples	1119	4.90	4.98	2.07	6.56	0.62
Energy	3404	4.78	4.83	1.36	6.44	0.54
Financials	296	4.68	4.88	1.81	5.80	0.72
Health care	506	4.86	4.92	2.51	6.45	0.55
Industrials	1632	5.03	5.07	2.13	6.63	0.56
Information technology	187	5.19	5.23	3.25	6.30	0.57
Materials	1482	4.91	5.01	1.81	6.41	0.61
Real estate	1386	5.14	5.20	2.34	6.88	0.61
Utilities	262	4.78	4.79	1.99	6.00	0.51
Total	11,826	4.90	4.96	1.36	6.88	0.60
Panel B. Country						
Bahrain	315	4.89	4.98	2.72	6.06	0.53
Egypt	1955	4.86	4.97	1.81	6.13	0.56
Iraq	73	4.49	4.69	2.30	5.36	0.71
Jordan	1393	4.78	4.84	2.03	6.26	0.57
Kuwait	1331	5.01	5.04	2.62	6.29	0.49
Lebanon	72	4.88	4.90	3.45	5.71	0.43
Morocco	737	5.47	5.54	1.81	6.88	0.61
Oman	1038	4.84	4.95	1.81	6.00	0.63
Palestine	355	4.66	4.74	1.36	6.27	0.67
Qatar	480	4.62	4.69	2.01	5.86	0.57
Saudi Arabia	2050	4.70	4.78	2.34	5.88	0.55
Sudan	15	5.52	5.56	4.95	5.95	0.31
Tunisia	705	5.38	5.48	2.25	6.61	0.64
UAE	1307	4.97	5.01	2.50	6.41	0.45
Total	11,826	4.90	4.96	1.36	6.88	0.60
Panel C. Year						
2013	918	4.95	5.11	2.44	6.45	0.71
2014	949	4.86	4.96	2.29	6.22	0.57
2015	970	5.00	5.16	1.36	6.88	0.77
2016	1001	5.13	5.23	2.30	6.56	0.62
2017	1023	4.96	5.03	2.49	6.18	0.50
2018	1060	4.97	4.97	3.54	6.28	0.41
2019	1127	4.66	4.70	2.91	6.31	0.53
2020	1152	4.87	4.92	2.34	6.60	0.59
2021	1178	4.70	4.74	2.61	6.24	0.58
2022	1211	4.90	4.95	2.69	6.63	0.60
2023	1237	4.96	5.03	2.95	6.42	0.58
Total	11,826	4.90	4.96	1.36	6.88	0.60

Notes: This table reports descriptive statistics on *CDS* across GICS sectors (Panel A), countries (Panel B), and years (Panel C). See Table 2.2 for details on the variable

Table 2.12 Descriptive statistics—Z-score (*ZSCORE*) in MENA countries

	N	Mean	Median	Min	Max	Std dev.
Panel A. Industry						
Communication services	363	38.04	7.38	0.00	2516.60	189.29
Consumer discretionary	1050	58.55	8.05	0.00	5371.81	328.77
Consumer staples	1076	212.83	8.25	0.00	107,689.70	3382.52
Energy	266	90.40	6.03	0.00	6387.70	644.16
Financials	3051	455.57	5.41	0.00	236,307.70	8482.08
Health care	527	2947.52	10.24	0.00	485,351.60	34,707.19
Industrials	1572	25.27	7.48	0.00	5253.09	171.67
Information technology	194	29.89	10.52	0.00	340.62	57.59
Materials	1450	620.43	8.25	0.00	193,067.30	8128.99
Real estate	1202	667.67	7.50	0.00	142,914.50	7244.65
Utilities	273	18.24	4.57	0.00	1439.83	117.21
Total	11,024	455.76	7.27	0.00	485,351.60	9659.41
Panel B. Country						
Bahrain	289	53.06	7.90	1.63	4256.95	294.91
Egypt	1782	742.36	8.08	0.00	193,067.30	8979.29
Iraq	56	9610.78	43.60	5.66	173,462.40	38,857.30
Jordan	1128	253.41	8.38	0.00	189,617.50	5683.98
Kuwait	1223	49.14	6.23	0.00	14,555.62	539.12
Lebanon	78	21.15	5.03	3.63	1056.50	119.26
Morocco	725	2580.41	7.34	0.00	485,351.60	30,043.46
Oman	940	26.23	6.01	0.00	1439.83	100.46
Palestine	299	56.27	7.60	0.00	9925.71	581.16
Qatar	456	21.48	6.65	0.00	1059.83	73.75
Saudi Arabia	2183	44.49	7.58	0.00	11,396.84	328.53
Sudan	11	4.53	4.70	3.04	6.24	1.09
Tunisia	656	77.12	5.77	0.00	6387.70	540.48
UAE	1198	609.70	7.56	0.00	236,307.70	9222.29
Total	11,024	455.76	7.27	0.00	485,351.60	9659.41
Panel C. Year						
2013	847	684.11	7.05	0.00	485,351.60	16,720.67
2014	876	800.60	7.17	0.00	440,153.50	15,760.47
2015	898	653.95	7.21	0.00	420,796.40	14,278.17
2016	910	136.17	7.30	0.00	59,801.59	2076.29
2017	927	541.47	7.29	0.00	173,462.40	7730.91
2018	963	603.88	6.96	0.00	236,307.70	9916.12
2019	1093	556.56	7.15	0.00	178,169.70	8355.28
2020	1139	351.59	7.19	0.00	197,481.80	6595.87
2021	1161	253.88	7.54	0.00	193,067.30	5710.82
2022	1139	285.48	7.35	0.00	180,863.10	5594.82
2023	1071	298.94	7.74	0.00	189,617.50	5980.42
Total	11,024	455.76	7.27	0.00	485,351.60	9659.41

Notes: This table reports descriptive statistics on *ZSCORE* across GICS sectors (Panel A), countries (Panel B), and years (Panel C). See Table 2.2 for details on the variable

variability in credit risk profiles. Consumer staples and consumer discretionary follow with mean *DTD* scores of 71.52 and 61.49, respectively, also suggesting relatively lower credit risk compared to other sectors. In contrast, sectors such as communication services and information technology have mean *DTD* scores of 10.96 and 7.99, respectively, indicating comparatively higher credit risk exposure.

At the country level, the UAE exhibits the highest mean *DTD* score of 262.47, highlighting significantly lower credit risk and substantial variability, with a maximum *DTD* value of 26,334.91. Oman also shows a high mean *DTD* score of 201.64, suggesting lower credit risk but with a large standard deviation of 932.54, indicating considerable variability. However, countries with relatively fewer observations, such as Sudan, Lebanon, and Iraq, show mean *DTD* scores of 3.22, 6.81, and 10.42, respectively. Due to their smaller sample sizes, these scores are not directly comparable to those of other countries, and conclusions drawn from these values should be approached with caution.

The temporal analysis reveals fluctuating mean *DTD* scores over the years, reflecting varying credit risk exposure. From 2013 to 2014, the mean *DTD* rose significantly from 14.06 to 39.64, indicating reduced credit risk during this period. The peak was observed in 2021, with a mean *DTD* score of 105.81, suggesting a period of relatively lower credit risk, likely due to economic conditions at the time. However, mean *DTD* scores saw a sharp decline in 2022, dropping to 20.27, before slightly increasing to 35.80 in 2023, indicating a partial stabilization in credit risk exposure.

The descriptive statistics of the natural log of CDS spreads across sampled MENA countries show notable variations across sectors, countries, and over time. Among sectors, the information technology sector has the highest mean *CDS* score of 5.19, suggesting relatively higher credit risk premiums compared to other sectors, with a maximum value of 6.30 and a standard deviation of 0.57. Real estate and industrials also display high mean scores of 5.14 and 5.03, respectively, indicating elevated credit risk. In contrast, the communication services sector has the lowest mean score of 4.54, reflecting comparatively lower perceived credit risk, with a standard deviation of 0.66.

Country-level statistics indicate that Tunisia and Morocco exhibit the highest mean *CDS* scores of 5.38 and 5.47, respectively, indicating higher credit risk premiums in these markets. Kuwait also shows a relatively high mean score of 5.01, suggesting notable credit risk exposure. On the other hand, Iraq, Lebanon, and Sudan have relatively fewer observations, making their scores less reliable and not directly comparable to other countries. Despite this limitation, Sudan shows a high mean *CDS* score of 5.52, reflecting significant credit risk exposure, though caution should be exercised in interpreting these results due to the smaller sample sizes.

The temporal analysis reveals fluctuations in mean *CDS* scores over the years. The highest mean was observed in 2016 at 5.13, with a median of 5.23, indicating heightened credit risk during this period. The lowest mean occurred in 2019 at 4.66, suggesting a relative decrease in credit risk premiums. Over the years, the mean *CDS* scores generally oscillated around 4.90, with a slight increase to 4.96 in 2023, reflecting some stabilization in credit risk perceptions toward the latter years of the period.

The descriptive statistics of the Z-score among sampled MENA countries show notable differences across sectors, countries, and over time. In this context, a higher *ZSCORE* indicates lower credit risk. Among sectors, the health care sector stands out with the highest mean *ZSCORE* of 2947.52, suggesting significantly lower credit risk, although it also exhibits substantial variability, with a maximum value of 485,351.60 and a standard deviation of 34,707.19. The materials and real estate sectors also show high mean *ZSCORE* of 620.43 and 667.67, respectively, reflecting relatively lower credit risk compared to other sectors. Conversely, sectors such as communication services and utilities have much lower mean Z-scores of 38.04 and 18.24, respectively, indicating comparatively higher credit risk exposure.

At the country level, Morocco and the UAE show the highest mean *ZSCORE*s of 2580.41 and 609.70, respectively, suggesting significantly lower credit risk in these markets. Egypt also has a high mean *ZSCORE* of 742.36, reflecting relatively lower credit risk exposure. However, countries with smaller sample sizes, such as Sudan, Lebanon, and Iraq, exhibit lower mean Z-scores of 4.53, 21.15, and 9610.78, respectively. Due to the limited number of observations, the Z-scores for these countries are not directly comparable to those of other nations and should be interpreted with caution.

The temporal analysis of *ZSCORE* reveals substantial fluctuations over the years, indicating varying levels of credit risk exposure. The mean *ZSCORE* peaked in 2014 at 800.60, suggesting a period of relatively low credit risk. However, this trend reversed in subsequent years, with the mean score dropping to 136.17 in 2016, reflecting a period of increased risk. In recent years, the mean *ZSCORE* has generally stabilized, with values ranging between 253.88 in 2021 and 298.94 in 2023, indicating a moderate level of credit risk exposure.

Across Tables 2.7–2.9, there is moderate variability in credit risk across different sectors, countries, and years. Some sectors and countries face higher credit risk premiums, reflecting heightened market concerns, while others maintain relatively lower risk levels, suggesting more stable credit profiles. The data highlights that these fluctuations in credit risk perceptions are influenced by broader economic and market conditions in the MENA region. The observed deviations among *DTD*, *CDS*, and *ZSCORE* can be attributed to the differing foundations and methodologies of these measures. The *ZSCORE* relies solely on firm-specific financial data, reflecting a more internal assessment of credit risk. In contrast, *DTD*, incorporating both accounting and market data, and *CDS*, a purely market-based measure, are aligned in reflecting the market's perception of credit risk. Additionally, the temporal discrepancies among these measures contribute to the variability in their assessments of credit risk.

2.5 Conclusion

This chapter provides a comprehensive analysis of the carbon performance, climate change commitment, and credit risk of 1519 listed firms from 14 MENA countries over the period 2013–2023. We evaluate firms' carbon emissions (*CE*) as raw total, scope 1, scope 2, and scope 3 GHG emissions disclosed by companies and assess carbon performance (*CP*) by considering firm size, industry sector, and total, scope 1, scope 2, and scope 3 GHG emissions. Climate change commitment is measured using two indicators: *CC_RF*, a composite score derived by summing four components—SDG 13 on climate action, Climate Change Commercial Risk Opportunities, Scope 3 CO_2 emissions, and Emission Reduction Targets, and *CC_BB*, which utilizes a consolidated CDP metric to quantify a company's commitment to climate change mitigation, adaptation, and transparency. To assess credit risk, we employ the *DTD*, CDS spreads, and *ZSCORE* as key indicators. Existing literature suggests that effective carbon management (Capasso et al., 2020; Kabir et al., 2021; Nguyen et al., 2023; Zhang & Zhao, 2022) and stronger climate change commitment (Kabir et al., 2021) contribute to reducing corporate credit risk.

The analysis of carbon emissions and carbon performance among MENA firms reveals not only substantial variability in emissions across sectors and countries but also significant data limitations, with 463 firm-year observations disclosing emissions. Utilities, materials, and energy consistently emerged as the most carbon-intensive sectors, while information technology and financials reported the lowest emissions across all scopes. Although sectors such as communication services and real estate report high carbon performance (*CP*) scores, they also have inherently low absolute emissions. This suggests that strong relative performance does not necessarily reflect a significant contribution to reducing the overall carbon footprint. Country-level patterns show that while Saudi Arabia consistently leads in absolute emissions due to its industrial base, countries such as Lebanon, Bahrain, and Qatar report high *CP* scores, often coupled with low emissions, which may reflect either efficient management or disclosure biases. These findings highlight the need for more consistent and transparent emissions reporting across the region.

At the country level, Saudi Arabia contributed the highest absolute emissions across all scopes, driven by the dominance of energy and heavy industrial firms. Meanwhile, Lebanon, Bahrain, and Qatar displayed high carbon performance scores, but these were accompanied by low raw emissions, suggesting either effective management within lower-emission sectors or selective disclosure among less carbon-intensive firms. These findings underscore that while carbon performance (*CP*) provides insight into relative efficiency, it must be interpreted alongside absolute emissions to assess the true carbon footprint. Moreover, the overall low levels of reported climate commitment, particularly in the form of target setting or public disclosure, indicate limited strategic alignment with global climate goals in many MENA firms. However, gaps in data, especially in countries such as Iraq, Palestine, Sudan, and Tunisia, highlight the need for comprehensive reporting across the region. Countries such as Saudi Arabia, Morocco, and the UAE lead in emissions, underscoring the significant impact of large-scale industrial activities on national carbon footprints. However, gaps in data, especially in countries such as Iraq, Palestine, Sudan, and Tunisia, highlight the need for comprehensive reporting across the region.

While the data indicate a modest upward trend in climate change commitment from 2017 onward, overall climate engagement remains low. Most firms in the sample do not disclose detailed climate strategies or emission reduction targets. Countries such as the UAE and Qatar show relatively higher scores on both indices, suggesting stronger engagement with climate-related disclosures and frameworks. However, in many other countries, particularly Iraq, Lebanon, and Sudan, climate commitment is negligible, further compounded by large data gaps. These limitations in disclosure not only constrain comparative analysis but also reflect structural challenges across the region, including limited regulatory pressure and insufficient incentives for firms to adopt climate-aligned strategies. Therefore, while the findings suggest growing awareness, they stop short of indicating widespread, firm-level commitment to long-term climate objectives.

The analysis highlights patterns suggesting potential links between stronger environmental performance and lower credit risk—echoing trends identified in existing literature—but cannot establish causality due to the presence of confounding macroeconomic and regulatory factors. For instance, firms with lower emissions and higher *CP* scores may also benefit from stronger governance, better capital structures, or greater investor confidence, which independently reduce credit risk. Moreover, firms that disclose emissions and climate data may be systematically different from non-disclosing firms, potentially biasing risk estimates. These limitations suggest that while environmental metrics may offer useful signals to investors and credit analysts, further research using longitudinal or experimental methods is needed to establish clear causal pathways.

2.6 Implications

This chapter also reflects on the evolving, yet uneven, regulatory landscape in the MENA region. While initiatives such as the Saudi Green Initiative and the expansion of voluntary carbon markets demonstrate positive momentum (United Nations, 2023; Middle East Institute, 2023), these efforts are largely concentrated in a few high-capacity states. The findings of this chapter suggest that policy responses should be tailored to the region's heterogeneity: for resource-rich countries with high industrial emissions, policy efforts might include sector-specific emissions caps, scaled-up investment in clean energy, and gradually introducing carbon pricing mechanisms. However, for countries with weaker institutional capacity or less diversified economies, softer instruments, such as voluntary disclosure frameworks, capacity-building programs, and international climate finance, may be more feasible entry points.

Barriers to implementation persist. Political resistance, fiscal limitations, and competing development priorities often hinder regulatory enforcement, particularly among SMEs, which face structural and financial constraints (World Economic

Forum, 2024). Bridging this gap requires a coordinated effort among multiple stakeholders. Investors can play a critical role by integrating carbon risk metrics into capital allocation decisions, while international organizations can support harmonized disclosure standards and technical assistance. Civil society and advocacy groups may also strengthen accountability by demanding greater transparency from firms and governments alike.

By providing firm-level evidence on carbon emissions, carbon performance, climate commitment, and credit risk, this chapter contributes to a more nuanced understanding of where progress is being made and where critical gaps remain. The findings offer a data-driven foundation for designing pragmatic and differentiated policy tools, while also encouraging broader stakeholder engagement to support an inclusive transition toward climate-aligned economic development in the MENA region.

References

Agrawal, A. K., & Matsa, D. A. (2013). Labor unemployment risk and corporate financing decisions. *Journal of Financial Economics, 108*(2), 449–470. https://doi.org/10.1016/j.jfineco.2012.11.006

Albaker, A., Abbasi, K. R., Haddad, A. M., Radulescu, M., Manescu, C., & Bondac, G. T. (2023). Analyzing the impact of renewable energy and green innovation on carbon emissions in the MENA region. *Energies, 16*(16), 6053. https://doi.org/10.3390/en16166053

Albitar, K., Al-Shaer, H., & Liu, Y. S. (2023). Corporate commitment to climate change: The effect of eco-innovation and climate governance. *Research Policy, 52*(2), 104697. https://doi.org/10.1016/j.respol.2022.104697

Albitar, K., Hussainey, K., Kolade, N., & Gerged, A. M. (2020). ESG disclosure and firm performance before and after IR: The moderating role of governance mechanisms. *International Journal of Accounting & Information Management, 28*(3), 429–444. https://doi.org/10.1108/IJAIM-09-2019-0108

Aldy, J. E., & Gianfrate, G. (2019). Future-proof your climate strategy. *Harvard Business Review, 97*(4), 16–86.

Alsaifi, K., Elnahass, M., & Salama, A. (2020). Carbon disclosure and financial performance: UK environmental policy. *Business Strategy and the Environment, 29*(2), 711–726. https://doi.org/10.1002/bse.2426

Altman, E. I. (1968). Financial ratios, discriminant analysis and the prediction of corporate bankruptcy. *The journal of finance, 23*(4), 589–609. https://doi.org/10.1111/j.1540-6261.1968.tb00843.x

Altman, E. I., Iwanicz-Drozdowska, M., Laitinen, E. K., & Suvas, A. (2017). Financial distress prediction in an international context: A review and empirical analysis of Altman's Z-score model. *Journal of International Financial Management & Accounting, 28*(2), 131–171. https://doi.org/10.1111/jifm.12053

Amit, R., & Schoemaker, P. J. (1993). Strategic assets and organizational rent. *Strategic Management Journal, 14*(1), 33–46. https://doi.org/10.1002/smj.4250140105

Anand, A., Vanpée, R., & Lončarski, I. (2023). Sustainability and sovereign credit risk. *International Review of Financial Analysis, 86*, 102494. https://doi.org/10.1016/j.irfa.2023.102494

Ang, A., & Longstaff, F. A. (2013). Systemic sovereign credit risk: Lessons from the US and Europe. *Journal of Monetary Economics, 60*(5), 493–510. https://doi.org/10.1016/j.jmoneco.2013.04.009

Bae, S. C., Chang, K., & Yi, H. C. (2018). Corporate social responsibility, credit rating, and private debt contracting: New evidence from syndicated loan market. *Review of Quantitative Finance and Accounting, 50*, 261–299. https://doi.org/10.1007/s11156-017-0630-4

Battiston, S., Mandel, A., Monasterolo, I., Schütze, F., & Visentin, G. (2017). A climate stress-test of the financial system. *Nature Climate Change, 7*(4), 283–288. https://doi.org/10.1038/nclimate3255

Ben Cheikh, N., & Ben Zaied, Y. (2021). A new look at carbon dioxide emissions in MENA countries. *Climatic Change, 166*(3), 27. https://doi.org/10.1007/s10584-021-03126-9

Bendig, D., Schäper, T., & Erbar, F. (2024). Revealing the truth: The moderating role of internal stakeholders in sustainability communication. *Journal of Cleaner Production, 434*, 139969. https://doi.org/10.1016/j.jclepro.2023.139969

Bento, N., Borello, M., & Gianfrate, G. (2020). Market-pull policies to promote renewable energy: A quantitative assessment of tendering implementation. *Journal of Cleaner Production, 248*, 119209. https://doi.org/10.1016/j.jclepro.2019.119209

Birindelli, G., Bonanno, G., Dell'Atti, S., & Iannuzzi, A. P. (2022). Climate change commitment, credit risk and the country's environmental performance: Empirical evidence from a sample of international banks. *Business Strategy and the Environment, 31*(4), 1641–1655. https://doi.org/10.1002/bse.2974

Boiral, O., Henri, J. F., & Talbot, D. (2012). Modeling the impacts of corporate commitment on climate change. *Business Strategy and the Environment, 21*(8), 495–516. https://doi.org/10.1002/bse.723

Bui, B., Houqe, M. N., & Zaman, M. (2020). Climate governance effects on carbon disclosure and performance. *The British Accounting Review, 52*(2), 100880. https://doi.org/10.1016/j.bar.2019.100880

Capasso, G., Gianfrate, G., & Spinelli, M. (2020). Climate change and credit risk. *Journal of Cleaner Production, 266*, 121634. https://doi.org/10.1016/j.jclepro.2020.121634

CDP. (2023). *Scoring Introduction 2023*. https://cdn.cdp.net/cdp-production/cms/guidance_docs/pdfs/000/000/233/original/Scoring-Introduction.pdf?1639144388

Chelli, M., Durocher, S., & Richard, J. (2014). France's new economic regulations: Insights from institutional legitimacy theory. *Accounting, Auditing & Accountability Journal, 27*(2), 283–316. https://doi.org/10.1108/AAAJ-07-2013-1415

Chen, X. H., Tee, K., Elnahass, M., & Ahmed, R. (2023). Assessing the environmental impacts of renewable energy sources: A case study on air pollution and carbon emissions in China. *Journal of Environmental Management, 345*, 118525. https://doi.org/10.1016/j.jenvman.2023.118525

Cho, C. H., & Patten, D. M. (2007). The role of environmental disclosures as tools of legitimacy: A research note. *Accounting, Organizations and Society, 32*(7–8), 639–647. https://doi.org/10.1016/j.aos.2006.09.009

Christy, A., Elnahass, M., Amezaga, J., Browne, A., & Heidrich, O. (2024). A dynamic framework to align company climate reporting and action with global climate targets. *Business Strategy and the Environment, 33*(4), 3103–3128. https://doi.org/10.1002/bse.3635

Clarkson, P. M., Li, Y., Pinnuck, M., & Richardson, G. D. (2015). The valuation relevance of greenhouse gas emissions under the European Union carbon emissions trading scheme. *European Accounting Review, 24*(3), 551–580. https://doi.org/10.1080/09638180.2014.927782

Costola, M., & Vozian, K. (2025). Pricing climate transition risk: Evidence from European corporate CDS. Energy Economics, 143, 108248. https://doi.org/10.1016/j.eneco.2025.108248

Cotter, J., & Najah, M. M. (2012). Institutional investor influence on global climate change disclosure practices. *Australian Journal of Management, 37*(2), 169–187. https://doi.org/10.1177/0312896211423945

De Greiff, K., Delis, M. D., & Ongena, S. (2018). *Being stranded on the carbon bubble? Climate policy risk and the pricing of bank loans*. Climate Policy Risk and the Pricing of Bank Loans (May 2018). CEPR Discussion Paper No. DP12928. https://cepr.org/publications/dp12928

Deegan, C. (2002). Introduction: The legitimising effect of social and environmental disclosures – A theoretical foundation. *Accounting, Auditing & Accountability Journal, 15*(3), 282–311. https://doi.org/10.1108/09513570210435852

Eliwa, Y., Aboud, A., & Saleh, A. (2021). ESG practices and the cost of debt: Evidence from EU countries. *Critical Perspectives on Accounting, 79*, 102097. https://doi.org/10.1016/j.cpa.2019.102097

Erdoğan, S., Yıldırım, S., Yıldırım, D. Ç., & Gedikli, A. (2020). The effects of innovation on sectoral carbon emissions: Evidence from G20 countries. *Journal of Environmental Management, 267*, 110637. https://doi.org/10.1016/j.jenvman.2020.110637

Esmaeili, B., Pereira, J. S., Malafront, I., & Nurullah, M. (2024). *The effect of carbon performance on firm credit risk.*

Freeman, R. E. (1984). *Strategic management: A stakeholder approach.* Cambridge University Press.

Gilchrist, S., & Zakrajšek, E. (2012). Credit spreads and business cycle fluctuations. *American Economic Review, 102*(4), 1692–1720. https://doi.org/10.1257/aer.102.4.1692

Hahn, T., Pinkse, J., Preuss, L., & Figge, F. (2015). Tensions in corporate sustainability: Towards an integrative framework. *Journal of Business Ethics, 127*(2), 297–316. https://doi.org/10.1007/s10551-014-2047-5

Harangozo, G., & Szigeti, C. (2017). Corporate carbon footprint analysis in practice – With a special focus on validity and reliability issues. *Journal of Cleaner Production, 167*, 1177–1183. https://doi.org/10.1016/j.jclepro.2017.07.237

Hart, S. L. (1995). A natural-resource-based view of the firm. *Academy of Management Review, 20*(4), 986–1014. https://doi.org/10.5465/amr.1995.9512280033

Hassan, T., Khan, Y., He, C., Chen, J., Alsagr, N., & Song, H. (2022). Environmental regulations, political risk and consumption-based carbon emissions: Evidence from OECD economies. *Journal of Environmental Management, 320*, 115893. https://doi.org/10.1016/j.jenvman.2022.115893

International Renewable Energy Agency. (2023). *MENA's transforming role in an evolving energy landscape.* https://www.irena.org/News/expertinsights/2023/Oct/MENAs-Transforming-Role-in-an-Evolving-Energy-Landscape

Ji, J., Li, T., & Yang, L. (2023). Pricing and carbon reduction strategies for vertically differentiated firms under Cap-and-Trade regulation. *Transportation Research Part E: Logistics and Transportation Review, 171*, 103064. https://doi.org/10.1016/j.tre.2023.103064

Jiang, M., Gao, X., Guan, Q., Hao, X., & An, F. (2019). The structural roles of sectors and their contributions to global carbon emissions: A complex network perspective. *Journal of Cleaner Production, 208*, 426–435. https://doi.org/10.1016/j.jclepro.2018.10.127

Jung, J., Herbohn, K., & Clarkson, P. (2018). Carbon risk, carbon risk awareness, and the cost of debt financing. *Journal of Business Ethics, 150*, 1151–1171. https://doi.org/10.1007/s10551-016-3207-6

Kabir, M. N., Rahman, S., Rahman, M. A., & Anwar, M. (2021). Carbon emissions and default risk: International evidence from firm-level data. *Economic Modelling, 103*, 105617. https://doi.org/10.1016/j.econmod.2021.105617

Kinley, R. (2017). Climate change after Paris: From turning point to transformation. *Climate Policy, 17*(1), 9–15. https://doi.org/10.1080/14693062.2016.1191009

Kleimeier, S., & Viehs, M. (2018, January 7). *Carbon disclosure, emission levels, and the cost of debt.* https://doi.org/10.2139/ssrn.2719665

Koten, H. (2018). Hydrogen effects on the diesel engine performance and emissions. *International Journal of Hydrogen Energy, 43*(22), 10511–10519. https://doi.org/10.1016/j.ijhydene.2018.04.146

Lemma, T. T., Lulseged, A., & Tavakolifar, M. (2021). Corporate commitment to climate change action, carbon risk exposure, and a firm's debt financing policy. *Business Strategy and the Environment, 30*(8), 3919–3936. https://doi.org/10.1002/bse.2849

Liêu, M. L., Dao, T., Nguyen, T. H., & Trinh, V. Q. (2024). Climate governance and carbon risk in the global energy sector: Insights into corporate environmental initiatives. *Energy Economics, 137*, 107782. https://doi.org/10.1016/j.eneco.2024.107782

Liu, Y. S., Zhou, X., Yang, J. H., Hoepner, A. G. F., & Kakabadse, N. (2023). Carbon emissions, carbon disclosure and organizational performance. *International Review of Financial Analysis, 90*, 102846. https://doi.org/10.1016/j.irfa.2023.102846

Longstaff, F. A., Pan, J., Pedersen, L. H., & Singleton, K. J. (2011). How sovereign is sovereign credit risk? *American Economic Journal: Macroeconomics, 3*(2), 75–103. https://doi.org/10.1257/mac.3.2.75

MacKie-Mason, J. K. (1990). Do taxes affect corporate financing decisions?. *The journal of finance, 45*(5), 1471–1493. https://doi.org/10.2307/2328746

Mariani, M., Pizzutilo, F., Caragnano, A., & Zito, M. (2021). Does it pay to be environmentally responsible? Investigating the effect on the weighted average cost of capital. *Corporate Social Responsibility and Environmental Management, 28*(6), 1854–1869. https://doi.org/10.1002/csr.2164

Merton, R. C. (1974). On the pricing of corporate debt: The risk structure of interest rates. *The Journal of Finance, 29*(2), 449–470. https://doi.org/10.2307/2978814

Middle East Institute. (2023). *Carbon trading in the MENA region: Opportunities and challenges*. https://www.mei.edu/publications/carbon-trading-mena-region-opportunities-and-challenges

Misani, N., & Pogutz, S. (2015). Unraveling the effects of environmental outcomes and processes on financial performance: A non-linear approach. *Ecological Economics, 109*, 150–160. https://doi.org/10.1016/j.ecolecon.2014.11.010

Nguyen, Q., Diaz-Rainey, I., & Kuruppuarachchi, D. (2023). In search of climate distress risk. *International Review of Financial Analysis, 85*, 102444. https://doi.org/10.1016/j.irfa.2022.102444

Nordhaus, W. (2017). Integrated assessment models of climate change. *NBER Reporter, 3*, 16–20. https://www.econstor.eu/bitstream/10419/178753/1/2017-no3-3.pdf

O'Donovan, G. (2002). Environmental disclosures in the annual report: Extending the applicability and predictive power of legitimacy theory. *Accounting, Auditing & Accountability Journal, 15*(3), 344–371. https://doi.org/10.1108/09513570210435870

Omri, A., & Saidi, K. (2022). Factors influencing CO_2 emissions in the MENA countries: The roles of renewable and non-renewable energy. *Environmental Science and Pollution Research, 29*(37), 55890–55901. https://doi.org/10.1007/s11356-022-19727-5

Owolabi, A., Mousavi, M. M., Gozgor, G., & Li, J. (2024). The impact of carbon risk on the cost of debt in the listed firms in G7 economies: The role of the Paris Agreement. *Energy Economics, 139*, 107925. https://doi.org/10.1016/j.eneco.2024.107925

Parmar, B. L., Freeman, R. E., Harrison, J. S., Wicks, A. C., Purnell, L., & de Colle, S. (2010). Stakeholder theory: The state of the art. *Academy of Management Annals, 4*(1), 403–445. https://doi.org/10.5465/19416520.2010.495581

Patten, D. M. (2002). The relation between environmental performance and environmental disclosure: A research note. *Accounting, Organizations and Society, 27*(8), 763–773. https://doi.org/10.1016/S0361-3682(02)00028-4

Pereira, J., Sorwar, G., & Nurullah, M. (2018). What drives corporate CDS spreads? A comparison across US, UK, and EU firms. *Journal of International Financial Markets, Institutions and Money, 56*, 188–200. https://doi.org/10.1016/j.intfin.2018.02.002

Qiu, Y., Shaukat, A., & Tharyan, R. (2016). Environmental and social disclosures: Link with corporate financial performance. *The British Accounting Review, 48*(1), 102–116. https://doi.org/10.1016/j.bar.2014.10.007

Roberts, R. W. (1992). Determinants of corporate social responsibility disclosure: An application of stakeholder theory. *Accounting, organizations and society, 17*(6), 595–612. https://doi.org/10.1016/0361-3682(92)90015-K

Rodríguez, I. M., Dandapani, K., & Lawrence, E. R. (2019). Measuring sovereign risk: Are CDS spreads better than sovereign credit ratings?. *Financial Management, 48*(1), 229–256. https://doi.org/10.1111/fima.12223

Safiullah, M., Phan, D. H. B., & Kabir, M. N. (2024). Green innovation and corporate default risk. *Journal of International Financial Markets, Institutions and Money, 95*, 102041. https://doi.org/10.1016/j.intfin.2024.102041

Sarumpaet, S., Nelwan, M. L., & Dewi, D. N. (2017). The value relevance of environmental performance: Evidence from Indonesia. *Social Responsibility Journal, 13*(4), 817–827. https://doi.org/10.1108/SRJ-01-2017-0003

Suchman, M. C. (1995). Managing legitimacy: Strategic and institutional approaches. *Academy of Management Review, 20*(3), 571–610. https://doi.org/10.5465/amr.1995.9508080331

United Nations. (2015). *The Paris Agreement*. https://unfccc.int/sites/default/files/english_paris_agreement.pdf

United Nations Climate Change. (2022). *MENA Climate Week opens with aim to accelerate climate action*. https://unfccc.int/news/mena-climate-week-opens-with-aim-to-accelerate-climate-action

United Nations. (2023). *Saudi Arabia leading efforts in afforestation and climate change adaptation in the MENA region*. https://saudiarabia.un.org/en/249596-saudi-arabia-leading-efforts-afforestation-and-climate-change-adaptation-mena-region

Vassalou, M., & Xing, Y. (2004). Default risk in equity returns. *The Journal of Finance, 59*(2), 831–868. https://doi.org/10.1111/j.1540-6261.2004.00650.x

Vershinina, N., Rodgers, P., Tarba, S., Khan, Z., & Stokes, P. (2020). Gaining legitimacy through proactive stakeholder management: The experiences of high-tech women entrepreneurs in Russia. *Journal of Business Research, 119*, 111–121. https://doi.org/10.1016/j.jbusres.2018.12.063

Wiedmann, T. (2009). Editorial: Carbon footprint and input–output analysis—An introduction. *Economic Systems Research, 21*(3), 175–186. https://doi.org/10.1080/09535310903541256

World Economic Forum. (2024). *How MENA's biggest actors can help the region's suppliers and SMEs to decarbonize*. https://www.weforum.org/agenda/2024/05/mena-climate-action-decarbonization-smes/

WRI, & WBCSD. (2020). *Corporate value chain (scope 3) standard*. https://ghgprotocol.org/sites/default/files/standards/Corporate-Value-Chain-Accounting-Reporing-Standard_041613_2.pdf

Zhang, Z., & Zhao, R. (2022). Carbon emission and credit default swaps. *Finance Research Letters, 50*, 103286. https://doi.org/10.1016/j.frl.2022.103286

Chapter 3
The Impact of Natural Disasters on Credit Creation: Evidence from Middle East and North African (MENA) Countries

Rasim Simsek, Ramshah Rashid Lone, Md Tanvir Hamim, and Shakila Halim

Abstract This study investigates the impact of climate-related natural disasters on credit creation in Middle East and North African (MENA) countries compared to non-MENA countries. Using a fixed-effects panel regression model and a dataset covering 22 MENA and 200 non-MENA countries from 1974 to 2022, we find that natural disasters lead to increased credit creation in MENA countries, whereas their impact on credit creation is statistically insignificant in non-MENA countries. Further analysis reveals that the relationship between natural disasters and credit creation varies by income level: high-income MENA countries experience increased credit creation following natural disasters, while low-income non-MENA countries show a negative relationship. Our findings are robust across various econometric specifications, including a two-step system generalized method of moments regression and propensity score matching. This study contributes to the ongoing debate on the economic implications of natural disasters in the MENA region by providing empirical evidence on their effect on credit creation. The results offer important policy implications for financial regulators, policymakers, and institutions in

R. Simsek (✉)
Faculty of Economics and Administrative Sciences, Bilecik Seyh Edebali University, Bilecik, Turkey
e-mail: rasim.simsek@bilecik.edu.tr

R. R. Lone
Faculty of Business and Law, Anglia Ruskin University, Chelmsford, UK
e-mail: ramsha.Lone@aru.ac.uk

M. T. Hamim
Department of Banking and Insurance, University of Dhaka, Dhaka, Bangladesh
e-mail: tanvir.hamim@du.ac.bd

S. Halim
Finance and Economics Department, Business and Law Faculty, Manchester Metropolitan University, Manchester, UK
e-mail: s.halim@mmu.ac.uk

M. Elnahass et al. (eds.), *A Greenprint for Net Zero*, Gulf Studies 15,
https://doi.org/10.1007/978-981-92-1762-5_3

MENA, highlighting the need for effective risk management strategies to mitigate the effects of such disasters on the credit creation mechanism.

Keywords Natural disasters · Financial institutions · Credit creation · MENA

3.1 Introduction

Over the past 30 years, approximately 10,000 natural disasters have occurred globally, impacting over 7 billion people and causing an estimated US$2 trillion in losses—a trend that continues to rise (EM-DAT, 2013). Regulatory bodies increasingly require banks and financial institutions to maintain adequate capital buffers and account for systemic environmental risks that may threaten financial stability (Alexander, 2014). The increasing risk of natural disasters is largely driven by population growth, urbanization, and unpredictable weather patterns linked to climate change (Knutson et al., 2020). Moreover, environmental degradation has exacerbated these risks, contributing to economic instability (CISL & UNEPFI, 2015). Studies indicate that the frequency, intensity, and costs associated with natural disasters—key manifestations of environmental risk—are increasing (Freeman et al., 2003; Gramlich et al., 2023). Between 1980 and 2012, economic losses from catastrophes reached US$3.8 trillion, with 74% attributed to adverse weather events (Bettencourt et al., 2013). Each major disaster disrupts significant portions of regional financial infrastructure (Klomp, 2014). Climate-related disasters such as droughts, floods, and heatwaves continue to stress financial intermediaries and economic stability.

However, the impact of such shocks differs between high- and low-income countries (Alalmaee, 2024). High-income nations typically suffer fewer severe consequences due to stronger infrastructure, better credit ratings, and established insurance systems. Existing literature suggests that natural disasters affect financial system resilience by increasing non-performing loans and triggering deposit withdrawals (Anh et al., 2022; Walker et al., 2023). Firms and households often accumulate more debt post-disaster (Keerthiratne & Tol, 2017).

Credit facilitates post-crisis recovery by allowing businesses to replace damaged assets and resume profit-generating activities (Collier & Babich, 2019). Immediate access to finance is crucial for clean-up, rehabilitation, and rebuilding (Brei et al., 2024). In such scenarios, both households and firms rely on credit to fund recovery (Berg & Schrader, 2012; Del Ninno et al., 2003). Financial institutions not directly impacted by disasters may benefit by providing loans to affected areas (Koetter et al., 2020). In emerging markets with weak insurance and social safety nets, credit becomes essential for managing catastrophic losses (Collier & Babich, 2019). In line with this reasoning, it is anticipated that people in calamity-stricken areas may attempt to obtain loans and advances, particularly immediately following a natural disaster.

The Middle East and North African (MENA) region is highly susceptible to natural disasters, including floods, droughts, and sandstorms. Countries such as Yemen, Egypt, the UAE, and Saudi Arabia frequently face flooding due to heavy rainfall, urbanization, and poor drainage. Droughts have harmed agriculture in Qatar, Kuwait, Tunisia, and Morocco, jeopardizing food security. Sandstorms disrupt daily life and worsen health issues across the region. MENA's high population density further increases its climate vulnerability.[1] Despite these challenges, most literature focuses on natural disasters in developed countries, leaving MENA relatively understudied.

This study aims to provide an understanding of how natural disasters affect credit creation in the MENA region. We also seek to highlight potential policy implications for both the public and private sectors, drawing upon empirical evidence. In this study, we examine the impact of natural disasters on credit creation, with a particular focus on the MENA region. Our primary objective is to assess the extent to which natural disasters influence the scope of credit creation across the 22 MENA countries. The sample period spans from 1974 to 2022, with a final sample size of 192 countries, including 22 from the MENA region. After applying a fixed-effects model to estimate the regression coefficients, our analysis reveals that natural disasters lead to an increase in credit creation in the MENA countries. This finding suggests that there is an increased demand for credit, as individuals, businesses, and communities tend to deal with the damage and losses caused by natural disasters within the region. To ensure the robustness of our baseline estimates, we run a two-step generalized method of moments (GMM) regression. Furthermore, we employ a propensity score matching technique to mitigate any potential endogeneity bias associated with the baseline findings. Finally, to gain further insights, we split the full sample into high- and low-income countries in both the MENA and non-MENA regions to perform sub-sample analyses.

The available literature generally agrees that natural calamities have a detrimental impact on a country's short-term macroeconomic performance (Felbermayr & Groschl, 2014; Klomp, 2016; Sseruyange & Klomp, 2021). Hence, it is commonly believed that natural disasters are likely to impede credit creation. Our findings from non-MENA countries are in line with the relevant literature. However, the current understanding of the scope and likelihood of credit creation may not hold for countries in the MENA region, as the empirical results indicate that a higher ratio of the population affected by natural disasters is associated with greater credit demand. We argue that this positive association is plausible because households and businesses within the MENA region seek loans to manage the aftermath of natural disasters, whereas financial intermediaries assist in the process of providing credit facilities. Furthermore, despite MENA being one of the most severely affected by climate change, there is still an absence of empirical investigation into how natural disasters might affect credit creation within this region (Gebrael et al., 2024).

[1] https://www.iemed.org/publication/climate-change-in-the-mena-region-environmental-riskssocioeconomiceffectsandpolicychallengesforthefuture

Therefore, we are among the first to fill this vacuum and offer novel empirical evidence that natural disasters stimulate credit creation in MENA countries. This study contributes to the ongoing debate on natural disasters and their economic implications in the MENA region.

The remainder of this chapter is organized as follows: Section 3.2 reviews the relevant literature. Sections 3.3 and 3.4 present the data and empirical results, while Sect. 3.5 discusses the findings. Section 3.6 offers recommendations for policymakers and practitioners. Finally, Sect. 3.7 concludes the chapter with a summary and suggestions for future research.

3.2 Literature Review

Countries worldwide face an increased risk from natural disasters due to urbanization, population growth, and erratic weather patterns associated with climate change (Knutson et al., 2020). While immediate and unrestricted access to finance must be available for recovery, natural disasters create uncertainty and ambiguity, resulting in shortages in credit and funding markets (Brei et al., 2024). Any delays in the availability of funds can hinder pre-disaster economic growth, potentially resulting in a recession and slowing the implementation of future technologies (Noy, 2009). Thus, a country's ability to quickly mobilize finance plays a vital role in mitigating the impacts of disasters and averting systemic macroeconomic consequences. Banks play an essential role in this context as insurance policies and government disaster relief measures often fail to cover the full extent of damages (Barth et al., 2022; Mejia et al., 2019). Garmaise and Moskowitz (2009) provide a detailed overview of how imperfections in the supply of catastrophic insurance can distort real estate markets by limiting the provision of bank credit and preventing positive NPV projects from being undertaken. This study finds that banks' monitoring problems (as lenders) may make them unwilling to extend credit in the absence of insurance, regardless of the promised interest rate. Alternatively, a lack of insurance may raise the interest rate to a level that borrowers find unacceptable. As a result, underdeveloped insurance markets can restrict bank lending, hampering neighborhood revitalization in disadvantaged areas. The authors further argue that terrorism and political risk share the central features of natural disaster risk and may be subject to an insufficient supply of insurance. Thus, the absence of terrorism insurance and the restricted supply of political risk insurance discourage foreign investment and depress emerging market capital markets. This is particularly important for financial institutions as terrorism risk increases credit, insolvency, and systemic risk, which negatively affect financial stability (Elnahass et al., 2022).

Although the drivers of credit availability have been analyzed from various perspectives, it remains important to address the question of whether and to what extent access to credit changes after natural disasters. It is important to find answers to these issues, as it is often assumed that the existence of credit markets can dilute the effects of shocks, for example, through the provision of emergency loans. The determinants of access to credit depend on a wide range of factors; however, important aspects include, among others, the institutional and regulatory environment, clients' financial means, and qualitative information on the potential borrower, which is particularly important in the case of relationship lending. Berg and Schrader (2012) used a natural experiment approach to identify the determinants of credit availability in response to unpredictable aggregate shocks. This study uses a dataset that includes loan applications and subsequent approvals for a self-sustainable Ecuadorian microfinance institution (MFI) from 2002 to 2007. The empirical results show that credit demand increased significantly after volcanic eruptions, suggesting that there is a need for additional financing in the aftermath of shocks. When analyzing access to credit on the level of the individual client, the results indicate that a history of natural disasters in the affected area before the credit application leads to a significant decrease in the probability of loan approvals. However, being a repeat borrower lowers these constraints. More specifically, the results show that returning clients not only have a higher probability of receiving a loan in general, but they are approximately equally likely to be approved for a loan after volcanic shocks occur. However, new clients do not have this advantage; therefore, in times of crisis, it becomes even more difficult for them to receive financing. This study shows that bank–borrower relationships are important determinants for increasing access to finance in times of unpredictable aggregate shocks.

While well-functioning credit markets are crucial for the effective allocation of resources, economic growth and financial intermediaries are also affected by exogenous shocks. Thus, the supply of credit after a disaster is shaped by financial institutions' strategies. Analyzing how these potential shocks affect financial intermediation, Choudhary and Jain (2022) explore banks' exposure to unprecedented flooding in Pakistan. This study shows how a change in a bank's funding cost after a disaster influences its lending decisions. The study finds that banks adopt credit rationing following a funding shock after a natural disaster; specifically, a 1% point increase in the funding shock leads to a 1.5% point decrease in the likelihood that a bank will provide a loan to a given borrower 2 years after the flood in non-flooded areas. Second, banks disproportionately reduce credit to certain borrowers; specifically, consumers with little education and credit history are rationed the most. Third, the reduction in credit is not compensated for by increased aggregate lending from less-affected banks. The study suggests that a reduction in banks' incentives to monitor loans is the most likely cause of the disproportionate fall in lending to individuals with low education and seasonal occupations. Specifically, in the non-flooded area after the floods, more-affected banks reduced lending more on loans with higher initial overdue rates than less-affected banks. Moreover, more-affected banks reduced lending for unsecured loans more than less-affected banks. Finally, the study shows that public banks are less affected by funding shocks resulting from natural disasters and are more likely to survive negative shocks. The findings suggest a possible role for policy interventions in limiting credit rationing to marginalized, economically vulnerable borrowers.

Koetter et al. (2022) analyze the role of regional banks in absorbing regional economic shocks from natural disasters. Specifically, they examine how banks lend to disaster-struck firms and how these firms borrow. The study focuses on the 2013 flooding of the River Elbe and its tributaries as a natural experiment. The study finds three main results: first, local banks in unaffected areas increased their lending by 3% after the flood compared to unexposed banks. These banks were better capitalized but had reduced liquidity buffers. Second, banks within diversified banking groups—able to collect deposits and originate loans across regions—were better able to withstand adverse conditions. These banks provided recovery lending without taking additional risks. Finally, firms in flooded areas with ties to lenders in unaffected regions benefited from increased borrowing, while firms without such ties experienced a decline in credit after the flood. These results suggest that local banks play a crucial role in providing recovery lending, especially for smaller firms affected by regional shocks. Diversified networks of local banks are more effective than isolated regional banking systems in buffering such shocks. Therefore, these dynamics indicate that an overall increase or decrease in credit creation following a natural disaster may mask deeper shifts in borrowing composition and access. Lee et al. (2022) examine the relationship between climate change and bank liquidity creation in 56 countries over the period 1995–2012. Specifically, they investigate whether the relationship between climate change and bank liquidity creation varies by bank and country. The results reveal that climate sensitivity and exposure negatively impact overall liquidity creation by banks. On the demand side, banks tend to reduce lending during uncertain times, thus decreasing asset-side liquidity creation. Conversely, on the supply side, banks may hold more liquid liabilities for precautionary reasons, thus increasing supply-side liquidity creation. Additionally, banks shift away from businesses that contribute to climate risk and focus more on environmentally friendly investments, which can enhance demand-side liquidity creation. This study finds that liquidity effects are more pronounced for larger banks with lower capital, banks located in lower gross domestic product (GDP) and developing countries, and those in Asia. The results suggest that policymakers should exercise caution when formulating and implementing climate-related strategies, as these can influence liquidity creation, which, in turn, can affect macroeconomic stability.

While supporting economic recovery and growth, banks are also affected by natural disasters. Natural disasters affect the repayment capabilities of both individuals and small firms. Following a catastrophe, bank insolvency risk increases because of various factors, such as immediate deposit withdrawal, an increase in borrowing demand, lower creditworthiness, and an increase in loan loss provisions (Alalmaee, 2024; Do et al., 2022). However, banks adopt different risk management strategies to protect their business performance. Collier et al. (2011) assess the effect of flooding disasters in Peru and report a 3.6% increase in restructured loans as a percentage of total loans. Berg and Schrader (2012) analyze the after-effects of natural calamities on credit access and demand for loans and find that although loan demand increases due to natural disasters, the probability of loan approvals declines. Do et al. (2022) conduct a study on bank vulnerability to natural disasters, primarily due to liquidity and volatility in bank deposits. They further argue that banks increase their loan loss provisions to minimize volatility and liquidity problems, resulting in a loss of competitiveness. However, they suggest that maintaining an appropriate level of loan loss provisions before a disaster can be an effective strategy to minimize the impact of disaster risk without weakening bank capital. Dal Maso et al. (2024) also emphasize that banks should cater to risk arising from natural disasters by following risk management strategies through loan loss provisions. Their findings suggest a positive relationship between natural disasters and loan loss provisions, with a 5.5–7% increase in loan loss provisions due to a one standard deviation change in disaster risk. Chamberlain et al. (2019) study the relationship between natural disasters and bank liquidity and find a decrease in bank deposits following the disaster, affecting the supply of funds necessary to facilitate loan growth. Their study suggests that banks need more traditional policies regarding loan loss provisions and that banks that make prudent or timely provisioning decisions in the period preceding disasters can respond more quickly to new loan demands and experience more significant growth in their loan portfolios. Furthermore, Albuquerque and Wassim (2019) explore the relationship between natural disasters and bank stability, stating that the effect of natural disasters is interim and disadvantageous only to non-performing loans.

While extensive literature exists on how natural disasters affect the supply of credit, relatively few studies have examined the effect of calamities on bank deposits or how they can affect loan demand. Skidmore and Toya (2002) explore the saving tendencies of households in the event of natural catastrophes. The results suggest a higher savings rate among households in disaster-stricken areas. They argue that households adopt this behavior as a self-insurance mechanism and that this happens when government policies and insurance claims do not provide enough security against unforeseen circumstances. Brei et al. (2019) argue that, after a natural catastrophe, banks face an increased number of deposit withdrawals. The authors argue that this occurs because firms and households try to hold liquid funds to overcome losses in income and cash flow shortages due to natural calamities.

An increase in credit creation following natural disasters is often interpreted as a sign of economic resilience and a responsive financial sector. However, higher credit demand and supply may also reflect a failure in public disaster relief mechanisms, potentially exacerbating existing inequalities within affected communities. Berg and Schrader (2012) show that increased credit provision after disasters tends to benefit borrowers with pre-existing relationships with financial institutions, while new applicants are more likely to be excluded. Ratcliffe et al. (2020) find that individuals who were already financially struggling before a disaster suffer more severe financial consequences afterward, particularly in areas lacking sustained public recovery support. Their findings suggest that post-disaster credit expansion does not necessarily imply inclusive recovery or long-term financial stability; rather, it may disproportionately burden financially vulnerable populations. Government financial assistance plays a critical role in mitigating the negative distributional effects of unequal credit access by reducing households' immediate reliance on private borrowing. Public loan initiatives can protect marginalized groups from immediate financial limitations and pave the way for future private financing by improving borrower creditworthiness and minimizing risk and uncertainty within the financial sector (Collier et al., 2024).

Therefore, whether post-disaster credit expansion constitutes a positive financial response or a symptom of institutional failure remains a contentious issue in the literature. Our study contributes to this discussion by empirically examining heterogeneity in credit creation across regional contexts. In particular, the observed increase in credit creation may reflect the presence of a stronger financial infrastructure that enables post-disaster lending. However, such expansion may also stem from ineffective or insufficient government financial responses, ultimately exacerbating the vulnerability of financially excluded groups who are forced to rely more heavily on private debt in the absence of robust public financial support mechanisms.

3.2.1 Developed and Developing Countries

While the literature on the effect of natural disasters on the economy has been growing in the past two decades (Klomp & Valckx, 2014), further research is required to investigate the impact of natural disasters by exploring the dissimilarities between developed and developing countries, with developing countries being more exposed to natural disasters (Noy, 2009). Empirical evidence exists that shows that developing countries are more exposed to natural calamities, as their loan portfolios are more concentrated in specific sectors and locations (BCBS, 2021). Banks adopt a concentration strategy regarding loan portfolios to overcome the issue of imperfect information; however, this strategy fails when such sectors or locations are more exposed to natural calamities (Collier et al., 2011). Developing countries further increase the credit risk of banks due to limited insurance options available to individuals and businesses, poor infrastructure, and lack of social support from the governments of these countries (Bueno et al., 2008; Lashley & Warner, 2013; Pelham et al., 2011).

Cavallo et al. (2013) discuss that it is not always the natural calamities that affect the economic growth of developing countries; other internal factors also have an important role in affecting these countries' financial stability. The result of their study suggests that natural disaster events do not necessarily have a damaging effect on the economy of all developing countries unless they are combined with political distress. Caprio and Klingebiel (1996) show that droughts, combined with poor governance and lending standards by banks, resulted in a banking crisis in Kenya and Senegal, where many financial institutions had to be liquidated or rescued by the government. After the 1998 floods in Bangladesh, Dowla (2018) reports that Grameen Bank encountered increased credit risk as 25% of the bank's borrowers defaulted on their loans. Massive deposit withdrawals further complicated the credit crisis, leading to government interventions that helped with the bank's liquidity problem. A similar study on the 2010 floods in Pakistan reports threats to bank capital and liquidity, resulting in reduced lending to new borrowers (Choudhary & Jain, 2022). Siamwalla et al. (1990), conducting a study on floods in Thailand, find the inability of rural borrowers to obtain funds from rural lenders or any other financial institution.

Another noticeable aspect of natural disasters is the time horizon over which disaster risks materialize and require critical consideration. Boustan et al. (2020) link it to the maturity profile of the loan book, disaster relief measures, and the capacity of the private sector to withstand adverse shocks. Brei et al. (2024) focus on continental countries and report that banks' funding is massively affected in the first 2 years following the disaster, which leads to a decrease in bank lending and net income over 4–6 years. Further results also highlight an increase in banks' non-performing loan (NPL) ratios in these countries. Berg and Schrader (2009) explore the relationship between natural disasters and bank stability in the Latin America and Caribbean regions. The authors show that natural disasters increased the default risk of first-time borrowers compared to existing or old borrowers. The reason for this difference is that banks charge lower interest rates in the recovery phase to borrowers with whom they have long-term relationships. In a follow-up study, Berg and Schrader (2012) document that credit applications increase with higher loan approvals for firms with existing bank relationships. A study by Collier et al. (2011) concludes that microfinance institutions in Peru become more vulnerable following natural disasters, leading to a high level of loan restructuring.

Based on the literature review, it is evident that natural disasters have a variety of effects on the economy and financial sector. These effects are contingent upon the development status of a country.

3.3 Data

We collect natural disaster information for 222 countries, comprising 22 MENA[2] and 200 non-MENA countries, covering the period from 1974 to 2023, from EM-DAT, the International Disaster Database from the Centre for Research on the Epidemiology of Disasters. While EM-DAT includes a range of different types of natural disasters categorized by their sources, such as meteorological, geophysical, technological, hydrological, biological, or climatological, we only focus on climate-related natural disasters categorized under meteorological, hydrological, and climatological events. We restrict our sample to climate-related natural disasters that had observable human impacts—such as fatalities, injuries, homelessness, or displacement—as certain events may occur without directly affecting individuals. Table 3.1 summarizes the number of climate-related natural disasters occurring in MENA and non-MENA countries over the last 50 years, from 1974 to 2023.

[2] Following the IMF classification, this study considers MENA countries to include Algeria, Bahrain, Djibouti, Egypt, the Islamic Republic of Iran, Iraq, Jordan, Kuwait, Lebanon, Libya, Mauritania, Morocco, Oman, Qatar, Saudi Arabia, Somalia, Sudan, the Syrian Arab Republic, Tunisia, the United Arab Emirates, the West Bank and Gaza, and Yemen.

Table 3.1 Distribution of the natural disasters

	MENA		Non-MENA		Total	
	No. of events	Occurrence rate	No. of events	Occurrence rate	No. of events	Occurrence rate
1. Meteorological	70	19.61%	2918	33.17%	2988	32.65%
1.1. Storm	55	15.41%	2741	31.16%	2796	30.55%
1.2. Extreme temperature	15	4.20%	177	2.01%	192	2.10%
2. Hydrological	259	72.55%	5101	57.99%	5360	58.56%
2.1. Mass movement	8	2.24%	457	5.20%	465	5.08%
2.2. Flood	251	70.31%	4644	52.80%	4895	53.48%
3. Climatological	28	7.84%	777	8.83%	807	8.82%
3.1. Drought	21	5.88%	475	5.40%	496	5.42%
3.2. Wildfire	7	1.96%	302	3.43%	309	3.38%
Total	357	100.00%	8796	100.00%	9153	100.00%

The distribution of natural disaster types reveals similarities and differences between MENA and non-MENA countries. Floods, categorized under hydrological disasters, are the most prevalent in both groups. The occurrence rate of floods is significantly higher in MENA countries, accounting for 70% of all climate-related natural disasters, compared to 58.5% in non-MENA countries. On the other hand, the MENA region experiences storms categorized under Meteorological disasters far less frequently than non-MENA countries, with a 15.4% occurrence rate in MENA compared to 31.1% of climate-related natural disasters caused by storms in non-MENA countries. Finally, climatological disasters display a similar pattern in both MENA and non-MENA countries, with occurrence rates of 7.8% and 8.8%, respectively, among climate-related natural disasters. In the last 50 years, the MENA region has experienced 357 climate-related natural disasters that have impacted human life, while non-MENA countries have experienced 8796 such disasters. To assess disaster severity, we construct the variable *AffectedRatio*, defined as the total number of individuals affected by climate-related natural disasters relative to the total population.

We utilize the World Bank DataBank database to collect credit creation data. We select three different metrics to measure credit creation from the Global Financial Development database: the ratio of private credit by deposit money banks to GDP (*Credit1*), the ratio of domestic credit to the private sector to GDP (*Credit2*), and the ratio of domestic credit to the private sector by banks to GDP (*Credit3*).

To capture the control variables, we also rely on data from the World Bank DataBank. These control variables include GDP per capita (*GDPPC*), inflation (*Inflation*), CO_2 emissions per capita (*Emission*), foreign direct investment (*FDI*) as a percentage of GDP, and trade openness (*Trade*), measured as the sum of exports and imports as a percentage of GDP. Higher *GDPPC* typically indicates a more developed economy with better financial infrastructure, which can facilitate higher levels of credit creation (Chang & Zhang, 2020). Wealthier countries tend to have more robust banking systems (Claessens & Laeven, 2003), higher savings rates (Dynan et al., 2004), and greater demand for credit (Beck et al., 2012), all of which result in higher credit demand, leading to increased credit creation. Furthermore, countries with higher *GDPPC* may have better financial and institutional mechanisms to mitigate the immediate negative effects of disasters, such as insurance systems, government relief programs, and more diversified economies. Moderate inflation might stimulate borrowing by reducing the real cost of debt, but high inflation can deter credit creation by increasing uncertainty and eroding the value of collateral (Fouejieu, 2017). Higher CO_2 emissions might indicate a more industrialized and energy-dependent economy, which could correlate with greater credit demand for industrial expansion and energy-related projects (Ali et al., 2021). However, high emissions also suggest potential environmental risks, which could impact the stability of credit markets in the long term (Nasreen et al., 2017). FDI can stimulate credit creation by increasing the demand for credit to finance projects linked to foreign investments. It also enhances financial stability and access to capital by bringing in foreign capital and expertise (Pek et al., 2018). Countries with higher FDI often have more developed financial markets and institutions, which can support higher levels of credit creation (Carbonell & Werner, 2018). Finally, higher trade openness is positively associated with greater economic activity, which leads to better access to international financial markets and can enhance credit creation (Ashraf, 2018; Niepmann & Schmidt-Eisenlohr, 2017).

3.4 Method

We conduct a fixed-effect regression analysis to examine the impact of climate-related natural disasters on credit creation across MENA and non-MENA countries. The baseline model is specified as follows:

$$CreditCreation_{i,t} = \alpha_0 + \beta_1 AffectedRatio_{i,t-1} + \beta_2 CreditCreation_{i,t-1} + \Upsilon X_{i,t} + \varepsilon_{i,t}$$

where $CreditCreation_{i,t}$ represents the credit creation measures for country i at time t, $AffectedRatio_{i,t}$ captures the ratio of the population affected by climate-related natural disasters to the total population, and $X_{i,t}$ denotes the vector of control variables, including *GDPPC*, inflation, CO_2 emissions, FDI, and trade openness, α is the constant, and $\varepsilon_{i,t}$ is the error term. We employ the 1-year lagged ranks of the independent and dependent variables in the model. This helps to address potential endogeneity concerns by reducing the possibility of reverse causality. Additionally, the use of lagged variables allows us to capture the delayed effects of natural disasters on credit creation, as the impact of such events on financial systems and credit markets may not be immediate. This methodology also helps to control for any temporal autocorrelation that might exist in the data, thereby providing more robust and reliable estimates of the relationship between climate-related natural disasters and credit creation. This study employs a fixed-effects model for these estimations, following the statistically significant results from the Hausman test.

3.5 Empirical Results

3.5.1 Descriptive Statistics

Table 3.2 presents the summary statistics of the variables. All financial variables are winsorized at the 1st and 99th percentiles to deal with extreme outliers or any data coding errors. Table 3.2 presents the number of observations, means, standard deviations, medians, minimums, and maximums. The mean (median) values of the dependent variables for the full sample are 0.442 (0.321) for *Credit1*, 0.492 (0.331) for *Credit2*, and 0.431 (0.312) for *Credit3*. The mean (median) value of *AffectedRatio* is 0.029 (0.002) for the full sample. For the MENA and non-MENA subgroups, the mean (median) values are 0.339 (0.301) and 0.448 (0.322) for *Credit1*, 0.334 (0.263) and 0.502 (0.335) for *Credit2*, and 0.323 (0.301) and 0.438 (0.313) for *Credit3*, respectively. In terms of disaster exposure, the mean (median) values of *AffectedRatio* are 0.012 (0.000) for MENA countries and 0.030 (0.002) for non-MENA countries. The t-test statistics indicate that the mean values of *Credit1*, *Credit2*, *Credit3*, and *AffectedRatio* are statistically different between MENA and non-MENA countries.

Table 3.3 reports the results of the correlation analysis. As expected, the correlations between *AffectedRatio* and all credit creation proxies are negative and statistically significant at the 1% level. In line with the literature, credit creation measures are positively and significantly correlated with *GDPPC*, *Emissions*, *FDI*, and *Trade*, while *Inflation* is negatively and significantly associated with credit creation variables.

Furthermore, the four graphs provided in Fig. 3.1 illustrate the trends of *AffectedRatio*, *Credit1*, *Credit2*, and *Credit3* for the MENA and Non-MENA regions from 1974 to 2022. Graph (a) under Fig. 3.1 clearly demonstrates that MENA countries exhibit a higher ratio of affected population during natural disasters, indicating a more severe impact relative to their population. In contrast, non-MENA countries maintain a lower and more stable ratio, suggesting either better disaster preparedness or differences in the occurrence and scale of such events. This graph highlights the disparity in disaster impact ratios between the MENA and non-MENA regions, with MENA countries appearing more vulnerable to natural disasters.

In Graphs (b), (c), and (d) under Fig. 3.1, all three credit creation measures exhibit a similar upward trend over time. The most notable observation is that non-MENA countries experienced steadier growth with less volatility compared to MENA countries. Non-MENA countries display smoother growth patterns with fewer pronounced dips, indicating more stable credit expansion over the long term. In contrast, MENA countries underwent aggressive credit creation between 1980

Table 3.2 Descriptive statistics

	Panel A: MENA countries						Panel B: Non-MENA countries						Panel C
	Obs.	Mean	S.D.	50%	Min	Max	Obs.	Mean	S.D.	50%	Min	Max	t-test
Credit 1	115	0.339	0.234	0.301	0.019	0.960	1913	0.448	0.378	0.322	0.019	1.691	−3.071***
Credit 2	102	0.334	0.265	0.263	0.020	1.006	1637	0.502	0.448	0.335	0.020	1.829	−3.730***
Credit 3	115	0.323	0.220	0.301	0.013	0.962	1870	0.438	0.375	0.313	0.012	1.679	−3.249***
AffectedRatio	115	0.012	0.060	0.000	0.000	0.574	1913	0.030	0.085	0.002	0.000	1.014	−2.141**
GDPPC	115	7.968	0.863	7.920	6.299	10.670	1913	7.977	1.504	7.938	4.856	11.244	−0.062
Inflation	115	0.090	0.104	0.050	−0.026	0.532	1913	0.089	0.229	0.043	−0.026	2.549	0.046
Emissions	115	11.262	1.359	11.290	6.028	13.317	1913	10.072	2.458	9.853	4.182	14.845	5.146***
FDI	115	20.240	2.353	20.830	6.908	23.812	1913	20.740	2.582	20.788	9.210	25.500	−2.025**
Trade	115	0.711	0.420	0.624	0.292	3.209	1913	0.688	0.356	0.594	0.155	3.209	0.669
	Panel D: Full sample												
	Obs.	**Mean**	**S.D.**	**50%**	**Min**	**Max**							
Credit1	2028	0.442	0.372	0.321	0.019	1.691							
Credit2	1739	0.492	0.441	0.331	0.020	1.829							
Credit3	1985	0.431	0.369	0.312	0.012	1.679							
AffectedRatio	2028	0.029	0.083	0.002	0.000	1.014							
GDPPC	2028	7.976	1.475	7.931	4.856	11.244							
Inflation	2028	0.089	0.224	0.044	−0.026	2.549							
Emissions	2028	10.139	2.424	10.044	4.182	14.845							
FDI	2028	20.712	2.572	20.790	6.908	25.500							
Trade	2028	0.689	0.360	0.596	0.155	3.209							

Note: *Credit1* represents the ratio of private credit provided by deposit money banks to GDP. *Credit2* is the ratio of domestic credit extended to the private sector to GDP, while *Credit3* denotes the ratio of domestic credit to the private sector provided specifically by banks to GDP. *AffectedRatio* is the ratio of the total number of people affected by natural disasters to the total population. *GDPPC* stands for GDP per capita, *Inflation* refers to the year-on-year change in the Consumer Price Index (CPI), and *Emission* indicate CO_2 emissions per capita. *FDI* represents the ratio of foreign direct investment to GDP, and *Trade* is the sum of exports and imports expressed as a percentage of GDP. ***, **, and * denote statistical significance at the 1%, 5%, and 10% levels, respectively

Table 3.3 Correlation matrix

	(1)	(2)	(3)	(4)	(5)	(6)	(7)	(8)	(9)
(1)	1.000								
(2)	0.910***	1.000							
(3)	0.957***	0.929***	1.000						
(4)	–0.129***	–0.130***	–0.134***	1.000					
(5)	0.534***	0.554***	0.539***	–0.258***	1.000				
(6)	–0.288***	–0.292***	–0.291***	0.072***	–0.279***	1.000			
(7)	0.417***	0.443***	0.425***	–0.216***	0.440***	–0.084***	1.000		
(8)	0.432***	0.444***	0.435***	–0.203***	0.502***	–0.156***	0.537***	1.000	
(9)	0.132***	0.107***	0.127***	0.033***	0.121***	–0.084***	–0.131***	–0.015	1.000

Note: (1) *Credit1* represents the ratio of private credit provided by deposit money banks to GDP. (2) *Credit2* is the ratio of domestic credit extended to the private sector to GDP. (3) *Credit3* is the ratio of domestic credit to the private sector provided specifically by banks to GDP. (4) *AffectedRatio* is the ratio of the total number of people affected by natural disasters to the total population. (5) *GDPPC* stands for GDP per capita. (6) *Inflation* refers to the year-on-year change in the Consumer Price Index (CPI). (7) *Emissions* indicate CO_2 emissions per capita. (8) *FDI* represents the ratio of FDI to GDP. (9) *Trade* is the sum of exports and imports expressed as a percentage of GDP. ***, **, and * denote statistical significance at the 1%, 5%, and 10% levels, respectively

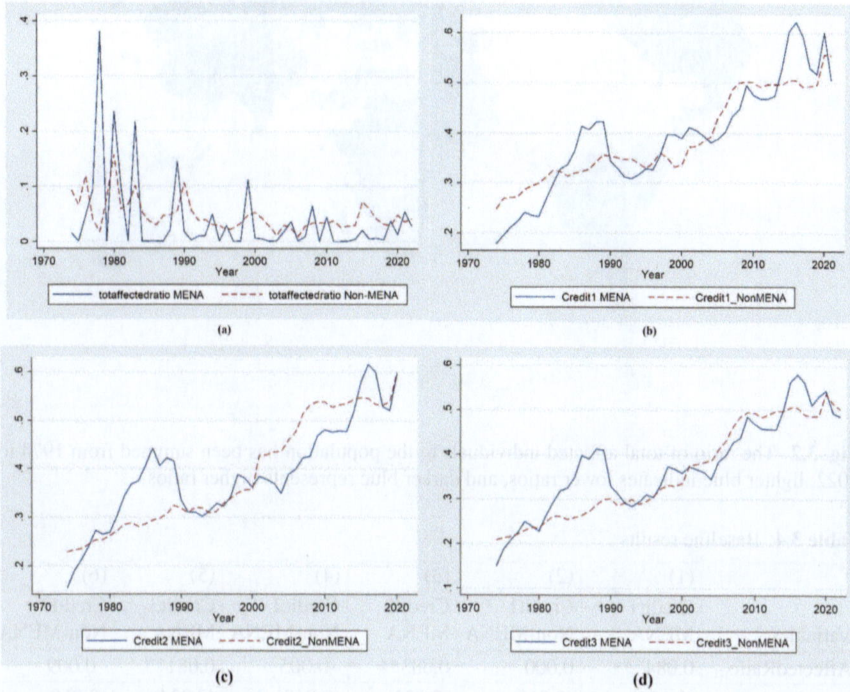

Fig. 3.1 Time-series plots of the *TotalAffected* (**a**), *Credit1* (**b**), *Credit2* (**c**), and *Credit3* (**d**) in MENA and Non-MENA countries from 1974 to 2022

and 1990, followed by a trajectory resembling that of non-MENA countries up until the end of the 2010s.

Figure 3.2 is based on historical data of the countries. The map illustrates the ratio of the total population affected across different regions, represented using a gradient of blue hues. Lighter shades of blue correspond to regions with a lower ratio of affected individuals, while darker shades indicate regions with a higher ratio.

3.5.2 Baseline Results

Table 3.4 presents the results of the baseline models. Models 1 and 2 examine the effect of the ratio of the population affected by natural disasters on the ratio of private credit by deposit money banks to GDP in MENA countries and the rest of the world, respectively. Models 3 and 4 assess the same independent variable's impact on credit creation, measured by the ratio of domestic credit to the private sector to GDP, again for MENA and non-MENA countries, respectively. Finally, Models 5 and 6 focus on domestic credit to the private sector by banks as a share of GDP in both groups of countries.

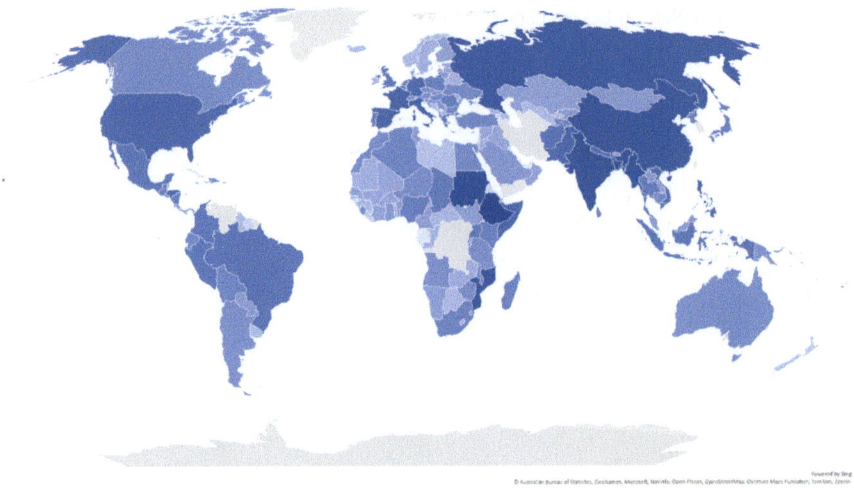

Fig. 3.2 The ratio of total affected individuals to the population has been summed from 1974 to 2022; lighter blue indicates lower ratios, and darker blue represents higher ratios

Table 3.4 Baseline results

Variables	(1) Credit1 MENA	(2) Credit1 Non-MENA	(3) Credit2 MENA	(4) Credit2 Non-MENA	(5) Credit3 MENA	(6) Credit3 Non-MENA
$AffectedRatio_{t-1}$	0.084***	0.000	0.083**	–0.005	0.081**	–0.009
	(0.031)	(0.010)	(0.033)	(0.012)	(0.034)	(0.010)
$Credit1_{t-1}$	0.929***	0.983***				
	(0.033)	(0.009)				
$Credit2_{t-1}$			0.965***	0.983***		
			(0.020)	(0.009)		
$Credit3_{t-1}$					0.947***	0.976***
					(0.024)	(0.010)
GDPPC	1.557*	0.147	1.134	0.320	1.321	0.357
	(0.896)	(0.225)	(0.865)	(0.249)	(0.848)	(0.228)
Inflation	–14.835***	–2.151**	–13.436**	–1.259*	–14.311**	–1.960**
	(5.640)	(0.954)	(5.716)	(0.716)	(5.653)	(0.964)
Emission	–0.866	0.144	–0.517	0.169	–0.666	0.176*
	(0.770)	(0.098)	(0.897)	(0.121)	(0.779)	(0.103)
FDI	–0.019	0.074	–0.066	0.048	–0.077	0.015
	(0.126)	(0.103)	(0.113)	(0.105)	(0.115)	(0.101)
Trade	–0.044**	0.009	–0.029	0.002	–0.032	0.006
	(0.020)	(0.006)	(0.025)	(0.007)	(0.020)	(0.006)
Constant	5.123	–2.765*	3.280	–3.316**	4.401	–3.086*
	(7.442)	(1.547)	(8.768)	(1.550)	(7.413)	(1.675)
Observations	113	1,863	100	1,589	113	1,841
No. of countries	15	135	15	132	15	135
$chi^2(1)$	34.06***		23.61***		26.21***	
Prob > chi^2	0.0000		0.0000		0.0000	
Hausman test	76.07***		68.69***		63.37***	

Note: ***, **, and * denote statistical significance at the 1%, 5%, and 10% levels, respectively

In Models 1, 3, and 5, the coefficient for *AffectedRatio* is positive and statistically significant, suggesting that in MENA countries, an increase in the ratio of the population affected by natural disasters leads to an increase in credit demand. This implies that natural disasters may stimulate credit creation in these regions, possibly due to the increased demand for credit in the aftermath of disasters. In addition to this, a significant relationship between the ratio of the population affected by natural disasters and credit creation does not appear in the non-MENA countries. In Models 2, 4, and 6, the coefficient for *AffectedRatio* is either small or negative and statistically insignificant, indicating that natural disasters do not have a significant effect on credit creation in non-MENA countries as they have in MENA countries.

On the one hand, this may reflect a responsive and resilient financial sector that facilitates access to funds for recovery—enabling households and businesses to finance reconstruction, replace lost assets, and stabilize their operations (Duqi et al., 2021; Koetter et al., 2020). Increased lending can act as an economic stabilizer by stimulating consumption and investment in the aftermath of disasters. From this perspective, higher credit creation may be interpreted as a sign of financial system adaptability and a contributor to economic resilience. On the other hand, reliance on credit-based recovery may introduce longer-term risks. Gallagher and Hartley (2017) and Ratcliffe et al. (2020) show that affected individuals and firms may incur substantial debt burdens that are not always matched by income-generating recovery, leading to increased financial vulnerability, especially among already disadvantaged groups. If borrowed funds are used to meet basic consumption or cover uninsured losses rather than for productive investment, the accumulation of debt may exacerbate inequality and delay recovery (Fengler et al., 2008; Ratcliffe et al., 2019). Therefore, while credit creation can offer short-term relief and liquidity, it may also generate unsustainable financial stress when public safety nets and insurance systems are weak or absent.

One potential problem that exists in our results is the disparity in the number of observations and countries between MENA and non-MENA regions. MENA models (1), (3), and (5) include substantially fewer observations than their non-MENA counterparts (Models 2, 4, and 6), reflecting the smaller sample size available for MENA countries. To address the possibility that these differences might influence our findings, we perform a coefficient comparison between each model for MENA and non-MENA countries. The results yielded a significant chi-square test, suggesting that the observed relationships are robust and not driven by differences in sample size between MENA and non-MENA countries.

The two-step system GMM regression results, presented in Table 3.5, suggest that the model is not biased, as the residuals in the first difference show serial correlation as indicated by a statistically significant AR(1), while the residuals in the second difference do not exhibit serial correlation as indicated by a statistically insignificant AR(2). The model includes valid instruments, as the Hansen J-statistics of over-identifying restrictions are statistically insignificant, indicating the validity of the instrument. Despite variations in the significance levels of the relationships, the findings from both the fixed-effects panel and the two-step system GMM regressions demonstrate consistency; an increase in the ratio of the population affected by natural disasters leads to a statistically significant increase in credit creation in MENA countries.

Table 3.5 System GMM estimation results

Variables	(1) Credit1 MENA	(2) Credit1 Non-MENA	(3) Credit2 MENA	(4) Credit2 Non-MENA	(5) Credit3 MENA	(6) Credit3 Non-MENA
$AffectedRatio_{t-1}$	0.111***	–0.013	0.097***	–0.010	0.106***	–0.019
	(0.038)	(0.019)	(0.027)	(0.017)	(0.033)	(0.018)
$Credit1_{t-1}$	0.991***	1.015***				
	(0.011)	(0.004)				
$Credit2_{t-1}$			1.003***	1.007***		
			(0.007)	(0.003)		
$Credit3_{t-1}$					1.004***	1.012***
					(0.010)	(0.003)
GDPPC	0.000	–0.003***	–0.001	–0.001	–0.001	–0.002**
	(0.004)	(0.001)	(0.002)	(0.001)	(0.003)	(0.001)
Inflation	–0.087***	–0.018***	–0.078***	–0.013**	–0.083***	–0.017***
	(0.025)	(0.005)	(0.018)	(0.005)	(0.022)	(0.004)
Emission	–0.002	–0.000	–0.001	0.000	–0.001	0.000
	(0.003)	(0.001)	(0.002)	(0.001)	(0.003)	(0.001)
FDI	0.000	0.001	0.000	0.001	0.000	0.001
	(0.001)	(0.001)	(0.001)	(0.001)	(0.001)	(0.001)
Trade	–0.033***	0.005	–0.024***	0.004	–0.027***	0.005*
	(0.010)	(0.003)	(0.008)	(0.003)	(0.009)	(0.003)
Constant	0.047	0.002	0.048**	–0.009	0.054*	0.003
	(0.031)	(0.010)	(0.024)	(0.010)	(0.028)	(0.009)
Observations	113	1863	100	1589	113	1,841
No. of countries	15	135	15	132	15	135
AR(1)	0.009	0.000	0.078	0.000	0.046	0.000
AR(2)	0.416	0.186	0.293	0.960	0.299	0.403
Hansen J-Statistics	0.807	0.910	0.426	0.221	0.665	0.858

Note: ***, **, and * denote statistical significance at the 1%, 5%, and 10% levels, respectively

3.5.3 High- and Low-Income Countries

Examining countries based on their income levels can yield important insights into the subject. For instance, Elnahass et al. (2022) investigate the effects of terrorism in high- and low-income MENA countries and find that countries from different income groups exhibit variation in their financial outcomes. In this regard, Table 3.6 reveals the baseline results using the subsamples of high- and low-income countries in MENA and non-MENA regions to test the relationship between natural disasters and credit creation in these regions. World Bank DataBank classifies the income of the countries into four categories: low income, lower middle income, upper middle income, and high income. Following Chang and Zhang (2020) and Elnahass et al. (2022), we consider low-income and lower-middle-income countries as low-income

Table 3.6 Baseline results from high- and low-income countries

Variables	(1) MENA High income	(2) MENA Low income	(3) Non-MENA High income	(4) Non-MENA Low income
Panel A—Dependent variable: Credit1				
$AffectedRatio_{t-1}$	0.105***	–0.139	0.020	–0.017**
	(0.018)	(0.126)	(0.029)	(0.007)
$Credit1_{t-1}$	1.001***	0.871***	0.977***	1.014***
	(0.027)	(0.065)	(0.011)	(0.016)
GDPPC	0.004	0.025	0.003	–0.005*
	(0.012)	(0.020)	(0.004)	(0.003)
Inflation	–0.173*	–0.186*	–0.028	–0.010*
	(0.098)	(0.099)	(0.017)	(0.006)
Emission	0.005	–0.004	0.000	0.002*
	(0.018)	(0.011)	(0.002)	(0.001)
FDI	–0.001	0.005	0.001	0.000
	(0.002)	(0.007)	(0.002)	(0.001)
Trade	–0.027	–0.035**	–0.002	0.023***
	(0.044)	(0.017)	(0.006)	(0.007)
Constant	–0.031	–0.138	–0.022	0.001
	(0.136)	(0.100)	(0.036)	(0.014)
Observations	66	47	1,020	843
No. of countries	9	6	78	57
$chi^2(1)$	11.53***		1.48	
$Prob>chi^2$	0.000		0.224	
Panel B—Dependent variable: Credit2				
$AffectedRatio_{t-1}$	0.109***	-0.129**	0.042	–0.025***
	(0.018)	(0.051)	(0.038)	(0.008)
$Credit2_{t-1}$	0.990***	0.949***	0.981***	1.031***
	(0.028)	(0.029)	(0.009)	(0.024)
GDPPC	0.005	0.025**	0.005	–0.007**
	(0.012)	(0.011)	(0.004)	(0.003)
Inflation	–0.159*	–0.102	–0.015	–0.011
	(0.090)	(0.091)	(0.010)	(0.009)
Emission	0.005	0.007	0.002	0.000
	(0.017)	(0.006)	(0.002)	(0.001)
FDI	–0.001	–0.003	0.000	0.001
	(0.002)	(0.004)	(0.002)	(0.001)
Trade	–0.024	–0.004	–0.006	0.019***
	(0.045)	(0.021)	(0.006)	(0.006)
Constant	–0.031	–0.151*	–0.037	0.019
	(0.137)	(0.086)	(0.031)	(0.015)
Observations	66	34	840	749
No. of countries	9	6	77	55

(continued)

Table 3.6 (continued)

Variables	(1) MENA High income	(2) MENA Low income	(3) Non-MENA High income	(4) Non-MENA Low income
$chi^2(1)$	12.48***		0.44	
Prob>chi^2	0.000		0.507	
Panel C—Dependent variable: Credit3				
$AffectedRatio_{t-1}$	0.109***	–0.129*	0.015	–0.024***
	(0.018)	(0.077)	(0.028)	(0.008)
$Credit3_{t-1}$	0.992***	0.917***	0.976***	1.024***
	(0.028)	(0.052)	(0.011)	(0.022)
GDPPC	0.005	0.020	0.004	–0.005*
	(0.012)	(0.014)	(0.004)	(0.003)
Inflation	–0.158*	–0.141*	–0.025	–0.006
	(0.092)	(0.075)	(0.017)	(0.006)
Emission	0.005	0.002	0.000	0.001
	(0.018)	(0.006)	(0.002)	(0.001)
FDI	–0.001	–0.001	0.001	0.000
	(0.002)	(0.004)	(0.002)	(0.001)
Trade	–0.021	–0.016	–0.004	0.021***
	(0.047)	(0.010)	(0.006)	(0.005)
Constant	–0.033	–0.092	–0.023	0.010
	(0.139)	(0.080)	(0.037)	(0.014)
Observations	66	47	986	855
No. of countries	9	6	78	57
$chi^2(1)$	17.68***		1.28	
Prob>chi^2	0.000		0.258	

Note: ***, **, and * denote statistical significance at the 1%, 5%, and 10% levels, respectively

countries, while upper-middle-income and high-income countries are categorized as high-income countries.

After classifying countries into two main income groups, the number of observations and countries in the high- and low-income MENA subsamples is relatively limited compared to the non-MENA counterparts. While the number of countries and observations in Table 3.6 are consistent with our baseline results in Table 3.4, the sample sizes of the MENA subgroups are relatively smaller compared to the non-MENA subgroups, reflecting the limited number of countries, the availability of long-term consistent data, and the frequency of climate-related disaster reporting. Therefore, to address internal variation and sample size concerns, we follow Simsek et al. (2024) and conduct a coefficient comparison between high- and low-income groups within MENA and non-MENA countries.[3] The results show that the

[3] We also conduct coefficient comparisons between high-income MENA and high-income non-MENA countries, as well as between low-income MENA and low-income non-MENA subgroups. The statistically significant chi-squared test for the high-income comparison indicates that regional differences persist after controlling for income levels.

differences in the impact of natural disasters on credit creation between income groups are statistically significant between high- and low-income countries in the MENA region, supporting the conclusion that these differences are not driven by sample size variation. This suggests that income-based heterogeneity in post-disaster credit dynamics is more evident in MENA countries, likely due to sharper contrasts in financial infrastructure, credit access, and institutional resilience.

A positive and statistically significant coefficient, *AffectedRatio*, indicates a rise in credit creation in high-income MENA countries. This finding suggests that natural disasters may lead to increased lending activity. Higher credit creation following natural disasters implies that businesses and individuals may resort to debt to finance recovery and reconstruction. Furthermore, the accumulation of debt might undermine long-term recovery, particularly if the borrowed funds do not translate into sustainable income or assets. Duqi et al. (2021) find that natural disasters may stimulate lending activity in countries with resilient banking sectors, where the financial infrastructure can absorb risk and sustain credit intermediation in times of crisis, especially by increasing the supply and refinancing the existing mortgage. Similarly, Koetter et al. (2020) find that financial institutions in non-affected areas increase the supply of credits for the companies operating in affected areas, thereby capitalizing on reconstruction demands. Although an increase in credit creation after a disaster may reflect strong support from the financial sector, it can also result in adverse consequences by adding financial stress to already indebted individuals and firms, particularly when the newly incurred debt does not yield productive returns or contribute to income generation.

Conversely, the findings show a negative but insignificant coefficient of the *AffectedRatio* in low-income MENA countries, suggesting that natural disasters may not impact or even decrease credit creation. This could be due to either efficient government financial assistance, which minimizes the need for additional borrowing, or the limited financial resources available in these regions. Dafermos et al. (2018) demonstrate that climate-related shocks exacerbate financial system fragility by limiting credit availability in less developed financial systems. Similarly, Collier and Babich (2019) emphasize that in the absence of robust financial markets, low-income populations often lack access to credit. In the same vein, Berg and Schrader (2012) show that the probability of loan approval significantly declines for low-income borrowers, although demand for loans increases after natural disasters.

For non-MENA regions, the results reveal a contrasting pattern based on income levels compared to MENA countries. In high-income non-MENA countries, the effect of natural disasters on credit creation is generally small and statistically insignificant, suggesting that these countries may not experience a significant shift in credit dynamics following natural disasters. On the other hand, in low-income non-MENA countries, the effect of natural disasters on credit creation is negative and statistically significant. This negative relationship could occur because of stronger institutional frameworks, efficient government financial assistance measures to alleviate the effects of disasters, and diversified financial markets that can absorb shocks more effectively, mitigating the immediate need for additional credit (Brei et al., 2019; Cavallo & Noy, 2011; Klomp, 2014; Noy, 2009).

3.5.4 Propensity Score Matching

The descriptive statistics of the sample presented in Table 3.7 indicate that there are differences between MENA and non-MENA countries from various perspectives. While the mean comparisons indicate no significant differences in credit creation measures and the population affected by natural disasters between MENA and non-MENA countries, certain control variables exhibit statistically significant differences. Accordingly, CO_2 emissions are significantly higher in MENA countries compared to non-MENA countries, whereas FDI is lower. Combined with the substantial difference in the number of observations between the two sample groups, these disparities could cause biased results, such as sample selection bias. To overcome this problem, we employ a propensity score matching approach to construct a balanced sample by matching MENA (treatment group) with non-MENA (control group) based on observed covariates. We then run our baseline model using the matched sample, which is constructed using one-to-one nearest neighbor matching without replacement, utilizing *GDPPC*, inflation, CO_2 emissions, FDI, and trade openness as matching variables. Panel A of Table 3.7 presents the descriptive statistics of the matched sample. The matching sample consists of 230 matched observations (115 MENA, 115 non-MENA) for *Credit1*, 204 matched observations (102 MENA, 102 non-MENA) for *Credit2*, and 232 matched observations (116 MENA, 116 non-MENA) for *Credit3*.

Panel B of Table 3.7 presents the results of the baseline regression using the propensity score-matched sample; the results for our dependent variable, which is the lagged ratio of the population affected by natural disasters, remain consistent. Specifically, the coefficient for the lagged ratio is positive and statistically significant for all credit creation measures in MENA countries, indicating that natural disasters continue to be associated with increased credit creation in these regions. Conversely, the coefficient for the same variable is insignificant or negative for non-MENA countries, suggesting a weaker or negligible impact on credit creation. These results confirm the robustness of the baseline findings, as the matching procedure effectively balances the covariates between MENA and non-MENA countries, minimizing sample selection bias.

3.6 Policy Implication

The findings of this study carry several important policy implications, particularly regarding the role of credit markets and public institutions in responding to climate-related natural disasters. Our findings reveal that in MENA nations, credit creation experiences a notable increase after natural disasters. In contrast, this relationship is insignificant in countries outside the MENA region.

Although it is crucial to address the economic impacts of natural disasters, nations should initially focus on enacting policies for climate change mitigation and

Table 3.7 Baseline results with PSM

Panel A: Balancing tables for propensity score matching

	Treatment		Control				Treatment		Control				Treatment		Control		
	N	Mean	N	Mean	t-test		N	Mean	N	Mean	t-test		N	Mean	N	Mean	t-test
Credit1	115	0.338	115	0.394	−1.65	Credit2	102	0.334	102	0.390	−1.40	Credit3	116	0.326	116	0.319	0.26
AffectedRatio	115	0.012	115	0108	0.25	AffectedRatio	102	0.013	102	0.014	−0.10	AffectedRatio	116	0.012	116	0.010	0.28
GDPPC	115	7.967	115	8.058	−0.57	GDPPC	102	8.011	102	8.082	−0.44	GDPPC	116	7.983	116	7.772	1.37
Inflation	115	0.089	115	0.073	1.43	Inflation	102	0.096	102	0.098	−0.06	Inflation	116	0.089	116	0.091	−0.17
Emissions	115	11.262	115	11.508	−1.05	Emissions	102	11.370	102	11.282	0.35	Emissions	116	11.261	116	11.296	−0.15
FDI	115	20.240	115	20.514	−0.79	FDI	102	20.204	102	20.359	−0.41	FDI	116	20.250	116	20.465	−0.66
Trade	115	0.711	115	0.673	0.74	Trade	102	0.718	102	0.803	−1.32	Trade	116	0.713	116	0.715	−0.04

Panel B: Baseline regressions using propensity score matching samples

	(1)	(2)	(3)	(4)	(5)	(6)
Variables	**Credit1 MENA**	**Credit1 Non-MENA**	**Credit2 MENA**	**Credit2 Non-MENA**	**Credit3 MENA**	**Credit3 Non-MENA**
$AffectedRatio_{t-1}$	0.098***	−0.005	0.094***	−0.004	0.096***	0.083
	(0.008)	(0.039)	(0.011)	(0.035)	(0.010)	(0.103)
$Credit1_{t-1}$	0.943***	0.970***				
	(0.042)	(0.025)				
$Credit2_{t-1}$			0.993***	0.941***		
			(0.010)	(0.023)		
$Credit3_{t-1}$					1.000***	0.920***
					(0.011)	(0.042)
GDPPC	−0.013	0.006	−0.015	0.014***	−0.013	0.004
	(0.009)	(0.004)	(0.009)	(0.004)	(0.009)	(0.005)
Inflation	−0.144**	−0.036	−0.114*	0.005	−0.116*	−0.099***

(continued)

Table 3.7 (continued)

Panel A: Balancing tables for propensity score matching																	
	Treatment		Control				Treatment		Control				Treatment		Control		
	N	Mean	N	Mean	t-test		N	Mean	N	Mean	t-test		N	Mean	N	Mean	t-test
	(0.062)	(0.062)	(0.069)	(0.085)	(0.070)	(0.029)											
Emission	0.016***	–0.001	0.014**	0.002	0.015***	0.008											
	(0.006)	(0.003)	(0.006)	(0.004)	(0.004)	(0.005)											
FDI	0.001	–0.001	0.001	–0.002	0.000	-0.006											
	(0.002)	(0.002)	(0.002)	(0.004)	(0.002)	(0.004)											
Trade	0.015	0.005	0.020	–0.007	0.022	0.035											
	(0.018)	(0.013)	(0.020)	(0.009)	(0.014)	(0.024)											
Constant	–0.079	0.011	–0.052	–0.061	–0.067	0.008											
	(0.095)	(0.030)	(0.094)	(0.041)	(0.079)	(0.041)											

Note: ***, **, and * denote statistical significance at the 1%, 5%, and 10% levels, respectively

adaptation. This is particularly urgent for MENA countries, given their significant environmental footprint. According to the EDGAR 2024 report, nine MENA countries (Qatar, Kuwait, Bahrain, UAE, Oman, Saudi Arabia, Libya, Iran, and Iraq) rank within the top quartile globally for per capita greenhouse gas emissions, with the first six of these given MENA countries are among the top 12 worldwide.[4] Furthermore, over 90% of the primary energy supply in many MENA countries comes from fossil fuels, far exceeding the global average, while the region also accounts for nearly one-third of global crude oil exports.[5] Despite abundant solar potential, the share of renewables in the energy mix remains below 5% in most MENA countries.[6] These figures highlight the region's significant environmental impact and an urgent and comprehensive need for climate policies. Therefore, politicians and environmentalists must prioritize the widespread adoption of renewable energy sources to mitigate the severity of natural disasters in the MENA region. By transitioning from fossil fuels to clean energy, renewable technologies significantly reduce carbon emissions and address climate risks, such as extreme weather events and rising sea levels. The use of renewable energy minimizes negative environmental impact and strengthens resilience to climate-related challenges (Chen et al., 2023). Furthermore, investing in disaster risk reduction and climate change adaptation should be prioritized. To mitigate the effects of disasters, this effort should involve constructing resilient infrastructure, establishing early warning systems, and promoting sustainable land use practices.

Second, given the importance of strengthening disaster risk financing and insurance mechanisms, it should receive special attention from the government and public entities. Promoting public–private partnerships is, therefore, necessary to create contemporary financial products like parametric insurance and catastrophe bonds. Policymakers and regulators should also focus on reinforcing a more supportive regulatory environment for the insurance industry, aiming to expand access to micro-insurance for vulnerable populations and small businesses.

Third, it is always crucial to ensure the stability of the financial sector. Christy et al. (2024) highlight that banks exposed to higher climate risks tend to build stronger capital buffers, particularly in more stringent environments. In this regard, establishing explicit guidance for financial institutions' asset-liability management, implementing stress testing, and encouraging prudent risk management strategies can guarantee that financial institutions are better prepared for potential shocks. Fourth, targeted fiscal and monetary policies are required for economic recovery and supporting credit growth following disasters. For example, implementing fiscal stimulus measures, easing monetary policies, and creating disaster relief funds can address immediate necessities after a natural disaster. Finally, maintaining international cooperation and enhancing knowledge sharing are notable components of

[4] https://edgar.jrc.ec.europa.eu/report 2024?vis=ghgpop#emissions table

[5] https://www.iea.org/regions/middle-east

[6] https://www.irena.org/-/media/Files/IRENA/Agency/Publication/2024/Jul/Renewable energy highlights FINAL July 2024.pdf

effective disaster response. Regional collaboration, technical assistance, and the exchange of information resources and research are vital to building capacity and ensuring coordinated actions across borders. By adopting the aforementioned comprehensive policy recommendations and related strategies, MENA countries can greatly reduce the adverse effects of natural disasters on their financial systems and overall economies.

3.7 Conclusion

This research provides insight into the link between climate-related natural disasters and credit creation in MENA and non-MENA countries over the period from 1974 to 2023. The findings show that climate-related natural disasters have a positive impact on credit generation in MENA countries. The study reveals an association between the proportion of the affected population and credit generation in MENA countries, suggesting that these regions experience increased credit demand in the aftermath of disasters, likely because of the urgent need for financial resources to support recovery efforts. On the other hand, the results suggest that more resilient financial systems or the presence of alternative recovery mechanisms that do not heavily rely on credit expansion exist in non-MENA countries, in which the impact of natural disasters on credit creation appears negative or insignificant.

The disparity between MENA and non-MENA countries highlights the importance of considering regional differences in economic and institutional structures when assessing the financial consequences of natural disasters. In MENA countries, credit creation following disasters may indicate a reliance on financial markets to manage the economic disruptions caused by such events. This reliance, however, also raises concerns about the long-term sustainability of credit expansion as a disaster response strategy, particularly in regions with less developed financial systems and higher levels of economic vulnerability.

Although this study provides significant policy implications, it has certain limitations that could be addressed in future research. Future research could explore the nature and consequences of credits and loans following natural disasters to determine whether increased lending translates into productive investment or exacerbates household and firm indebtedness. This analysis would also shed light on whether specific groups—such as small businesses, low-income households, or informal sector workers—have equitable access to credit and whether they are disproportionately burdened by debt accumulation in the aftermath of natural disasters. Future studies could examine banks' behavioral responses to natural disasters, including changes in risk preferences, shifts in lending to disaster-affected sectors, and modifications in collateral requirements when the pertinent data is available. Future research might explore how financial instruments such as disaster and green bonds, other state-sponsored incentive packages, and diversification strategies influence banks' capacity to maintain credit supply after natural disasters. Another potential avenue for further research is to investigate how global and regional climate change mitigation and adaptation policies affect banks' credit creation, particularly within sectors vulnerable to environmental risks, and how these policies or legislation influence banks' lending decisions and risk management. Similarly, it would also be intriguing to analyze how the ESG or sustainability performance of commercial banks operating in the MENA region affects credit creation efficiency. Finally, future research could study the adoption of technological tools by banks, such as big data analytics, fintech, and artificial intelligence, to evaluate their impact on credit risk management following a natural disaster and on credit creation efficiency with a particular emphasis on the MENA region.

Appendix

Definitions of the Variables

Variables	Definitions
Credit 1	Private credit by deposit money banks (% of GDP). The financial resources provided to the private sector by domestic money banks as a share of GDP. Domestic money banks comprise commercial banks and other financial institutions that accept transferable deposits, such as demand deposits. *Source: World Bank*
Credit 2	Domestic credit to private sector (% of GDP). Refers to financial resources provided to the private sector, such as through loans, purchases of nonequity securities, and trade credits and other accounts receivable, that establish a claim for repayment. For some countries, these claims include credit to public enterprises. *Source: World Bank*
Credit 3	Domestic credit to private sector by banks (% of GDP). Refers to financial resources provided to the private sector by deposit-taking corporations (excluding central banks), such as through loans, purchases of nonequity securities, and trade credits and other accounts receivable, that establish a claim for repayment. For some countries, these claims include credit to public enterprises. *Source: World Bank*
AffectedRatio	The ratio of the total number of people affected (fatalities, injuries, homelessness, displacement and other forms of effects) by natural disasters to the total population. *Source: EM-DAT*
GDPPC	The natural logarithm of GDP per capita (current US$). *Source: World Bank*
Inflation	Inflation, consumer prices (annual %). *Source: World Bank*
Emissions	The natural logarithm of total greenhouse gas emissions (kt of CO_2 equivalent). *Source: World Bank*
FDI	The natural logarithm of net inflows of foreign direct investment. *Source: World Bank*
Trade	The ratio of imports and exports of goods and services to GDP. *Source: World Bank*

References

Alalmaee, H. (2024). Natural disasters and banking stability. *Economies, 12*(2), 31. https://doi.org/10.3390/economies12020031

Albuquerque, P., & Wassim, R. (2019). Banking stability, natural disasters, and state fragility: Panel VAR evidence from developing countries. *Research in International Business and Finance, 50*, 430–443. https://doi.org/10.1016/j.ribaf.2019.06.001

Alexander, K. (2014). *Stability and sustainability in banking reform: Are environmental risks missing in Basel III?* CISL and UNEP FI, University of Cambridge, Institute for Sustainability Leadership, Cambridge and Geneva.

Ali, S., Ying, L., Anjum, R., Nazir, A., Shalmani, A., Shah, T., & Shah, F. (2021). Analysis on the nexus of CO2 emissions, energy use, net domestic credit, and GDP in Pakistan: An ARDL bound testing analysis. *Environmental Science and Pollution Research, 28*, 4594–4614. https://doi.org/10.1007/s11356-020-10763-7

Anh, Q., Phan, V., & Nguyen, D. T. (2022). How do local banks respond to natural disasters? *European Journal of Finance, 29*(7), 754–779. https://doi.org/10.1080/1351847x.2022.2055969

Ashraf, B. N. (2018). Do trade and financial openness matter for financial development? Bank-level evidence from emerging market economies. *Research in International Business and Finance, 44*, 434–458. https://doi.org/10.1016/j.ribaf.2017.07.115

Barth, J. R., Miller, S. M., Sun, Y., & Zhang, S. (2022). Natural disaster impacts on US Banks. *American Business Review, 25*(2), 452–487. https://doi.org/10.37625/abr.25.2.452-487

BCBS. (2021). *Climate-related risk drivers and their transmission channels*. Basel Committee on Banking Supervision, Bank for International Settlements.

Beck, T., Buyukkarabacak, B., Rioja, F. K., & Valev, N. T. (2012). Who gets the credit? And does it matter? Household vs. firm lending across countries. *The B.E. Journal of Macroeconomics, 12*(1), 1–44. https://doi.org/10.1515/1935-1690.2262

Berg, G., & Schrader, J. (2009). *Relationship lending in times of crises: What about default and interest rates*. Available at SSRN 1497506.

Berg, G., & Schrader, J. (2012). Access to credit, natural disasters, and relationship lending. *Journal of Financial Intermediation, 21*(4), 549–568. https://doi.org/10.1016/j.jfi.2012.05.003

Bettencourt, U. M. D., Ebinger, T. S., Fay, J. O., Ghesquiere, M., Gitay, F., Krausing, H., Kull, J., Mccall, D. W., Reid, K., Simpson, R. C. J., & Leigh, A. (2013). *Building resilience: Integrating climate and disaster risk into development the World Bank Group experience: Main report*. No: 82648. The World Bank.

Boustan, L. P., Kahn, M. E., Rhode, P. W., & Yanguas, M. L. (2020). The effect of natural disasters on economic activity in U.S. counties: A century of data. *Journal of Urban Economics, 118*, 103257. https://doi.org/10.1016/j.jue.2020.103257

Brei, M., Mohan, P., Barahona, A. P., & Strobl, E. (2024). Transmission of natural disasters to the banking sector: Evidence from thirty years of tropical storms in the Caribbean. *Journal of International Money and Finance, 141*, 103008. https://doi.org/10.1016/j.jimonfin.2023.103008

Brei, M., Mohan, P., & Strobl, E. (2019). The impact of natural disasters on the banking sector: Evidence from hurricane strikes in the Caribbean. *The Quarterly Review of Economics and Finance, 72*, 232–239. https://doi.org/10.1016/j.qref.2018.12.004

Bueno, R., Herzfeld, C., Stanton, E. A., & Ackerman, F. (2008). *The Caribbean and climate change: The costs of inaction*. Stockholm Environment Institute—US Center and Global Development and Environment Institute, Tufts University.

Caprio, G., & Klingebiel, D. (1996). *Bank insolvencies: Cross country experience*. Policy Research Working Paper Series, 1620, The World Bank.

Carbonell, J., & Werner, R. (2018). Does foreign direct investment generate economic growth? A new empirical approach applied to Spain. *Economic Geography, 94*(4), 425–456. https://doi.org/10.1080/00130095.2017.1393312

Cavallo, E., Galiani, S., Noy, I., & Pantano, J. (2013). Catastrophic natural disasters and economic growth. *The Review of Economics and Statistic, 95*(5), 1549–1561.

Cavallo, E., & Noy, I. (2011). *The economics of natural disasters: A survey*. IDB Working Paper Series, IDB-WP-124. https://doi.org/10.18235/0010924

Chamberlain, S., Vijayaraghavan, R., & Zheng, Y. (2019). *Natural disasters, loan loss accounting and subsequent lending*. University of British Columbia.

Chang, C. P., & Zhang, L. W. (2020). Do natural disasters increase financial risks? An empirical analysis. *Bulletin of Monetary Economics and Banking, 23*, 62–86. https://doi.org/10.21098/bemp.v23i0.1258

Chen, X. H., Tee, K., Elnahass, M., & Ahmed, R. (2023). Assessing the environmental impacts of renewable energy sources: A case study on air pollution and carbon emissions in China. *Journal of Environmental Management, 345*, 118525. https://doi.org/10.1016/j.jenvman.2023.118525

Choudhary, M. A., & Jain, A. K. (2022). Finance and inequality: The distributional impacts of bank credit rationing. *Journal of Financial Intermediation, 52*, 100997. https://doi.org/10.1016/j.jfi.2022.100997

Christy, A., Elnahass, M., Amezaga, J., Browne, A., & Heidrich, O. (2024). A dynamic framework to align company climate reporting and action with global climate targets. *Business Strategy and the Environment, 33*(4), 3103–3128. https://doi.org/10.1002/bse.3635

CISL, & UNEPFI. (2015). *Banking & sustainability Time for convergence: A policy briefing on the links between financial stability and environmental sustainability*. University of Cambridge, Institute for Sustainability Leadership, Cambridge and UNEPFI.

Claessens, S., & Laeven, L. (2003). *What drives bank competition? Some international evidence*. Policy Research Working Paper, No. 3113. World Bank.

Collier, B., & Babich, V. O. (2019). Financing recovery after disasters: Explaining community credit market responses to severe events. *Journal of Risk and Insurance, 86*(2), 479–520. https://doi.org/10.1111/jori.12221

Collier, B., Katchova, A. L., & Skees, J. R. (2011). Loan portfolio performance and El Nino, an intervention analysis. *Agricultural Finance Review, 71*(1), 98–119. https://doi.org/10.1108/00021461111128183

Collier, B. L., Howell, S. T., & Rendell, L. (2024). After the storm: How emergency liquidity helps small businesses following natural disasters. *NBER Working Paper Series, 32326*, 1–86. https://doi.org/10.3386/w32326

Dafermos, Y., Nikolaidi, M., & Galanis, G. (2018). Climate change, financial stability and monetary policy. *Ecological Economics, 152*, 219–234. https://doi.org/10.1016/j.ecolecon.2018.05.011

Dal Maso, L., Kanagaretnam, K., Lobo, G. J., & Mazzi, F. (2024). Does disaster risk relate to banks' loan loss provisions? *European Accounting Review, 33*(3), 825–854. https://doi.org/10.1080/09638180.2022.2120513

Del Ninno, C., Dorosh, P. A., & Smith, L. (2003). Public policy, markets and household coping strategies in Bangladesh: Avoiding a food security crisis following the 1998 floods. *World Development, 31*(7), 1221–1238. https://doi.org/10.1016/s0305-750x(03)00071-8

Do, Q. A., Phan, V., & Nguyen, D. T. (2022). How do local banks respond to natural disasters? *European Journal of Finance, 29*(7), 754–779. https://doi.org/10.1080/1351847X.2022.2055969

Dowla, A. (2018). Climate change and microfinance. *Business Strategy and Development, 1*(2), 78–87. https://doi.org/10.1002/bsd2.13

Duqi, A., McGowan, D., Onali, E., & Torluccio, G. (2021). Natural disasters and economic growth: The role of banking market structure. *Journal of Corporate Finance, 71*, 102101. https://doi.org/10.1016/j.jcorpfin.2021.102101

Dynan, K. E., Skinner, J., & Zeldes, S. P. (2004). Do the rich save more? *Journal of Political Economy, 112*(2), 397–444. https://doi.org/10.1086/381475

Elnahass, M., Marie, M., & Elgammal, M. (2022). Terrorist attacks and bank financial stability: Evidence from MENA economies. *Review of Quantitative Finance and Accounting, 59*(1), 383–427. https://doi.org/10.1007/s11156-022-01043-1

EM-DAT. (2013). *The OFDA/CRED international disaster database*. Universite Catholique de Louvain.

Felbermayr, G., & Groschl, J. (2014). Naturally negative: The growth effects of natural disasters. *Journal of Development Economics, 111*, 92–106. https://doi.org/10.1016/j.jdeveco.2014.07.004

Fengler, W., Ihsan, A., & Kaiser, K. (2008). *Managing post-disaster reconstruction finance: International experience in public financial management* (pp. 1–35). Policy Research Working Paper, 4475. World Bank.

Fouejieu, A. (2017). Inflation targeting and financial stability in emerging markets. *Economic Modelling, 60*, 51–70. https://doi.org/10.1016/j.econmod.2016.08.020

Freeman, P. K., Keen, M., & Mani, M. (2003). *Dealing with increased risk of natural disasters: Challenges and options*. IMF Working Paper, No. 197, Washington, DC.

Gallagher, J., & Hartley, D. (2017). Household finance after a natural disaster: The case of Hurricane Katrina. *American Economic Journal. Economic Policy, 9*(3), 199–228. https://doi.org/10.1257/pol.20140273

Garmaise, M. J., & Moskowitz, T. J. (2009). Catastrophic risk and credit markets. *The Journal of Finance, 64*(2), 657–707. https://doi.org/10.1111/j.1540-6261.2009.01446.x

Gebrael, K., Mitri, G., & Kalantzi, O. (2024). Overview of nature-based solutions for climate resilience in the MENA region. *Nature-Based Solutions, 6*, 100159. https://doi.org/10.1016/j.nbsj.2024.100159

Gramlich, D., Walker, T., Zhao, Y., & Bitar, M. (2023). After the storm: Natural disasters and bank solvency. *International Journal of Central Banking, 19*(2), 199–249.

Keerthiratne, S., & Tol, R. S. J. (2017). Impact of natural disasters on financial development. *Economics of Disasters and Climate Change, 1*(1), 33–54. https://doi.org/10.1007/s4188501700025

Klomp, J. (2014). Financial fragility and natural disasters: An empirical analysis. *Journal of Financial Stability, 13*, 180–192. https://doi.org/10.1016/j.jfs.2014.06.001

Klomp, J. (2016). Economic development and natural disasters: A satellite data analysis. *Global Environmental Change, 36*, 67–88. https://doi.org/10.1016/j.gloenvcha.2015.11.001

Klomp, J., & Valckx, K. (2014). Natural disasters and economic growth: A meta-analysis. *Global Environmental Change, 26*, 183–195. https://doi.org/10.1016/j.gloenvcha.2014.02.006

Knutson, T., Camargo, S. J., Chan, J. C. L., Emanuel, K., Ho, C.-H., Kossin, J., Mohapatra, M., Satoh, M., Sugi, M., Walsh, K., & Wu, L. (2020). Tropical cyclones and climate change assessment: Part II: Projected response to anthropogenic warming. *Bulletin of the American Meteorological Society, 101*(3), 303–322. https://doi.org/10.1175/BAMSD180194.1

Koetter, M., Noth, F., & Rehbein, O. (2020). Borrowers under water! Rare disasters, regional banks, and recovery lending. *Journal of Financial Intermediation, 43*, 100811. https://doi.org/10.1016/j.jfi.2019.01.003

Koetter, M., Noth, F., & Rehbein, O. (2022). Borrowers under water! Rare disasters, regional banks, and recovery lending. *Journal of Financial Intermediation, 43*, 100811. https://doi.org/10.1016/j.jfi.2019.01.003

Lashley, J., & Warner, K. (2013). Evidence of demand for microinsurance for coping and adaptation to weather extremes in the Caribbean. *Climatic Change, 133*, 101–112. https://doi.org/10.1007/s10584-013-0922-1

Lee, C.-C., Wang, C.-W., Thinh, B. T., & Xu, Z.-T. (2022). Climate risk and bank liquidity creation: International evidence. *International Review of Financial Analysis, 82*, 102198. https://doi.org/10.1016/j.irfa.2022.102198

Mejia, S. A., Baccianti, C., Mrkaic, M., Novta, N., Pugacheva, E., & Topalova, P. (2019). *Weather shocks and output in low-income countries: Adaptation and the role of policies*. IMF Working Paper, No. 178, Washington, DC.

Nasreen, S., Anwar, S., & Ozturk, I. (2017). Financial stability, energy consumption and environmental quality: Evidence from South Asian economies. *Renewable and Sustainable Energy Reviews, 67*, 1105–1122. https://doi.org/10.1016/j.rser.2016.09.021

Niepmann, F., & Schmidt-Eisenlohr, T. (2017). International trade, risk and the role of banks. *Journal of International Economics, 107*, 111–126. https://doi.org/10.1016/j.jinteco.2017.03.007

Noy, I. (2009). The macroeconomic consequences of disasters. *Journal of Development Economics, 88*(2), 221–231. https://doi.org/10.1016/j.jdeveco.2008.02.005

Pek, S., Oh, C. H., & Rivera, J. (2018). MNC foreign investment and industrial disasters: The moderating role of technological, safety management, and philanthropic capabilities. *Strategic Management Journal, 39*(2), 502–526. https://doi.org/10.1002/smj.2737

Pelham, L., Clay, E., & Braunholz, T. (2011). *Natural disasters: What is the role of social safety nets?* Social Protection Discussion Paper, No. 3113. World Bank.

Ratcliffe, C., Congdon, W., Teles, D., Stanczyk, A., & Martın, C. (2020). From bad to worse: Natural disasters and financial health. *Journal of Housing Research, 29*(1), 25–53. https://doi.org/10.1080/10527001.2020.1838172

Ratcliffe, C., Congdon, W. J., Stanczyk, A., Teles, D., Martin, C., & Kotapati, B. (2019). Insult to injury: Natural disasters and residents' financial health. *Urban Institute*, 1–82.

Siamwalla, A., Pinthong, C., Poapongsokorn, N., Satsanguan, P., Nettayarak, P., Mingmaneenakin, W., & Tubpun, Y. (1990). The Thai rural credit system: Public subsidies and segmented markets. *The World Bank Economic Review, 4*(3), 271–295.

Simsek, R., Mollah, S., & Tunyi, A. (2024). Corporate governance structure and climate-related financial disclosure: Conventional banks versus Islamic banks. *Business Strategy and the Environment, 33*(6), 5503–5528. https://doi.org/10.1002/bse.3753

Skidmore, M., & Toya, H. (2002). Do natural disasters promote long-run growth? *Economic Inquiry, 40*(4), 664–687. https://doi.org/10.1093/ei/40.4.664

Sseruyange, J., & Klomp, J. (2021). Natural disasters and economic growth: The mitigating role of microfinance institutions. *Sustainability, 13*(9), 5055. https://doi.org/10.3390/su13095055

Walker, T., Xu, Y., Gramlich, D., & Zhao, Y. (2023). The impact of natural disasters on the performance and solvency of US banks. *International Journal of Managerial Finance, 19*(1), 136–154. https://doi.org/10.1108/ijmf-08-2020-0406

Chapter 4
Scaling Up Sustainable Finance in the Middle East and North African Region

Lucia Alessi, Annette Becker, Mélina London, and Caterina Rho

Abstract The Middle East and North African (MENA) region is particularly vulnerable to the effects of climate change, has a key role to play in global climate goals, and faces a series of challenges related to the social and governance dimension. Against this background, this chapter analyzes where we stand with respect to the development of sustainable finance in the region and how it could be scaled up. We start by looking at the broader macrofinancial context and then focus on sustainable financial market segments, also in comparison with other areas. Then, we discuss lessons learned from relevant initiatives in and outside of the region, tying these lessons to the MENA context. We conclude by providing an overview of innovative tools such as blended finance, catastrophe bonds, and securitization, which could unleash their potential, particularly in a region where derisking mechanisms are needed to attract investors.

Keywords Sustainable finance · ESG risks · Climate adaptation strategies · Climate resilience · Natural disasters · Risk mitigation · Capacity building

4.1 Introduction

Sustainable finance is a critical tool for fostering economic growth and addressing environmental and social challenges through the integration of environmental, social, and governance (ESG) criteria into financial decision-making processes. As it aims to promote long-term sustainable development, sustainable finance has a key

L. Alessi (✉) · M. London · C. Rho
European Commission, Joint Research Centre, Ispra, Italy
e-mail: Lucia.alessi@ec.europa.eu; Melina.london@ec.europa.eu; Caterina.rho@ec.europa.eu

A. Becker
Deutsche Bundesbank, Frankfurt, Germany
e-mail: Annette.becker@bundesbank.de

M. Elnahass et al. (eds.), *A Greenprint for Net Zero*, Gulf Studies 15,
https://doi.org/10.1007/978-981-92-1762-5_4

role to play, particularly in low- and middle-income countries (LMICs), including the Middle East and North African (MENA) region.

One of the main reasons why it is crucially important to scale up sustainable finance in the MENA region is that this region is highly vulnerable to climate-related physical risks, both acute and chronic. On the night of September 10–11, 2023, the city of Derna, in north-eastern Libya, was devastated by a catastrophic flood following heavy rainfall from Storm Daniel.[1]

On top of some parts of the MENA region being prone to flash floods, the region experiences some of the highest temperatures globally and is warming nearly twice as fast as the global average, with temperatures expected to rise by 4–5 °C by the end of the century in the Middle East (Zittis et al., 2022). This can lead to heatwaves and exacerbate the already critical issue of water scarcity. Prolonged periods of drought and water shortages are indeed becoming more common and are coupled with desertification, which progressively reduces arable land. Finally, coastal areas, particularly in North Africa, are at risk from rising sea levels, which can lead to coastal erosion and saltwater intrusion, endangering fluvial ecosystems. All these hazards have huge implications in terms of health issues, food security, population displacement, as well as overall quality of life, and need to be addressed via coordinated efforts in climate change adaptation and mitigation.

With reference to the social and governance dimensions, investments in sustainable projects can improve social outcomes by providing better access to healthcare, education, and clean water. This, in turn, can reduce poverty and inequality, contributing to more inclusive growth. Improving on these fronts is crucial in a region that faces significant economic disparities, with wealth concentrated in a few oil-rich countries, while others struggle with poverty. At the same time, the MENA region is characterized by a large young population, which, however, suffers from high rates of youth unemployment. Gender inequality remains a significant issue, as women in many MENA countries face barriers to education, employment, and political participation. Finally, while effective governance is essential for implementing policies that promote social welfare and economic growth, weak governance and corruption generally hinder social development in the region.

In particular, based on the Worldwide Governance Indicators by the World Bank, corruption remains a significant issue in many MENA countries, impacting economic development and public trust in institutions. Furthermore, many MENA countries face challenges in the areas related to the rule of law, due to issues related to judicial independence and the enforcement of laws, and suffer from political instability due to ongoing conflicts, political unrest, and security issues. MENA countries typically score lower compared to other regions in the area related to political freedoms and media restrictions, while they show varied performance with respect to government effectiveness and regulatory quality.

[1] Reprofiling the Derna river and reconstructing dams are vital adaptation measures, see Annunziato et al. (2024) and European Commission Joint Research Centre (JRC) (2023).

Against this background, this chapter analyzes challenges and opportunities related to scaling up sustainable finance in the MENA region. The contribution of this chapter to the academic literature on the MENA region is twofold. First, it provides quantitative evidence on the development of the financial system and of sustainable finance in particular. Quantitative financial studies focusing on this region are not overly numerous, also owing to data availability issues. In particular, reliable, consistent, and standardized ESG data are often lacking in the region. At the same time, available papers often look at only particular countries or cases. This chapter provides the big picture of the MENA region as a whole, systematically analyzing available data sources. The second contribution relates to the forward-looking part of the chapter, where we discuss relevant initiatives, tools, and actions to scale up sustainable finance. While these initiatives, tools, and actions have normally been developed in other contexts than the MENA region or with a wider scope than the MENA region, we tie broader lessons to the MENA region and discuss their potential relevance and implications for this peculiar group of countries.

The chapter is structured as follows. The following section provides an overview of sustainable finance in the region, starting from the general macro-financial context and comparing developments across regions. Section 4.3 discusses lessons from existing regional initiatives and the international context. Section 4.4 covers available tools to increase the financing of sustainable activities, including catastrophe bonds for climate adaptation, blended finance and derisking schemes, and securitization. Section 4.5 concludes.

4.2 Overview of Sustainable Finance in the MENA Region

In this section, we provide a comprehensive overview of the sustainable finance landscape in the MENA region. We will first explore the financing of sustainable investment and the macro-financial context in the region, looking at the development of the banking sector and financial markets (Section 4.2.1.1), and the patterns of climate finance with a focus on public versus private investment divides (Section 4.2.1.2). Then, we will examine sustainable financial flows and regional climate finance patterns (Section 4.2.2), alongside the challenges that hinder sustainable finance developments in the region (Section 4.2.3).

4.2.1 *Financing of Sustainable Investment Needs and Their Macro-Financial Context*

4.2.1.1 Quantitative Evidence on Regional Banking and Financial Markets, and Their Maturity

In this section, we present the domestic banking and financial sector of the region of interest, with the help of some financial indicators and statistics widely used in the literature. We also present some descriptive statistics about international banking activity in the countries of interest.

To this end, we refer to two main sources of data: the World Development Indicators (WDI) and the Global Financial Development Database (GFDD) by the World Bank.

The WDI database is the primary World Bank database of development indicators compiled from officially recognized international sources. It presents the most current and accurate global development data available and includes national, regional, and global estimates. We select several indicators on banking resilience and stability from this database, as well as on the development of the domestic banking sector and on financial markets in the MENA countries of interest.

The GFDD is an extensive collection of financial system characteristics for 214 countries. Time series have an annual frequency and cover the period from 1960 to 2022. This database collects many financial sector indicators covering both financial markets and the banking sector.

A general caveat on the data applies to all our results. Although we are referring to reliable data providers that apply the highest statistical standards, data are not always available for our selection of countries and for all data series.[2] There are several technical reasons why data may not be available for a part of the indicators in some countries or for a couple of time periods. In some cases, indicators are derived from sporadic surveys and are only available every few years. In other cases, datasets or indicators are available only from the year they were initiated in the country. Some countries do not regularly report data due to conflict, lack of statistical capacity, or other reasons. Finally, the databases are not published annually: the most recent version of the GFDD is from September 2022, the WDI database is updated annually, but its data sources may be surveys updated every 2–4 years.[3]

[2] We use the IMF definition of the MENA region, including the following countries (22): Algeria, Bahrain, Djibouti, Egypt, the Islamic Republic of Iran, Iraq, Jordan, Kuwait, Lebanon, Libya, Mauritania, Morocco, Oman, Qatar, Saudi Arabia, Somalia, Sudan, the Syrian Arab Republic, Tunisia, the United Arab Emirates, the West Bank and Gaza, and Yemen.

[3] This is the case of some of the financial inclusion data mentioned in this chapter, which are extracted from the 2021 edition of the Global Findex Database (Demirgüç-Kunt et al., 2021).

Dimension and Development of the Banking Sector and Financial Markets in the MENA Region

At lower levels of economic development, commercial banks tend to dominate the financial system, while at higher levels, domestic stock markets tend to become more active and efficient. Hence, the amount of credit issued by the formal banking sector, or by the non-banking financial sector, can be used as an indicator of the degree of development of the financial sector in a given country or region.

To compare the size of the banking sector and the stock market within our sample of countries, we refer to the level of bank deposits, domestic credit to the private sector, and market capitalization of listed domestic companies (expressed as a percentage of gross domestic product (GDP)). Bank deposits include the total value of demand, time, and saving deposits at domestic commercial banks and other financial institutions that accept transferable deposits. Listed companies, including foreign companies, are those that have shares listed on an exchange at the end of the year. Market capitalization is the share price multiplied by the number of shares outstanding for listed companies on an exchange.

The indicator we use to describe the state of the credit market is domestic credit to the private sector, provided by both the banking sector and the non-banking sector, expressed as a share of GDP. We focus on the time dimension, presenting the indicators as aggregate at the level of the MENA region or at the country level, and particularly on the period from 2012 to 2021.[4]

Panel (a) of Fig. 4.1 shows the development of banking activity and of the stock market, considering the average for the 22 countries of the MENA region.[5] In particular, the chart reports bank deposits and domestic credit to the private sector, as well as stock market capitalization, all taken as a percentage of GDP. Domestic credit and bank deposits have a similar trend through the years, increasing from 2012 to 2016 and declining afterward, standing at around 60% of GDP based on the latest available data. Stock market capitalization remains constant over time from 2012 to 2018, increasing from 40% to 100% of GDP from 2018 to 2020. This sudden rise is due to a stock exchange bubble in Iran in 2019–2020 that burst during the COVID-19 crisis, and to the opening of the stock market to foreign investors in Saudi Arabia in August 2019, after having been segregated from the global stock market for a long time (Alshammari & Goto, 2022). As a reference, in the European Union, where the banking sector is still the main actor in the financial market, bank deposits reached 85% of GDP in 2019 before the COVID-19 pandemic, and 95% in 2020, while bank credit stands at 75% of GDP. Pre-COVID-19 market capitalization in the EU was around 50% of GDP.

[4] The data coverage of the GFDD stops at 2021 for this time series.

[5] Domestic credit data for Somalia and Syria are missing. Stock market data for Djibouti, Iraq, Libya, Mauritania, Somalia, Sudan, Syria, and Yemen are also missing.

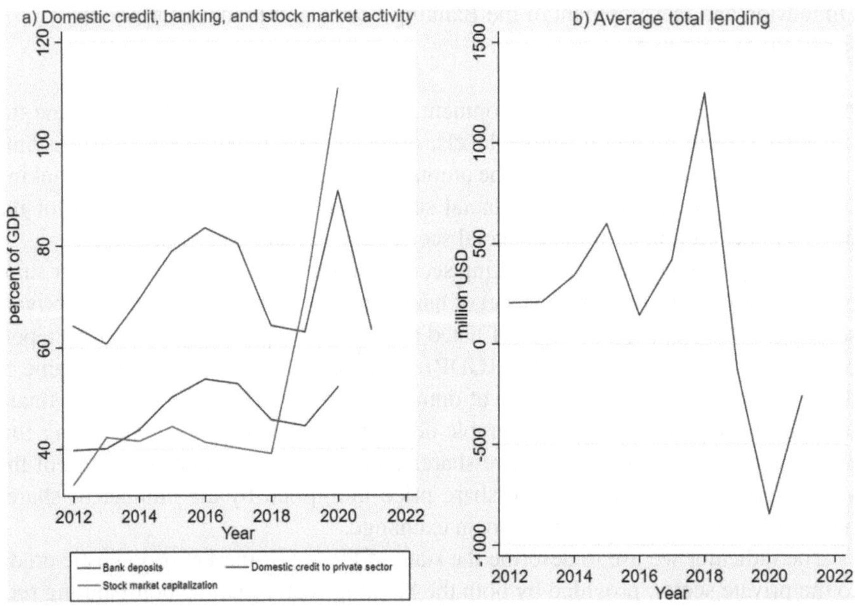

Fig. 4.1 Banking sector and financial markets in the MENA region, 2012–2021. Source: World Bank, GFDD

Panel (b) of Fig. 4.1 shows the amount of total lending to the private sector in the MENA region from 2012 to 2021.[6] The sudden decrease after 2018 is due to the economic crisis in Lebanon, with lending in the other countries remaining quite stable.

Access to Finance and Banking Concentration

Access to finance is the first step to deepen and strengthen the credit market and the financial sector in a country: a stable financial system that promotes efficient savings and investment is crucial for economic growth and development. Availability and costs of funding are among the factors that define access to finance. If credit services are not widespread, or if the cost of credit is too high, financial development will suffer, and this will translate into a lower level of investment and consumption in the country, as well as slower economic development. Another factor that influences access to finance and the diversification of financial services available in a country is bank concentration. Scientific evidence shows that higher levels of bank concentration may increase access to finance, provided that the market power of banks is limited by adequate regulation (Owen & Pereira, 2018). Bank

[6] Data for Bahrain, Kuwait, Libya, Mauritania, Oman, Qatar, Saudi Arabia, Somalia, United Arab Emirates, and West Bank and Gaza are missing.

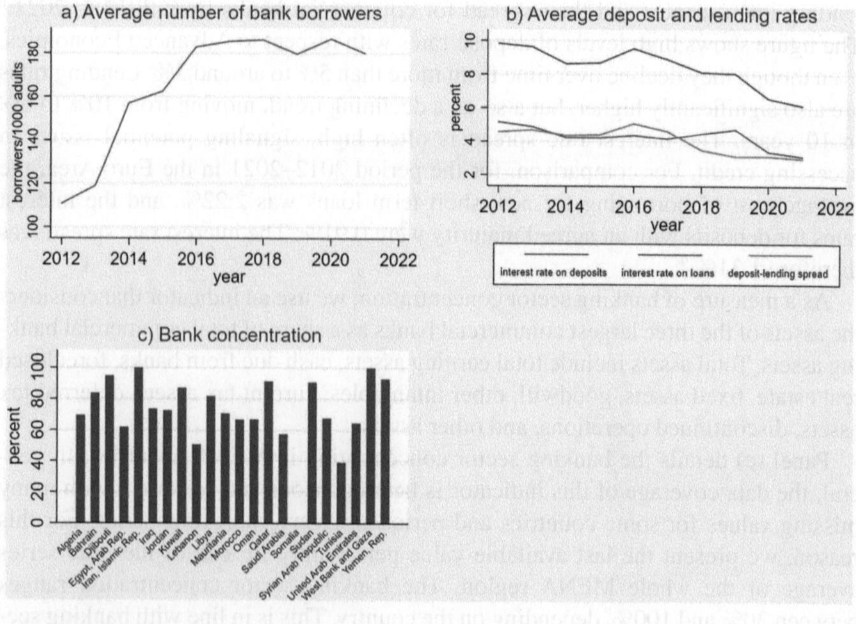

Fig. 4.2 Access to finance and banking concentration, MENA region, 2012–2021. Source: World Bank, WDI

concentration in a proper regulatory environment may also help to stabilize the financial system and prevent bank crises (Beck et al., 2005).

As an indicator of access to finance, we use the average number of borrowers from commercial banks per 1000 adults. Borrowers are defined as resident customers that are non-financial corporations and households. Panel (a) of Fig. 4.2 shows that there have been positive developments in the banking sector of the MENA countries during the observation period with respect to this indicator.[7] Indeed, the average number of bank borrowers per 1000 persons increased from 100 units in 2012 to around 180 units after 2016.

Deposits and lending interest rates, and their difference (interest rate spread), are useful indicators of financing costs. The deposit rate is the interest that the bank pays to its depositors. The lending rate is the interest that the bank charges to its borrowers and is normally differentiated according to the creditworthiness of borrowers, while the deposit rate depends on the financing strategy of the credit institution. The interest rate spread measures financial sector efficiency in intermediation. A lower spread means lower transaction costs and a lower cost of funds for investment. A higher spread indicates a higher cost of lending, that is, a potential impediment to investment and economic growth. Panel (b) reports the level of deposit and

[7] Data for Bahrain, Iran, Iraq, Jordan, Libya, Morocco, Oman, Somalia, and Sudan are missing.

lending interest rates and their spread for commercial banks from 2012 to 2021.[8] The figure shows high levels of deposit rates with respect to Advanced Economies, even though they decline over time from more than 5% to around 3%. Lending rates are also significantly higher, but also on a declining trend, moving from 10% to 6% in 10 years. The interest rate spread is often high, signaling potential issues in accessing credit. For comparison, for the period 2012–2021 in the Euro Area, the average cost of borrowing for new short-term loans was 2.22%, and the interest rates for deposits with an agreed maturity were 0.91%. The interest rate spread was therefore 1.31%.[9]

As a measure of banking sector concentration, we use an indicator that considers the assets of the three largest commercial banks as a share of total commercial banking assets. Total assets include total earning assets, cash due from banks, foreclosed real estate, fixed assets, goodwill, other intangibles, current tax assets, deferred tax assets, discontinued operations, and other assets.

Panel (c) details the banking sector concentration in MENA countries. In general, the data coverage of this indicator is heterogeneous and scattered, with many missing values for some countries and periods, or very short time series. For this reason, we present the last available value per country instead of the time series average of the whole MENA region. The banking sector concentration ranges between 30% and 100%, depending on the country. This is in line with banking sector concentration levels in Advanced Economies (75% in the Euro Area, and 38% in the United States in 2021).[10]

Banking Sector Resilience

One of the possible causes of a financial crisis is excessive leverage for the banking sector and decreasing quality and quantity of the capital base. These factors were both key in the case of the Global Financial Crisis of 2008, for example. Insufficient liquidity buffers are a second factor of risk for the banking sector. Low capitalization and liquidity may decrease investors' confidence in the banking sector because the banking system would not be able to absorb losses after a negative shock. To limit the contagion to the real economy during a financial crisis, the public sector may step in with injections of liquidity, capital support, and guarantees, if fiscal space allows. Often, LMICs do not have the necessary fiscal space to bail out their financial sectors, so over time, most of them adopted and maintained prudent regulatory and supervision systems. Moreover, banks located in countries with high

[8] Data for Djibouti, Morocco, Saudi Arabia, Somalia, Sudan, Syria, Tunisia, and United Arab Emirates are missing.

[9] Data extracted from the European Central Bank (ECB) Data Portal.

[10] Sources: ECB Data Portal, Federal Reserve Economic Data.

exposure to conflict or terrorist attacks may exhibit low financial stability, due to high credit or insolvency risk (Elnahass et al., 2022).

To measure bank solvency and resilience, we refer to the bank regulatory capital to risk-weighted assets ratio (expressed as a percentage rate). This indicator shows the extent to which banks can deal with unexpected losses. As a measure of bank liquidity, we refer to the liquid assets to deposits and short-term funding ratio (expressed as a percentage rate). Liquidity risk is a structural risk for a bank, deriving from its intermediation and maturity transformation activity. If bank liquidity is too low with respect to its assets, a bank may not be able to face unexpected withdrawals from its investors, triggering a liquidity crisis or a "bank run." Finally, we look at the ratio between non-performing loans (NPLs) and gross loans (expressed as a percentage rate). The ratio of NPLs to total gross loans measures bank profitability and efficiency by identifying problems with asset quality in the loan portfolio. NPLs may originate from exogenous macro-financial shocks (e.g., negative shocks to export volumes or prices), a weak domestic economy and/or a fragile business environment, foreign exchange crises, or weather-related shocks (Eyraud et al., 2021).

Panel (a) of Fig. 4.3 shows the bank capital to asset ratio for those MENA countries for which data are available. Despite missing data, one can still get a general picture of the strength and stability of the banking sector of the region. The Basel III standard requires that the minimum regulatory capital be at least 9% of the risk-weighted assets of a bank. In our sample, on average, all countries respect this

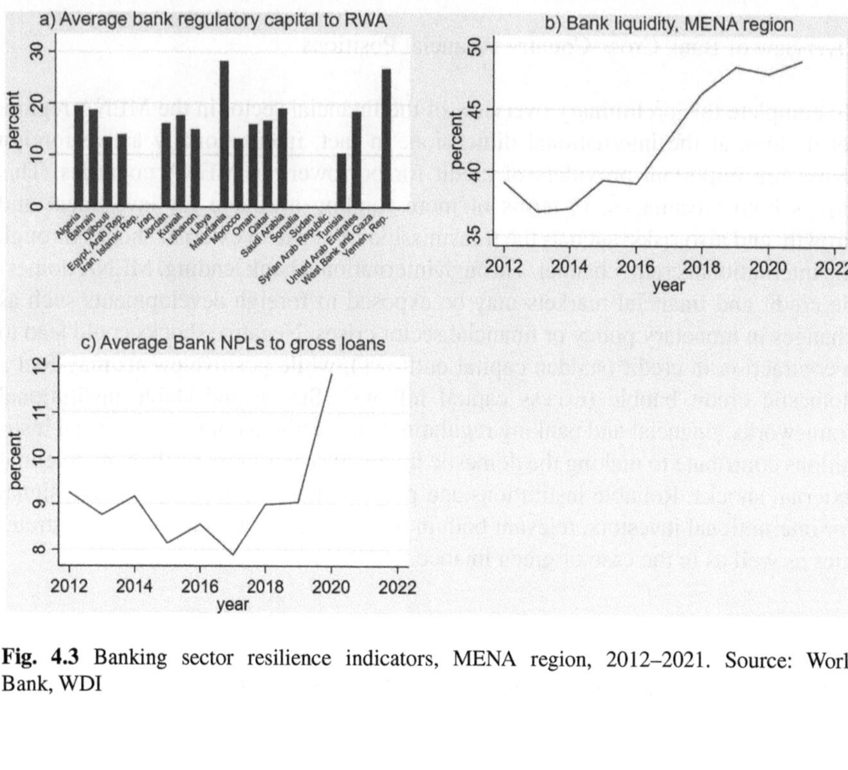

Fig. 4.3 Banking sector resilience indicators, MENA region, 2012–2021. Source: World Bank, WDI

requirement, showing higher levels of capitalization than the required minimum, going from 10% of Tunisia to almost 30% of Mauritania. Emerging Market economies may tend to overcapitalize their banking sector to maintain a more conservative regulatory stance. This approach may help to maintain financial stability and financial sector confidence, but may also impede the development of the credit market, as banks are overly cautious in opening new credit lines. As alternative indicators of bank resilience, panel (b) presents the development over time of the average ratio of bank liquid assets to deposits and short-term funding,[11] while panel (c) presents the regional average ratio of NPLs to the total loans of the bank. Banks in the MENA region have a growing liquidity coverage that has increased from less than 40% in 2012 to almost 50% in 2021. As a reference, the minimum liquidity coverage ratio that banks must have under the Basel III standards was phased in at 70% in 2016 and steadily increased to 100% by 2019. In line with international banking standards, the data show a sudden increase in the liquidity ratio after 2016.

Average NPLs in the MENA region (panel (c)) are in line with historical trends for developing countries, as presented in the Global Financial Development Report for 2019–2020 (The World Bank, 2020).[12] On average, the ratio of NPLs to gross loans is 8 percentage points higher in developing countries than in high-income countries (13% versus 5%). In our sample, we observe similar values, but also a certain level of heterogeneity. On average, NPLs grew from 9% in 2021 to 12% in more recent years. However, data are scattered: one-third of the countries miss data, and some of them have data coverage only for 2–4 years.

Overview of Bank Cross-Country Financial Positions

To complete this preliminary overview of the financial sector in the MENA region, let us look at the international dimension. In fact, internationally active foreign banks are important providers of credit for borrowers in MENA countries. This brings both advantages, in terms of more funding available for investment and growth, and also risks, such as the transmission of negative external shocks through the international credit channel. Through international bank lending, MENA domestic credit and financial markets may be exposed to foreign developments such as changes in monetary policy or financial sector crises. Negative shocks could lead to a contraction in credit (sudden capital outflows), while positive events may fuel a domestic credit bubble (excess capital inflows). Strong and stable institutional frameworks, financial and banking regulation, and resilient domestic financial institutions contribute to making the domestic financial sector more resilient to potential external shocks. Reliable institutions and proper regulation are a reassuring signal for international investors, relevant both in the case of traditional investment strategies as well as in the case of green finance.

[11] Data for Somalia are missing.

[12] Data for Iran, Iraq, Libya, Somalia, Sudan, Syria, West Bank, and Gaza are missing.

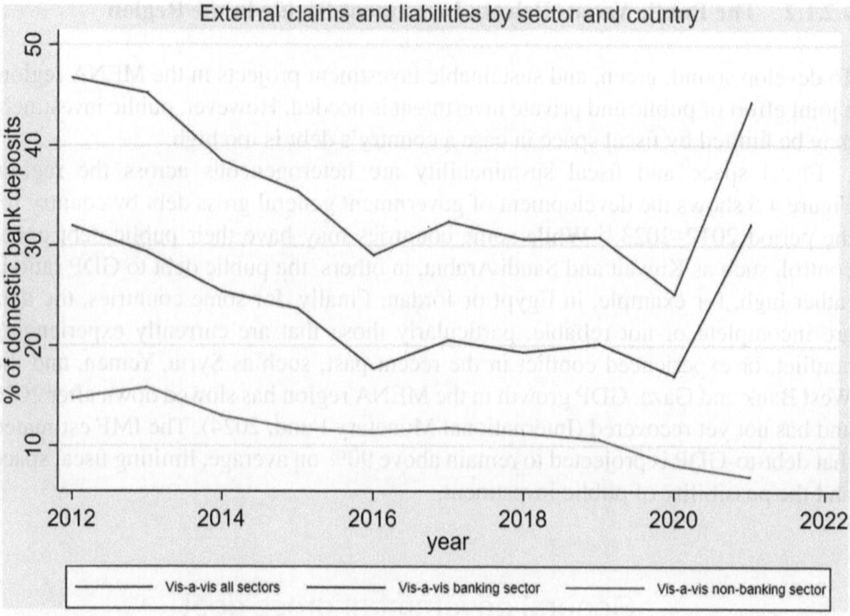

Fig. 4.4 Average external claims and liabilities by sector and country, MENA region, 2012–2021. Source: World Bank, GFDD

Figure 4.4 focuses on international banking flows.[13] In particular, the chart presents some evidence regarding the destination of capital flows over time, looking at total external claims and liabilities for the banking sector and the non-banking sector, as a percentage of domestic bank deposits. The non-banking sector includes all counterparties that are not banks: financial institutions different from banks and the non-financial sector (households, general government, and others). The results depend on the openness of the economies in the MENA region and the dimension of the domestic market. We find that the MENA region as a whole has an average level of total claims and liabilities comparable with those of some European countries (41% in France, and 44% on average in the Euro Area in 2021, according to GFDD data). We find that, also at the international level, the banking sector plays a more significant role than the non-banking sector, and this is reflected in the higher level of claims and liabilities directed to the banks of the MENA region rather than the non-banking sector.

[13] Data for Somalia and Syria are missing.

4.2.1.2 The Public Versus Private Investment Divide in the Region

To develop sound, green, and sustainable investment projects in the MENA region, a joint effort of public and private investment is needed. However, public investment may be limited by fiscal space in case a country's debt is too high.

Fiscal space and fiscal sustainability are heterogeneous across the region. Figure 4.5 shows the development of government general gross debt by country for the period 2012–2023.[14] While some countries may have their public debt under control, such as Kuwait and Saudi Arabia, in others, the public debt to GDP ratio is rather high, for example, in Egypt or Jordan. Finally, for some countries, the data are incomplete or not reliable, particularly those that are currently experiencing conflict, or experienced conflict in the recent past, such as Syria, Yemen, and the West Bank and Gaza. GDP growth in the MENA region has slowed down after 2020 and has not yet recovered (International Monetary Fund, 2024). The IMF estimated that debt-to-GDP is projected to remain above 90% on average, limiting fiscal space and the possibility of public investment.

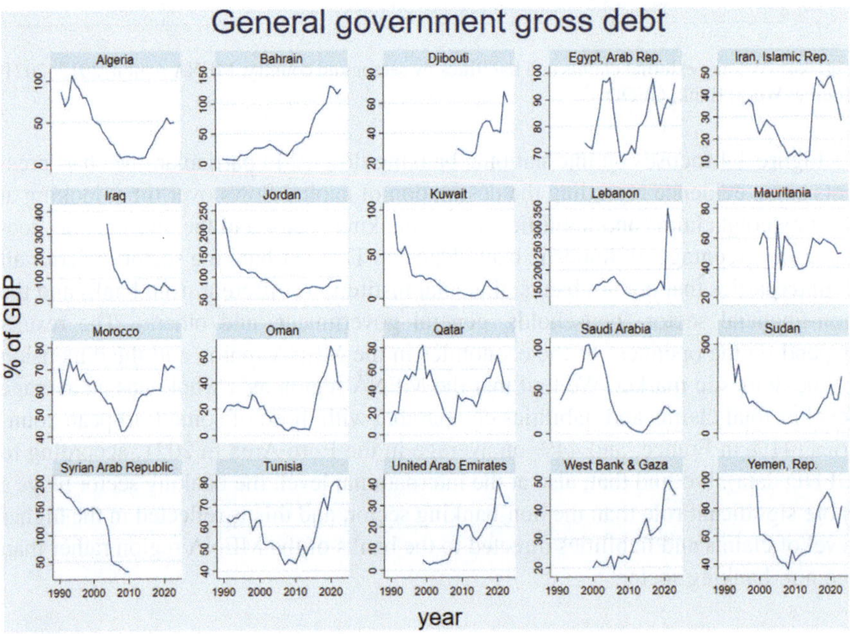

Fig. 4.5 General government gross debt in percentage of GDP by country, 2012–2023. Source: IMF, WEO database

[14] Data extracted from the IMF World Economic Outlook Database, April 2024 update.

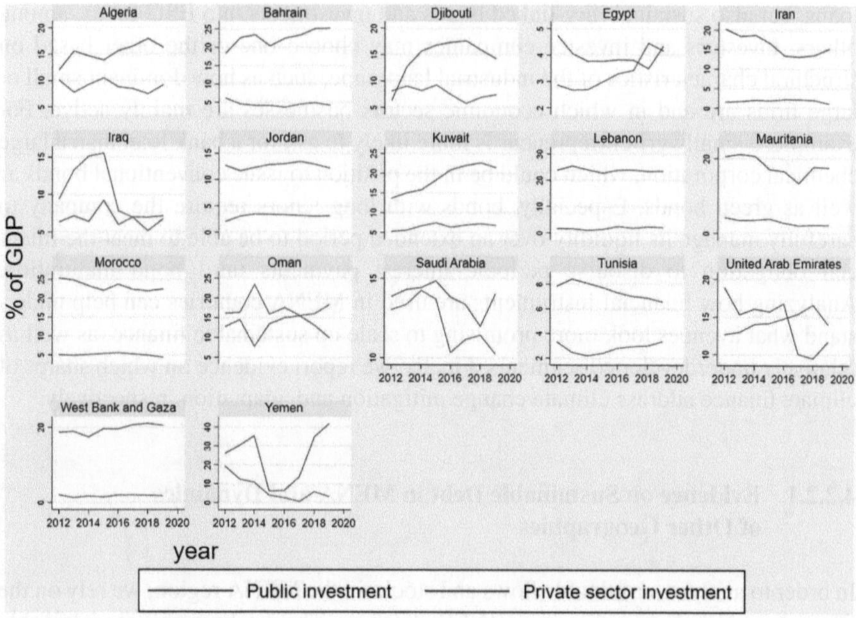

Fig. 4.6 Public and private investment in percentage of GDP by country, 2012–2019. Source: IMF, ICSD

We find significant heterogeneity also in the level of public and private investment with respect to GDP (Fig. 4.6).[15] The figure depicts general investment for the private and public sectors in selected MENA countries. The gap between public and private investment varies over time and among countries, going from less than 5% to more than 30%. In some cases, for instance, in countries experiencing conflict, or at high geopolitical risk, data are to be taken with caution, since GDP also experiences unexpected shocks, and it may not always be reliable.[16]

4.2.2 *Development of Sustainable Flows and Regional Patterns of Climate Finance*

Sustainable financial flows are defined as funds for long-term investments in activities with a focus on the ESG dimensions. There are several financial instruments that may be used to channel funding to ESG activities, for example, green bonds and

[15] Data extracted from the Investment and Capital Stock Dataset by the IMF. The last available data are from 2019.

[16] This is the most complete metric for accounting investment among those available, the dataset does not report the level of investment in nominal or real terms.

loans, but also sustainability-linked bonds and investments into ESG funds, among others. Investors and investee companies may choose one or the other based on structural characteristics of the industrial landscape, such as how dominant small or large firms are and in which economic sectors businesses are mainly active. For example, a small agri-entrepreneur is more likely to ask for a bank loan than a large chemical corporation, which could be in the position to issue conventional bonds as well as green bonds. Especially, bonds with long tenors require the company to carefully manage its liquidity over an extended period to be able to meet the financial obligation to disburse periodic interest payments until bond maturation. Analyzing how financial instruments are used in MENA countries can help understand what avenues look more promising to scale up sustainable finance, as well as relatively underdeveloped segments. Finally, we report evidence on which shares of climate finance address climate change mitigation and adaptation, respectively.

4.2.2.1 Evidence on Sustainable Debt in MENA and Dynamics of Other Geographies

In order to analyze sustainable flows and stocks in the MENA region, we rely on the Bloomberg New Energy Finance (BNEF) database. The database covers sustainable fixed-income securities and syndicated loans of the sustainable segment. It includes green, social, sustainability, sustainability-linked, and transition debt types. Companies issuing green, social, and sustainability bonds make a binding commitment in the "Use of proceeds" to use these funds for sustainable projects and activities. While a large fraction of sustainable financial products does not offer specific information on the degree of sustainability of the activities they finance, the financial products considered in this section adhere to the International Capital Markets Association bond principles and the Loan Market Association loan principles for green, social, sustainability, and sustainability-linked types.[17] Following this independent and globally acknowledged certification scheme allows us to: (i) exclude from the analysis financial products at higher risk not to finance sustainable activities and (ii) compare results across geographies, as standards do not depend on local regulation.

Narrowing the spatial examination down to locations of relevance, countries are selected by the domicile of the issuer's or borrower's headquarters.[18] To put sustainable debt in the MENA region in perspective, we make a comparison with global developments, as well as with the evolution in Asian countries. Vintages for which data have been collected include pre-2012 (jointly due to marginal amounts) and distinct year-by-year statistics from 2012 to 2024, with significantly growing amounts from 2021 onward. We provide flow data, such as annual issuances, as a

[17] For example, https://www.icmagroup.org/sustainable-finance/the-principles-guidelines-and-handbooks/green-bond-principles-gbp/, https://www.lsta.org/content/green-loan-principles/

[18] The domicile country is where the company's senior management is located and generally applies for tax purposes (https://lei.bloomberg.com/docs/faq)

time series to illustrate structural changes and the recent emergence of instrument types.

Sustainable Debt Worldwide

Looking at the composition of sustainable debt worldwide, the most sizeable source of funding stems from green bonds, sustainability-linked loans (SLLs), and sustainability bonds (Fig. 4.7). Considering 2023, out of the 1452.4 bn$ of sustainable debt issued, almost half of the funds (44%) are backed by green bonds (646.5 bn$), trailed by 14.6% of the total from SLL (212.3 bn$), with smaller contributions from sustainability bonds, social bonds, as well as green loans. The same proportions hold in previous years, as well as for the first 7 months of 2024, with 45% of the total 929.9 bn$ issued through green bonds (422.6 bn$), and another 15% by SLL (138.2 bn$).

Analyzing the global balance of bonds and loans, bonds have featured a dominance over loans in the past years when it comes to contribution size (green, sustainability, and social bonds). Compared to loans, this market segment, consisting of more novel debt instruments, could profit from the general perception that bonds offer a safer investment opportunity. With respect to the developments over time, green bonds were the first type of financial instrument to be launched in the sustainable segment, foremost by the public sector,[19] long before other instruments.

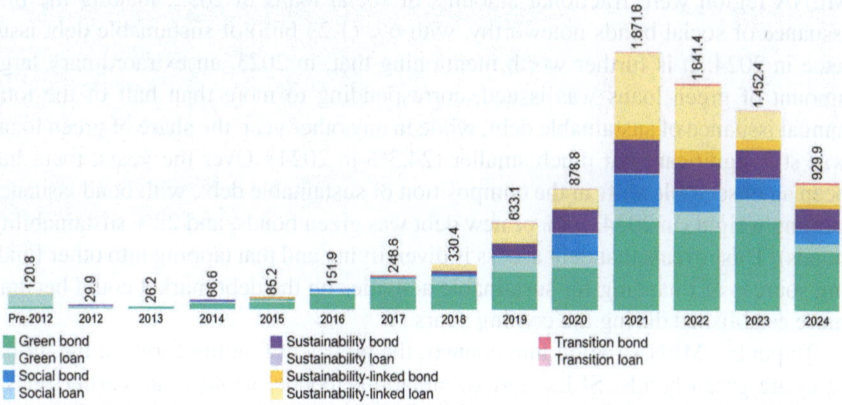

Fig. 4.7 Worldwide—annual issuance of sustainable debt types, bn$, pre-2012–2024 (cut-off date 31.07.24). Source: BNEF

[19] The EIB issued the first green bond in 2007, named the "Climate Awareness Bond."

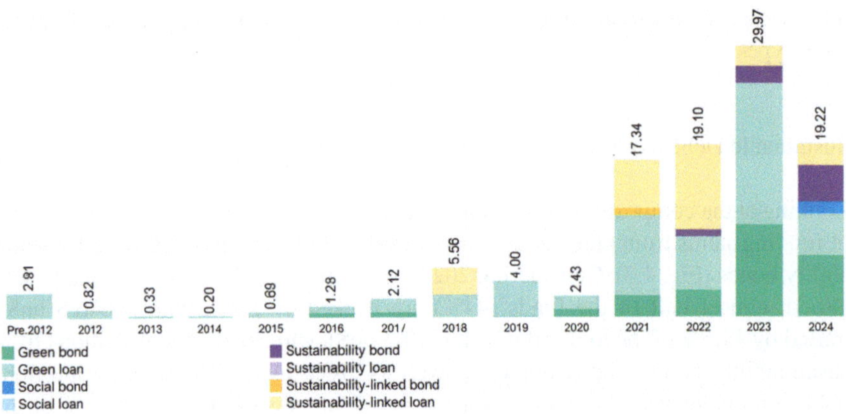

Fig. 4.8 MENA—annual issuance of sustainable debt types, bn$, pre-2012–2024 (cut-off date 31.07.24). Source: BNEF

Sustainable Debt in MENA Countries

Sustainable debt for the MENA region[20] is more loan-based than at the global level. Regarding sustainable debt flows for 2023 (Fig. 4.8), the overall amount sums up to 29.97 bn$, with the balance tilting toward green loans, making up more than half of the volume with 52% (15.56 bn$), while 34% are green bonds (10.27 bn$). The remainder is split into sustainability bonds (6% with 1.72 bn$), SLL, and sustainability and social loans. The first social debt instruments entering the stage in the MENA region were fractional amounts of social loans in 2022, making the first issuance of social bonds noteworthy, with 6% (1.23 bn$) of sustainable debt issuance in 2024. It is further worth mentioning that, in 2023, an extraordinary large amount of green loans was issued, corresponding to more than half of the total annual issuance of sustainable debt, while in any other year, the share of green loans was still significant but much smaller (24.3% in 2024). Over the years, there has been an observable shift in the composition of sustainable debt, with bond issuance gaining weight (in 2024, 35% of new debt was green bonds, and 22% sustainability bonds). This reveals that debt access is diversifying and that tapping into other funding sources of financing for sustainable activities on the debt market could become more established during the coming years.

To put the MENA region into context, the prevalent funding forms at the global level are green bonds, SLLs, and sustainability bonds, while in an earlier phase, green loans were somewhat more relevant. With the exception of green loans, which are still relatively rather important in the MENA region, the pattern worldwide is therefore not very different from that in the MENA region. As its market matures,

[20]The sample available in BNEF consists of ten countries: Bahrain, Egypt, Iraq, Israel, Jordan, Lebanon, Morocco, Palestine, Qatar, Saudi Arabia, and the UAE.

and businesses search for alternative funding sources, minor sustainable bond and loan types might gain momentum and grow in size in the MENA region as well.

Sustainable Debt in Asian Countries

In Asian countries,[21] the composition of new sustainable debt is different compared to global and MENA ones.

In particular, the period 2021–2023 is characterized by annual issuances of sustainable debt types remaining in stable proportions (Fig. 4.9). While 2023 totaled 351.3 bn$ of new sustainable debt, 44% was green bonds (155.2 bn$), 19% social bonds (65.25 bn$), 12% SLL (41.29 bn$), 11% sustainability bonds (39.32 bn$), and 10% green loans (34.17 bn$). SLB, social and sustainability loans, and transition bonds and loans account for a marginal fraction.

This illustrates that the flow composition of sustainable debt in the observed Asian countries features a proportional growth of all represented types until 2023, with the onset of transition bonds in 2024 introducing a nascent segment. Since a more diversified use of the instruments has emerged in 2021, no type has been set back or substituted by another. A novelty emerging in the first 7 months of 2024 is the visible rise of transition bonds with 9% of new sustainable debt (18.24 bn$), showing that further market capacity in Asian countries exists to absorb securities devoted to the decarbonization of firms and industrial sectors.

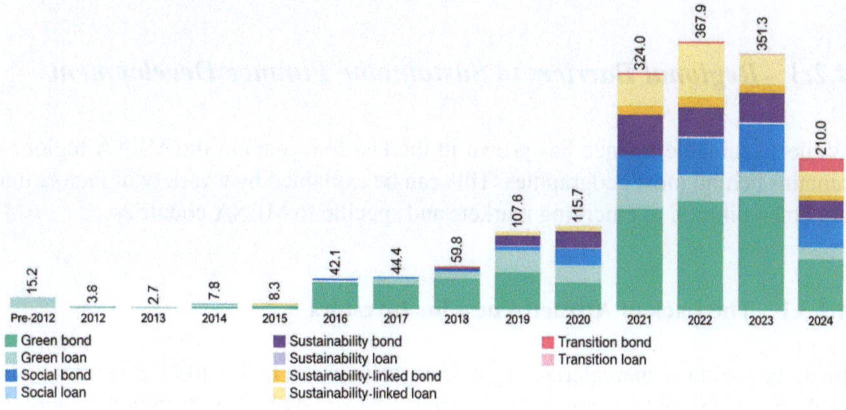

Fig. 4.9 Asia—annual issuance of sustainable debt types, bn$, pre-2012–2024 (cut-off date 31.07.24). Source: BNEF

[21]We consider the following 18 countries: China, Hong Kong, India, Indonesia, Japan, Laos, Macao, Malaysia, Myanmar, Nepal, Pakistan, Philippines, Singapore, South Korea, Sri Lanka, Taiwan, Thailand, and Vietnam.

Compared to the MENA region, the social segment has been increasing steadily in the Asian market over the past few years, while it is currently less represented in the MENA region. Another difference is that while sustainable activities in the MENA region are mostly financed by (green and SLL) loans, all the sustainable bond segments are more in use at the global level as well as in the Asian countries under study, proving a more advanced diversification in the instrumental landscape. Financial progress and convergence in standardization across regions could accelerate the absorption of novel debt types in the MENA region, too.

4.2.2.2 Evidence on the Mitigation Focus in the Region

According to Climate Policy Initiative data for the years 2021 and 2022, only 5 bn$ were dedicated to adaptation finance projects in the region,[22] compared to 28 bn$ for climate mitigation and 4 bn$ for projects with multiple objectives. Hence, adaptation represents only 14% of all climate flows in the region. While this is higher than the global average at 5%, it means that regional climate finance flows are still strongly oriented toward mitigation projects, despite the stronger adaptation needs in the region.

Moreover, 100% of these flows are financed by the public sector, and mostly internationally through multilateral and bilateral development finance institutions, and therefore not by local actors (only a small minority from governments).

Looking at sectors, a quarter of adaptation flows was directed toward water and wastewater projects, followed by agriculture, forestry, and other land use.

4.2.3 Regional Barriers to Sustainable Finance Development

While sustainable finance has grown in the last few years in the MENA region, it remains behind most geographies. This can be explained by a variety of factors that are both common to emerging markets and specific to MENA countries.

4.2.3.1 The Lack of Attractiveness for Investors

First, the region is characterized by a heavy reliance on high-carbon assets for some of the countries, with oil rents representing 12% of GDP on average in the region in 2021, and reaching 56% for Libya, according to World Bank data.[23] Oil and gas

[22] Climate Policy Initiatives provide aggregate data on the MENA region defined as including the following countries: Algeria, Bahrain, Egypt, Iraq, Islamic Republic of Iran, Israel, Jordan, Kuwait, Lebanon, Libya, Morocco, Oman, Qatar, Saudi Arabia, State of Palestine, Syrian Arab Republic, Tunisia, UAE, and Yemen.

[23] https://databank.worldbank.org/source/adjusted-net-savings/Series/NY.GDP.PETR.RT.ZS

production have been the main sources of government income for several decades in oil-exporting countries and have been both very attractive assets for local and foreign investors and cheap energy sources for local companies. Given this dependence, divesting from high-carbon assets in the MENA region is much more costly than in other regions. This is fully in line with predictions from the financial literature, according to which risk diversification considerations impede total divestment from high-carbon assets (Steuer & Tröger, 2022). Moreover, because of the lack of a commonly shared narrative on climate risks in the region, local investors still often restrict climate change risks to physical risks. Hence, they perceive that climate change will affect green and high-carbon assets in a symmetrical way through natural disasters. Because they do not see high-carbon assets being primarily affected by the green transition, they do not divest from those sectors.

Linked to this, as described in the report of the High-Level Expert Group on scaling up sustainable finance in LMICs (2024) established by the European Commission, investors struggle to find investments with the required levels of risk-adjusted return, size, liquidity, and sustainability. Investors see returns as insufficient compared to the level of risk, particularly for corporates in countries with non-investment sovereign grade.[24] In the MENA region, this problem particularly affects corporates outside of the Gulf countries, Egypt, and Morocco.

Investors also highlight the low level of returns compared to the cost of research, mostly for foreign investors that lack expertise on the countries they are investing in (High-Level Expert Group on scaling up sustainable finance in low- and middle-income countries, 2024).

For European investors, there is an additional challenge to find projects matching the sustainability criteria in place at home. With reference to the EU Regulation, the "do no significant harm" criteria and the governance criteria of Article 9 of the Sustainable Finance Disclosure Regulation are the hardest to fulfill. Finally, investors often highlight the lack of bankable projects in emerging countries (High-Level Expert Group on scaling up sustainable finance in low- and middle-income countries, 2024).

Regarding local investors, in the MENA region, environmental objectives are less of a priority than other social needs, such as poverty or unemployment (Al-Hamadeen, 2021). Sustainable investments are also seen as a philanthropic matter rather than a business. This general lack of interest in sustainable investments makes such assets quite illiquid, adding an extra constraint on investors.

[24] In these countries, the sovereign rating caps the risk assessment of all corporates in the country, which are therefore assessed as highly risky.

4.2.3.2 Lack of Transparent Data and Uncertain Regulatory Frameworks

In order to take informed investment decisions, professional investors need concise, comprehensive, comparable, and credible non-financial information (Cohen et al., 2015). Furthermore, institutional investors value and demand information on firms' climate risk exposure (Ilhan et al., 2023). However, in developing countries, corporates' disclosures are less extensive and credible compared to more advanced economies (Choi & Meek, 2014). This is reinforced by the lower attention given to the investigation of environmental disclosure practices in the MENA region compared to developed countries (Al-Hamadeen, 2021; Al-Hamadeen & Badran, 2014; Ku Ismail & Ibrahim, 2008; Luo et al., 2013; Said Suwaidan et al., 2004). Moreover, most studies have focused on disclosure in oil-related sectors, adopting a clear mitigation stance, and leaving aside all the other sectors. Finally, the lack of specialized databases and research centers in the region tampers with academic enthusiasm to conduct empirical and theoretical research in this area (Al-Hamadeen, 2021).

While corporate disclosures in the MENA region are improving, there is still a strong heterogeneity in the quantity and quality of environmental disclosure across companies as well as across countries in the region (Gerged et al., 2018). Such heterogeneity is linked to the different ambitions of environmental regulations across the region. Based on a comprehensive review of the different sustainable strategies deployed in the region (Beyer & Bayouni, 2022; UNEP Finance Initiative, 2021), Morocco, Egypt, the United Arab Emirates (UAE), and Bahrain are the only ones in the region to have developed sustainable regulations for the bond and stock markets (only the stock market for Bahrain). However, existing strategies still often exclude a net-zero pledge (Beyer & Bayouni, 2022). For those that take a strong commitment, such as the UAE with a net-zero target by 2050, Bahrain and Saudi Arabia by 2060, or Morocco with a reinforced United Nations Framework Convention on Climate Change (UNFCCC) strategy in 2021 (Climate Action Tracker[25]), there is still a need to translate it into a financial strategy and build a sustainable finance framework with appropriate instruments.

Most countries in the region provide guidelines for voluntary ESG disclosures but fail to implement mandatory disclosure (UNEP Finance Initiative, 2021). Moreover, the poor enforcement of environmental laws in the region slows the inclusion of environmental and social risks into the core operations and strategies of financial institutions, as it requires clients to abide by environmental and social procedures (UNEP Finance Initiative, 2021).

Finally, despite recent initiatives launched at the regional level (Arab Green and Sustainable Finance Network, launched in 2023, or participation in the Sustainable Baking Finance Network), no regional standards and guidelines have been developed yet. This constrains the mobility of investors across markets in the region and blocks larger investors from richer countries in the region from investing in less developed markets, most in need of foreign investors.

---

[25] https://climateactiontracker.org/countries/uae/

4.2.3.3 The Public Good Spell: Barriers for Adaptation Finance

Investments in climate change adaptation differ from those in mitigation in several aspects, which have a bearing on the incentives to raise private funding (Mullan & Ranger, 2022).

First, there are no unique standards and methodologies to assess physical risk and the reduction thereof from adaptation investments. This hinders the definition of any clearly quantified and internationally agreed targets for adaptation. It also leads to inadequate pricing in financial modeling and affects decision-making made based on such modeling (UNEP Finance Initiative, 2021). Furthermore, adaptation investment is context- and location-specific, and so are its benefits. Hence, not surprisingly, adaptation finance is characterized by the predominance of public flows, which represent 98% of the total adaptation expenditure in 2021/2022 (Buchner et al., 2023).

According to Stout (2022), the lack of private investments is due to three main barriers. First, market actors suffer from information asymmetries and knowledge gaps, which reduce their incentive and ability to invest in adaptation. Second, adaptation projects display positive externalities, in the sense of positive benefits for society as a whole. Those environmental and social benefits are not fully captured in the rates of return for the investment and, hence, generate unfavorable risk-return profiles for investors. Finally, private investors' and adaptation projects' horizons normally do not match. While private actors operate in the short to medium term, adaptation projects will likely bring benefits only in the long term, conditional upon uncertain climate outcomes.

4.3 Scaling up Sustainable Finance: Lessons from Existing Regional Initiatives and the International Context

Starting from the state of play described in the previous section, we now delve into existing regional initiatives and international examples to draw lessons on how to scale up sustainable finance in the MENA region. We highlight the benefits derived from regulatory standardization (Section 4.3.1), the need to address the balance between global standards and regional nuances (Section 4.3.2), and the crucial implementation strategies involving capacity building and stakeholder engagement (Section 4.3.3).

4.3.1 The Benefits from Regulation and Standardization

With the rise of ESG voluntary disclosures in the MENA region, the first evidence has become available on the benefits of reporting. Several articles have established the link between corporate sustainability and financial performance in the region. Some find a positive impact of firms' environmental performance on the cost of finance, for both debt and equity, in emerging markets (Temiz, 2022). Looking at financial firms in Oman, reporting on Sustainable Development Goals (SDGs), while generally poor in the country, improves financial performance (Al Lawati & Hussainey, 2022). In Egypt, sustainability performance positively affects financial performance, even more after the political events in the country, mirroring changes in the ESG perceptions in the society (Aboud & Diab, 2019). With Saudi firms, reporting environmental sustainability practices has a positive and significant impact on firm value (Ammer et al., 2020).

Beyond disclosure, the literature also highlights the need for common definitions of sustainability concepts to support investors' confidence and foster investments (Migliorelli, 2021). The definition of such standards and guidelines is the backbone of the sustainable bond markets (Saltane et al., 2023; Tyson, 2021). Looking at what happened in other regions can help in identifying the benefits of sustainable finance regulations and standardization. For example, with the implementation of the EU Taxonomy and the EU Green Bond Standard (EUGBS), the European Union has given to all investors a common definition of green activities with clear criteria, which they can refer to. As a result of implemented regulation, today, the EU27 market shows the highest issuance of sustainable instruments (37% of global issuances of sustainable financial instruments in 2023).[26] In the absence of such standards, as it is still the case for sustainability-linked instruments, the level of trust is much lower, hindering the growth of relevant market segments. Actually, from the peak in 2021, global issuances of sustainability-linked bonds decreased from 629 bn$ in 2021 to 282 bn$ in 2023. Such instruments have lost attractiveness as investors fear greenwashing practices and lack of ambitions in targets. Looking at the EU example, the development of the EUGBS brought into light the essential question of sustainability reporting. Under the EUGBS, issuers of a green bond have to regularly disclose the use of proceeds, to monitor how the money was ultimately invested. Such ex-post reporting, yet to be standardized at the EU27 level, will be the backbone of the fight against green washing practices and will foster a real greening of the economy as result of sustainable investments.

Developing such standards and guidelines will be the next challenge for MENA countries, with the need to balance global standardization requirements and regional specificities.

[26] BNEF data.

4.3.2 *Balancing Global Standardization and Regional Specificities*

A first step in the direction of a homogeneous standardization in the MENA region could be a wider introduction of sustainability reporting frameworks. According to a survey (KPMG, 2024) conducted with a sample of the largest 100 companies in 58 countries (the N100), only 14% of the 800 covered firms in Africa and the Middle East disclose sustainable practices. An available framework to be adopted is provided by the International Financial Reporting Standards (IFRS), which have introduced the IFRS Sustainability Disclosure Standards. The IFRS are a set of accounting rules for the financial statements of public companies, developed by the International Accounting Board, that are intended to make them consistent, transparent, and easily comparable around the world. Other than setting an international accounting standard, in 2021, the IFRS Foundation created the International Sustainability Standards Board (ISSB), which developed in 2023 the IFRS Sustainability Disclosure Standards. This new set of standards is designed to deliver a global baseline for sustainability disclosures to inform capital markets. Countries in the MENA region have applied the IFRS standards in a mixed way. More than half of them permit or require the adoption of the IFRS standards, but in most cases, only for banks or other financial institutions, not for listed firms or small and medium enterprises (SMEs). Moreover, a common disclosure standard for sustainability purposes is still missing (see Section 4.2.3.2). The ISSB sustainability standards were published in 2023, and consultations with several countries—such as the European Union, Canada, Japan, and Malaysia, among others—were closed in October 2024. Currently, none of the countries in the MENA region has opened a jurisdictional consultation at the national or supra-national level on sustainability-related disclosures.

Although common disclosure standards are still missing in the region, environmental risks and opportunities are widely analyzed at the country level by the region's policymakers. The UN reported (UNEP Finance Initiative, 2021) that several countries in the MENA region, in particular, Egypt, Jordan, Morocco, Bahrain, and the UAE, have developed specific ESG guidelines. Central banks and financial supervisors are taking steps to integrate climate-related risks into their analytical frameworks and decision-making processes to ensure the financial sector's stability (International Monetary Fund, 2024). The development of models for the evaluation of asset exposures to climate risks and the setup of contingency plans within financial institutions is still in its initial phase. However, there are notable examples that show progress in the field. Several central banks in the area developed indicators to identify environmental and climate risks and roadmaps or tackle such challenges, such as Oman and Morocco. Others, such as Egypt and the UAE, publish analyses of the consequences of climate shocks on the financial system based on climate shock scenarios in their Financial Stability Reviews, or, such as Saudi Arabia, are planning to include physical and transition risks analysis in their Financial Stability Review. Awareness also appears to be increasing within financial institutions.

4.3.3 *Implementation by Capacity Building and Stakeholder Interaction*

Activities of financial intermediaries and businesses in MENA countries are not homogeneous and, in part, strongly lag behind other macro regions (see Section 4.2.1.2). A way to foster financial development is to focus on capacities that can be developed in interaction with the government sector. Looking at climate finance in particular, an examination of seven case studies by the United Nations Environmental Program—Financial Institutions (UNEP Finance Initiative, 2022) points to blended finance as a promising avenue. Blended finance is intended as a mixture of financing, technical assistance, and financial incentives for which parties are complementarily involved according to dedicated areas of expertise. Citing, for instance, a Moroccan energy efficiency measure of 1.6 mn$, only the technical support was foreign (European Bank of Reconstruction and Development, EBRD), whereas the expertise on sustainability was facilitated by the Bank of Africa in unison with the National Sustainable Development Strategy from the government. Against this background, the UNEP recommends the development of a green taxonomy led by the finance sector and tailored to the national or regional context, as well as locally driven solutions, such as climate-smart agriculture, to counterbalance imported innovation such as large-scale photovoltaic systems.

As a country case study of industry formation, it is noteworthy that Morocco's renewable energy sector is maturing at a faster pace compared to its regional neighbors: In 2023, it showcased a more than 20% share of electricity production from renewable sources,[27] such as geothermal, hydropower, modern biomass, solar, tidal power, and wind power. Regarding the potential capacity to produce renewable energy, there has been an almost steady annual upward movement from 24.9% of total electricity capacity in 2013 to 34.3% in 2022 (International Renewable Energy Agency (IRENA), 2023). Among the conditions favorable to this growth have been a green energy ecosystem and multiple waves of start-up development: for example, the city of Rabat offered proactive assistance to micro firms and SME for tapping into innovation financing (Tanchum, 2024). Being aware of the financial restrictions of particularly smaller businesses, Morocco's Ministry of Economy and Finance has collaborated with the Central Guarantee Fund to establish the "Innov Invest Fund" in 2016, which hosts a series of business foundation funding, that is, seed, early-stage, and venture capital financing services. These dedicated and cross-sectoral efforts to overcome entrepreneurial constraints allow the development of novel, path-independent economic activities with a less attractive risk-return profile in the beginning, such as renewable energy.

The roadmap for climate action in the MENA region by the World Bank (The World Bank Group, 2022) involves the consultation of national institutions and private investors together with other supranational financial intermediaries

[27] https://ourworldindata.org/energy/country/morocco#how-much-of-the-country-s-electricity-comes-from-renewables, 06.04.2025

(International Finance Corporation and Multilateral Investment Guarantee Agency). Their engagement is aimed at strengthening the role of treasury markets and commercial banks to improve access to finance for firms. Lifting some of the financial constraints and a part of the regulatory barriers in green industries, jointly with the implementation of reforms giving priority to businesses tapping into climate-aligned funds, is expected to widen the private sector participation in sustainable finance.

Considering that planning certainty is a driving force for investor decisions, the inclusion of third-party certifiers and data vendors is an essential element when scaling up sustainable finance. As emphasized by the Bundesbank (Mauderer, 2024), credit rating agencies and ESG data providers are indispensable for attracting private capital to emerging markets and developing economies, specifically in the MENA region and Central Asia. Their consistent integration into national adoptions of legislation on corporate data collection (High-Level Expert Group on scaling up sustainable finance in low- and middle-income countries, 2024) will pave the way for improved data quality.

A last reflection from banking practitioners refers to enabling SME's access to sustainable finance (High-Level Expert Group on scaling up sustainable finance in low- and middle-income countries, 2024). Looking at blended finance practice in the MENA region, the lack of financial intermediation for small enterprises hampers investments (Convergence Blended Finance, 2024). By nature, many SDG projects are of reduced size, whereas donors and investors are in pursuit of channeling large capital amounts to sustainable activities. The mismatch between investment strategies and supplied project leads to suboptimal engagements by foreign and private investors. As sovereign wealth funds have high political visibility in some of the MENA countries, they could take up financial intermediation and lower the risk exposure of investments. Such an engagement, as well as the promotion of public-private partnerships, can be vehicles to derisk foreign inflows by institutional investors. According to the United Nations Economic Commission for Africa (2024), leveraging foreign direct investments and official development aid has to occur more structurally in combination with private sector supply. A targeted coordination of funds already in place can unlock further investments and seize their full desired impact.

4.4 Financial Levers for the MENA Region: How Can the EU and External Parties Help?

Against this background comes the question of tools and levers to address the specific needs of the MENA region regarding sustainable finance. In this section, we delve into different tools that could be leveraged for both the public and private sectors and see how the European Union can help. We will start by exploring catastrophe bonds as a tool to increase MENA countries' resilience to natural disasters

(Section 4.4.1). Then we will examine examples of derisking tools (Section 4.4.2) before addressing sustainable assets recycling (Section 4.4.3).

4.4.1 A Potential Tool to Increase Financing of Climate Resilience: Catastrophe Bonds

In the first sections of this chapter, we discussed the high exposure of MENA countries to physical risk. This implies acute adaptation needs to cope with the increasing occurrence of disasters (acute risks), as well as with long-lasting climate shifts (chronic risks) affecting the countries' growth model. We also described the prevalence of climate mitigation rather than adaptation as an objective of regional sustainable financial flows. To address this double challenge, new financial tools could be leveraged to finance climate resilience in the region.

Given the limited fiscal space in some MENA countries (see Section 4.1) and the large expected losses from natural disasters, governments need to finance the response to natural disasters ex ante to avoid unaffordable stress on public debt. Financial markets have developed a new tool that can be used for such purposes. Catastrophe bonds are insurance-linked securities that allow the cedent (the insured) to get funding from the capital market, if and only if catastrophic conditions, such as a flood or hurricane, occur (Ando et al., 2022). The catastrophe bond involves the creation of a special-purpose vehicle (SPV), acting as an intermediate between the issuer and the investor. The SPV is in charge of collecting the premium from the issuer, as for a classic insurer, while collecting the proceeds from investors. As long as a natural disaster does not occur, the SPV transfers principal and coupon to the investors. However, if a catastrophic event occurs, the SPV disburses the reimbursement to the issuer, allowing the government to access funding from the capital market at a moment where other traditional financing tools are unavailable due to the increase in risks following the disaster. Hence, here investors are betting against the odds of a natural disaster occurring (Ando et al., 2022). The existence of the SPV as an intermediary allows for stock funds to be held under safe assets while enabling trading of the bonds on the secondary market. Such a bond structure also protects the issuer against a default from the investors, as all the proceeds are collected at the time of the bond purchase. It also allows access private capital to finance climate resilience. On the investor side, catastrophe bonds are providing high returns due to the high level of risks, and they have low correlation with other asset classes, allowing for diversification (The World Bank, 2022).

Such catastrophe bonds are predominantly used on the U.S. market (more than 50% of outstanding bonds according to Artemis[28]), but thanks to the intermediation made by the World Bank, some emerging markets have been able to issue some: Mexico (in 2009, 2012, 2017, and 2020 for hurricanes and earthquake), Chile,

[28] https://www.artemis.bm/dashboard/cat-bonds-ils-by-risk-or-peril/

Colombia, Mexico and Peru (in 2018 for earthquakes), the Philippines (in 2019 for tropical cyclones and earthquakes), and Jamaica (in 2021 for hurricanes). Such intermediation by the World Bank or other multilateral lenders is particularly useful to lower the spread (Ando et al., 2022). Indeed, in theory, the bond's pricing should be independent of the issuer's characteristics, given the existence of the SPV as an intermediary. However, scholars have shown that this is not the case, neither on the primary nor on the secondary market (Chatoro et al., 2023; Götze & Gürtler, 2020). They demonstrate that on both markets, the risk premium depends on the issuer's credit rating and length of presence on the market.

This type of intermediation by a multilateral lender would be particularly useful for MENA countries with limited fiscal space and high levels of risk. The European Union could also play a role in fostering the use of such instruments (High-Level Expert Group on scaling up sustainable finance in low- and middle-income countries, 2024), providing aid to low-income countries that do not have the capacity to pay the premium and promoting such instruments in its policy.

For MENA countries with greater financing capacity—such as Gulf states—investing in catastrophe bonds issued by other countries would also be a way to show commitment by internalizing climate risk happening elsewhere while earning a positive return on average (Ando et al., 2022).

4.4.2 Examples of Derisking and Sustainable Funds

Beyond the instrument level, risk reduction in private investments is a general topic ingrained into approaches by the European Union, global alliances, and bilateral partnerships. A public party can provide risk adjustment to corporate investors and financiers by absorbing first or specific losses, hence making the risk-return more acceptable to private sector investors. Blended finance should be encouraged specifically for investments in climate-resilient infrastructure (High-Level Expert Group on scaling up sustainable finance in low- and middle-income countries, 2024): the "Infrastructure Climate Resilient Fund" in cooperation with the "Africa Finance Corporation" delivers a prominent example of this kind. The European Fund for Sustainable Development Plus (EFSD+) is an instance of a more general, flexible device to promote risk sharing for sustainable investments in EU partner countries.[29] The EFSD+ offers a range of derisking tools reaching a maximum of 40 bn€ to crowd in private capital, thus targeting large projects toward a sum of 135 bn€, backed by the EIB as a major implementation partner. To give an example from the publicly accessible "EFSD+ Funding Database" from September 2024, the

[29] https://international-partnerships.ec.europa.eu/funding-and-technical-assistance/funding-instruments/european-fund-sustainable-development-plus_en

instrument of choice to mitigate risk for independent power producers in a long list of participating African states is a counter-guarantee.[30]

At the global level, the "Global Shield against Climate Risks" bases the deployment of its financial vehicles on a technical risk analysis, adjusted to the national or regional profile. After identifying the layers of climate risk, for example, in specific locations according to the severity of events, residual risk is managed by adequate and cost-effective instruments suited to the respective layer of risk (Global Shield, 2024). For instance, in October 2024, a country-led initiative by the Government of Ghana, jointly with the Global Shield and African Risk Capacity Ltd., launched a sovereign drought insurance policy to adapt to climate-related food insecurity. While the German "Kreditanstalt für Wiederaufbau" provided the funding, it was a joint effort to develop suitable risk management and sectoral measures to pursue agricultural resilience and livelihood protection for vulnerable populations in Ghana.[31] Despite the high number of Global Shield projects at the national level, the regional programs do not address the MENA region yet.

Taken together, the current derisking schemes dedicated to sustainable purposes cover a wide range, from concessional finance for climate-resistant infrastructure projects to counter-guarantees and insurance policies. Among them, the degree to which risk is identified, distributed among parties, and managed varies greatly in sophistication. Most financing cases available, like the sample presented here, focus on climate-vulnerable countries in Africa, whereas MENA as a region has not received particular attention so far.

4.4.3 Sustainable Asset Recycling and Securitization in LMICs

Asset recycling is a process that allows existing public infrastructure to be monetized through partial divestion, lease or concession, or the creation of a joint venture between the relevant public authority and private parties (The World Bank, 2023). First, the government needs to monetize value from revenue-generating public assets, so it can use the proceeds to invest in existing or new infrastructure assets. Public assets that may be eligible for asset recycling are toll roads, bridges, airports, transit systems, power generation and grids, warehouses, and so on. Usually, the public sector continues to own the asset (except in the case of sale, or full privatization). In the case of concession or lease, the selected private sector investors form an SPV to enter a long-term lease or concession agreement with the public authority to operate the asset. In the case of a joint venture, the relevant authority and the selected

[30] https://www.get-invest.eu/fund/european-fund-for-sustainable-development-plus-efsd/#:~:text=The%20EFSD%2B%20provides%20partial%20risk,the%20host%20government%20is%20available.&text=Yes%2C%20there%20are%20limits%20on,exposure%20that%20can%20be%20assumed, 26.09.2024

[31] https://www.globalshield.org/news/ghana-purchases-its-first-drought-insurance-policy-from-arc/, 17.10.24

private sector investors enter a joint venture to jointly operate the asset under a long-term lease or concession.

Securitization is an effective option to allow public entities to transfer their illiquid assets to the private sector through an ad hoc-created SPV. The SPV raises funds from investors through the issuance of Asset-Backed Securities that are issued in different tranches. Once illiquid assets, such as infrastructure, are pooled and securitized, they become more liquid and tradable, becoming accessible to a larger investor base.

Securitization, however, may be complex and costly to structure. It needs high-quality underlying assets, appropriate regulatory infrastructure, transparency backed by strong regulation and monitoring systems, and relatively developed capital markets and experienced institutional investors.

Asset recycling may be used for financing any kind of infrastructure project, but it can be a particularly effective tool for financing new sustainable infrastructure projects, such as sustainable energy projects, or for greening current infrastructure, such as increasing the energy efficiency of existing buildings (green use-of-proceeds). It suits sustainable finance initiatives well in those countries with high levels of public debt, as it allows the public financing of infrastructural projects at virtually no cost for the taxpayers, since the source of funding for the new infrastructure would be the revenues of the recycling process and not new public debt.

Among LMICs, there are examples of securitization, which could be replicated in the MENA region for green or sustainable objectives. In Uruguay, the Corporación Vial del Uruguay was set up to recycle toll road assets through the issuance of project bonds in local capital markets to finance new transport infrastructure (The World Bank, 2023). To advance universal access to electricity in Côte d'Ivoire, the Emerging Africa Infrastructure Fund is committing up to 48 mn$ to launch the first-ever social asset-backed security in West Africa. The bond, valued at 96 mn$, will be issued by an SPV fund in the local currency (Emerging Africa & Asia Infrastructure Fund, 2023). Finally, India has created Infrastructure Investment Trusts, pooled investment vehicles that buy, operate, and manage operational infrastructure assets, thus helping developers release their invested equity and deploy capital in new projects (Sanghai, 2024). In the context of Islamic finance, Malaysia, Indonesia, and Saudi Arabia have already issued green sukuks.[32] In 2014, Malaysia approved a solar sukuk program, the Quantum Solar Park (Semenanjung) Sdn Bhd. This SPV issued the largest solar project linked to a green sukuk. In 2018, Indonesia issued the first sovereign green sukuk, while in 2019, the UAE listed the world's first benchmark corporate green sukuk and first green sukuk issued by a corporate in the Middle Eastern region (Garcia-Kilroy et al., 2023).

[32] A sukuk is an Islamic financial certificate that complies with Islamic religious law. It is similar to a bond in Western finance.

4.5 Conclusion

Sustainable finance in the MENA region holds great potential for driving economic diversification, enhancing climate resilience, and improving social outcomes. This chapter proposed actionable takeaways, including recommendations for policymakers, financial institutions, and private sector stakeholders, to scale up sustainable finance in the region, while showing that several MENA countries have already made significant strides in sustainable finance. In particular, concrete proposals include the following:

- The development of ad-hoc ESG guidelines, the integration of climate-related risks in their supervisory frameworks, and the development of competitive renewable energy sectors. In this area, to name a few, Egypt, Jordan, Morocco, and the UAE constitute seminal case studies for the region.
- The development of sustainable finance tools for easing green investment in the region. On this, MENA countries may take inspiration from the experience of advanced economies such as the United States (catastrophe bonds to finance climate resilience) and the European Union (derisking schemes, at both the regional and country levels). Also, other emerging market countries constitute interesting case studies for the MENA region, both in the Western Hemisphere (e.g., the experience of Chile, Mexico, Peru, and Uruguay with catastrophe bonds and sustainable asset recycling), Africa (the Global Shield initiative in Ghana, Emerging Africa and Asia Infrastructure Fund (EIAF) SPV fund in Côte d'Ivoire), and in Asia (the Quantum Solar Park project in Malaysia).

In fact, concerted efforts from governments, financial institutions, and the private sector are needed to overcome existing barriers and leverage opportunities for sustainable growth. Green bonds, SLLs, and blended finance are some of the innovative financial instruments that can mobilize private capital for sustainable projects, as they help derisk investments and attract a broader range of investors. In this respect, further research is warranted to assess the impact of these financial instruments in particular MENA countries or regions. For example, we know that green bond issuers in the European Union benefit from a negative *greenium*; however, this only happens if they are transparent and credible enough, and it would be interesting to know whether issuers in the MENA region would enjoy the same conditions.[33]

Finally, as developing coherent regulatory frameworks is essential for attracting private investments, governments need to create policies that incentivize sustainable investments and ensure transparency and accountability. For example, MENA countries could build on the experience of Morocco, the UAE, and Bahrain to introduce harmonized sustainable regulations for the bonds and stock markets. The Moroccan green energy ecosystem can also inspire others in the region to implement industrial policies fostering innovation for domestic firms, including SMEs. Further research

[33] See, for example, Fatica et al. (2021).

is also needed to explore how country-specific policy frameworks could be developed, taking into consideration the specificities of MENA countries.

References

Aboud, A., & Diab, A. (2019). The financial and market consequences of environmental, social and governance ratings: The implications of recent political volatility in Egypt. *Sustainability Accounting, Management and Policy Journal, 10*(3), 498–520. https://doi.org/10.1108/SAMPJ-06-2018-0167

Al Lawati, H., & Hussainey, K. (2022). Does sustainable development goals disclosure affect corporate financial performance? *Sustainability, 14*(13), 1–14. https://doi.org/10.3390/su14137815

Al-Hamadeen, R. (2021). Environmental accounting and reporting: Evidence from the MENA region. In *Routledge handbook of environmental accounting* (pp. 300–314). Routledge. https://doi.org/10.4324/9780367152369-26

Al-Hamadeen, R., & Badran, S. (2014). Nature and determinants of CSR disclosure: Experience of the Jordanian public shareholding companies. *Europian Journal of Business and Management, 6*(13), 18–35.

Alshammari, S., & Goto, S. (2022). What factors drive Saudi stock markets? Firm characteristics that attract retail trades. *International Review of Economics and Finance, 80*, 994–1011. https://doi.org/10.1016/j.iref.2022.02.004

Ammer, M. A., Aliedan, M. M., & Alyahya, M. A. (2020). Do corporate environmental sustainability practices influence firm value? The role of independent directors: Evidence from Saudi Arabia. *Sustainability, 12*(22), 1–21. https://doi.org/10.3390/su12229768

Ando, S., Fu, C., Roch, F., & Wiriadinata, U. (2022, July 1–28). *Sovereign climate debt instruments: An overview of the green and catastrophe bond markets*. IMF staff climate note 2022/004, International Monetary Fund, Washington, DC.

Annunziato, A., Santini, M., Proietti, C., de Girolamo, L., Lorini, V., Gerhardinger, A., & Tucci, M. (2024). Modelling and validation of the Derna dam break event. *GeoHazards, 5*(2), 504–529. https://doi.org/10.3390/geohazards5020026

Beck, T., Demirgüç-Kunt, A., & Levin, R. (2005). *Bank concentration and fragility: Impact and mechanics* (no. 11500; NBER working paper series). http://www.nber.org/papers/w11500

Beyer, J., & Bayouni, M. (2022). *Financing a green transition in the Middle East*.

Buchner, B., Baysa, N., Padmanabhi, R., Stout, S., Strinati, C., Wignarajah, D., Miao, G., Jake, C., & Marini, N. (2023). *Global landscape of climate finance* (issue November). doi:https://doi.org/10.1201/9781003396949-21

Chatoro, M., Mitra, S., Pantelous, A. A., & Shao, J. (2023). Catastrophe bond pricing in the primary market: The issuer effect and pricing factors. *International Review of Financial Analysis, 85*. https://doi.org/10.1016/j.irfa.2022.102431

Choi, F. D. S., & Meek, G. (2014). *International accounting* (7th ed.). Pearson Education.

Cohen, J. R., Holder-Webb, L., & Zamora, V. L. (2015). Nonfinancial information preferences of professional investors. *Behavioral Research in Accounting, 27*(2), 127–153. https://doi.org/10.2308/bria-51185

Convergence Blended Finance. (2024). *Blended finance in the Middle East & North Africa* (issue 17).

Demirgüç-Kunt, A., Klapper, L., Singer, D., & Ansar, S. (2021). *The global findex database 2021: Financial inclusion, digital payments, and resilience in the age of COVID-19*.

Elnahass, M., Marie, M., & Elgammal, M. (2022). Terrorist attacks and bank financial stability: Evidence from MENA economies. In *Review of quantitative finance and accounting* (Vol. 59, Issue 1). Springer US. https://doi.org/10.1007/s11156-022-01043-1

Emerging Africa & Asia Infrastructure Fund. (2023). *EAIF commits up to 30 billion to launch West Africa first social asset-backed security, targeting 100% electrification in Cote d'Ivoire*. EAIF.

European Commission Joint Research Centre (JRC). (2023). *Flood risk and mitigation strategies in Derna (Libya)* [Dataset]. http://data.europa.eu/89h/65909314-0273-575c-9a7e-9126289b8af3

Eyraud, L., Bunda, I., Jack, J., Jardak, T., Ouedraogo, R., Wang, Z., & Wezel, T. (2021). *Resolving nonperforming loans in Sub-Saharan Africa in the aftermath of the COVID-19 Crisis* (International Monetary Fund. African Department (Series)). International Monetary Fund.

Fatica, S., Panzica, R., & Rancan, M. (2021). The pricing of green bonds: Are financial institutions special? *Journal of Financial Stability, 54*(C), 100873.

Garcia-Kilroy, C., Juan, E., Satheesh, S., & Martinez Torres, G. (2023). *Institutional investors and sustainable infrastructure. A global review of case studies to finance the infrastructure gap*.

Gerged, A. M., Cowton, C. J., & Beddewela, E. S. (2018). Towards sustainable development in the Arab Middle East and North Africa region: A longitudinal analysis of environmental disclosure in corporate annual reports. *Business Strategy and the Environment, 27*(4), 572–587. https://doi.org/10.1002/bse.2021

Götze, T., & Gürtler, M. (2020). Hard markets, hard times: On the inefficiency of the CAT bond market. *Journal of Corporate Finance, 62*. https://doi.org/10.1016/j.jcorpfin.2019.101553

High-Level Expert Group on scaling up sustainable finance in low- and middle-income countries. (2024). *Final recommendations* (issue April).

Ilhan, E., Krueger, P., Sautner, Z., & Starks, L. T. (2023). Climate risk disclosure and institutional investors. *Review of Financial Studies, 36*(7), 2617–2650. https://doi.org/10.1093/rfs/hhad002

International Monetary Fund. (2024). Regional economic outlook. Middle East and Central Asia: An uneven recovery amid high uncertainty. In *Regional outlook: Southeast Asia 2012–2013*.

International Renewable Energy Agency (IRENA). (2023). *Renewable energy statistics 2023*. https://www.irena.org/-/media/Files/IRENA/Agency/Publication/2023/Jul/IRENA_Renewable_energy_statistics_2023.pdf

KPMG. (2024). *The move to mandatory reporting survey of sustainability reporting 2024*. https://assets.kpmg.com/content/dam/kpmgsites/xx/pdf/2024/11/the-move-to-mandatory-reporting--executive-summary.pdf.coredownload.inline.pdf

Ku Ismail, K. N. I., & Ibrahim, A. H. (2008). Social and environmental disclosure in the annual reports of Jordanian companies. *Issues in Social and Environmental Accounting, 2*(2), 198. https://doi.org/10.22164/isea.v2i2.32

Luo, L., Tang, Q., & Lan, Y. C. (2013). Comparison of propensity for carbon disclosure between developing and developed countries: A resource constraint perspective. *Accounting Research Journal, 26*(1), 6–34. https://doi.org/10.1108/ARJ-04-2012-0024

Mauderer, S. (2024). *Unlocking potential: Green finance in the Middle East, North Africa and Central Asia*.

Migliorelli, M. (2021). What do we mean by sustainable finance? Assessing existing frameworks and policy risks. *Sustainability, 13*(2), 1–17. https://doi.org/10.3390/su13020975

Mullan, M., & Ranger, N. (2022). *Finance and investment: Framing paper* (no. 196; OECD environment working papers). doi:https://doi.org/10.1787/223ad3b9-en.

Owen, A. L., & Pereira, J. M. (2018). Bank concentration, competition, and financial inclusion. *Review of Development Finance Journal, 8*(1), 1–17.

Said Suwaidan, M., Moh'd Al-omari, A., & Hanna Haddad, R. (2004). Social responsibility disclosure and corporate characteristics: The case of Jordanian industrial companies. *International Journal of Accounting, Auditing and Performance Evaluation, 1*(4), 432–447. https://doi.org/10.1504/ijaape.2004.006355

Saltane, V., Amin, A., & Farazi, S. (2023). *Does sustainable finance regulation play a role in the "greening" of capital markets?, 22*. doi:https://doi.org/10.1596/40148.

Sanghai, S. (2024). Why REITS InvITs occupy significant mindspace in capital markets. *Economic Times*. https://economictimes.indiatimes.com/markets/stocks/news/why-reits-invits-occupy-significant-mindspace-in-capital-markets/articleshow/106832475.cms?from=mdr

Steuer, S., & Tröger, T. H. (2022). The role of disclosure in green finance. *Journal of Financial Regulation, 8*(1), 1–50. https://doi.org/10.1093/jfr/fjac001

Stout, S. (2022). Unlocking private sector adaptation finance. *Climate Policy Initiative*. https://www.climatepolicyinitiative.org/unlocking-private-sector-adaptation-finance/

Tanchum, M. (2024). *Renewable energy and Morocco's new green industries: How Morocco's green energy ecosystem can expand women and youth employment through sustainable development*. https://mei.edu/sites/default/files/2024-10/Renewable-Energy-and-Moroccos-New-Green-Industries-How-Moroccos-Green-Energy-Ecosystem-Can-Expand-Women-and-Youth-Employment-Through-Sustainable-Development.pdf

Temiz, H. (2022). Environmental performance and cost of finance: Evidence from emerging markets. *Sustainability Accounting, Management and Policy Journal, 13*(5), 1229–1250. https://doi.org/10.1108/SAMPJ-12-2021-0537

The World Bank. (2020). *Bank regulation and supervision a decade after the global financial crisis*.

The World Bank. (2022). *Building sovereign financial resilience in middle-income countries: Factsheet 5*.

The World Bank. (2023). *Guidelines for implementing asset recycling rights and permissions*. www.worldbank.org

Tyson, J. E. (2021). *Developing green bond markets for Africa*.

UNEP Finance Initiative. (2021). *Promoting sustainable finance and climate finance in the arab region* (issue January).

UNEP Finance Initiative. (2022). *Scaling private sector climate finance in the MENA region* (issue December).

United Nations Economic Commission for Africa. (2024). *Economic report on Africa 2024: Investing in a just and sustainable transition in Africa*.

Zittis, G., Almazroui, M., Alpert, P., Ciais, P., Cramer, W., Dahdal, Y., Fnais, M., Francis, D., Hadjinicolaou, P., Howari, F., Jrrar, A., Kaskaoutis, D. G., Kulmala, M., Lazoglou, G., Mihalopoulos, N., Lin, X., Rudich, Y., Sciare, J., Stenchikov, G., et al. (2022). Climate change and weather extremes in the eastern Mediterranean and Middle East. *Reviews of Geophysics, 60*(3). https://doi.org/10.1029/2021RG000762

Chapter 5
Impact of Green Finance and Sustainable Natural Resources on Climate Change: A Study of the Middle East and North African Region

Sana Saleem and Gamini Premaratne

Abstract This study examines the relationship between carbon emissions, reliance on natural resources, and green finance in 23 Middle East and North African (MENA) countries from 2000 to 2022, using the generalized method of moments to obtain robust estimates. The results indicate that green finance significantly reduces carbon emissions, especially in more urbanized contexts, where greener financial flows are more effective. However, environmental deterioration is made worse by the exploitation of natural resources. Our findings emphasize the importance of integrating green policies with targeted green investment plans to achieve climate resilience and long-term prosperity in resource-dependent nations.

Keywords Green finance · Net natural rent · CO_2 emission · Climate change · GMM · MENA region

5.1 Introduction

A swift transition to sustainable business practices is essential in response to the dual challenges of climate change and environmental degradation. A key component of this transition is green finance, which integrates environmental consideration into financial decision-making and promotes the reallocation of capital toward low-carbon, climate-resilient investments. While much of the global discourse on green finance has concentrated on developed economies, this chapter focuses on the Middle East and North African (MENA) region, where environmental risks are significant, yet the adoption of green finance remains uneven and underexplored.

S. Saleem · G. Premaratne (✉)
UBD School of Business and Economics, Gadong, Brunei
e-mail: gamini.premaratne@ubd.edu.bn

M. Elnahass et al. (eds.), *A Greenprint for Net Zero*, Gulf Studies 15,
https://doi.org/10.1007/978-981-92-1762-5_5

Despite its paradoxical dependence on fossil fuel-based economies, the MENA region is particularly vulnerable to the impacts of climate change, including desertification, rising temperatures, and water scarcity. Innovative funding mechanisms that promote long-term growth are urgently needed, considering these environmental challenges and structural economic constraints. However, the region continues to lag global standards in implementing climate-aligned fiscal policies, integrating environmental risk in financial regulation, and developing green financial products.

This chapter examines the relationship between natural resource dependence, environmental sustainability, and green finance in the MENA region. Due to the limited availability of data on green finance, we use renewable energy (RE) production as a proxy to capture environmentally conscious financial efforts. In addition, we incorporate net natural resource (NNR) rents to assess the role of resource dependence in shaping environmental outcomes. A central motivation of this chapter is to explore how the financial and energy sectors influence carbon emissions in the MENA. Accordingly, we pose the following research questions:

1. How does financial development in the MENA region influence environmental outcomes, particularly carbon emissions?
2. How can resource-dependent economies transition to RE and implement sustainable environmental policies with the support of green finance?
3. Can environmental stress and prospects for green finance be accurately predicted using NNR rents?

These questions are addressed using a panel data study of selected MENA countries from 2014 to 2022. Proxies for investment trends and resource utilization patterns, such as RE generation and NNR, are utilized.

5.1.1 Contributions of Theory and Empiricism

This work contributes to the growing body of research on green finance and environmental economics in key areas:

This study expands the framework of the environmental Kuznets curve (EKC) by incorporating measures of financial development and natural resource rent (NRR). Using updated data from IRENA and World Bank datasets, it provides a comparative assessment of financial indicators and RE in the MENA region. From the policy perspective, the chapter highlights the need for the MENA countries to shift from fossil fuel dependence toward growth based on renewable resources, while also offering region-specific insights that can inform sustainable development strategies.

We empirically investigate the research questions using panel data econometric techniques, including fixed-effects and dynamic generalized method of moments (GMM) estimators. Specifically, we study the relationship among NRRs, financial development, and environmental degradation (as measured by CO_2 emissions). The main sources of data for this study include the International Renewable Energy Agency (IRENA, 2023), the World Bank (2024), and the Global Carbon Atlas.

RE output serves as a proxy for green finance expenditures, while NNR rents provide insight into the sustainability of current resource extraction practices. We assess ecological stress and green investment initiatives using these metrics. The selection of MENA countries is guided by data availability and the economic significance of the energy transformation in the region.

According to the first findings, financial development helps the economy, but it usually makes environmental harm worse when there are inadequate regulatory frameworks in place. The rise in RE, especially solar energy, throughout the MENA suggests a possible shift toward sustainability, albeit with notable regional variation. Some nations, such as Morocco and the United Arab Emirates, have made significant investments in RE, while others still rely on fossil fuels. The resource curse hypothesis states that when NNR rents are not utilized to fund sustainable infrastructure or education, they are linked to increased environmental deterioration (Danish et al., 2019; Tufail et al., 2021). However, if the right governance structures and environmental laws are in place, NNR may be used to finance the green transition, develop human capital, and enhance institutional quality (Jiang & Li, 2022; Wan et al., 2022).

These observations have several policy implications. Green regulatory frameworks need to be strengthened to ensure funds are allocated to environmentally sustainable projects. To enhance income mobilization and distribution, resource-dependent MENA countries should adopt fiscal policies such as sovereign green bonds and environmental taxes. Education and innovation should be promoted as long-term strategies for enhancing institutional capacity and advancing sustainable development.

The remaining study is structured as follows: Section 5.2 presents theoretical and empirical evidence. Section 5.3 compiles the data and methodology. Section 5.4 provides an analysis, justification, and discussion of the findings, and Section 5.5 concludes the investigation.

5.2 Literature Review

This section provides a summary of the literature to support the analytical framework and empirical arguments of the chapter. To improve the clarity and emphasis, we divided the review into two sections: studies that provide empirical evidence, particularly from emerging and developing countries, and studies that support the theoretical foundation of fiscal multipliers.

5.2.1 Theoretical Framework

This chapter explores the relationship between CO_2 emissions, NNRs, and green finance in the MENA region. The term *green finance* refers to financial investments aimed at reducing CO_2 emissions and mitigating climate change (Anwar et al., 2022b; Chen et al., 2021; Fullman et al., 2017; Zheng et al., 2023).

Green finance contributes to emissions reduction in several ways. Its primary objective is to promote the development and adoption of RE technologies, including solar, wind, hydro, and geothermal energy. These sustainable energy sources help diversify national energy portfolios and reduce reliance on fossil fuels, thereby lowering carbon emissions. (Abbasi et al., 2022; Habiba et al., 2022; Lei et al., 2022; Su et al., 2021; Umar et al., 2021).

Moreover, green finance supports energy efficiency initiatives across a variety of sectors—such as buildings, transportation, and industry—by encouraging the adoption of energy-saving practices and technologies (Huang et al., 2023a, 2023b; Rosa et al., 2023). Examples include retrofitting buildings, promoting eco-friendly urban transportation systems, and offering financial incentives for the purchase of energy-efficient equipment.

Furthermore, green investment is essential for enhancing sustainable infrastructure, particularly in the MENA region, which is experiencing rapid urbanization. This includes financing for waste management, public transportation, RE grids, and climate-resilient infrastructure (De Lucia et al., 2020). These initiatives make green finance a vital tool for accelerating the transition to a low-carbon economy. However, the extent and effectiveness of green finance's impact on carbon reduction remain context-dependent. By examining this relationship, policymakers can better understand the trade-offs and synergies needed to balance economic development with environmental objectives.

The concept of NNRs—defined as the revenue generated from the extraction and use of natural resources such as minerals and fossil fuels—provides an alternative approach. While it can spur economic growth, NNR is often linked to environmental degradation due to its heavy reliance on resource extraction, which significantly increases CO_2 emissions (Dai et al., 2023; Durani et al., 2023; Zhengxia et al., 2023). The extraction and export of fossil fuels form a major part of the economies of several MENA countries. This dependence contributes to higher carbon emissions, given the carbon-intensive nature of their energy and industrial systems (Ai et al., 2020; Karmaker et al., 2020; Liu et al., 2015). Moreover, NNR may encourage further industrialization, thereby increasing demand for fossil fuels and accelerating emission growth (Wan et al., 2022).

Governance and institutional frameworks are also critically important. Emissions are exacerbated by lax regulatory environments in countries that prioritize economic gain over environmental stewardship (Candra et al., 2023; Nchofoung & Ojong, 2022; Skowron et al., 2023). Even with the availability of renewable technologies, the transition to greener energy systems remains challenging due to infrastructure constraints and financial lock-ins related to NNR extraction (Xin et al., 2022; Zhengxia et al., 2023). Understanding the relationship between NNR and green finance is essential. By facilitating sustainable investments, green finance has the potential to reverse the trend of rising CO_2 emissions in resource-dependent economies. Policymakers must rigorously analyze these dynamics in order to develop comprehensive policies that support decarbonization without compromising economic resilience (Anwar & Malik, 2021; Anwar et al., 2022a; Chen et al., 2021).

5.2.2 *Empirical Evidence*

This section highlights empirical research on the relationship between carbon emissions, NNRs, and green finance with a particular focus on developing countries, including those in the MENA region.

5.2.2.1 Green Finance and Carbon Emissions

Numerous studies show how crucial green finance is in enabling the transition from non-renewable to RE, which reduces carbon emissions. For instance, green finance promotes technological advancements that reduce emissions and hasten the use of RE (Ben Belgacem et al., 2023; Muazu et al., 2023). Research on the Brazil, Russia, India, China, and South Africa (BRICS) economies (2000–2018) shows that green financing (GF) enhances environmental quality despite rising FDI and economic growth (Fakher & Murshed, 2023; Xia et al., 2023).

Green finance may seem promising, but its effectiveness depends on institutional preparedness and policy alignment. The conflict between energy consumption and environmental objectives in a growing economy is emphasized by Gong et al. (2023). China's central bank defines green finance as a financial system that balances environmental protection with economic growth. Between 2005 and 2018, Chai et al. (2021) assessed green finance across 30 Chinese provinces, demonstrating its efficacy in reducing carbon intensity. Similar results show that Chinese provinces with a high penetration of green finance have superior environmental outcomes (Bai et al., 2022; Xiong & Sun, 2023). Scholars have used three perspectives to analyze how green finance addresses the tension between economic growth and environmental protection. Green finance was initially described as a specialized financial instrument intended to reduce pollution and boost energy efficiency (Anwar et al., 2023; Esmaeili et al., 2023; Jahanger et al., 2023; Xu et al., 2022).

The strategic integration perspective emphasizes integrating green finance into corporate and governmental objectives, supported by indicators derived from institutional and firm-level financial data (Debrah et al., 2022; Lim et al., 2016). The societal communication perspective examines how public perception and awareness affect the demand for environmentally friendly investments. As public interest in sustainability grows, financial institutions are facing increasing pressure to adhere to green principles (Chauvy et al., 2019; Herrmann et al., 2020). These findings show that although green finance can significantly reduce emissions, its effectiveness depends on systemic integration and strong public engagement.

5.2.2.2 NNRs and Carbon Emissions

Higher carbon emissions are generally associated with reliance on NNRs rather than green finance. Empirical studies have shown that environmental degradation is driven by resource-intensive activities such as fossil fuel extraction. In a global analysis of 208 countries from 1990 to 2018, Wang et al. (2023) found a significant correlation between NNR and CO_2 emissions. Similar findings were reported by Zhang et al. (2022b) for Morocco.

According to studies conducted in Ghana by Kwakwa et al. (2020) and Colombia by Awosusi et al. (2022), resource-driven economic expansion raises emissions. Furthermore, Hassan et al. (2022) demonstrate that although NNR and energy consumption increase emissions, green energy expenditures can offset these effects, highlighting the need for a hybrid approach. Kwakwa (2021) found that resource extraction significantly increases emissions in South Africa. In some countries, NNR contributes to higher emissions, as demonstrated by Adebayo et al. (2022) using the MMQR model and ecological footprint frameworks.

However, results are not always consistent. High levels of governance may have different effects compared to NNR. For example, Gyamfi et al. (2021) discovered that while NNR influenced emissions based on output, it had no significant effect on emissions based on consumption. On the other hand, Tufail et al. (2021) found that NNR reduced emissions in Organisation for Economic Co-operation and Development (OECD) countries due to the influence of institutional quality and technical innovation. Similarly, Ibrahim and Ajide (2021) showed that financial efficiency lowers the emissions impact of NNR, particularly in the context of the BRICS. Yang et al. (2021) also noted that financial structure and urbanization affect the link between NNR and emissions. Recent literature underscores the critical role of aligning energy transition policies and standardized carbon accounting practices—at both national and sectoral levels—to effectively advance toward carbon neutrality and enhance climate accountability (Chen et al., 2023; Christy et al., 2024).

According to theoretical and empirical studies, green finance can be an effective instrument for reducing CO_2 emissions if implemented properly. However, research indicates that in economies reliant on natural resources, such as those in the MENA region, continued use of such resources typically leads to higher emission levels. These processes are complex and influenced by the interaction of governance quality, economic structures, and institutional competency. This chapter adds to the existing body of knowledge by undertaking specific empirical research on the development and climate policy trajectories of MENA nations. It aims to deepen our understanding of these complex linkages by providing evidence-based insights to support and enhance regional policy.

5.3 Methodology

5.3.1 Data and Variables

This chapter employs a balanced panel dataset covering 23 MENA countries from 2000 to 2020. The selection of these countries and this period was influenced by their high reliance on natural resources, as well as an acute vulnerability to climate-related challenges. The MENA region presents a compelling case for investigating the relationship between institutional resilience, environmental sustainability, and economic growth, as extensively discussed in the literature. In addition to ensuring the availability and reliability of critical data in the fields of economics, environment, and energy, the selected time frame aligns with a period of heightened global climate debate and policy action. The empirical investigation presented in this chapter is grounded in a solid foundation of data quality and regional relevance. To maintain internal consistency and cross-economy comparison, our selection favors data coverage and quality within the MENA region.

Additionally, there is a deliberate methodological divergence from a broader worldwide sample. Most of the existing research on the relationship between green finance and environmental degradation is conducted in affluent economies or globally. The MENA region, however, presents a unique conundrum: it is a relatively new but growing player in RE development, while simultaneously remaining a major emitter due to its reliance on fossil fuels. As a result, the sample is intended to generate insights that are useful for region-specific policymaking, rather than to draw broad global conclusions.

This study uses CO_2 emissions per capita as a proxy for climate impact and environmental degradation. The primary independent variables include RE consumption, which serves as a proxy for green finance. While this proxy offers a narrower interpretation than global definitions, which may encompass green bonds, sustainable investments, and environmental, social, and governance (ESG) financing, it was adopted due to data limitations and a lack of comprehensive green finance statistics in many MENA countries. In this context, investing in RE is a sensible course of action, demonstrating a genuine commitment to the region's green transformation agenda. The degree of economic dependence on extractive industries is measured by NRRs (as a percentage of gross domestic product (GDP)), which are closely linked to both fiscal policy space and environmental impact in several MENA countries. By combining these variables, the research seeks to demonstrate how resource dependence and RE investments interact to influence environmental outcomes in the MENA region. The specific variables examined in this study, along with their measure definitions, are detailed in Table 5.1.

Table 5.1 Variables and description

Variable	Symbol	Measure	Source
Carbon emissions	CO_2	Metric tons per capita	World Bank
Green finance	GF	% of total electricity output	Our World Data
Natural resource rent	NRR	Percentage of GDP	World Bank
GDP per capita	GDP	Current US$	World Bank
Urban population	URB	% of the total population	World Bank

5.3.2 Selection of the MENA Region

The MENA region is a critical area of study for researchers examining climate change due to its diverse environmental and economic challenges. The region is characterized by a heavy reliance on NRRs, particularly those from oil and gas, which play a crucial role in the global energy market. At the same time, the MENA region is significantly vulnerable to the adverse effects of climate change, including intense heat waves, water scarcity, and desertification.

Given these challenges, examining green finance and sustainable natural resource management becomes imperative. The MENA region includes both oil-importing and oil-exporting countries, resulting in distinct economic dynamics across the bloc. In this study, we categorize Algeria, Bahrain, Iran, Iraq, Israel, Kuwait, Libya, Oman, Qatar, Saudi Arabia, and the United Arab Emirates as oil-exporting nations while Armenia, Azerbaijan, Djibouti, Egypt, Jordan, Lebanon, Mauritania, Morocco, Sudan, Syria, Tunisia, and Turkey are classified as oil-importing nations. We also examine how the effects of RE on CO_2 emissions vary between these two groups of countries.

5.3.3 Econometric Methodology

5.3.3.1 Model Specification

This study uses a two-step system GMM estimator to address potential endogeneity, unobserved heterogeneity, and autocorrelation in the panel dataset. GMM is preferred over more straightforward estimating methods, such as fixed effects or ordinary least squares, for several compelling reasons:

- Dynamic nature of emissions: A dynamic model framework is necessary, as carbon emissions are likely influenced by their own historical values.
- Endogeneity issues: Environmental factors and policy decisions may simultaneously affect variables such as investments in RE, creating potential reverse causality.
- Measurement errors: The internal instrumentation of GMM is better equipped to handle variations in environmental and economic variables obtained from various sources.

System GMM (Blundell & Bond, 1998) performs better in finite samples compared to alternative estimators, particularly when analyzing the dependent variable with a moderate time dimension.

Justification of Control Variables

The use of urban population and GDP per capita as control variables is generally supported by research on environmental economics:

- The EKC theory, which posits that environmental degradation initially increases with economic growth but eventually declines as countries become more developed and adopt cleaner technologies, is evaluated using GDP per capita (Grossman & Krueger, 1995)
- The urban population serves as a proxy for energy consumption, infrastructure intensity, and population density—all of which significantly influence emissions (De Lucia et al., 2020).

The Green Finance Proxy and Its Limitations

RE investments serve as a practical and policy-relevant proxy for green finance, despite having a more limited conceptual scope than green finance, which may include instruments such as green bonds, ESG funds, or climate-smart investments. This is particularly relevant in MENA countries, where comprehensive green finance data are either scarce or inconsistent. This approach aligns with earlier empirical research conducted in developing and resource-intensive regions (Zhao et al., 2023).

In this study, we examine the long-term persistence of carbon emissions while accounting for potential economic influences using the following model specifications:

$$CO2_{it} = \beta_0 + \beta_1 CO2_{it-1} + \beta_2 GF_{it} + v_{it} + \epsilon_{it} \tag{5.1}$$

$$CO2_{it} = \beta_0 + \beta_1 CO2_{it-1} + \beta_2 NNR_{it} + v_{it} + \epsilon_{it} \tag{5.2}$$

$$CO2_{it} = \beta_0 + \beta_1 CO2_{it-1} + \beta_2 GF_{it} + \beta_3 NRR_{it} + v_{it} + \epsilon_{it} \tag{5.3}$$

$$CO2_{it} = \beta_0 + \beta_1 CO2_{it-1} + \beta_2 GF_{it} + \beta_3 NNR_{it} + \beta_4 GDP_{it} + \beta_5 URB_{it} + v_{it} + \epsilon_{it} \tag{5.4}$$

$$CO2_{it} = \beta_0 + \beta_1 CO2_{it-1} + \beta_2 GF_{it} + \beta_3 GDP_{it} + \beta_4 URB_{it} + v_{it} + \epsilon_{it} \tag{5.5}$$

$$CO2_{it} = \beta_0 + \beta_1 CO2_{it-1} + \beta_2 NRR_{it} + \beta_3 GDP_{it} + \beta_4 URB_{it} + v_{it} + \epsilon_{it} \tag{5.6}$$

Where the dependent variable, $CO2_{it}$, measures carbon emissions (in metric tons per capita), for the country i at time t. The $CO2_{it-1}$ represents the 1-year lag value of carbon emissions, capturing the persistence of emissions over time. GF_{it} denotes the GF for the country i at time t defined as the total annual investment in RE initiatives. NRR_{it} measures the NRRs for a country i at time t expressed as the percentage of GDP derived entirely from natural resources. GDP_{it} represents GDP per capita (in constant 2010 US dollars) for country i at time t, reflecting the level of economic development. URB_{it} accounts for the degree of urbanization. v_i captures unobserved time-invariant country-specific traits that may affect carbon emissions. Finally, ϵ_{it} represents the error term accounting for stochastic shocks and omitted variables.

In the next stage, this study examines the dynamic interactions between carbon emissions and other key explanatory variables, with particular emphasis on the differences in impacts between oil-importing and oil-exporting countries. A dummy variable, DOE_i, is introduced, which takes the value 1 for oil-importing countries and 0 for oil-exporting countries:

$$CO2_{it} = \beta_0 + \beta_1 CO2_{it-1} + \beta_2 GF_{it} + \beta_3 NRR_{it} + \beta_4 DOE_i + v_{it} + \epsilon_{it} \quad (5.7)$$

$$\begin{aligned} CO2_{it} = {} & \beta_0 + \beta_1 CO2_{it-1} + \beta_2 GF_{it} + \beta_3 GDP_{it} + \beta_4 URB_{it} + \beta_5 DOE_i \\ & + \beta_6 RE_{it} \cdot DOE_i + \beta_7 GDP_{it} \cdot DOE_i + \beta_8 URB_{it} \cdot DOE_i + v_{it} + \epsilon_{it} \end{aligned} \quad (5.8)$$

$$\begin{aligned} CO2_{it} = {} & \beta_0 + \beta_1 CO2_{it-1} + \beta_2 NRR_{it} + \beta_3 GDP_{it} + \beta_4 URB_{it} + \beta_5 DOE_i \\ & + \beta_6 NRR_{it} \cdot DOE_i + \beta_7 GDP_{it} \cdot DOE_i + \beta_8 URB_{it} \cdot DOE_i + v_{it} + \epsilon_{it} \end{aligned} \quad (5.9)$$

Model 7 accounts for the direct effect of DOE_i on CO_2 emissions. Model 8, the extended version of Model 7, incorporates the interaction effect of DOE_i and key variables—GDP per capita, urbanization, RE consumption, and carbon emissions—to examine whether the effect of these variables on carbon emissions differs between oil-importing and oil-exporting countries.

Model 9 further explores the influence of NRRs, GDP per capita, and urbanization on CO_2 emissions, again distinguishing between oil-exporting and oil-importing countries by interacting these variables with DOE_i to capture their differential effects. By integrating these interaction terms, the models provide a comprehensive assessment of how the determinants of carbon emissions differ according to a country's energy dependence and economic structure. The findings offer valuable insights for policymakers to design targeted strategies for reducing carbon emissions under different economic models.

5.4 Results and Discussion

5.4.1 *Data and Descriptive Statistics*

This study utilizes panel data from 23 MENA countries for the period 2000–2020, based on data availability, to examine the relationship between green finance, NRRs, and climate change. Table 5.2 presents the descriptive statistics of the selected variables. RE investment data are used as a proxy for GF.

Table 5.2 Descriptive statistics

Variable	Obs	Mean	Std. dev.	Min	Max	Skewness	Kurtosis
GF	483	0.780	2.0542	−5.55	4.417	−0.609	2.931
CO2	483	1.486	1.2903	−1.60	4.212	−0.068	2.357
NNR	483	1.515	2.6349	−6.95	4.190	−1.775	5.021
GDP	483	8.692	1.2393	5.935	11.49	0.173	2.202
URB	483	4.235	0.2745	3.481	4.605	−0.933	3.357

NNR rents, measured as a percentage of GDP derived from total NRRs, are used to represent natural resource dependence. Data on NRRs and carbon emissions are obtained from the World Development Indicators, while data on renewable electricity are sourced from Our World Data. CO_2 emissions per capita are used as a proxy for climate change. Additionally, urbanization and GDP per capita are introduced as control variables in this study.

The logarithm of CO_2 emissions ($\ln CO_2$) for oil-importing and oil-exporting countries in the MENA region is represented in the two box plots. Fig. 5.1(a) shows a variation of CO_2 emissions in MENA countries. Sudan, Djibouti, and Mauritania

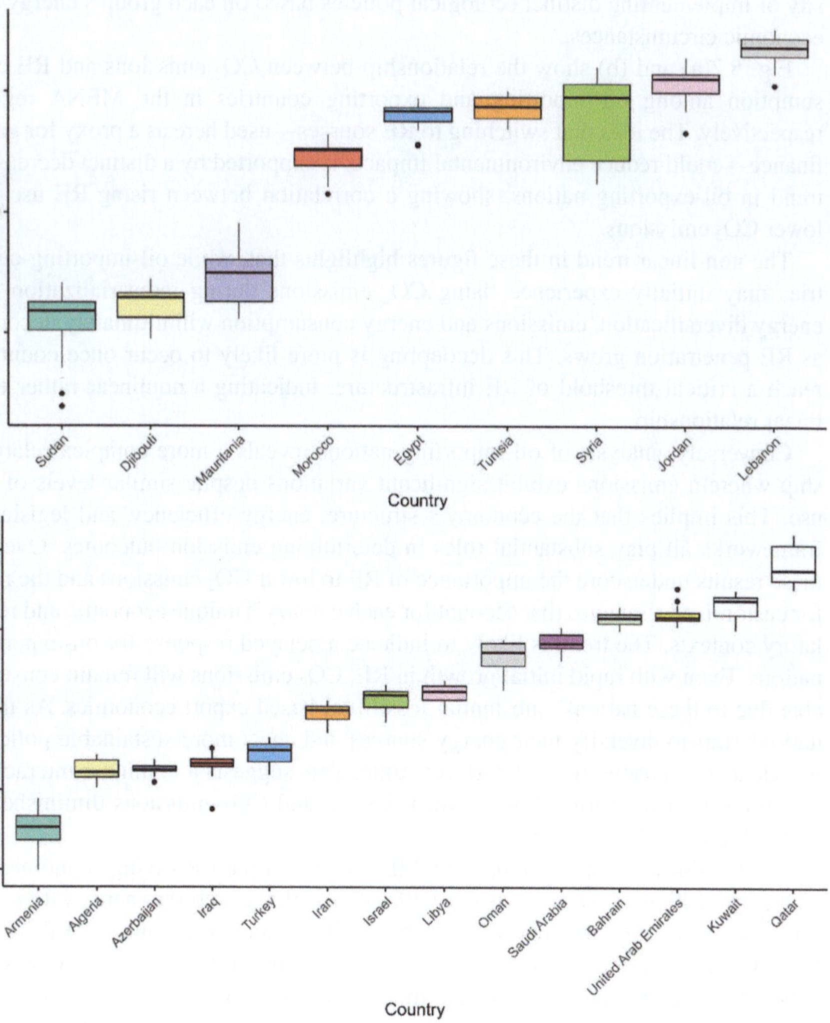

Fig. 5.1 (**a**) Distribution of CO_2 emissions across oil-importing MENA countries. (**b**) Distribution of CO_2 emissions across oil-exporting MENA countries

have the lowest amounts of CO2 emissions, while Lebanon and Jordan have the greatest CO_2 emissions among the oil-importing countries. This variability shows that these nations have different levels of industrialization, energy use, and environmental regulations. Fig. 5.1(b) also exhibits a significant rise in CO_2 emissions per capita, especially in the oil-exporting nations such as Kuwait, Qatar, and the United Arab Emirates, which have the highest levels of emissions.

Reliance on oil production and exporting is linked to increased CO_2 emissions. In contrast, oil-importing countries tend to have lower emissions, likely due to more diversified economic activity and lower fossil fuel dependency. The box plots depict the differences in emissions between these two categories, highlighting the necessity of implementing distinct ecological policies based on each group's energy and economic circumstances.

Fig. 5.2(a) and (b) show the relationship between CO_2 emissions and RE consumption among oil-importing and exporting countries in the MENA region, respectively. The idea that switching to RE sources—used here as a proxy for green finance—could reduce environmental impacts is supported by a distinct decreasing trend in oil-exporting nations, showing a correlation between rising RE use and lower CO_2 emissions.

The non-linear trend in these figures highlights that, while oil-importing countries may initially experience rising CO_2 emissions during industrialization and energy diversification, emissions and energy consumption will ultimately decouple as RE penetration grows. This decoupling is more likely to occur once countries reach a critical threshold of RE infrastructure, indicating a nonlinear rather than linear relationship.

Conversely, analysis of oil-importing nations reveals a more complex relationship wherein emissions exhibit significant variations despite similar levels of RE use. This implies that the economy's structure, energy efficiency, and legislative frameworks all play substantial roles in determining emission outcomes. Overall, these results underscore the importance of RE to lower CO_2 emissions and the need for customized strategies that account for each country's unique economic and regulatory contexts. The trend is likely to indicate a delayed response for oil-exporting nations. Even with rapid initial growth in RE, CO_2 emissions will remain considerable due to these nations' substantial fossil fuel-based export economies. As these nations start to diversify their energy sources and enact more sustainable policies, we see a slow change in the trend over time. This suggests a nonlinear interaction where the positive correlation between RE use and CO_2 emissions diminishes at higher levels of RE adoption.

The nonlinear trend illustrates how MENA countries are evolving economically as they transition to RE. Consistent with empirical research on energy efficiency, findings suggest that the marginal return on CO_2 reduction per unit of RE investment decreases after reaching a certain threshold in the energy mix (Jaffe & Stavins, 1994). Policymakers must recognize this non-linear relationship, as RE adoption does not always lead to proportional emission reductions, particularly in countries that still largely rely on fossil fuels.

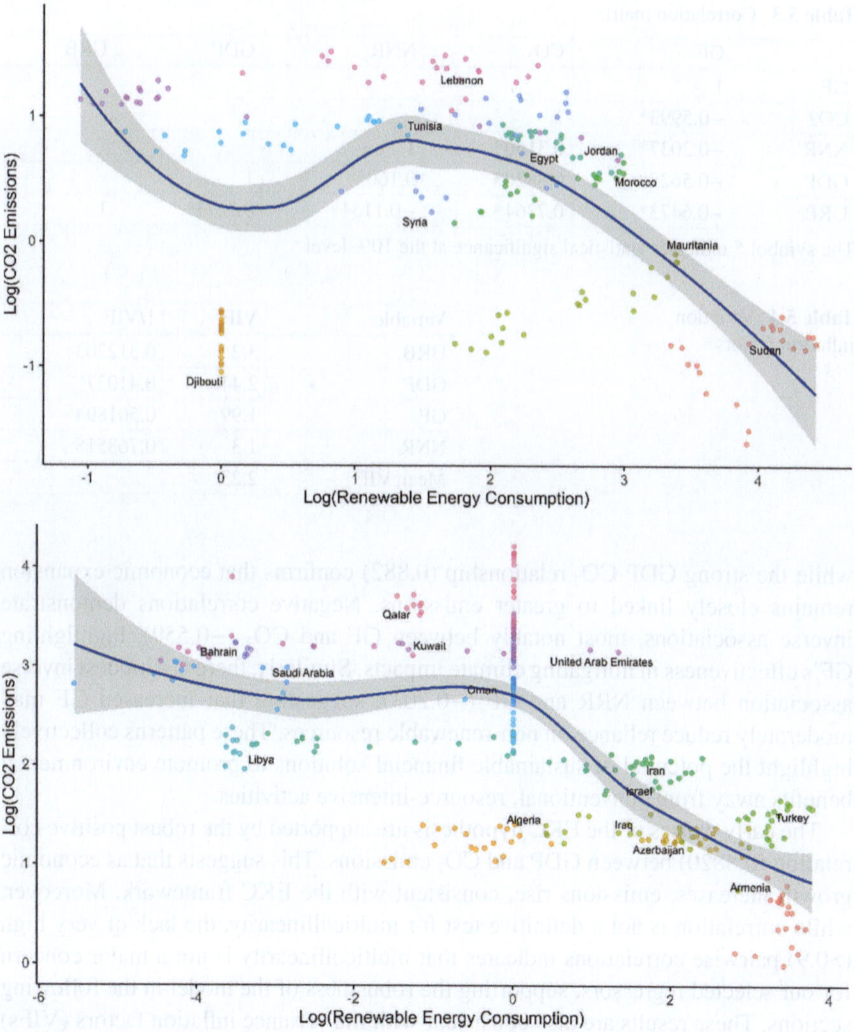

Fig. 5.2 (**a**) Relationship between CO_2 emissions and RE consumption in oil-importing MENA countries. (**b**) Relationship between CO_2 emissions and RE consumption in oil-exporting MENA countries

5.4.2 Empirical Results and Discussion

A Pearson correlation matrix in Table 5.3 reveals the fundamental relationships among key variables, GF, carbon emissions (CO_2), NNRs, GDP per capita (GDP), and urban population (URB). The weak correlation between GDP-NRR (0.160) indicates limited dependence of economic growth on non-renewable resources,

Table 5.3 Correlation matrix

	GF	CO_2	NNR	GDP	URB
GF	1				
CO2	−0.5993*	1			
NNR	−0.2037*	0.3160*	1		
GDP	−0.5627*	0.8820*	0.1605*	1	
URB	−0.6473*	0.7364*	−0.1131*	0.7273*	1

The symbol * indicates statistical significance at the 10% level

Table 5.4 Variation inflation factors

Variable	VIF	1/VIF
URB	3.2	0.312203
GDP	2.44	0.41027
GF	1.99	0.501894
NNR	1.3	0.768518
Mean VIF	2.23	

while the strong GDP-CO_2 relationship (0.882) confirms that economic expansion remains closely linked to greater emissions. Negative correlations demonstrate inverse associations, most notably between GF and CO_2 (−0.559), highlighting GF's effectiveness in mitigating climate impacts. Similarly, there is a modest inverse association between NRR and GF (−0.203), suggesting that increased GF may moderately reduce reliance on non-renewable resources. These patterns collectively highlight the potential of sustainable financial solutions to promote environmental benefits away from conventional, resource-intensive activities.

The early phases of the EKC hypothesis are supported by the robust positive correlation (0.8820) between GDP and CO_2 emissions. This suggests that as economic growth increases, emissions rise, consistent with the EKC framework. Moreover, while correlation is not a definitive test for multicollinearity, the lack of very high (>0.9) pairwise correlations indicates that multicollinearity is not a major concern for our selected regressors, supporting the robustness of the model in the following sections. These results are also consistent with the variance inflation factors (VIFs) reported in Table 5.4, all of which are below 5, indicating no evidence of multicollinearity.

The negative relationships between GF and other economic indicators, such as GDP per capita and urbanization, suggest the presence of structural barriers. More affluent or highly urbanized MENA countries may not yet be effectively directing their resources into RE initiatives. These trends highlight the need for further research using dynamic modeling approaches, particularly in light of the differing growth trajectories of oil-exporting and oil-importing countries.

Table 5.5 presents the estimated results of two-step GMM models (Models 1–6), with the dependent variable as log(CO_2), and independent variables lnGF, lnNNR, lnGDP, lnURB, and lnCO_2(−1). The table reports the coefficients, standard errors, and level of statistical significance for each variable across all the models.

Table 5.5 Two-step system GMM estimates for Models 1–6

Variable	Two-step system GMM					
	Model 1	Model 2	Model 3	Model 4	Model 5	Model 6
lnGF	−0.31*** 0.02		−0.26*** 0.01	−0.058 0.057	−0.04*** 0.009	
lnNRR		0.14*** 0.01	0.38*** 0.03	0.15*** 0.052		0.12*** 0.01
lnGDP	–	–	–	0.42*** 0.13	0.68*** 0.04	0.46*** 0.04
lnURB	–	–	–	1.62 1.21	1.49*** 0.28	1.16 0.75
$\ln CO_2(-1)$	0.20*** 0.01	0.36*** 0.008	0.34*** 0.04	0.44*** 0.138	0.49*** 0.02	0.48*** 0.02
Constant	1.73*** 0.08	1.25*** 0.10	1.13*** 0.07	−9.26** 4.45	−10.69*** 0.86	−7.64** 3.00
AR (1)	0.30	0.005	0.83	1.54	0.01	0.01
AR (2)	0.97	0.70	0.65	−0.18	0.37	0.68
Hansen test	0.63	0.35	0.23	0.304	0.27	0.30

The symbols ***, **, and * indicate statistical significance at the 1%, 5%, and 10% levels, respectively. Standard errors are reported in parentheses

The two-step system GMM results show that while greater resource exploitation and economic expansion result in higher emissions, green finance consistently reduces CO_2 emissions. According to Model 1, GF significantly reduces emissions, and historical emissions continue to influence present levels. Model 2 highlights the long-term persistence of CO_2 emissions and demonstrates how the use of natural resources significantly increases emissions. Model 3 shows that although green finance reduces emissions, natural resources have a greater positive influence on CO_2 emissions when both variables are included in the model simultaneously. Moreover, according to the study's findings, emissions behave as they do in various scenarios because of intricate relationships between environmental conditions, urbanization, and economic development. MENA's affluent or highly urbanized countries may not yet be effectively allocating their resources to RE, as seen by the negative relationships found between green funding and other economic variables such as GDP per capita and urban population. These tendencies underscore the need for more dynamic modeling research, given the differences in growth paths between oil-importing and oil-exporting countries.

GF continuously exhibits a strong negative correlation with CO2 emissions across all models, although the *lnGF* coefficient in Model 4 is statistically insignificant. This suggests that increased funding for environmentally friendly initiatives can effectively reduce CO_2 emissions. On the other hand, NNRs have a positive and substantial effect on CO_2 emissions, suggesting that higher levels of resource exploitation are linked to higher emissions because of the extraction operations' environmental impact (Zhang et al., 2022a).

GDP-based measures of economic growth also show a positive correlation with emissions, supporting the notion that although economic progress is important, it may result in environmental deterioration if not properly managed. Similarly, urbanization contributes to higher emissions; however, its effect becomes less pronounced when other variables are considered. The substantial coefficients for delayed emissions across all models demonstrate the persistence of CO_2 emissions and show that previous emissions have a major impact on current levels. This emphasizes how crucial it is to start early and maintain consistent effort to reduce emissions to lessen long-term environmental damage. The diagnostic tests demonstrate no substantial second-order autocorrelation and validate the instruments utilized, confirming the resilience of the GMM models. To effectively reduce CO_2 emissions and promote sustainable development, our findings underline the need for a balanced approach incorporating GF, sustainable resource management, and regulated urban expansion.

According to Model 5, the economy's structural features and continued reliance on conventional energy sources are to blame for the emissions' continuous long-term trajectory. This persistence implies that, absent significant changes in technology or policy, emissions typically stay high after they reach a certain level. Urbanization and economic expansion, as measured by GDP, are major contributors to rising emissions because of the energy requirements of growing urban centers, infrastructural development, and industrial operations. These elements have historically been associated with increased energy and fossil fuel use, which explains why there is a strong positive correlation between GDP and urbanization.

Model 6 provides crucial new insights into the variables affecting CO_2 emissions in the MENA region. The study demonstrates that both natural resource exploitation (NRR) and economic growth (GDP) significantly and positively affect emissions. Specifically, the coefficient of lnNRR is 0.12 (significant at the 1% level), suggesting that increased resource extraction increases emissions, while the coefficient of lnGDP is 0.46 (also significant at the 1% level), indicating that economic development contributes to emissions through increased energy consumption. Urbanization (lnURB) and emissions have a positive association with a coefficient of 1.16; however, this link is not statistically significant, suggesting that the effects of urbanization may vary by context.

The $\ln CO_2(-1)$ variable, which captures the influence of past emissions, highlights their persistence over time with a value of 0.48 (significant at the 1% level). Its constant term is −7.64 (significant at the 5% level); however, its practical application is limited. Diagnostic tests reveal no substantial autocorrelation, and the Hansen test ($p = 0.30$) validates the model's instrumentation. The model underscores the necessity of addressing both economic development and natural resources dependence in CO_2 emission reduction strategies due to the emissions' extraordinary persistence over time.

Table 5.6 provides valuable insights into the relationship between GF, NNR, urbanization (URB), and CO_2 emissions in the MENA region, highlighting the differences between oil-importing countries (DOE = 1) and oil-exporting countries (DOE = 0). The inclusion of the DOE variable allows us to assess how these factors influence emissions across two groups of countries.

Table 5.6 Two-step system GMM estimates with country dummies for oil importers/oil exporters

Variable	Two-step system GMM		
	Model 7	Model 8	Model 9
lnGF	−0.064*** 0.007	−0.03** 0.01	–
lnNNR	0.28*** 0.006	–	0.07** 0.02
DOE	3.40*** 0.16	43.60** 18.33	30.70** 11.7
lnGDP	–	0.04 0.06	−0.14** 0.06
lnURB	–	7.13** 2.80	5.97*** 1.70
lnGF_DOE	–	0.39*** 0.09	–
lnNNR_DOE		–	−0.10*** 0.06
lnGDP_DOE	–	0.15 0.09	0.51*** 0.11
lnURB_DOE	–	−11.4** 4.14	−8.85*** 2.74
$lnCO_2(-1)$	0.128*** 0.15	0.19*** 0.04	0.21*** 0.01
Constant	−0.72*** 0.14	−28.82** 12.7	−22.24*** 7.62
AR 1	0.44	0.36	0.31
AR 2	0.67	0.29	0.45
Hansen test	0.34	0.52	0.35

The symbols ***, **, and * indicate statistical significance at the 1%, 5%, and 10% levels, respectively. Standard errors are reported in parentheses

According to the coefficient of green finance (lnGF) in Model 7, which is significant and negative (−0.064), CO_2 emissions in MENA countries are significantly reduced by promoting green investments. The positive coefficient for NNR (lnNNR) (0.28) suggests the environmental cost of resource dependence, indicating that more resource extraction is associated with higher emissions in MENA countries. The positive and significant coefficient of DOE (3.40) in Model 7 suggests that, overall, being an oil-importing country contributes to higher CO_2 emissions compared to oil-exporting countries.

However, with the introduction of the DOE variable as an interaction term with lnGF in Model 8, the total effect of *lnGF* on CO_2 becomes positive for oil-importing countries. This may be because oil-importing countries, which rely heavily on fossil fuels, have not yet scaled green initiatives sufficiently to offset their carbon intensity. The positive coefficient for NNR (lnNNR) (0.28) further supports the environmental cost of resource reliance.

In Model 8, the coefficient of urbanization (lnURB) is positive and significant (7.13), indicating that urbanization in oil-exporting countries is associated with increased CO_2 emissions. This suggests that urban growth in oil-exporting nations is energy-intensive, leading to higher CO_2 emissions due to the availability of low-cost energy. The lagged CO_2 emissions term ($lnCO_2(-1)$) is also positive and significant in all three models (Models 7–9), demonstrating the persistence of emissions over time: past emissions levels continue to influence current emissions in MENA countries.

However, the interaction between urbanization and the importing dummy variable $ln(URB) * DOE$ in Model 8 is negative and significant. This implies that, in oil-importing countries, urbanization contributes less to emissions compared to oil-exporting countries. This may be attributed to more efficient infrastructure and transportation systems in oil-importing nations, which help mitigate fuel consumption and associated emissions.

The information shows a distinct dynamic among oil-exporting nations. The DOE variable's very significant correlation of 43.60 indicates that nations that export oil emit much more CO_2 than those that import it. This makes sense given that these countries' heavy reliance on the exploitation and export of fossil fuels greatly increases their carbon footprints. The positive coefficient for NNR (lnNNR) (0.16) further suggests that emissions are higher in oil-exporting countries due to intensive fossil fuel extraction, supporting this conclusion even more.

The adoption of sustainable urban development strategies, such as the installation of energy-efficient technologies and legislation intended to limit emissions in urban areas, may be the reason why oil-exporting nations have a negative coefficient (−11.4) for urbanization (lnURB).

Some nations may be implementing cleaner technologies or more energy-efficient infrastructure despite growing urbanization, which would lessen some of the emission pressures often associated with urban expansion. The delayed CO_2 emissions ($lnCO_2(-1)$) exhibit a positive correlation (0.19), indicating the persistence of emissions, since the previous carbon footprint still affects the current emissions levels in oil-exporting nations.

According to the results of Model 9, countries that import oil emit substantially less CO_2 than those that export it. The oil-exporter dummy variable (DOE) has a positive and statistically significant coefficient of 30.70. This finding illustrates how carbon-intensive some nations are due to their dependence on fossil fuel production and exports. This link is reinforced by the positive effect of NNR (lnNNR), which has a value of 0.07 and indicates that greater levels of resource extraction directly result in higher emissions.

Nonetheless, there is a distinct trend among oil-importing nations. The slightly negative coefficient (−0.03) effect of the green factor (lnGF) on emissions indicates that green initiatives and sustainable policies are beginning to reduce CO_2 emissions in these countries. Additionally, there is a negative interaction between urbanization and the DOE dummy variable (lnURB_DOE = −8.85), suggesting that urbanization increases emissions in oil-exporting nations more than in oil-importing nations.

This might be due to better energy efficiency and infrastructure in oil-importing nations, which help reduce emissions associated with urban growth.

In nations that export oil, there is a positive correlation between urbanization and CO_2 emissions, as seen by the significant coefficient for lnURB (5.97). This indicates that these countries have greater carbon footprints as a result of rising urbanization, which is frequently accompanied by higher energy demand. However, in some oil-exporting countries, the mitigating effect of urbanization on emissions may be explained by the adoption of sustainable urban development practices and the use of more energy-efficient technologies, as reflected in the interaction term with the DOE dummy (lnURB_DOE).

Different trends in CO_2 emissions between MENA oil-importing and oil-exporting countries are evident when comparing Model 9 with the other models in Tables 5.5 and 5.6. The results indicate that oil-exporting nations have significantly higher CO_2 emissions, largely due to their dependence on the production and sale of fossil fuels. This is supported by the positive coefficient for NNR (lnNNR), which shows that greater levels of resource extraction lead to increased emissions.

Incorporating green finance (lnGF) into Model 9 appears to reduce emissions in oil-importing countries, indicating the potential benefits of sustainable policies, although this effect is less noticeable in oil-exporting countries. In countries that export oil, urbanization contributes more substantially to emissions, especially when coupled with rising energy demand. The interaction variables, however, suggest that in oil-importing countries, the impact of urbanization on emissions is less significant, most likely due to improved infrastructure and greater energy efficiency.

Overall, the results show that emissions are persistent in both groups, with past emissions continuing to influence current levels. They also emphasize the importance of a comprehensive strategy to address the environmental consequences of economic growth—one that integrates GF, sustainable resource management, and effective urbanization.

5.5 Conclusion

This study examines the impact of NNR and green finance on carbon emissions in the MENA with a particular focus on the role of RE in mitigating climate change. The results of the two-step GMM technique show that green finance significantly lowers carbon emissions, while the exploitation of natural resources increases them. This emphasizes how crucial it is to foster economic expansion while practicing responsible resource management.

The study also emphasizes the potential contribution of RE to the reduction of CO_2 emissions, particularly when combined with financial expenditures in eco-friendly technologies. According to our research, switching to RE is essential for reducing climate change, enhancing environmental quality, and promoting regional economic growth in the MENA region. Green finance plays a major role in promoting the use of RE sources and reducing dependency on nonrenewable ones.

These findings, which suggest that reducing carbon emissions requires greater investment in RE sources and the sustainable use of natural resources, align with the global perspectives on the significance of GF in tackling climate change. By examining the effects of these variables in various regions and accounting for factors such as globalization, industrialization, and income disparities, future research could build on these findings to provide a more thorough understanding of the role of natural resources and green finance in mitigating climate change.

5.5.1 Policy Implications and Future Directions

The findings of this study highlight the policy implications regarding green finance and its impact on CO_2 emissions in oil-importing and oil-exporting MENA countries. In oil-importing nations, green finance appears to be effective, as evidenced by a notable reduction in emissions associated with investments in RE and sustainable projects. This underscores the need to prioritize policies that support green finance initiatives—such as public–private partnerships and RE investments—in these countries.

In contrast, oil-exporting countries face greater challenges, as their heavy reliance on fossil fuel revenues makes it more difficult to decouple emissions from economic growth. For these nations, policy measures should focus on carbon pricing, economic diversification, and the gradual removal of fossil fuel subsidies. Additionally, green finance should be strategically directed toward the development of RE technologies and infrastructure.

To create a business-friendly environment and foster a low-carbon transition, governments across the MENA region must also enhance regulatory frameworks, invest in RE infrastructure, and reform energy markets. These actions are critical for ensuring long-term fiscal stability and reducing dependence on fossil fuel in the context of the global energy transition.

The conclusions discussed above lead to several important policy considerations. Businesses should actively seek top talent and advanced technologies, adopt sustainable resource extraction methods, promote awareness of green exploration and development of natural resources, and work to minimize pollution and environmental damage.

In parallel, governments should strictly regulate the issuance of mining licenses and enforce formal registration procedures for resource-extracting entities. They should promote transparency in how natural resources are used and exploited, and establish accountability mechanisms to monitor and control emissions. As regulators, governments must fully embrace their role at the macro level by strengthening legal and regulatory frameworks related to the exploration, extraction, transportation, and utilization of natural resources. They must also apply strict penalties for violations. Financial institutions, as third-party actors, should proactively support the expansion of green finance and contribute to building a robust green economic system, including instruments such as green credit, bonds, funds, and insurance.

Overall, the results underscore the need for a comprehensive multifaceted strategy that addresses the environmental impact of urbanization and economic growth, promotes GF, and regulates natural resource use to reduce emissions and mitigate climate change. Stakeholders must be innovative and reform-oriented in shaping a green financial management system. This includes gradually advancing environmental information disclosure, increasing market transparency, and enhancing regulatory oversight of financial institutions. At the same time, they should improve the system of incentives and sanctions and actively provide funds to develop green and low-carbon sectors in the MENA region.

References

Abbasi, K. R., Shahbaz, M., Zhang, J., Irfan, M., & Alvarado, R. (2022). Analyze the environmental sustainability factors of China: The role of fossil fuel energy and renewable energy. *Renewable Energy, 187*, 390–402. https://doi.org/10.1016/J.RENENE.2022.01.066

Adebayo, T. S., Onifade, S. T., Alola, A. A., & Muoneke, O. B. (2022). Does it take international integration of natural resources to ascend the ladder of environmental quality in the newly industrialized countries? *Resources Policy, 76*, 102616. https://doi.org/10.1016/J.RESOURPOL.2022.102616

Ai, H., Hu, S., Li, K., & Shao, S. (2020). Environmental regulation, total factor productivity, and enterprise duration: Evidence from China. *Business Strategy and the Environment, 29*(6), 2284–2296. https://doi.org/10.1002/BSE.2502

Anwar, A., Chaudhary, A. R., & Malik, S. (2023). Modeling the macroeconomic determinants of environmental degradation in E-7 countries: The role of technological innovation and institutional quality. *Journal of Public Affairs, 23*(1), e2834. https://doi.org/10.1002/pa.2834

Anwar, A., & Malik, S. (2021). Cogitating the role of technological innovation and institutional quality on environmental degradation in G-7 countries. *International Journal of Green Economics, 15*(3), 213–232. https://doi.org/10.1504/IJGE.2021.120871

Anwar, A., Malik, S., & Ahmad, P. (2022a). Cogitating the role of technological innovation and institutional quality in formulating the sustainable development goal policies for E7 countries: Evidence from quantile regression. *Global Business Review, 26*(6), 1532–1555. https://doi.org/10.1177/09721509211072657

Anwar, A., Sinha, A., Sharif, A., Siddique, M., Irshad, S., Anwar, W., & Malik, S. (2022b). The nexus between urbanization, renewable energy consumption, financial development, and CO2 emissions: Evidence from selected Asian countries. *Environment, Development and Sustainability, 24*(5), 6556–6576. https://doi.org/10.1007/S10668-021-01716-2

Awosusi, A. A., Mata, M. N., Ahmed, Z., Coelho, M. F., Altuntaş, M., Martins, J. M., Martins, J. N., & Onifade, S. T. (2022). How do renewable energy, economic growth and natural resources rent affect environmental sustainability in a globalized economy? Evidence from Colombia based on the gradual shift causality approach. *Frontiers in Energy Research, 9*. https://doi.org/10.3389/FENRG.2021.739721/FULL

Bai, J., Chen, Z., Yan, X., & Zhang, Y. (2022). Research on the impact of green finance on carbon emissions: Evidence from China. *Economic Research-Ekonomska Istraživanja, 35*(1), 6965–6984. https://doi.org/10.1080/1331677X.2022.2054455

Ben Belgacem, S., Khatoon, G., & Alzuman, A. (2023). Role of renewable energy and financial innovation in environmental protection: Empirical evidence from UAE and Saudi Arabia. *Sustainability, 15*(11), 8684. https://doi.org/10.3390/SU15118684

Blundell, R., & Bond, S. (1998). Initial conditions and moment restrictions in dynamic panel data models. *Journal of Econometrics, 87*(1), 115–143. https://doi.org/10.1016/S0304-4076(98)00009-8
Candra, O., Chammam, A., Alvarez, J. R. N., Muda, I., & Aybar, H. (2023). The impact of renewable energy sources on the sustainable development of the economy and greenhouse gas emissions. *Sustainability, 15*(3), 2104–2104. https://doi.org/10.3390/SU15032104
Chai, K. C., Li, Q., Bao, X. L., Zhu, J., & He, X. X. (2021). An empirical study of economic cycle, air quality, and national health since reform and opening up. *Frontiers in Public Health, 9*, 706955. https://doi.org/10.3389/FPUBH.2021.706955/BIBTEX
Chauvy, R., Meunier, N., Thomas, D., & De Weireld, G. (2019). Selecting emerging CO_2 utilization products for short- to mid-term deployment. *Applied Energy, 236*, 662–680. https://doi.org/10.1016/J.APENERGY.2018.11.096
Chen, X., Chen, Z., Sharifi, A., Feng, C., & Yang, J. (2021). Can green finance development reduce carbon emissions? Empirical evidence from 30 Chinese provinces. *Sustainability, 13*(21), 12137. https://doi.org/10.3390/SU132112137
Chen, X. H., Tee, K., Elnahass, M., & Ahmed, R. (2023). Assessing the environmental impacts of renewable energy sources: A case study on air pollution and carbon emissions in China. *Journal of Environmental Management, 345*, 118525. https://doi.org/10.1016/J.JENVMAN.2023.118525
Christy, A., Elnahass, M., Amezaga, J., Browne, A., & Heidrich, O. (2024). A dynamic framework to align company climate reporting and action with global climate targets. *Business Strategy and the Environment, 33*(4), 3103–3128. https://doi.org/10.1002/BSE.3635;PAGE:STRING:ARTICLE/CHAPTER
Dai, J., Ahmed, Z., Sinha, A., Pata, U. K., & Alvarado, R. (2023). Sustainable green electricity, technological innovation, and ecological footprint: Does democratic accountability moderate the nexus? *Utilities Policy, 82*, 101541. https://doi.org/10.1016/J.JUP.2023.101541
Danish, Baloch, M. A., Mahmood, N., & Zhang, J. W. (2019). Effect of natural resources, renewable energy and economic development on CO2 emissions in BRICS countries. *Science of the Total Environment, 678*, 632–638. https://doi.org/10.1016/J.SCITOTENV.2019.05.028
De Lucia, C., Pazienza, P., & Bartlett, M. (2020). Does good ESG lead to better financial performances by firms? Machine learning and logistic regression models of public enterprises in Europe. *Sustainability, 12*(13). https://doi.org/10.3390/SU12135317
Debrah, C., Chan, A. P. C., & Darko, A. (2022). Green finance gap in green buildings: A scoping review and future research needs. *Building and Environment, 207*, 108443. https://doi.org/10.1016/J.BUILDENV.2021.108443
Durani, F., Bhowmik, R., Sharif, A., Anwar, A., & Syed, Q. R. (2023). Role of economic uncertainty, financial development, natural resources, technology, and renewable energy in the environmental Phillips curve framework. *Journal of Cleaner Production, 420*, 138334. https://doi.org/10.1016/J.JCLEPRO.2023.138334
Esmaeili, P., Balsalobre Lorente, D., & Anwar, A. (2023). Revisiting the environmental Kuznetz curve and pollution haven hypothesis in N-11 economies: Fresh evidence from panel quantile regression. *Environmental Research, 228*, 115844. https://doi.org/10.1016/J.ENVRES.2023.115844
Fakher, H. A., & Murshed, M. (2023). Does financial and economic expansion allow for environmental sustainability? Fresh insights from a new composite index and PSTR analysis. *Journal of Environmental Planning and Management.* https://doi.org/10.1080/09640568.2023.2205997
Fullman, N., Barber, R. M., Abajobir, A. A., Abate, K. H., Abbafati, C., Abbas, K. M., Abd-Allah, F., Abdulle, A. M., Abera, S. F., Aboyans, V., Abu-Raddad, L. J., Abu-Rmeileh, N. M. E., Adedeji, I. A., Adetokunboh, O., Afshin, A., Agrawal, A., Agrawal, S., Kiadaliri, A. A., Ahmadieh, H., et al. (2017). Measuring progress and projecting attainment on the basis of past trends of the health-related sustainable development goals in 188 countries: An analysis from the global burden of disease study 2016. *The Lancet, 390*(10100), 1423–1459. https://doi.org/10.1016/S0140-6736(17)32336-X

Gong, Q., Ying, L., & Dai, J. (2023). Green finance and energy natural resources nexus with economic performance: A novel evidence from China. *Resources Policy, 84*, 103765. https://doi.org/10.1016/J.RESOURPOL.2023.103765

Grossman, G. M., & Krueger, A. B. (1995). Economic growth and the environment. *The Quarterly Journal of Economics, 110*(2), 353–377. https://doi.org/10.2307/2118443

Gyamfi, B. A., Onifade, S. T., Nwani, C., & Bekun, F. V. (2021). Accounting for the combined impacts of natural resources rent, income level, and energy consumption on environmental quality of G7 economies: A panel quantile regression approach. *Environmental Science and Pollution Research, 29*(2), 1–13. https://doi.org/10.1007/S11356-021-15756-8

Habiba, U., Xinbang, C., & Anwar, A. (2022). Do green technology innovations, financial development, and renewable energy use help to curb carbon emissions? *Renewable Energy, 193*, 1082–1093. https://doi.org/10.1016/J.RENENE.2022.05.084

Hassan, S. T., Batool, B., Sadiq, M., & Zhu, B. (2022). How do green energy investment, economic policy uncertainty, and natural resources affect greenhouse gas emissions? A Markov-switching equilibrium approach. *Environmental Impact Assessment Review, 97*, 106887. https://doi.org/10.1016/J.EIAR.2022.106887

Herrmann, A., Sauerborn, R., & Nilsson, M. (2020). The role of health in households' balancing act for lifestyles compatible with the Paris agreement—Qualitative results from Mannheim, Germany. *International Journal of Environmental Research and Public Health, 17*(4), 1297. https://doi.org/10.3390/IJERPH17041297

Huang, S. Z., Sadiq, M., & Chien, F. (2023a). The impact of natural resource rent, financial development, and urbanization on carbon emission. *Environmental Science and Pollution Research, 30*(15), 42753–42765. https://doi.org/10.1007/S11356-021-16818-7

Huang, Y., Abidin, S. Z. U., & Raza, S. M. F. (2023b). Investigation of resource curse hypothesis: The role of renewable energy and urbanization in realizing environmental sustainability in China. *Environmental Science and Pollution Research, 30*(37), 86927–86939. https://doi.org/10.1007/S11356-023-28300-7/METRICS

Ibrahim, R. L., & Ajide, K. B. (2021). The dynamic heterogeneous impacts of nonrenewable energy, trade openness, total natural resource rents, financial development and regulatory quality on environmental quality: Evidence from BRICS economies. *Resources Policy, 74*, 102251. https://doi.org/10.1016/J.RESOURPOL.2021.102251

Jahanger, A., Awan, A., Anwar, A., & Adebayo, T. S. (2023). Greening the Brazil, Russia, India, China and South Africa (BRICS) economies: Assessing the impact of electricity consumption, natural resources, and renewable energy on environmental footprint. *Natural Resources Forum, 47*(3), 484–503. https://doi.org/10.1111/1477-8947.12294

Jiang, Y., & Li, X. (2022). Sustainable development: 30 years since the Rio de Janeiro earth summit. *ACS Environmental Au, 2*(6), 480–481. https://doi.org/10.1021/ACSENVIRONAU.2C00065

Karmaker, A. K., Rahman, M. M., Hossain, M. A., & Ahmed, M. R. (2020). Exploration and corrective measures of greenhouse gas emission from fossil fuel power stations for Bangladesh. *Journal of Cleaner Production, 244*, 118645. https://doi.org/10.1016/J.JCLEPRO.2019.118645

Kwakwa, P. A. (2021). The effects of natural resource extraction and renewable energy consumption on carbon dioxide emissions in sub-Saharan Africa. The Journal of Energy and Development. 47(1/2), 195–222. Retrieved June 5, 2024, from https://www.jstor.org/stable/27188919

Kwakwa, P. A., Alhassan, H., & Adu, G. (2020). Effect of natural resources extraction on energy consumption and carbon dioxide emission in Ghana. *International Journal of Energy Sector Management, 14*(1), 20–39. https://doi.org/10.1108/IJESM-09-2018-0003/FULL/PDF

Lei, X., Yang, Y., Alharthi, M., Rasul, F., & Faraz Raza, S. M. (2022). Immense reliance on natural resources and environmental challenges in G-20 economies through the lens of COP-26 targets. *Resources Policy, 79*, 103101. https://doi.org/10.1016/J.RESOURPOL.2022.103101

Lim, S. S., Allen, K., Dandona, L., Forouzanfar, M. H., Fullman, N., Goldberg, E. M., Hay, S. I., Holmberg, M., Kutz, M. J., Larson, H. J., Lopez, A. D., McNellan, C. R., Mokdad, A. H., Mooney, M. D., Naghavi, M., Olsen, H. E., Pigott, D. M., Vos, T., Wang, H., et al. (2016).

Measuring the health-related sustainable development goals in 188 countries: A baseline analysis from the global burden of disease study 2015. *The Lancet, 388*(10053), 1813–1850. https://doi.org/10.1016/S0140-6736(16)31467-2

Liu, J., Mooney, H., Hull, V., Davis, S. J., Gaskell, J., Hertel, T., Lubchenco, J., Seto, K. C., Gleick, P., Kremen, C., & Li, S. (2015). Systems integration for global sustainability. *Science, 347*(6225). https://doi.org/10.1126/SCIENCE.1258832

Muazu, A., Yu, Q., & Alariqi, M. (2023). The impact of renewable energy consumption and economic growth on environmental quality in Africa: A threshold regression analysis. *Energies, 16*(11), 4533. https://doi.org/10.3390/EN16114533

Nchofoung, T. N., & Ojong, N. (2022). Natural resources, renewable energy, and governance: A path towards sustainable development. *Sustainable Development, 31*(3), 1553–1569. https://doi.org/10.1002/SD.2466

Rosa, D., Bianco, M., Barth, V., Pereira Da Silva, H., Salgado, V., Pallonetto, C., De Rosa, M., Bianco, V., Barth, H., Pereira Da Silva, P., Salgado, C. V., & Pallonetto, F. (2023). Technologies and strategies to support energy transition in urban building and transportation sectors. *Energies, 16*(11), 4317. https://doi.org/10.3390/EN16114317

Skowron, Ł., Chygryn, O., Gąsior, M., Koibichuk, V., Lyeonov, S., Drozd, S., & Dluhopolskyi, O. (2023). Interconnection between the dynamic of growing renewable energy production and the level of CO_2 emissions: A multistage approach for modeling. *Sustainability, 15*(12), 9473–9473. https://doi.org/10.3390/SU15129473

Su, C. W., Huang, S. W., Tao, R., & Haris, M. (2021). Does economic overheating provide positive feedback on population health? Evidence from BRICS and ASEAN countries. *Frontiers in Public Health, 9*, 661279. https://doi.org/10.3389/FPUBH.2021.661279/BIBTEX

Tufail, M., Song, L., Adebayo, T. S., Kirikkaleli, D., & Khan, S. (2021). Do fiscal decentralization and natural resources rent curb carbon emissions? Evidence from developed countries. *Environmental Science and Pollution Research, 28*(35), 49179–49190. https://doi.org/10.1007/S11356-021-13865-Y/METRICS

Umar, M., Ji, X., Mirza, N., & Rahat, B. (2021). The impact of resource curse on banking efficiency: Evidence from twelve oil producing countries. *Resources Policy, 72*, 102080. https://doi.org/10.1016/J.RESOURPOL.2021.102080

Wan, Q., Miao, X., & Afshan, S. (2022). Dynamic effects of natural resource abundance, green financing, and government environmental concerns toward the sustainable environment in China. *Resources Policy, 79*, 102954. https://doi.org/10.1016/J.RESOURPOL.2022.102954

Wang, Q., Zhang, F., & Li, R. (2023). Revisiting the environmental kuznets curve hypothesis in 208 counties: The roles of trade openness, human capital, renewable energy and natural resource rent. *Environmental Research, 216*, 114637. https://doi.org/10.1016/J.ENVRES.2022.114637

Xia, L., Liu, Y., & Yang, X. (2023). The response of green finance toward the sustainable environment: The role of renewable energy development and institutional quality. *Environmental Science and Pollution Research, 30*(21), 59249–59261. https://doi.org/10.1007/S11356-023-26430-6/TABLES/4

Xin, L., Sun, H., & Xia, X. (2022). Renewable energy technology innovation and inclusive low-carbon development from the perspective of spatiotemporal consistency. *Environmental Science and Pollution Research, 30*(8), 20490–20513. https://doi.org/10.1007/S11356-022-23556-X

Xiong, Q., & Sun, D. (2023). Influence analysis of green finance development impact on carbon emissions: An exploratory study based on fsQCA. *Environmental Science and Pollution Research, 30*(22), 61369–61380. https://doi.org/10.1007/s11356-021-18351-z

Xu, Y., Li, S., Zhou, X., Shahzad, U., & Zhao, X. (2022). How environmental regulations affect the development of green finance: Recent evidence from polluting firms in China. *Renewable Energy, 189*, 917–926. https://doi.org/10.1016/J.RENENE.2022.03.020

Yang, Y., Su, X., & Yao, S. (2021). Nexus between green finance, fintech, and high-quality economic development: Empirical evidence from China. *Resources Policy, 74*, 102445. https://doi.org/10.1016/J.RESOURPOL.2021.102445

Zhang, J., Muhammad, S., Raza, F., Huang, Y., & Wang, C. (2022a). What affect energy poverty in China? A path towards sustainable development. *Economic Research-Ekonomska Istraživanja, 36*(2). https://doi.org/10.1080/1331677X.2022.2111316

Zhang, Y., Khan, I., & Zafar, M. W. (2022b). Assessing environmental quality through natural resources, energy resources, and tax revenues. *Environmental Science and Pollution Research, 29*(59), 89029–89044. https://doi.org/10.1007/S11356-022-22005-Z/METRICS

Zhao, J., Taghizadeh-Hesary, F., Dong, K., & Dong, X. (2023). How green growth affects carbon emissions in China: The role of green finance. *Economic Research-Ekonomska Istraživanja, 36*(1), 2090–2111. https://doi.org/10.1080/1331677X.2022.2095522

Zheng, M., Feng, G. F., Jiang, R. A., & Chang, C. P. (2023). Does environmental, social, and governance performance move together with corporate green innovation in China? *Business Strategy and the Environment, 32*(4), 1670–1679. https://doi.org/10.1002/BSE.3211

Zhengxia, T., Batool, Z., Ali, S., Haseeb, M., Jain, V., Raza, S. M. F., & Chakrabarti, P. (2023). Impact of technology on the relation between disaggregated energy consumption and CO_2 emission in populous countries of Asia. *Environmental Science and Pollution Research, 30*(26), 68327–68338. https://doi.org/10.1007/S11356-023-26980-9/METRICS

Chapter 6
Rolling the Greenflation Dice: Will MENA Stock Markets Win or Lose Liquidity?

Abdessamad Raghibi, Anas Alaoui Mdaghri, and Karim Iddouch

Abstract This study investigates the impact of greenflation—the inflationary pressures from environmental policies—on stock market liquidity in the MENA region. Using daily market data from 823 publicly listed firms across 12 MENA countries from November 29, 2019 to December 29, 2023, we assess market liquidity through its tightness and depth dimensions. Our findings indicate that greenflation enhances market liquidity by reducing transaction costs and increasing market depth, driven by growing investor demand for sustainable assets. The impact, however, varies across regions: Middle Eastern markets show significant liquidity improvements, while North African markets exhibit a more limited response. Furthermore, the post-pandemic COVID-19 period has amplified greenflation's positive effects, which indicates increased investments in sustainable projects. These findings highlight the potential of greenflation to advance market liquidity and sustainability goals.

Keywords Greenflation · MENA · Market liquidity · COVID-19 · Sustainability

---

A. Raghibi
Ibnou Zohr University, Agadir, Morocco
e-mail: abdessamad.raghibi@edu.uiz.ac.ma

A. Alaoui Mdaghri (✉)
National School of Business and Management (ENCG), Ibnou Zohr University, Agadir, Morocco
e-mail: a.alaouimdaghri@uiz.ac.ma

K. Iddouch
Université Internationale de Casablanca, Business School, Casablanca, Morocco
e-mail: karim.iddouch@uic.ac.ma

M. Elnahass et al. (eds.), *A Greenprint for Net Zero*, Gulf Studies 15,
https://doi.org/10.1007/978-981-92-1762-5_6

6.1 Introduction

In the words of African environmental advocate Wangari Maathai, "*The environment and the economy are really two sides of the same coin. If we cannot sustain the environment, we cannot sustain ourselves.*" This perspective resonates deeply with the challenges and opportunities facing the MENA region in its pursuit of climate resilience. While climate change impacts are global, the MENA region's millennia-long battle with harsh environmental conditions uniquely positions it at the frontlines of climate vulnerability and adaptation.

Over the past decade, global awareness of climate issues has grown, making the transition to a greener economy increasingly urgent. This greener economy must provide enough energy to: (1) replace the 100 million barrels of oil currently consumed daily, and (2) meet the energy demands of a growing global population expected to reach 10 billion by 2050. The International Energy Agency (IEA) predicts that global energy demand could rise to between 800 and 1000 exajoules (approximately 220,000–280,000 TWh) under a "business-as-usual" scenario—representing a 60% increase compared to current consumption (IEA, 2024).

The global energy system is undergoing a significant transformation toward cleaner sources, especially since the adoption of the Paris Agreement in December 2015. This decade is pivotal, as the world must achieve nearly half of the emissions reductions required to meet the net-zero target by 2050 (Agbo-Bloua, 2023). Recent geopolitical developments have further accelerated the clean energy transition. Prominent policy initiatives include Europe's REPowerEU, which aims to reduce fossil fuel dependence and improve energy efficiency through a €210 billion investment over 5 years (European Commission, 2022). Similarly, the USA has renewed its commitment to climate action, allocating $369 billion under the Inflation Reduction Act (US Department of the Treasury, 2022).

In the MENA region, climate vulnerabilities and the need for clean and affordable energy are prompting governments to take action. While Western countries' climate fight is largely data-driven, the MENA region's fight stems from a deep historical understanding of the challenges ahead. Around 3500 years ago, the "Curse of Akkad" mythologized the real-world decline of the Akkadian empire, based south of Baghdad, which some scientists partly link to a climate-driven "mega-drought." This drought's effects on Akkadian society were worsened by poor governance, centralization, and crop mismanagement. "The great agricultural tracts produced no grain… The gathered clouds did not rain," laments the ancient text. Consequently, the climate fight in the MENA region is very much a "live or die" battle (Ersek, 2019).

Now that the climate fight is underway, the core challenge is how to finance, develop, and sustain clean energy initiatives. Critical minerals lie at the heart of this transition. A paradigm shift is expected between now and 2030, with strong demand growth for these minerals to support cleaner energy and transport systems. For example, a typical electric vehicle requires six times more mineral inputs than a conventional car, while an onshore wind plant requires nine times more than a gas-fired plant. Over 20 metals are essential to combating climate change, and cumulative demand for copper and nickel must more than double and quadruple, respectively, over the next 30 years to meet the Paris Agreement targets (Agbo-Bloua, 2023). The mismatch between supply and demand for key commodities will necessitate a 2.5-fold increase in investment (Agbo-Bloua, 2023). In other words, the green transition is likely to trigger a new commodity super-cycle, driving inflationary pressures—what is now referred to as greenflation. Greenflation is the inflation that arises from rising demand for, or constrained supply of, critical raw materials needed to build a greener global economy (Chung & Kim, 2024).

How significant is greenflation for the MENA region, and how likely is it to affect stock markets? Between 1992 and 2023, the MENA region received a cumulative total of $24.4 billion in climate-related finance—among the lowest regional shares globally (SRMG Think, 2023). Most of these investments were concentrated in Morocco and Egypt. However, we argue that the true drivers of the region's energy transition are its National Oil Companies (NOCs), such as Saudi Aramco and the Abu Dhabi National Oil Company (ADNOC). These economies are highly carbon-intensive and likely to be among the first to feel the impact of a global green shift. The region holds 57% of the world's proven oil reserves and 41% of its gas reserves (BP, 2021), though unevenly distributed. Thus, transitioning away from fossil fuels poses high economic costs, potentially depriving oil-producing nations of key revenue streams. These countries also maintain carbon-intensive lifestyles; per capita emissions in several MENA states are 60% higher than the average for emerging economies (SRMG Think, 2023).

Consequently, MENA stock markets—particularly in the Middle East—are expected to undergo structural shifts due to: (1) increased investment in green projects, which are often riskier and less profitable, and (2) decreased global appetite for fossil fuel investments. For example, the mining sector, historically underinvested in the MENA region, is beginning to flourish. Capital expenditure (CAPEX) ratios are at all-time lows, and few greenfield projects have been announced (OECD, 2021). Moreover, the supply of key minerals is heavily concentrated in a few countries, introducing further systemic risks. Simultaneously, declining investor interest in the fossil fuel sector could shift capital away from MENA markets. Between 2010 and 2019, publicly listed fossil fuel companies delivered lower and more volatile returns than the broader equity market. Moreover, institutional investors managing over $11 trillion have committed to divesting from fossil fuels (BCG, 2023). According to the Natural Resource Governance Institute (NRGI), more than $400 billion in NOC spending may not generate profit under a sustainable development scenario—$66 billion of that is in the MENA region. Algeria's Sonatrach, for instance, has over $12 billion—more than 25% of its CAPEX—invested in projects with breakeven prices between $40 and $70 per barrel (NRGI, 2021).

Amid conflicting views on the relationship between sustainable green investments and market liquidity, a key question emerges: What is the impact of greenflation on stock market liquidity in the MENA region? Market liquidity—defined as the ability to buy or sell assets quickly with minimal price impact and transaction costs (Chordia et al., 2005; Kyle, 1985)—is a vital concern for investors. Illiquid markets heighten risk and reduce portfolio efficiency, ultimately affecting investment returns (Amihud, 2002; Hendershott & Seasholes, 2014).

This chapter addresses this gap by investigating how green policies, proxied through a greenflation index, influence stock market liquidity across MENA countries. Specifically, we focus on the supply-side dynamics of the energy transition. While earlier studies have primarily explored demand-driven greenflation, we argue that the green transition's reliance on critical minerals generates short-term spillover effects in stock markets. These effects arise from investor behavior and asset reallocation in response to rising commodity prices.

As the global economy pivots toward sustainability, the demand for critical minerals will intensify, pushing greenflation upward. Assuming greenflation serves as a reliable proxy for the economic pressures of the green transition, stock markets in the MENA region are likely to experience one or both of the following impacts: (1) increased investment in green sectors, and/or (2) a declining appetite for fossil fuel-linked assets. By focusing on these short-term market responses, this chapter complements existing supply-side narratives and sheds light on the broader financial market implications of the green transition in MENA stock exchanges.

Using a panel dataset of 823 listed firms from MENA countries—comprising over 325,000 firm-day observations between November 29, 2019 and December 29, 2023—we find that greenflation improves market liquidity by reducing transaction costs and deepening market activity. However, regional heterogeneity emerges: stock markets in the Middle East show significant improvements in liquidity, while North African markets respond more weakly. These differences likely reflect varying regulatory frameworks, market depths, and structural factors. Furthermore, post-COVID investments in sustainability appear to have amplified greenflation's influence, presenting both a challenge and an opportunity for MENA capital markets.

Our findings provide valuable insights for policymakers. First, MENA governments must foster strong and resilient financial markets that can absorb greenflation-related shocks. This requires diversifying investments, especially into green technologies, and reducing dependence on fossil fuel revenues. Second, regulatory frameworks must incentivize private sector contributions to climate goals, while maintaining overall market stability. Third, regional cooperation is crucial in addressing disparities in greenflation's impact across Middle Eastern and North African markets. Coordinated investments in infrastructure and technology-sharing can enhance regional resilience. Finally, aligning national policies with global climate finance initiatives will help MENA countries secure a greater share of international green funding.

The structure of this chapter is as follows: Sect. 6.2 reviews the literature and develops hypotheses. Section 6.3 discusses the data, variables, and methodology employed in the study. Section 6.4 presents empirical findings. Section 6.5 delves into further sub-analyses. Finally, Sect. 6.6 concludes the chapter.

6.2 Literature Review and Hypotheses Development

6.2.1 Greenflation

The first logical question to address is how energy transition could prompt an increase in long-term price expectations. One way to capture this effect is through the demand-output equilibrium. The green transition is costly in terms of GDP, as the introduction of emission taxes or caps compels firms to reduce production and incur abatement costs (Diluiso et al., 2021; Ferrari Minesso & Pagliari, 2023). From a price dynamics perspective, introducing carbon taxes or fiscal disincentives might potentially increase long-term inflation expectations. However, our article specifically aims to capture the short-term dynamics of the energy transition by examining how heightened demand for raw materials can drive prices upward, thereby spilling over into the broader economy.

Given the scarcity of academic research on greenflation, as the concept is still in its infancy, industry-related analogies seem more practical when elaborating our theoretical framework. First, accurately capturing greenflation requires clearly defining the concept. Sharma (2021) notes that new government-directed spending is driving up demand for materials needed to build a cleaner economy, while simultaneously, tightening regulations restrict supply by discouraging investment in mines, smelters, or other carbon-intensive sources. The unintended consequence, according to the author, is "greenflation," characterized by rising prices for metals and minerals, such as copper, aluminum, and lithium, essential for solar and wind power, electric vehicles, and other renewable technologies. Alternatively, Olovsson and Vestin (2023) describe greenflation as the upward inflationary pressure resulting from the transition toward clean energy production. These definitions align with the European Central Bank's (ECB) statement that "the introduction of new technologies to cope with climate change would entail substantial expenditures, investments, innovations, and changes in relative prices, with far-reaching economic consequences" (Andersson et al., 2021). Accordingly, price increases for critical minerals needed in green transition projects constitute one manifestation of these economic consequences.

Nevertheless, one notable observation from the existing literature review is that most studies emphasize inflation induced by either rising energy prices or carbon taxes. For instance, Heyen et al. (2020) highlight the inflationary effects of the green transition, attributing them either to rising carbon taxes associated with green pathways or to changes in the energy mix. We consider these perspectives as primarily emphasizing the supply side of the equation—namely, the reduced availability of fossil energy due to underinvestment—or institutional approaches where governments actively discourage fossil-fuel consumption through taxes and regulations. In contrast, our approach to capturing greenflation focuses on the demand side, specifically the increased demand for green minerals that drives industry adaptation through reinvestment in clean energy-related projects. We believe these demand-driven supply shocks are more relevant and forward-looking than traditional energy-related inflation measures or government-based policy metrics.

6.2.2 Stock Market Liquidity

Market liquidity is a primary concern for investors, as it significantly influences portfolio returns. Illiquid assets often involve higher transaction costs and increase the risk of substantial losses during trading. Numerous studies have attempted to price illiquidity and explore its relationship with expected stock returns (Amihud, 2002; Amihud & Mendelson, 1986; Hendershott & Seasholes, 2014). Despite extensive research, liquidity remains an elusive concept (Amihud, 2002)—intuitive to sense but challenging to define and even more difficult to quantify (Lesmond, 2005; O'Hara, 2004). Its multifaceted nature resists a universally accepted definition that encompasses all dimensions of liquidity.

Kyle (1985) defines market liquidity as the ability to trade an asset quickly with minimal value loss, while Chordia et al. (2005) describe it as "the ability to buy or sell large quantities of an asset quickly and at low cost." Moreover, Wyss (2004) highlights four main characteristics of liquid markets: depth, the ability to trade significant quantities without major price impact; tightness, low transaction costs enabling trades at near-identical prices; immediacy, the speed at which trades can be executed; and resiliency, the rapid adjustment of markets to new orders that restore balance. However, no single liquidity measure can comprehensively capture these diverse aspects.

6.2.3 Hypotheses Development

Greenflation is an emerging area of study, and the global transition toward net-zero emissions is expected to bring about significant and unprecedented changes in capital markets. Given the uncertainty surrounding its effects, adopting a bi-directional hypothesis development approach is essential to comprehensively explore its potential impact.

6.2.3.1 Greenflation Supports MENA Countries' Stock Markets Liquidity

The objective of a utility-maximizing investor is to maximize wealth, specifically by maximizing the present value of their investments (Auerbach, 1979). This goal aligns closely with maintaining robust market liquidity, as any disruption to liquidity may jeopardize investors' ability to convert wealth into realized returns.

The traditional shareholder wealth maximization perspective views environmental activities primarily as costs without corresponding financial benefits. Consequently, rational investor behavior dictates minimizing these costs as much as possible (Feldman et al., 1997). However, an alternative viewpoint suggests that environmental investments and related activities can enhance financial performance due to their positive impact on a firm's operating efficiency and long-term profitability (Feldman et al., 1997).

Accordingly, given that the threat of climate change is particularly significant for MENA countries, investors might anticipate a substantial influx of financial investment into green projects, thus attracting more funds to the region. Nonetheless, existing theory offers two distinct perspectives influencing investor perceptions of green investments. The shunned-stock hypothesis argues that socially responsible or green investors are value-driven; following Merton's (1987) incomplete information model, these investors prioritize non-pecuniary benefits rather than financial returns. As a result, socially responsible investments trade at a premium (i.e., lower expected returns) relative to irresponsible investments, primarily due to limited risk-sharing opportunities (Derwall et al., 2011). Conversely, the errors-in-expectation hypothesis suggests that socially responsible stocks are systematically undervalued, thus generating higher returns for investors who recognize their true value (Derwall et al., 2011).

Interestingly, we could also argue that an increase in greenflation might exert an "insurance effect" by helping listed MENA firms reduce their business risks. Specifically, rising greenflation, signaling a more rapid transition to green energy, could motivate MENA countries to proactively engage in green investments. This dynamic could subsequently influence stock markets either by encouraging investments in green-oriented firms or by reducing overall market risk, thereby supporting greater market liquidity.

***H1:** Greenflation improves market liquidity in MENA countries.*

6.2.3.2 Greenflation Reduces Market Liquidity in MENA Countries

Alternatively, one might consider that the shift toward greener projects—and potentially greener economies—could disrupt stock markets in the MENA region. One channel through which green investment affects market liquidity is controversy. Green investments are considered "green" due to their strong alignment with global climate initiatives. From a capital markets perspective, this alignment is validated through reporting and disclosure requirements. As a result, increased exposure to green investments renders stock markets more sensitive to controversies. Such controversies can significantly reduce liquidity and increase the risk profile of "green" stocks, as evidenced by wider bid-ask spreads (Kong & Narayanan, 2023).

Furthermore, Acemoglu et al. (2014) discuss the concept of path dependence in innovation, which refers to the tendency of firms to continue innovating in areas where they have previously succeeded. In the case of MENA National Oil Companies

(NOCs), which have historically innovated in "brown" or carbon-intensive technologies, this path dependence may persist due to their existing expertise and comparative advantage. From a market liquidity standpoint, if investors perceive that these firms will continue prioritizing brown innovation over green alternatives, a gradual shift of capital toward greener markets may occur. Such a shift could adversely impact regional market liquidity as the green transition gains momentum.

Therefore, we propose the following hypothesis:

***H2:** Greenflation reduces market liquidity in MENA countries.*

6.3 Methodology and Data

6.3.1 Data

Our unbalanced panel dataset includes 823 listed firms from the MENA region, covering 325,093 firm-day observations from November 29, 2019 to December 29, 2023. The 12 MENA countries include Saudi Arabia, Qatar, Tunisia, Kuwait, Egypt, UAE, Morocco, Bahrain, Lebanon, Jordan, Iraq, and Oman. Daily market data was sourced from Refinitiv Eikon, while the greenflation index was obtained from Bloomberg, with 2019 selected as the starting point to align with the inception of the index.

6.3.2 Stock Market Liquidity

We employ several liquidity measures, which covers both tightness (transaction costs proxies) and *depth* dimensions of market liquidity (volume-based proxies). These measures include the *Quoted Spread (QS), Effective Spread (ES), Imputed Roundtrip Cost (IRC), Amivest Liquidity Ratio (AV), Turnover (TN), and Amihud Illiquidity Ratio (AH)*. Each measure is briefly described below with its formulation and relevant literature references.

Quoted Spread (QS): The Quoted Spread, introduced by Stoll (1989), reflects the cost of crossing the spread, providing a measure of market tightness. It is calculated as:

$$QS_t = \frac{Ask_t - Bid_t}{\frac{Ask_t + Bid_t}{2}} \tag{6.1}$$

where Ask_t and Bid_t are the ask and bid prices at time t. This measure is widely used in liquidity studies for corporate stocks, as seen in works by Schestag et al. (2016).

Effective Spread (ES): Defined by Lee (1993), the Effective Spread provides a more precise measure of the trading cost by considering the trade direction. It is given by:

$$ES_k = 2 \cdot D_k \cdot \left(\ln(p_k) - \ln(m_k)\right) \tag{6.2}$$

where D_k is the trade direction indicator, p_k is the trade price, and m_k is the midpoint price. The Effective Spread uses transaction data, including quote prices and trade signs, and has been utilized in studies such as Beber et al. (2008) and Hasbrouck (2009).

Imputed Roundtrip Cost (IRC): Developed by Dick-Nielsen et al. (2012), this measure evaluates the roundtrip transaction cost by capturing the range of transaction prices over a specific period. The formula is:

$$IRC_t = \frac{p_{\max,t} - p_{\min,t}}{p_{\max,t}} \tag{6.3}$$

where $p_{max,\,t}$ and $p_{min,\,t}$ denote the maximum and minimum transaction prices within a given period ttt. This measure, applied in studies like Goldstein et al. (2007) and Bao et al. (2011), provides insight into market immediacy and tightness, particularly for corporate bonds.

Amihud Illiquidity Ratio (AH): Proposed by Amihud (2002), this measure assesses the price impact of trading volume, thus capturing market depth. It is calculated as:

$$AH_t = \frac{1}{K_t} \sum_{k=1}^{D_t} \frac{|r_{t,k}|}{p_{t,k} \cdot TV_{t,k}} \tag{6.4}$$

where D_t is the number of transactions on day t, $r_{t,\,k}$ is the return, $p_{t,\,k}$ is the price, and $TV_{t,\,k}$ is the trading volume for each transaction k. This ratio has been used in liquidity assessments in studies such as Mahanti et al. (2008) and Han and Zhou (2016).

Turnover (TN): Turnover, as defined by Gravelle (1999), represents the frequency of trading in relation to the amount outstanding. It is formulated as:

$$TN_t = \frac{TV_t}{AmOut} \tag{6.5}$$

where TV_t is the total traded volume and *AmOut* is the amount outstanding. Turnover is a common depth measure in liquidity research, indicating the frequency of transactions. It has been used in studies like Downing et al. (2005) and Sadka (2006).

Amivest Liquidity Ratio (AV): Introduced by Cooper et al. (1985), the Amivest Liquidity Ratio measures depth by analyzing the trade volume's impact on returns. The formula is:

$$AV_t = \frac{1}{K_t} \sum_{k=1}^{D_t} \frac{p_{t,k} \cdot TV_{t,k}}{|r_{t,k}|} \tag{6.6}$$

This ratio captures the market's ability to absorb large trades without significant price changes, and it has been applied in liquidity studies for corporate and treasury stocks, such as Custódio et al. (2011) and Díaz and Escribano (2017).

6.3.3 Greenflation Index

To capture greenflation, our study uses an equally weighted price index based on the quoted prices of 24 critical metals used in renewable technologies, utilizing daily data. The metals included in the index are as follows: copper, aluminum, nickel, zinc, cobalt, lithium, graphite, manganese, uranium, platinum, silver, palladium, chromium, molybdenum, vanadium, dysprosium, neodymium, praseodymium, silicon, terbium, cadmium, gallium, iridium, and tellurium.

$$Greenflation_t = \sum_{j=1}^{J} P_{j,t}\omega_{j,t} \tag{6.7}$$

where $P_{j,\,t}$ is the logarithm price of the metal j in period t. Whereas $\omega_{j,\,t}$ are the respective weights.

6.3.4 Control Variables

We also include key market-level control variables to account for factors that influence firms' stock market liquidity. Stock daily returns (Return) are included to capture investor sentiment and trading motivation; positive returns are generally associated with improved liquidity, as they reduce adverse selection and encourage trading activity, while negative or volatile returns can widen spreads due to increased perceived risk (Chordia et al., 2000; Foucault et al., 2013). Daily value traded (Value Traded), a proxy for trading activity and market depth, is positively correlated with liquidity; higher trading volumes are associated with tighter bid-ask spreads and lower price impact (Chordia et al., 2001; Hasbrouck, 2009). Daily market capitalization (Capitalization)—calculated as the natural logarithm of the daily closing price multiplied by the number of outstanding shares—serves as a proxy for firm size. Larger firms typically enjoy higher liquidity due to greater analyst coverage, investor following, and deeper order books, while smaller firms often face thinner markets and higher transaction costs (Alaoui Mdaghri et al., 2021; Amihud, 2002; Lesmond et al., 1999). Finally, stock daily volatility (Volatility) is measured using the Garman-Klass estimator, which accounts for both opening and closing prices as well as intra-day highs and lows (Garman & Klass, 1980). Volatility represents market uncertainty and risk; higher volatility is typically associated with reduced liquidity, wider spreads, and lower trading volume due to the increased cost of market making (Acharya & Pedersen, 2005; Glosten & Milgrom, 1985). See Appendix A for detailed variable definitions.

6.3.5 Model Specification

To investigate the impact of greenflation on stock market liquidity across listed firms in the MENA region, we employ an econometric model that accounts for daily variations in market liquidity measures as a function of the greenflation index and other control variables. The baseline model specification is given as:

$$Market\ Liquidity_{i,t} = \alpha + \beta_1\, Greenflation_t + \beta_2\, Controls_{i,t} + \gamma_i + \delta_i + \epsilon_{i,t} \quad (6.8)$$

where *Market Liquidity*$_{i,\,t}$ represents the market liquidity measure for stock i on day t, with separate regressions for each proxy. α is the constant term. *Greenflation*$_t$ denotes the greenflation index on day t. *Controls*$_{i,\,t}$ represents a vector of control variables. γ_i and δ_i represents, respectively, firm- and country-fixed effects to control for heterogeneous characteristics across MENA firms and countries. $\epsilon_{i,\,t}$ is the error term.

To estimate Eq. (6.1) and to follow prior studies (e.g., Alaoui Mdaghri et al., 2021; Narayan et al., 2021), we use the ordinary least square (OLS) model with robust standard errors clustered at the firm level to control for heteroskedasticity and autocorrelation issues.

6.4 Empirical Analyses

6.4.1 Descriptive Statistics

Table 6.1 provides the descriptive statistics for the key variables used in this study, highlighting the characteristics of liquidity measures, the greenflation index, and control variables. The Quoted Spread (QS), with a mean of 0.025, indicates moderate transaction costs, while Effective Spread (ES) and Imputed Roundtrip Cost (IRC), averaging 0.001 and 0.024 respectively, reflect low trading costs across MENA markets. Depth measures such as the Amivest Liquidity Ratio (AV, mean 7.185) and Turnover (TN, mean 0.623) show high market depth and active trading, though variability is evident, particularly with extreme observations. The Amihud Illiquidity Ratio (AH) suggests generally low illiquidity, but its range (−29.458 to

Table 6.1 Descriptive statistics

Variable	Obs	Mean	Std. Dev.	Min	Max
QS	325,093	0.025	0.025	0	1.292
ES	325,093	0.001	0.024	−0.751	0.649
IRC	325,093	0.024	0.023	0	0.785
AV	277,149	7.185	1.199	−4.995	12.845
TN	325,082	0.623	0.218	0	2.07
AH	324,995	0.002	0.292	−29.458	147.742
Greenflation	325,093	7.669	0.4	6.36	8.095
Return	325,009	0.001	0.124	−6.685	6.78
Value traded	325,082	14.124	3.828	−4.605	26.759
Capitalization	325,093	21.267	2.675	11.234	30.014
Volatility	324,086	0.014	0.015	0	0.942

This table presents descriptive statistics for the full sample over the period from November 29, 2019 to December 29, 2023. See Appendix A for variables' definitions

Table 6.2 Correlation matrix

	Greenflation	Return	Value traded	Capitalization	Volatility
Greenflation	1				
Return	0.003624	1			
Value traded	0.004026	0.002012	1		
Capitalization	0.007965	0.000684	0.006835	1	
Volatility	−0.027219	0.016524	0.006569	−0.019885	1

This table reports Pearson correlation matrix of all independent variables employed in our empirical models

147.742) highlights disparities across firms. The Greenflation index (mean 7.669, standard deviation 0.4) shows consistent inflationary pressures linked to environmental policies, while stock returns (mean 0.001) indicate typical market performance, albeit with notable volatility. Daily Value Traded (mean log value 14.124) points to active trading, while Market Capitalization (mean log value 21.267) highlights the mix of small- and large-cap firms. Volatility is generally low (mean 0.014) but varies significantly across stocks.

Table 6.2 presents the correlation matrix of all independent variables for our full sample of firms in MENA region. The results exhibit very low accepted correlation coefficients (less than 0.5), which alleviates major concerns for multicollinearity.

6.4.2 Baseline Findings

Table 6.3 presents the baseline results of the impact of greenflation on stock market liquidity, with market tightness measures displayed in columns (1)–(3) and *market depth measures* in columns (4)–(6). Across all specifications for market tightness, the coefficients are negative and strongly significant, indicating that as greenflation

Table 6.3 Impact of greenflation on market liquidity in MENA region

	Tightness			Depth		
	(1)	(2)	(3)	(4)	(5)	(6)
	QS	ES	IRC	AV	TN	AH
Greenflation	−0.000***	−0.000**	−0.000***	0.217***	0.000***	−0.000
	(0.000)	(0.000)	(0.000)	(0.018)	(0.000)	(0.000)
Return	0.001***	0.022***	0.001***	−12.820	0.005**	0.277***
	(0.000)	(0.000)	(0.000)	(92.252)	(0.002)	(0.004)
Value traded	0.000	−0.000**	0.000	0.000***	0.000***	−0.000
	(0.000)	(0.000)	(0.000)	(0.000)	(0.000)	(0.000)
Capitalization	−0.000*	0.000***	−0.000***	0.000***	−0.000***	−0.000
	(0.000)	(0.000)	(0.000)	(0.000)	(0.000)	(0.000)
Volatility	1.600***	0.176***	1.510***	−56551.083***	2.805***	−0.033
	(0.001)	(0.003)	(0.001)	(903.048)	(0.016)	(0.038)
Constant	0.002***	0.000	0.002***	1631.091***	0.528***	0.003
	(0.000)	(0.000)	(0.000)	(105.479)	(0.002)	(0.004)
Firms FE	Yes	Yes	Yes	Yes	Yes	Yes
Countries FE	Yes	Yes	Yes	Yes	Yes	Yes
Number of Obs.	324,002	324,002	324,002	277,240	323,991	323,988
R-squared	0.918	0.034	0.917	0.035	0.689	0.014

This table displays regression results on the effect of greenflation on various market liquidity measures in the MENA region, covering transaction cost indicators like Quoted Spread (QS), Effective Spread (ES), and Imputed Roundtrip Cost (IRC), as well as volume-based measures like Amivest Liquidity Ratio (AV), Turnover (TN), and Amihud Illiquidity Ratio (AH). The models include controls for Return, Value Traded, Capitalization, and Volatility, with country fixed effects. Robust standard errors are shown in parentheses, and significance levels are marked by ***$p < 0.01$, **$p < 0.05$, and *$p < 0.1$. See Appendix A for variables' definitions

rises, bid-ask spreads in MENA stock markets narrow. In other words, the cost of transactions in MENA stock markets decreases as greenflation intensifies. This result suggests that increasing environmental and sustainability pressures might actually reduce transaction costs, potentially due to heightened trading activity or investor focus on sustainable investments within the MENA region. In other words, as greenflation intensifies, it likely stimulates increased trading activity in sustainable assets. Investors align portfolios with ESG trends, raising demand for these stocks and fostering active trading. Similarly, the coefficients for market depth measures are positive and significant (with the exception of the Amihud measure, AH, which does not show significance). This finding implies that as greenflation escalates, the depth of MENA stock markets improves, enabling the absorption of a higher volume of trade without significant price fluctuations. These results collectively indicate that rising greenflation enhances liquidity in MENA stock markets, both by lowering transaction costs and by increasing market depth.

Our findings are particularly interesting, as one might initially expect the opposite—that greenflation, with its associated costs and economic pressures, would dampen market liquidity by increasing transaction costs and reducing market depth.

However, the observed improvements in liquidity as greenflation intensifies could reflect an adaptive market response where investors and institutions actively pursue sustainable assets, driving up demand and trading volumes. This shift might be attributed to a growing preference for environmentally responsible investments, which could foster greater liquidity as investors seek to align their portfolios with sustainability trends. These insights highlight a potential positive feedback loop between environmental sustainability initiatives and stock market performance in the MENA region. As greenflation continues to shape market dynamics, it may create opportunities for liquidity enhancement, particularly in sectors aligned with sustainable practices.

6.5 Extensions and Robustness Checks

6.5.1 Sub-Sample Analysis: Middle Eastern vs. North African Countries

In Table 6.4, we present a regional sub-sample analysis, with Panel A displaying results for Middle Eastern countries and Panel B showing results for North African countries. Our findings reveal a clear divergence in how greenflation impacts stock market liquidity across these two regions. In Panel A, we observe that greenflation significantly enhances both market tightness and depth in Middle Eastern markets. Conversely, in Panel B, greenflation shows no consistent or statistically significant effect on liquidity measures in North African markets. This discrepancy can be explained by a range of institutional, structural, and economic differences.

We argue that Middle Eastern countries—particularly those in the Gulf Cooperation Council (GCC)—have benefited from accelerated developments in green finance and ESG adoption. With strong policy backing from initiatives such as Saudi Arabia's Vision 2030 and the UAE's Green Agenda, these countries have fostered mature financial systems that are more resilient and responsive to environmental shocks. These reforms have enabled Middle Eastern markets to attract sustainable capital, deepen liquidity, and reduce transaction costs as greenflation rises. Our results suggest that these markets are leveraging greenflation as an opportunity, reflecting their strategic alignment with global ESG investment trends. On the other hand, we find that North African markets exhibit weaker transmission channels between greenflation and liquidity, likely due to their less developed green finance infrastructure, voluntary or limited ESG disclosure requirements, and smaller, less liquid stock markets. These conditions may prevent greenflation from meaningfully influencing trading behavior or attracting sustainable investment flows.

Additionally, many North African countries face fiscal constraints and rely heavily on external financing for climate projects, further slowing the integration of ESG practices into their markets. We believe these regional disparities reflect broader economic and institutional differences. While Middle Eastern economies are

Table 6.4 Greenflation impact on market liquidity: Middle Eastern vs. North African firms

	Tightness			Depth		
	(1)	(2)	(3)	(4)	(5)	(6)
	QS	ES	IRC	AV	TN	AH
Panel A: Middle East						
Greenflation	−0.000***	−0.000***	−0.000***	0.238***	0.000***	−0.000***
	(0.000)	(0.000)	(0.000)	(0.019)	(0.000)	(0.000)
Constant	0.002***	0.001***	0.002***	1587.806***	0.529***	0.001***
	(0.000)	(0.000)	(0.000)	(92.077)	(0.002)	(0.000)
Controls	Yes	Yes	Yes	Yes	Yes	Yes
Firms FE	Yes	Yes	Yes	Yes	Yes	Yes
Countries FE	Yes	Yes	Yes	Yes	Yes	Yes
Number of Obs.	192,159	192,159	192,159	166,394	192,148	192,147
R-squared	0.903	0.226	0.903	0.056	0.638	0.005
Panel B: North Africa						
Greenflation	0.000*	0.000	0.000	0.181*	0.000	−0.000
	(0.000)	(0.000)	(0.000)	(0.033)	(0.000)	(0.000)
Constant	0.000*	−0.008**	0.001***	3339.163***	0.557***	0.006
	(0.000)	(0.000)	(0.000)	(98.321)	(0.001)	(0.005)
Controls	Yes	Yes	Yes	Yes	Yes	Yes
Firms FE	Yes	Yes	Yes	Yes	Yes	Yes
Countries FE	Yes	Yes	Yes	Yes	Yes	Yes
Number of Obs.	131,843	131,843	131,843	110,846	131,843	131,841
R-squared	0.937	0.029	0.935	0.019	0.537	0.014

This table presents regression results on the impact of greenflation on market liquidity measures, comparing firms in the Middle East (Panel A) and North Africa (Panel B) regions. The dependent variables include liquidity measures categorized by transaction cost proxies (QS, ES, IRC) and volume-based proxies (AV, TN, AH). Each model includes controls for Return, Value Traded, Capitalization, and Volatility, with country fixed effects applied. Robust standard errors are in parentheses, and significance levels are denoted by ***$p < 0.01$, **$p < 0.05$, and *$p < 0.1$. See Appendix A for variables' definitions

actively diversifying away from oil dependence through large-scale investment in sustainable sectors, North African countries are still navigating foundational challenges in establishing green finance ecosystems.

6.5.2 *Pre-, During, and Post-COVID-19 Analysis*

We acknowledge that global sustainability efforts had already been unfolding before the COVID-19 pandemic. However, our decision to frame the analysis around the pre-COVID-19 (Panel A), during COVID-19 (Panel B), and post-COVID-19 (Panel C) periods is grounded in both conceptual reasoning and empirical observation. The pandemic constituted a structural break that reshaped fiscal priorities, accelerated sustainable investment trends, and triggered major shifts in public policy—especially in emerging and resource-dependent regions such as MENA. Globally, COVID-19 served as a turning point, pushing governments and financial institutions to incorporate environmental sustainability into economic recovery strategies. This was reflected in unprecedented stimulus packages, international calls for "green recovery," and the growing mainstream adoption of ESG investing.

In the MENA region specifically, the pandemic catalyzed a wave of policy reforms and investment initiatives that explicitly prioritized green infrastructure, renewable energy, and climate resilience. Countries such as the UAE and Saudi Arabia embedded green ambitions into national recovery plans, while Morocco and Egypt advanced green bond frameworks and climate-aligned finance strategies. These institutional developments contributed to a marked increase in sustainable capital flows across the region, and this trend is empirically captured by the sharp and sustained rise in the greenflation index during and after the pandemic, as shown in Fig. 6.1.

To empirically validate this framing, we conduct a period-specific analysis presented in Table 6.5, where we examine the relationship between greenflation and market liquidity across the three distinct phases of the pandemic. In Panel A (pre-COVID-19), the coefficients for greenflation are statistically insignificant across nearly all liquidity measures, indicating a weak or non-existent relationship during that time. This suggests that, prior to the pandemic, greenflation was not yet an influential force within MENA capital markets. However, in Panel B (during COVID-19), greenflation begins to exhibit a significantly positive relationship with both market tightness and depth, a trend that becomes even more pronounced in Panel C (post-COVID-19), where the coefficients are stronger and more consistently significant.

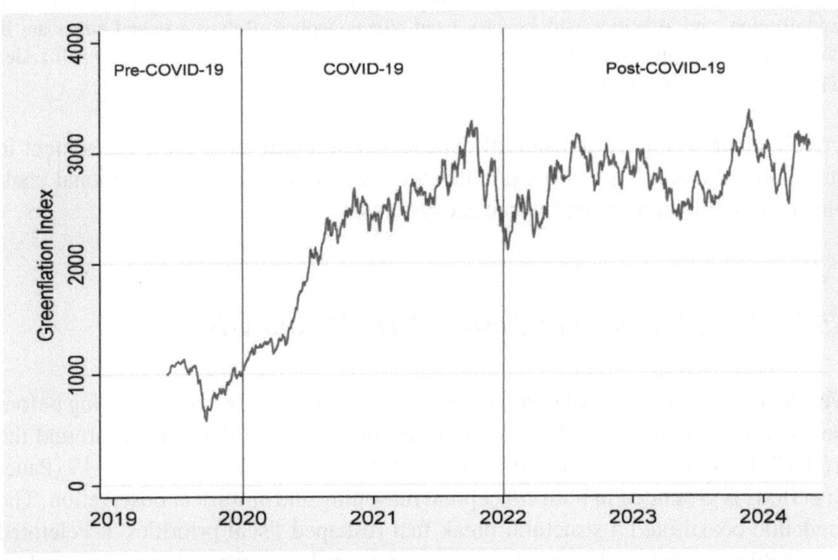

Fig. 6.1 Trend of greenflation index across Pre-COVID19, COVID, and Post-COVID periods

Table 6.5 Greenflation impact on market liquidity: pre-, during, and post-COVID-19 analysis

	Tightness			Depth		
	(1)	(2)	(3)	(4)	(5)	(6)
	QS	ES	IRC	AV	TN	AH
Panel A: Pre-COVID-19 pandemic						
Greenflation	0.000	0.000*	0.000	1.534	0.000	−0.000*
	(0.000)	(0.000)	(0.000)	(1.255)	(0.000)	(0.000)
Constant	−0.000	−0.031*	0.000	676.420	0.495***	0.023**
	(0.001)	(0.004)	(0.001)	(1400.641)	(0.023)	(0.008)
Controls	Yes	Yes	Yes	Yes	Yes	Yes
Firms FE	Yes	Yes	Yes	Yes	Yes	Yes
Countries FE	Yes	Yes	Yes	Yes	Yes	Yes
Number of Obs.	17,945	17,945	17,945	15,097	17,945	17,945
R-squared	0.901	0.048	0.897	0.033	0.690	0.604
Panel B: During COVID-19 pandemic						
Greenflation	−0.000***	−0.000***	−0.000***	0.590***	0.000***	−0.000
	(0.000)	(0.000)	(0.000)	(0.029)	(0.000)	(0.000)
Constant	0.003***	−0.001*	0.003***	1007.655***	0.521***	0.004
	(0.000)	(0.000)	(0.000)	(160.128)	(0.002)	(0.008)
Controls	Yes	Yes	Yes	Yes	Yes	Yes
Firms FE	Yes	Yes	Yes	Yes	Yes	Yes
Countries FE	Yes	Yes	Yes	Yes	Yes	Yes
Number of Obs.	131,843	131,843	131,843	110,846	131,843	131,841
R-squared	0.937	0.029	0.935	0.019	0.537	0.014
Panel C: Post-COVID-19 pandemic						
Greenflation	−0.000**	0.000***	−0.000**	−0.000**	0.201**	0.000***
	(0.000)	(0.000)	(0.000)	(0.000)	(0.068)	(0.000)
Constant	0.002***	−0.006**	0.002***	0.002***	1556.589***	0.505***
	(0.000)	(0.001)	(0.000)	(0.000)	(230.991)	(0.004)
Controls	Yes	Yes	Yes	Yes	Yes	Yes
Firms FE	Yes	Yes	Yes	Yes	Yes	Yes
Countries FE	Yes	Yes	Yes	Yes	Yes	Yes
Number of Obs.	173,130	173,130	173,130	148,084	173,126	173,124
R-squared	0.914	0.038	0.914	0.030	0.695	0.091

This table provides regression results on the impact of greenflation on market liquidity measures across three periods: pre-COVID-19 (Panel A), during COVID-19 (Panel B), and post-COVID-19 (Panel C). The dependent variables include liquidity measures categorized by transaction cost proxies (QS, ES, IRC) and volume-based proxies (AV, TN, AH). Each model incorporates controls for Return, Value Traded, Capitalization, and Volatility, with country fixed effects. Robust standard errors are in parentheses, and significance levels are denoted by ***p < 0.01, **p < 0.05, and *p < 0. See Appendix A for variables' definitions

This shift confirms that the pandemic acted as a turning point in shaping how financial markets internalize sustainability-related risks and opportunities. The segmentation into these three periods allows us to isolate and highlight this evolution in a meaningful way. Without this temporal disaggregation, the amplification of greenflation's effects on market liquidity—particularly in response to policy-driven sustainable investment efforts—would be obscured. Thus, the pre-, during-, and post-COVID framework is not only conceptually relevant but also essential to capturing the dynamic, time-sensitive nature of the relationship between environmental policy pressures and stock market behavior in the MENA region.

6.5.3 Robustness Checks

To address potential endogeneity concerns related to the greenflation variable, we conducted a 2SLS analysis as presented in Table 6.6. We employ the Price-to-Earnings (PE) ratio as an instrumental variable for greenflation.

The rationale behind this choice is twofold. First, the inclusion restriction is satisfied, as the PE ratio is empirically and conceptually correlated with greenflation:

Table 6.6 Impact of greenflation on market liquidity: 2SLS with IV regression

	2SLS regressions					
	(1)	(2)	(3)	(4)	(5)	(6)
	Tightness			Depth		
	QS	ES	IRC	AV	TN	AH
Greenflation	−0.000**	−0.000**	−0.000***	0.647**	−0.000	0.000**
	(0.000)	(0.000)	(0.000)	(0.710)	(0.000)	(0.000)
Return	0.001***	0.022***	0.001***	−35.020	0.005	0.277**
	(0.000)	(0.002)	(0.000)	(55.430)	(0.006)	(0.105)
Value traded	0.000	−0.000	0.000	0.000*	0.000*	−0.000*
	(0.000)	(0.000)	(0.000)	(0.000)	(0.000)	(0.000)
Capitalization	−0.000**	0.000	−0.000***	0.000***	−0.000**	−0.000***
	(0.000)	(0.000)	(0.000)	(0.000)	(0.000)	(0.000)
Volatility	1.571***	0.098***	1.489***	−52633.496***	4.765***	−0.035
	(0.004)	(0.012)	(0.003)	(1226.946)	(0.062)	(0.022)
Constant	0.013**	−0.007	0.011**	2234.605	0.787***	−0.010
	(0.005)	(0.012)	(0.004)	(1637.901)	(0.084)	(0.007)
Firms FE	Yes	Yes	Yes	Yes	Yes	Yes
Countries FE	Yes	Yes	Yes	Yes	Yes	Yes
Number of Obs.	324,002	324,002	324,002	277,240	323,991	323,988
R-squared	0.900	0.009	0.905	0.014	.	0.014
p-value	0.000	0.000	0.000	0.000	0.000	0.000

This table presents 2SLS regression results examining the impact of greenflation on market liquidity measures, using instrumental variables (IV) to address potential endogeneity. PE Ratio is the IV used. The dependent variables include transaction cost measures (QS, ES, IRC) and volume-based measures (AV, TN, AH), categorized under tightness and depth. Control variables are Return, Value Traded, Capitalization, and Volatility, with country fixed effects applied. Standard errors are in parentheses, and significance levels are indicated by ***$p < 0.01$, **$p < 0.05$, and *$p < 0.1$. See Appendix A for variables' definitions

firms with higher exposure to sustainable and green investment trends often command valuation premiums, reflected in elevated PE ratios. As investor sentiment increasingly shifts toward ESG-aligned firms, their PE ratios tend to rise, tracking the same sustainability-driven valuation dynamics that underlie movements in the greenflation index. Second, the exclusion restriction is justified on the basis that the PE ratio, while linked to firm valuation, does not directly influence stock market liquidity once standard firm-level controls such as returns, volatility, trading volume, and market capitalization are accounted for. Thus, its effect on liquidity operates solely through its influence on greenflation, making it a suitable instrument.

The 2SLS results confirm that greenflation continues to exert a significant effect on both market tightness and depth dimensions. The signs and statistical significance of the coefficients are largely consistent with those obtained from the baseline OLS estimations, suggesting that the relationship between greenflation and market liquidity is robust and not driven by endogeneity bias. These findings reinforce the credibility of our empirical strategy and validate our core conclusions.

6.6 Conclusion

Exploring the impact of greenflation on stock market liquidity in the MENA region sheds light on an emerging dynamic between environmental policies and financial market behaviors. As MENA countries gradually integrate sustainability practices, the influence of greenflation—a measure of inflation driven by environmentally responsible policies—becomes a crucial factor. Understanding this relationship is essential, given that the region is balancing its traditional reliance on natural resources with a strategic pivot toward sustainable investments. This study contributes to this understanding by examining whether rising greenflation affects the liquidity of MENA stock markets, potentially revealing a pathway for markets to adapt to global green finance trends.

Using a comprehensive daily market dataset of 823 listed firms from 12 MENA countries over the period from November 2019 to December 2023, we analyzed the relationship between greenflation and stock market liquidity. By computing multiple liquidity measures that capture both market tightness and depth, the study offers a multi-faceted view of market liquidity dynamics in response to greenflation. Our findings indicate that greenflation positively impacts MENA stock market liquidity. Contrary to expectations that increased costs from environmental policies might deter trading activity, we observe that greenflation is associated with narrower bid-ask spreads and improved market depth, allowing for higher trading volumes without significant price impacts. This suggests that investors are increasingly aligning their portfolios with ESG principles, perhaps driven by a preference for sustainable assets that mitigate long-term environmental risks. However, our sub-regional analysis reveals differences: while greenflation enhances liquidity in Middle Eastern markets—where sustainability practices are more established—it shows limited impact in North African markets, likely due to regulatory and structural barriers in green finance adoption. Moreover, post-COVID analysis indicates that greenflation's effect on liquidity intensified, reflecting a surge in sustainable investments following the pandemic.

These findings lead to several policy implications that could support sustainable liquidity growth in MENA stock markets. First, regulators should encourage the integration of ESG principles across markets by promoting transparency in reporting and providing incentives for sustainable investments. This could attract ESG-focused investors, bolstering market depth and transaction efficiency, particularly in regions like North Africa that are currently less impacted by greenflation. Second, developing a robust green finance framework in North Africa is essential. Regulatory reforms to lower entry barriers for green bonds, along with the establishment of local green funds, could facilitate the transition toward sustainable finance. This development would help bridge the gap between Middle Eastern and North African markets, allowing for more consistent liquidity responses to greenflation across the region. Third, policymakers should sustain the momentum of post-COVID sustainable initiatives, reinforcing long-term green policies that align with global ESG standards. Such policies could stabilize market reactions to greenflation and reduce volatility, contributing to steady liquidity growth even as environmental costs rise. Finally, monitoring the dynamics of transaction costs amid rising greenflation is important. Since our findings show that greenflation can reduce transaction costs, regulators could leverage this trend by encouraging investments in sustainable assets. This approach would support liquidity and help maintain market stability, benefiting both investors and broader financial ecosystems in the MENA region.

Our paper acknowledges a key limitation. The measure of greenflation used in this study is based on global commodity price data for critical minerals, which may not fully capture the region-specific dynamics of the green transition or the policy heterogeneity across MENA countries. While this global approach reflects broader supply-side pressures, it overlooks local differences in regulatory frameworks, climate policies, and inflationary responses tied to national sustainability efforts. Future research could address this gap by developing a more holistic and context-sensitive greenflation index, incorporating country-level indicators such as carbon pricing, green bond issuance, public environmental spending, and inflation components related to green sectors. Such an index would enable a more accurate assessment of how domestic green transitions shape financial market behavior across diverse MENA economies.

Appendix A: Variable Definitions and Data Sources

Variable	Abbreviation	Definition	Data source
Dependent variables			
Transaction cost proxies—Tightness liquidity measures			
Quoted spread	QS	$QS_t = \frac{Ask_t - Bid_t}{\frac{Ask_t + Bid_t}{2}}$	Refinitiv Eikon
Effective spread	ES	$ES_k = 2 \cdot D_k \cdot (\ln(p_k) - \ln(m_k))$	Refinitiv Eikon
Imputed roundtrip cost	IRC	$\mathrm{IRC}_t = \frac{p_{\max,t} - p_{\min,t}}{p_{\max,t}}$	Refinitiv Eikon
Volume-based proxies—Depth liquidity measures			
Amivest liquidity ratio	AV	$AV_t = \frac{1}{K_t} \sum_{k=1}^{Dt} \frac{p_{t,k} \cdot TV_{t,k}}{\lvert r_{t,k} \rvert}$	Refinitiv Eikon
Turnover	TN	$TN_t = \frac{TV_t}{AmOut}$	Refinitiv Eikon
Amihud illiquidity ratio	AH	$AH_t = \frac{1}{K_t} \sum_{k=1}^{Dt} \frac{\lvert r_{t,k} \rvert}{p_{t,k} \cdot TV_{t,k}}$	Refinitiv Eikon
Independent variable			
Greenflation global index	Greenflation	$Greenflation_{i,t} = \sum_{j=1}^{J} P_{j,t} \omega_{j,t}$	Bloomberg

(continued)

(continued)

Variable	Abbreviation	Definition	Data source
Control variables			
Dail returns	Return	$Return = \frac{Price_t - Price_{t-1}}{Price_{t-1}}$	Refinitiv Eikon
Daily value traded	Value traded	$Value\ Traded = Price_t \times Volume_t$	Refinitiv Eikon
Market capitalization	Capitalization	$Capitalization = \ln(Price_t \times \text{Outstanding Shares } t)$	Refinitiv Eikon
Market volatility	Volatility	$GK_{VOL} = \sqrt{\frac{1}{2} - \log\left(\frac{HP_{i,t}}{LP_{i,t}}\right) - (2\log(2)-1)\log\left(\frac{CP_{i,t}}{OP_{i,t}}\right)}$	Refinitiv Eikon

This table presents the variable definitions, and their data sources used in this empirical research

References

Acemoglu, D., Aghion, P., & Hémous, D. (2014). The environment and directed technical change in a North-South model. *Oxford Review of Economic Policy, 30*(3), 513–530. https://doi.org/10.1093/oxrep/gru031

Acharya, V. V., & Pedersen, L. H. (2005). Asset pricing with liquidity risk. *Journal of Financial Economics, 77*(2), 375–410.

Agbo-Bloua, K. (2023). *Invest without carbon: Greenflation, and turning costs into opportunities through ESG research*. Societe Generale, News and Insights.

Alaoui Mdaghri, A., Raghibi, A., Thanh, C. N., & Oubdi, L. (2021). Stock market liquidity, the great lockdown and the COVID-19 global pandemic nexus in MENA countries. *Review of Behavioral Finance, 13*(1), 51–68. https://doi.org/10.1108/RBF-06-2020-0132

Amihud, Y. (2002). Illiquidity and stock returns: Cross-section and time series effects. *Journal of Financial Markets, 5*, 31–56.

Amihud, Y., & Mendelson, H. (1986). Asset pricing and the bid-ask spread. *Journal of Financial Economics, 17*(2), 223–249. https://doi.org/10.1016/0304-405X(86)90065-6

Andersson, M., Morgan, J., & Baccianti, C. (2021). Climate change and the macro economy. *SSRN Electronic Journal*, 243. https://doi.org/10.2139/ssrn.3632832

Auerbach, A. J. (1979). Wealth maximization and the cost of capital. *Quarterly Journal of Economics, 93*(3), 433–446. https://doi.org/10.2307/1883167

Bao, J., Pan, J., & Wang, J. (2011). The illiquidity of corporate bonds. *Journal of Finance, 66*(3), 911–946.

Beber, A., Brandt, M. W., & Kavajecz, K. A. (2008). Flight-to-quality or flight-to-liquidity? Evidence from the euro-area bond market. *Review of Financial Studies, 22*(3), 925–957.

Boston Consulting Group. (2023). *Value creators report: Oil and gas in volatile times*. Retrieved from https://www.bcg.com/publications/2023/report-on-oil-and-gas-tsr-in-volatile-times

BP. (2021). *BP statistical review of world energy 2021: Middle East insights*. BP Global. Retrieved from https://www.bp.com/content/dam/bp/business-sites/en/global/corporate/pdfs/energy-economics/statistical-review/bp-stats-review-2021-middle-east-insights.pdf

Chordia, T., Roll, R., & Subrahmanyam, A. (2000). Commonality in liquidity. *Journal of Financial Economics, 56*(1), 3–28.

Chordia, T., Subrahmanyam, A., & Anshuman, V. R. (2001). Trading activity and expected stock returns. *Journal of Financial Economics, 59*(1), 3–32.

Chordia, T., Sarkar, A., & Subrahmanyam, A. (2005). An empirical analysis of stock and bond market liquidity. *Review of Financial Studies, 18*(1), 85–129. https://doi.org/10.1093/rfs/hhi010

Chung, C., & Kim, J. (2024). Greenflation, a myth or fact? Empirical evidence from 26 OECD countries. *Energy Economics, 139*, 107906. https://doi.org/10.1016/j.eneco.2024.107906

Cooper, S., Groth, K., & Avera, W. (1985). Liquidity, exchange listing and common stock performance. *Journal of Economics and Business, 37*, 19–33.

Custódio, C., Ferreira, M. A., & Laureano, L. (2011). *The declining corporate debt maturity: The impact of asymmetric information and new listings*. Working paper.

Derwall, J., Koedijk, K., & Ter Horst, J. (2011). A tale of values-driven and profit-seeking social investors. *Journal of Banking and Finance, 35*(8), 2137–2147. https://doi.org/10.1016/j.jbankfin.2011.01.009

Díaz, A., & Escribano, A. (2017). Liquidity measures throughout the lifetime of the US Treasury bond. *Journal of Financial Markets, 33*, 42–74.

Dick-Nielsen, J., Feldhütter, P., & Lando, D. (2012). Corporate bond liquidity before and after the onset of the subprime crisis. *Journal of Financial Economics, 103*(3), 471–492.

Diluiso, F., Annicchiarico, B., Kalkuhl, M., & Minx, J. C. (2021). Climate actions and macro-financial stability: The role of central banks. *Journal of Environmental Economics and Management, 110*(October), 102548. https://doi.org/10.1016/j.jeem.2021.102548

Downing, C., Underwood, S., & Xing, Y. (2005). *Is liquidity risk priced in the corporate bond market? Working paper*. Rice University.

Ersek, V. (2019). *How climate change caused the world's first ever empire to collapse*. The Conversation.

European Commission. (2022). *REPowerEU: A plan to rapidly reduce dependence on Russian fossil fuels and fast forward the green transition*. Retrieved from https://ec.europa.eu/commission/presscorner/detail/en/IP_22_3131

Feldman, S. J., Soyka, P. A., & Ameer, P. G. (1997). Does improving a firm's environmental management system and environmental performance result in a higher stock price? *The Journal of Investing, 6*(4), 87–97. https://doi.org/10.3905/joi.1997.87

Ferrari Minesso, M., & Pagliari, M. S. (2023). No country is an island. International cooperation and climate change. *Journal of International Economics, 145*(June). https://doi.org/10.1016/j.jinteco.2023.103816

Foucault, T., Pagano, M., & Röell, A. (2013). *Market liquidity: Theory, evidence, and policy*. Oxford University Press.

Garman, M. B., & Klass, M. J. (1980). On the estimation of security price volatilities from historical data. *Journal of Business, 53*, 67–78.

Glosten, L. R., & Milgrom, P. R. (1985). Bid, ask and transaction prices in a specialist market with heterogeneously informed traders. *Journal of Financial Economics, 14*(1), 71–100.

Goldstein, M. A., Hotchkiss, E. S., & Sirri, E. R. (2007). Transparency and liquidity: A controlled experiment on corporate bonds. *Review of Financial Studies, 20*(2), 235–273.

Gravelle, T. (1999). *Liquidity of the Government of Canada securities market: Stylized facts and some market microstructure comparisons to the United States Treasury market* (pp. 11–99). Bank of Canada Working Paper.

Han, S., & Zhou, H. (2016). Effects of bond liquidity on the non-default component of corporate yield spreads: Evidence from intraday transactions data. *Quarterly Journal of Finance, 6*(3), 1650012.

Hasbrouck, J. (2009). Trading costs and returns for US equities: Estimating effective costs from daily data. *Journal of Finance, 64*, 1445–1477.

Hendershott, T., & Seasholes, M. S. (2014). Liquidity provision and stock return predictability. *Journal of Banking and Finance, 45*(1), 140–151. https://doi.org/10.1016/j.jbankfin.2013.12.021

Heyen, D. A., Menzemer, L., & Wolff, F. (2020). Just transition in the context of EU environmental policy and the European green Deal environmental policy in the 21st century view project. March, 49. www.oeko.de

International Energy Agency. (2024). *Global energy review 2024*. Retrieved from https://www.iea.org/reports/global-energy-review-2024

Kong, X., & Narayanan, A. (2023). *Do controversies matter? A study of how stock markets react to ESG-related issues*. Asian Infrastructure Investment Bank.

Kyle, A. S. (1985). Continuous auctions and insider trading. *Econometrica, 53*(6), 1315–1335.

Lee, C. M. C. (1993). Market integration and price execution for NYSE-listed securities. *Journal of Finance, 48*, 1009–1038.

Lesmond, D. A. (2005). Liquidity of emerging markets. *Journal of Financial Economics, 77*(2), 411–452. https://doi.org/10.1016/j.jfineco.2004.01.005

Lesmond, D. A., Ogden, J. P., & Trzcinka, C. A. (1999). A new estimate of transaction costs. *Review of Financial Studies, 12*(5), 1113–1141.

Mahanti, S., Nashikkar, A., Subrahmanyam, M., Chacko, G., & Mallick, G. (2008). Latent liquidity: A new measure of liquidity, with an application to corporate bonds. *Journal of Financial Economics, 88*, 272–298.

Merton, R. C. (1987). A simple model of capital market equilibrium with incomplete information. *The Journal of Finance, 42*(3), 483–510. https://doi.org/10.1111/j.1540-6261.1987.tb04565.x

Narayan, P. K., Phan, D. H. B., & Liu, G. (2021). COVID-19 lockdowns, stimulus packages, travel bans, and stock returns. *Finance Research Letters, 38*(June 2020), 101732. https://doi.org/10.1016/j.frl.2020.101732

Natural Resource Governance Institute. (2021). *Risky bet: National oil companies in the energy transition*. Retrieved from https://resourcegovernance.org/sites/default/files/documents/risky-bet-national-oil-companies-in-the-energy-transition.pdf

O'Hara, M. (2004). Liquidity and financial market stability. In *The National Bank of Belgium (issue 55)*. https://doi.org/10.2139/ssrn.1691574

OECD. (2021). *Middle East and North Africa investment policy perspectives*. Organisation for Economic Co-operation and Development. Retrieved from https://www.oecd.org/content/dam/oecd/en/publications/reports/2021/03/middle-east-and-north-africa-investment-policy-perspectives_eed61198/6d84ee94-en.pdf

Olovsson, C., & Vestin, D. (2023). *Greenflation? 2023 (May)* (1–43). www.riksbank.se/en/research

Sadka, R. (2006). Momentum and post-earnings-announcement drift anomalies: The role of liquidity risk. *Journal of Financial Economics, 80*(2), 309–349.

Schestag, R., Schuster, P., & Uhrig-Homburg, M. (2016). Measuring liquidity in bond markets. *Review of Financial Studies, 29*(5), 1170–1219.

Sharma, R. (2021). 'Greenflation' Threatens to Derail Climate Change Action. Financial Times, August 2, 2021.

SRMG Think. (2023). *Mind the gap: Highlighting MENA's climate finance challenge*. Retrieved from https://www.srmgthink.com/highlighting-menas-climate-finance-challenge

Stoll, H. R. (1989). Inferring the components of the bid-ask spread: Theory and empirical tests. *Journal of Finance, 44*(1), 115–134.

U.S. Department of the Treasury. (2022). *Treasury announces guidance on inflation reduction act's strong labor protections*. Retrieved from https://home.treasury.gov/news/press-releases/jy1128

Wyss, R. Von. (2004). Measuring and predicting liquidity in the stock market.

Chapter 7
Analysis of Sustainability Practices Integration Among Publicly Listed Companies: Country-Specific Case Study

Ayodeji Michael Obadire and Tinotenda Douglas Hwara

Abstract The adoption of sustainability practices in emerging economies like Botswana remains under-researched. This study aims to fill this gap by examining the sustainability practices of selected listed companies in Botswana during 2022 and 2023. The research provides empirical insights into how industries in Botswana engage with sustainability frameworks and their impact on corporate performance, contributing valuable data to the existing literature on sustainability in the MENA region. Using a quantitative content analysis method, the study analyzed sustainability disclosures across environmental, social, and governance (ESG) pillars. The narrative approach enabled a focused examination of reporting trends within a short time frame, capturing changes across various selected sectors. Key findings reveal that social sustainability practices were the most reported, accounting for 52% of disclosures in 2023 and 51% in 2022. Governance practices followed at 30% in 2023 and 33% in 2022, while environmental disclosures were the least represented at 18% in 2023 and 16% in 2022. Sector analysis identified the mining industry as the leading contributor to sustainability disclosures, with agriculture also showing notable engagement. These findings offer practical recommendations and implications for firms looking to enhance their sustainability strategies, particularly in improving environmental and governance practices and policymakers aiming to develop effective regulations. Firms can use the research findings to strategically strengthen their sustainability initiatives especially by enhancing environmental and governance practices while regulators and policymakers can translate these insights into targeted policies that promote deeper, sector-wide integration of sustainability, ensuring both business competitiveness and alignment with broader development goals.

Keywords Environmental, social, and governance · Integrated report · Content analysis · Stakeholder engagement · Sustainable business practices

A. M. Obadire (✉) · T. D. Hwara
School of Finance and Professional Studies, Botswana School of Business Sciences (BAC), Gaborone, Botswana
e-mail: ayodejio@bac.ac.bw; tinotendah@bac.ac.bw

M. Elnahass et al. (eds.), *A Greenprint for Net Zero*, Gulf Studies 15,
https://doi.org/10.1007/978-981-92-1762-5_7

7.1 Introduction

The escalating global crises and the depletion of both natural and artificial resources have heightened the urgency for sustainability discourse in business practices (Piketty & Saez, 2014). In nations like Botswana, Libya, Morocco, and Tunisia, integrating sustainable business practices into operations is crucial for achieving long-term economic stability and growth (Radikoko et al., 2019). These countries, characterized by diverse economies and unique sociopolitical landscapes, increasingly recognize the importance of sustainability in enhancing firm performance and contributing to national development goals.

Singh and Rahman (2021) define sustainable business practices as specific actions, strategies, or initiatives that integrate sustainability into day-to-day operations, minimizing negative environmental and social impacts. These include improving quality of life, adopting energy-efficient technologies, reducing waste, and implementing fair labour practices. These practices go beyond regulatory compliance, rather fostering active engagement with stakeholders, including employees, communities, suppliers, customers, and investors.

Despite the recognized benefits, implementing sustainability practices in businesses in these nations faces challenges rooted in political and environmental contexts. For instance, political instability in Libya creates an unpredictable business environment, hindering long-term planning and investment (Mohamed et al., 2015). In Morocco and Tunisia, firms grapple with regulatory uncertainties and the slow pace of policy implementation. Moreover, the economic dependence on traditional industries, such as oil in Libya and agriculture in Morocco and Tunisia, makes transitioning to sustainable practices more challenging (Abdulaziz-al-Humaidan et al., 2022; Ibourk & Amaghouss, 2016). In Botswana, cultural factors limit sustainability awareness, leading to low prioritization in business strategies (Mbekomize & Wally-Dima, 2013).

Tilt et al. (2021) argue that sustainable business practices foster better organizational performance, enhanced risk management, and stakeholder trust. Transparent sustainability reporting is increasingly vital, helping firms build strong relationships with stakeholders in stable and volatile markets (Wells et al., 2020). As these nations align with global sustainable development goals, capital allocation towards sustainable companies will be crucial for driving economic growth and ensuring long-term prosperity. However, achieving this requires comprehensive approaches considering regulatory frameworks, stakeholder engagement, and industry collaboration (Geissdoerfer et al., 2016).

Prior studies, such as those by Garetti and Taisch (2012), Upward and Jones (2016), Abdelkafi and Täuscher (2016), Evans et al. (2017), and Geissdoerfer et al. (2016), have primarily focused on sustainable business models in developed nations, with little emphasis on sustainable business practices. Sustainable business models encompass the broader structure and approach of a business, while sustainable business practices involve specific actions within the existing business model to achieve targeted sustainability outcomes (Tilt et al., 2021). The integration of sustainable business practices and its level of adoption in Africa, particularly in countries such as Libya, Botswana, or Morocco requires further investigation.

This study uses firms listed on the Botswana Stock Exchange (BSE) as a case study to examine the extent to which sustainable business practices have been adopted by these listed firms. Botswana and countries in North Africa, particularly Libya, share similar socio-economic characteristics, including relatively small populations, comparable gross domestic product (GDP) and gross national product (GNP), and classification as upper middle-income countries in Africa (International Monetary Fund, 2023). These similarities make Botswana a logical and noteworthy choice for comparison. More so, the selection of Botswana as the case study is based on the availability and accessibility of reliable data, and other countries' fundamental differences, such as, a stable political environment and active national efforts towards sustainability. Botswana offers a unique regulatory landscape and practical accessibility, enabling deeper engagement with local stakeholders and institutions. Using Botswana as a context provides the necessary depth and feasibility to generate meaningful insights that can inform sustainable business practices in similar developing MENA economies such as Libya and Morocco.

The paper is organized as follows: Sect. 7.2 presents the review of relevant literature, Sect. 7.3 details the methodology, Sect. 7.4 provides the analysis and discussion of findings, and Sect. 7.5 concludes with recommendations and implications for the MENA region.

7.2 Literature Review

7.2.1 Conceptual Reviews

In understanding sustainability practices in business, the United Nations' Brundtland Commission defined sustainability as "meeting the needs of the present without compromising the ability of future generations to meet their own needs" (United Nations, 2023). Additionally, the International Sustainability Standards Board (ISSB) defines sustainability as "the ability for a company to sustainably maintain resources and relationships with and manage its dependencies and impacts within its whole business ecosystem over the short, medium and long term" (IFRS Foundation, 2022). The two definitions are consistent as they both identify a focus on the "future generations," on the one hand, and a business's "long-term dependencies" on the other. Both definitions point out the need for "resources" now and in the future, with the UN definition referring to these as "needs".

The ISSB definition tends to centre around what sustainability means in the business context, which further underscores the importance of sustainability in a business context by identifying it as a condition for a company to access over time the resources and relationships needed (such as financial, human, and natural), ensuring their proper preservation, development, and regeneration to achieve its goals (IFRS

Foundation, 2022). This furthers the view that sustainability is not an optional practice born out of the benevolence of organizations to the public but rather a necessity for an organization's long-term survival (Business and Sustainable Development Commission, 2017; Obadire & Moyo, 2024).

Based on the aforementioned, this study adopted the definition of sustainability practices as business practices that give an organization the ability to access the financial, manufactured, intellectual, human, natural, and social resources and relationships to meet their needs without compromising the ability of future generations to meet their needs (The International Integrated Reporting Council, 2021). Building upon the six capitals identified as necessary for value creation by the integrated reporting framework whose concepts influenced the ISSB definition of sustainability (Hwara et al., 2024; IFRS Foundation, 2022).

The study employed this definition to identify sustainability practices within the listed companies on the Botswana Stock Exchange by examining environmental impact mitigation measures (tied to Natural Capital disclosures), ethical business conduct (associated with human capital disclosures and corporate governance), and social responsibility initiatives (connected to social capital disclosures) through an archival study of integrated reports (The International Integrated Reporting Council, 2021). Thus, these factors were taken into consideration in assessing the presence of sustainability practices among BSE-listed companies.

7.2.2 Empirical Reviews

7.2.2.1 Sustainability Disclosures and Firm Characteristics

Initial empirical investigations into sustainability practices in Botswana were conducted by Mbekomize and Wally-Dima (2013), who examined the extent of social and environmental disclosures among parastatals and listed companies on the Botswana Stock Exchange. Employing regression analysis, they explored how firm-specific factors such as size, industry type, and ownership structure influenced disclosure levels. Their findings revealed no significant relationship between these variables and the extent of sustainability reporting. At the time, the absence of formalized sustainability guidelines in Botswana contributed to this outcome, as companies lacked clear regulatory direction on disclosure expectations and standards.

Since the study by Mbekomize and Wally-Dima (2013), the sustainability reporting landscape in Botswana has evolved considerably. The introduction of formal guidelines by the BSE in 2016 marked a turning point, providing a structured framework for corporate sustainability disclosures. This regulatory development has been complemented by the emergence of advanced reporting tools and frameworks, as highlighted by Hwara (2023) and Obadire et al. (2025), which collectively aim to enhance the quality and consistency of sustainability reporting. These changes have not only improved corporate transparency but have also created opportunities for researchers and practitioners to re-examine the dynamics between sustainability disclosures and firm performance with greater precision.

The evolution of reporting frameworks now enables a more sophisticated assessment of the role of sustainability practices in driving corporate outcomes. With clearer guidelines and improved disclosure mechanisms, companies are better positioned to align their sustainability initiatives with strategic objectives, potentially influencing operational efficiency and financial performance.

7.2.2.2 Empirical Perspectives on ESG Impact: Lessons from Global, MENA, and Emerging Markets

The empirical relationship between ESG practices and financial performance has been the subject of extensive global and regional investigation, offering a nuanced understanding of how sustainability initiatives shape corporate outcomes. A recent study by Buła et al. (2024) provides robust evidence from Poland, examining firms listed between 2013 and 2022. They found that ESG integration significantly enhances both financial performance and operational efficiency, reinforcing the strategic value of sustainability. This positive correlation echoes the findings of Whelan et al. (2021), who reported improvements in key operational metrics such as Return on Equity (ROE) and Return on Assets (ROA) among global corporations integrating ESG factors.

Similarly, Sinha Ray and Goel (2023), focusing on Indian firms from 2011 to 2019, observed that high ESG scores—particularly in environmental dimensions—positively impacted market valuations and financial performance, although these benefits often emerged over time. This pattern of time-lagged benefits underscores the long-term strategic importance of ESG investments, especially in markets where regulatory frameworks are evolving and investor expectations around sustainability are intensifying.

The Middle East and North Africa (MENA) region provides further insightful case studies reinforcing this global trend, particularly in Morocco and Tunisia. Ibourk and Amaghouss (2016) highlighted that Moroccan firms, following the introduction of formal sustainability reporting guidelines, have increasingly integrated environmental and social considerations into their core operations to enhance efficiency and meet regulatory expectations. Ahmad and Mokhchy (2023) deepened this understanding by demonstrating that Moroccan companies engaging actively in energy efficiency and waste management experienced notable improvements in financial performance and competitive positioning.

These findings align closely with earlier research by Khlif et al. (2015), which established positive relationships between sustainability initiatives and financial metrics such as ROA and ROE. Similarly, Mohamed et al. (2015) documented that Libyan companies adopting energy-efficient practices realized cost savings and operational improvements, providing further regional affirmation of sustainability's value in enhancing business outcomes. In Tunisia, Abdulaziz-al-Humaidan et al. (2022) found that responsible sourcing and environmental conservation contributed to operational efficiencies and enhanced market reputations, even amidst challenges like regulatory hurdles and organizational resistance to change. Complementing this, Chtourou and Triki (2017) confirmed that Tunisian firms with robust sustainability frameworks not only experienced stronger financial performance but also benefitted from improved stakeholder trust and corporate credibility.

On a broader global scale, comprehensive research by Chen et al. (2023) substantiated these positive associations through an analysis of over 3300 listed companies worldwide from 2011 to 2020. Their findings underscored that ESG performance significantly boosts corporate financial outcomes, particularly for larger firms operating in high-risk environments where reputational and operational risks are heightened. Ahmad et al. (2021), focusing on the UK's FTSE350 firms between 2002 and 2018, echoed these conclusions, highlighting that ESG adoption positively impacts financial performance, though the magnitude of this impact varies across the environmental, social, and governance pillars. Notably, their analysis found that larger corporations tend to experience a more pronounced benefit, likely due to their greater capacity for resource allocation and their heightened visibility among stakeholders and investors prioritizing sustainable practices.

However, while many studies affirm the financial advantages of ESG adoption, some research introduces a more complex perspective, particularly in emerging markets and within specific ESG dimensions. Strekalina et al. (2023), examining 257 companies across BRICS nations between 2017 and 2021, presented a more mixed picture. Although their study did not establish a significant relationship between overall ESG scores and ROA, it did highlight a negative association between governance-related disclosures and ROA, interpreted through the lens of legitimacy theory. This suggests that in certain contexts, firms may engage in governance disclosures more as a legitimacy-seeking mechanism rather than as a driver of financial performance, especially when facing institutional pressures without immediate economic rewards.

Rao et al. (2023), analysing India's Nifty 50 companies from 2015 to 2022, further complicated the narrative. Their fixed-effects panel quantile regression analysis revealed that ESG's impact on profitability varied across ROE distributions, with environmental and governance factors generally exerting negative influences, while the social component showed largely insignificant effects. These findings illustrate the contextual dependency of ESG's financial impacts, influenced by sectoral characteristics, firm-specific dynamics, and broader market conditions.

Turkey offers another example of this dynamic relationship. Saygili et al. (2022) studied Turkish listed companies for over a decade (2007–2017) and found that while social and governance practices—such as stakeholder engagement and sound board governance—were associated with improved financial performance, environmental disclosures were linked to negative financial outcomes. This disparity suggests that while firms benefit from transparent social practices and robust governance frameworks, environmental initiatives may incur significant short-term costs that outweigh immediate financial returns, particularly in industries with high compliance burdens or where environmental innovations are still in early stages of adoption. Such complexities highlight the importance of contextual and temporal considerations in evaluating the ESG–financial performance nexus.

In sum, the evolving body of empirical evidence paints a largely positive but complex picture of the relationship between ESG practices and corporate financial performance. Across diverse geographic and market contexts—from Poland and the UK to Morocco, Tunisia, and India—many firms adopting comprehensive sustainability strategies have realized operational efficiencies, enhanced reputations, and improved financial outcomes. Yet, this positive relationship is not universally guaranteed.

Factors such as firm size, industry type, market maturity, and regulatory environments significantly influence the extent and immediacy of ESG benefits. Importantly, while larger firms and those in more mature regulatory contexts tend to derive clearer financial gains, firms in emerging markets may encounter mixed results, particularly where short-term costs associated with environmental compliance are high or where ESG initiatives are pursued more for legitimacy than for strategic transformation.

Despite the global insights from the empirical discussions, there remains a gap in focused, up-to-date research within Botswana's evolving context. Given the country's advancing sustainability agenda post-2016 and increasing corporate attention to ESG, a detailed examination of Botswana's firms could enrich the literature by offering findings from a developing and resource-dependent economy that is transitioning towards sustainable development.

7.2.3 Theoretical Motivation

This study is grounded on four relevant theories that explain why businesses have shifted towards sustainability practices. The underpinning theories are the stakeholder theory, legitimacy theory, agency theory, and the resource-based view (Kuzey & Uyar, 2017; Haji et al., 2022; Barney, 1991).

Firstly, stakeholder theory posits that firms must consider the interests of all stakeholders—including shareholders, employees, customers, suppliers, communities, and the environment—rather than focusing solely on shareholder value (Freeman, 1999). This broader accountability encourages firms to adopt sustainability practices as a response to stakeholder pressures and expectations. Empirical evidence strongly supports the view that engaging stakeholders through sustainability initiatives enhances a firm's corporate reputation and performance (Hu et al., 2023; Talan et al., 2024; Wang, 2024). Companies that actively address environmental and social concerns are perceived as responsible and trustworthy, which strengthens their public image and differentiates them from competitors (Wang, 2024). A positive reputation not only fosters goodwill among communities and regulators but also serves as a buffer during crises, helping firms maintain their market position even in challenging times.

Moreover, sustainability initiatives contribute significantly to building customer loyalty. Consumers are increasingly favouring brands that demonstrate genuine commitment to sustainable practices (Talan et al., 2024). By aligning business operations with societal values, companies create emotional connections with customers, encouraging repeat purchases and positive word-of-mouth referrals. This loyalty translates into stable revenue streams, giving firms a competitive advantage in both mature and emerging markets (Hu et al., 2023). In addition to reputational benefits and customer loyalty, sustainable practices attract investors and enhance financial performance through improved operational efficiencies. Clark et al. (2015) argue that firms with robust environmental, social, and governance (ESG) performance often experience lower capital costs. Investors view such firms as lower risk due to their proactive risk management and long-term value orientation. Furthermore, operational improvements—such as energy efficiency and waste reduction—not only lower costs but also increase profitability, reinforcing the business case for sustainability. As such, the Stakeholder Theory underpins the strategic importance of responding to stakeholder expectations, suggesting that sustainability practices can enhance long-term financial performance through risk mitigation and value creation (Freeman, 1999).

Also, legitimacy theory posits that organizations actively engage in activities to align their behaviours and practices with societal norms, values, and expectations to gain and maintain legitimacy in the eyes of stakeholders. According to O'Donovan (2002), the legitimacy theory suggests that businesses must adhere to societal norms and values perceived as socially acceptable to achieve success. This theory explains the increasing trend in environmental disclosure, as organizations strive to align their actions and reporting with societal expectations (Archel et al., 2009; O'Donovan, 2002). Kuzey and Uyar (2017) support this perspective, highlighting the role of sustainability reporting in legitimizing companies and demonstrating compliance with societal norms. Additionally, Archel et al. (2009) emphasize the influence of the state in shaping legitimacy, as companies seek to align themselves with government expectations to gain societal acceptance. In the context of sustainability practices amongst listed firms, legitimacy theory suggests that firms adopt sustainability initiatives not only as a response to genuine concern for social and environmental issues but also to enhance their legitimacy and reputation (Archel et al., 2009; Kuzey & Uyar, 2017). More so, firms may perceive sustainability practices as a means to demonstrate their commitment to social responsibility and environmental stewardship, thereby enhancing their legitimacy in the eyes of stakeholders such as customers, investors, regulators, and the broader society. By aligning their activities with societal expectations regarding sustainability, firms aim to secure continued support, trust, and acceptance from these stakeholders, which can contribute to long-term organizational success and competitive advantage.

Based on this, an increase in the implementation of sustainability practices should increase revenue spurred by the legitimizing effects of demonstrating adherence to societal expectations. Hence, from a firm performance perspective, legitimacy theory postulates that adopting sustainability practices can positively influence financial and non-financial performance metrics. By addressing ESG concerns, firms may enhance operational efficiency, reduce risks, attract investment capital, improve brand reputation, and foster innovation. Moreover, by meeting or exceeding stakeholder expectations regarding sustainability, firms can mitigate the risk of legitimacy threats such as negative public perception, stakeholder activism, regulatory scrutiny, and legal sanctions, which could adversely affect firm performance.

Moreover, the agency theory of Ross (1973) proposes that an agency relationship emerges when one party represents another, known as the principal or stakeholder, while the other party acts on their behalf. This representation typically occurs in companies where management, including the board of directors, represents shareholders (Obadire et al., 2023). Jensen and Meckling (1976) extended this theory, asserting that optimal outcomes for principals are achieved when they incur agency costs. Agents, often not aligned with shareholders' interests, can be incentivized to better serve shareholders through these costs. By establishing agency costs, the principal and agent align their desired outcomes, fostering optimal results.

Noreen (1988) aligns with this perspective, connecting agency theory to the necessity of internal mechanisms to deter opportunistic behaviour and address the mutual distrust between managers (agents) and shareholders (principals). The agency theory is relevant in the context of sustainability as it supports the disclosure of information about the implementation of sustainability practices as a tool to reduce agency costs. Kuzey and Uyar (2017) highlighted sustainability reporting as a tool to mitigate information asymmetry and enhance transparency, thereby reducing agency costs. Moreover, agency theory posits that businesses benefit from disclosing information about their implementation of sustainability practice, a facet often covered in sustainability reports.

Lastly, the Resource-Based View (RBV) emphasizes the importance of internal capabilities as the foundation for achieving and sustaining competitive advantage. According to Barney (1991), firms that possess resources that are valuable, rare, inimitable, and non-substitutable (VRIN) are better positioned to outperform competitors. In this context, sustainability can be a source of such resources, particularly when integrated into core competencies like innovation, efficient supply chains, and human capital development. Supporting this perspective, Moussa et al. (2024) argue that robust environmental, social, and governance (ESG) practices enhance financial performance through more effective resource allocation and a heightened ability to respond to regulatory and market shifts. Firms with well-integrated ESG strategies are often more agile in navigating environmental regulations and societal expectations, giving them a dynamic edge. These companies are better positioned to anticipate and mitigate risks, seize emerging opportunities, and allocate capital towards initiatives that drive both sustainable impact and profitability.

Furthermore, companies that proactively invest in sustainable technologies and practices cultivate capabilities that are difficult for competitors to imitate. Such investments lead to operational efficiencies, cost reductions, and access to premium markets that value sustainable offerings (Bhandari et al., 2022; Kumar et al., 2024). Additionally, sustainability-oriented innovation fosters the development of new products and services that cater to evolving consumer preferences, opening up new revenue streams and further reinforcing financial performance (Sharma et al., 2019).

These innovations not only bolster profitability but also enhance a firm's long-term resilience in an increasingly sustainability-focused global economy.

However, RBV also cautions that sustainability initiatives must be strategically aligned with a firm's unique resources and capabilities to yield tangible benefits. Sheffi (2018) warns that without such alignment, sustainability efforts risk becoming superficial gestures or burdensome expenses that dilute competitive focus. Simply adopting sustainability practices without integrating them into the firm's core resource base may lead to inefficiencies and missed opportunities. Therefore, the RBV perspective underscores that sustainability, when embedded within a firm's distinctive capabilities, serves not only as a catalyst for improved financial performance but also as a foundation for enduring competitive advantage.

Together, these four theories offer a holistic understanding of how businesses adopt sustainability practices to balance stakeholder interests, maintain legitimacy, align managerial actions with shareholder goals, and leverage internal resources for enhanced financial performance.

7.3 Methodology

This section outlines the data utilized in the study and describes the quantitative content analysis approach employed.

7.3.1 Data and Variable Definitions

The target population comprised all companies listed on Botswana's domestic main board. Presently, there are 24 companies listed on this board (Appendix A). The Botswana Stock Exchange is a participant in the Sustainable Stock Exchange (SSE) Initiative, which underscores its commitment to improving performance on ESG issues, as well as promoting sustainable investment, compliance, and reporting.

Given the specific nature of the population in this study, a non-probability sampling technique was employed to select a sample of listed companies. Purposive non-probability sampling was deemed most appropriate because the researchers focused specifically on firms listed on the domestic main board of the BSE that publicly disclose information about their sustainable business practices in their integrated reports. According to Patton (2007), purposive sampling is widely used in research to identify and select information-rich cases related to the phenomenon of interest.

The total sample size consisted of six listed firms that prepare their annual reports in accordance with integrated reporting guidelines. Each firm represents a different sector across the domestic board, excluding banks and non-bank financial services firms. The firm chosen from each sector was selected based on its highest market capitalization and its demonstrated commitment to impactful sustainability practices and reporting. Firms that did not have 2023 annual integrated reports were excluded from the analysis, resulting in the exclusion of two industries from the sample, as the firms listed in those sectors did not have published integrated reports.

Banks and non-bank financial service firms were excluded from the sample due to their subjection to stringent regulatory requirements, particularly by bodies such as the Basel Committee on Banking Supervision. This introduces a level of homogeneity that may limit variability in financial fundamentals (Obadire et al., 2023). Furthermore, these sectors operate under unique fundamental structures, involving complex financial instruments, and the differences in risk and performance metrics between financial and non-financial sectors justify their exclusion, ensuring a more focused examination of listed non-financial services firms.

The integrated reports of the sampled firms were collected over a 2-year period, from 2022 to 2023, to capture recent trends in sustainability practices. Data from 2022 was included because prior to that year, most listed firms had not yet adopted integrated reporting guidelines, resulting in a lack of data on sustainability practices for analysis.

The study used the standardized audited integrated report of the listed companies which were obtained from the companies' and BSE websites. This process ensures that the data used for the analysis relies on current and accurate information on key financial contexts, contributing to the robustness and reliability of the research findings. To fully understand the sample selection, see Appendix A.

To achieve the study's aim of examining the prominent adoption of sustainability practices among listed companies in Botswana, the study analyzed the frequency of various sustainability-related terms within the companies' integrated reports by adopting the narrative data analysis approach. This approach was underpinned by using a quantitative content analysis technique, which involved systematically reviewing the integrated annual reports and company websites to identify and quantify the occurrences of these terms.

Content analysis is a widely used method in studies that seek to examine definitions and concepts such as sustainability models, social participation, green operations, corporate social responsibility, and green finance (Ahi & Searcy, 2013; Kozlowski et al., 2015; Moldavska & Welo, 2017; Nunes & Bennett, 2010). This approach allows researchers to interpret text data content through a systematic classification process, identifying patterns and themes within the data.

While content analysis is often regarded as a qualitative research method, it is equally valuable in quantitative research. In the context of this study, the content analysis was employed in its quantitative form to provide measurable results, specifically focusing on the frequency with which sustainability practices were mentioned in the integrated reports. The quantitative approach enabled the researchers to systematically quantify the occurrences of sustainability practices, offering a clear understanding of how often these practices are adopted by the listed firms in Botswana. By analyzing these frequencies, the study provided deeper insights into the sustainability practices that are mostly implemented by these companies, thereby offering a comprehensive assessment of the integration of sustainability practices in the corporate landscape of Botswana.

Importantly, the key terms for analysis were derived and chosen from the recurring SBP themes outlined in the BSE and UNDP (2018) report (Table 7.1). These terms served as a structured framework for identifying relevant sustainability business practices across firms and sectors. The study then applied a coding process that involved searching for these terms within each firm's integrated report, documenting their occurrence, and aggregating the results (Table 7.2).

Table 7.1 Relevant recurring sustainability practices under the ESG framework

Environmental	Social	Governance and ethics
Words/terms	Words/terms	Words/terms
1. Investment in local supplier 2. Energy-saving lights/bulbs 3. Rainwater harvesting 4. Measuring carbon footprints 5. Monitoring carbon footprints 6. Solar power usage 7. Green building funding 8. Waste separation and management 9. Water recycling 10. Water saving mechanism 11. Water-saving toilet flushing system 12. Grey water for garden use 13. Power saving cards 14. Renewable materials 15. Save water campaign 16. Promote showers instead of bathtubs 17. On-site water purifier 18. Separator pits for prevention of water pollution 19. Safe handling of used oil 20. Virtual meetings 21. Sharing telecommunication masts 22. E-invoicing 23. Fleet management to reduce carbon footprint 24. Electronic control system for refrigeration 25. Recyclable packaging 26. Encouraging suppliers to embrace responsible sustainability practices	1. Partnership with local communities 2. Educational support to vulnerable groups 3. Cancer awareness 5. Sport sponsorships and development 6. Outdoor public gym 7. Charity donations 8. Student work placement 9. Community education and training 10. Employee training 11. Tourism and excursions for students 12. Training and attachment 13. Provision of school supplies for student 14. Scholarships 15. Adopt a school initiative 16. Reward employees for good performance on sustainability issues 17. Tele-centre in underdeveloped area 18. e-learning platforms for employee training 19. Schools support 20. Provision of foodstuffs for less privileged members of the community 21. Employment policy for people living with disabilities 22. Provision of housing to the needy 23. Partnerships with farming communities to enhance food security	1. Integrated report 2. Measure of company sustainability performance 3. Sustainability champion team 4. Disclosure of sensitive information on products to protect consumers 5. Board-level leadership on sustainability 6. Green building accreditation 7. Participation in national sustainability initiatives 8. Local supply chain engaged to minimize environmental impact 9. Dedicated sustainable office 10. Brand champion 11. Social, ethics and sustainability committee under the board of directors 12. Demand employee accountability on sustainability issues 13. Management and staff held accountable for environmental performance 14. Rewarding/incentivizing excellence on sustainability issues 15. Environmental sustainability considered in selection of suppliers

Source: BSE and UNDP (2018) and Obadire et al. (2025)

Table 7.2 Frequency of occurrence of the sustainability practices adopted across the sampled industry

Sustainability aspect	Reference for sustainable business practices	Checklist/Measures	ICT		Agriculture		Mining		Tourism		Retail & wholesale		Property	
			2023	2022	2023	2022	2023	2022	2023	2022	2023	2022	2023	2022
Panel A														
Environmental impact mitigation measures (Natural Capital)	Investment in sustainable products and services	In-house waste reduction and pollution prevention mechanisms	0.00%	0.00%	0.11%	0.86%	0.75%	1.40%	0.11%	0.11%	0.43%	0.43%	0.00%	0.00%
	Innovation	Technology solutions that address sustainable development challenges in e.g. waste, water, energy	0.11%	5.38%	2.47%	1.29%	0.97%	0.65%	0.86%	0.11%	0.75%	0.22%	0.22%	5.38%
	Transportation	Use of fuel-efficient vehicle fleets. Buildings and facilities Design and construction of environmentally sustainable buildings	0.00%	0.00%	0.00%	0.00%	0.00%	0.00%	0.00%	0.00%	0.00%	0.00%	0.00%	0.00%
	Design	Integration of alternative, recycled, and recyclable materials in production and packaging design	0.00%	0.11%	0.11%	0.43%	0.00%	0.00%	0.00%	0.22%	0.65%	0.00%	0.00%	0.11%
	Certification/ Environmental Management Systems	ISO certification	0.11%	3.98%	3.98%	2.26%	0.97%	0.22%	0.11%	0.86%	0.22%	0.11%	0.11%	3.98%

(continued)

Table 7.2 (continued)

Sustainability aspect	Reference for sustainable business practices	Checklist/Measures	ICT		Agriculture		Mining		Tourism		Retail & wholesale		Property	
			2023	2022	2023	2022	2023	2022	2023	2022	2023	2022	2023	2022
	Supply chain management	Sustainably produced or harvested inputs or raw materials	0.00%	1.40%	1.40%	0.00%	0.00%	0.00%	0.00%	0.22%	0.00%	0.00%	0.00%	1.40%
Social responsibility initiatives	Partnerships with communities	Setting aside time for community projects Fair compensation for supplier communities Partnerships with research institutions, schools, etc.	1.94%	2.04%	2.90%	2.90%	6.02%	5.59%	5.59%	0.00%	3.76%	6.02%	0.75%	0.00%
	Investment in sustainability projects	Grants to environmental groups	0.32%	0.86%	0.00%	0.00%	0.00%	0.00%	0.11%	0.00%	0.00%	0.00%	0.00%	0.00%
	Investor dialogue	Engaging local communities to accelerate investment in sustainable livelihoods, e.g. farming	0.00%	0.00%	0.00%	0.00%	0.00%	0.00%	0.00%	0.00%	0.00%	0.00%	0.00%	0.00%
	Human rights	Combating inequalities and discrimination	0.65%	1.40%	1.51%	1.51%	1.29%	1.08%	0.00%	0.00%	0.86%	0.65%	0.11%	0.11%
	Investment in employee/ community social welfare	Employee education Gender balance Safety and health	4.52%	3.23%	6.45%	6.45%	5.70%	5.81%	0.86%	0.75%	2.37%	5.38%	0.11%	0.11%
	National aspirations	Contribution to nationally defined development targets e.g. poverty eradication poverty	0.43%	0.22%	1.18%	1.18%	2.26%	1.51%	0.54%	0.11%	0.43%	0.86%	0.22%	0.11%

Panel B														
Governance and ethical business conduct	Reporting	Integration of social, financial, and environmental goals in annual reports or separate sustainability reporting	1.18%	0.54%	1.18%	1.18%	1.18%	0.75%	0.22%	0.32%	2.37%	2.37%	0.22%	0.22%
	Support to teams or individuals facilitating sustainability in the company	Environmental champions/ green teams employee engagement employee education/awareness on sustainability issues	0.32%	0.75%	0.97%	0.97%	0.32%	0.32%	0.22%	0.00%	0.54%	0.65%	0.00%	0.00%
	Stakeholder engagement	Presentation of the company's sustainability strategy and goals in annual general meetings	1.08%	1.83%	0.97%	0.97%	4.62%	4.73%	0.11%	0.11%	0.97%	2.15%	1.40%	1.18%
	Supply chain management	Working with suppliers to establish GHG emissions reductions and energy efficiency targets	0.00%	0.00%	0.00%	0.00%	0.00%	0.00%	0.00%	0.00%	0.00%	0.00%	0.00%	0.00%
	Management accountability and executive compensation	Holding management accountable to sustainability and rewarded good sustainability performance	0.00%	0.00%	0.00%	0.00%	0.00%	0.00%	0.00%	0.00%	0.00%	0.00%	0.00%	0.00%
	Policies, plans and programmes (PPPs)	Inclusion of sustainability aspects in company PPPs	0.22%	0.00%	3.12%	3.12%	1.29%	0.86%	0.00%	0.11%	1.83%	0.54%	0.00%	0.22%

(continued)

Table 7.2 (continued)

Sustainability aspect	Reference for sustainable business practices	Checklist/Measures	ICT		Agriculture		Mining		Tourism		Retail & wholesale		Property	
			2023	2022	2023	2022	2023	2022	2023	2022	2023	2022	2023	2022
	Disclosure	Transparency in the use of environment-sensitive inputs	0.11%	0.00%	0.00%	0.00%	0.00%	0.00%	0.00%	0.00%	0.00%	0.00%	0.00%	0.00%
	Adoption of SDGs/ sustainable development	Priority SDGs that the company focuses on	0.00%	0.00%	0.97%	0.97%	2.58%	2.47%	0.00%	0.00%	0.32%	1.51%	0.00%	0.00%
	Incentivizing employees	Rewarding safety and environmental stewardship	0.00%	0.00%	0.00%	0.00%	0.00%	0.00%	0.00%	0.00%	0.00%	0.00%	0.00%	0.00%
	Ownership of sustainability principles	Board-level leadership on sustainability senior management leadership on sustainability	0.75%	0.65%	0.54%	0.65%	0.32%	0.11%	1.08%	0.32%	0.54%	0.32%	0.43%	0.32%

Source: Reference for sustainability practices and measures were adapted from the report of BSE and UNDP (2018), Adam et al. (2020), and Obadire et al. (2025)

Following the coding process, the identified terms were further classified and aggregated under ESG categories. Table 7.2 presents the summarized quantified results, showing the frequency of key SBP terms extracted through the coding and analysis process.

7.4 Analysis and Discussion of Findings

This section presents and discusses the result of the quantitative content analysis.

7.4.1 Content Analysis

The content analysis approach enabled the researchers to identify how many times the sustainability practices related terms occurred in the integrated reports of the sampled firms. The frequency of occurrence of the sustainability practices adopted across the sampled industry is presented in Tables 7.1, 7.2, and 7.3.

The analysis of integrated reports from the sampled firms listed on the BSE, as shown in Tables 7.1, 7.2, and 7.3, highlights the adoption of sustainable business practices across various industries. Table 7.1's content analysis identifies 26 environmental terms, 22 social terms, and 15 governance terms frequently mentioned in these reports. These terms were further examined and categorized into themes in Table 7.2 to assess their frequencies of occurrence.

Table 7.2 presents a comparative analysis of sustainability practices across various sectors, including ICT, agriculture, mining, tourism, retail and wholesale, and property, for the years 2022 and 2023. It examines different sustainability aspects such as environmental impact mitigation, social responsibility initiatives, and governance and ethical business conduct. The narrative data analysis approach yielded the elaborate discussion of findings below.

Table 7.3 Summarized frequency and percentage distribution of sustainability practices occurrences

	2023		2022		
Sustainability aspect	Frequency of occurrence	% of occurrence	Frequency of occurrence	% of occurrence	% change based on frequency of occurrence
Environmental impact mitigation measures (Natural Capital)	166	18.00%	145	16.00%	14.48%
Social responsibility initiatives	484	52.00%	445	51.00%	8.76%
Governance and ethical business conduct	280	30.00%	289	33.00%	−3.11%
Total	930	100%	879	100%	5.80%

Source: Author's Analysis (2024)

7.4.1.1 Environmental Impact Initiatives

The findings reveal a mixed picture of sustainability practices across sectors for environmental initiatives, with the mining and agriculture sectors showing slight improvements, while sectors such as ICT and transportation lag behind. Specifically, investment in sustainable products and services remains generally low, although agriculture saw an increase from 1.29% in 2022 to 2.47% in 2023, and mining reported incremental advancements. Conversely, the ICT sector experienced a dramatic decline in innovation in sustainable technologies from 5.38% in 2022 to 0.11% in 2023, and transportation maintained consistent zero values, indicating a sector-wide neglect of fuel-efficient technologies and sustainable building practices.

From the perspective of legitimacy theory, these findings suggest that sectoral engagement with sustainability initiatives is closely tied to societal expectations and external pressures. Mining, for example, faces intense scrutiny from regulators, local communities, and environmental activists due to its historically adverse environmental impact. To maintain its social licence to operate, the mining sector is compelled to invest in sustainability initiatives, not only as a genuine response to ecological concerns but also to align with societal norms and expectations, thus enhancing its legitimacy (Archel et al., 2009; O'Donovan, 2002). This explains why mining leads other sectors—its visibility and environmental footprint make sustainability reporting a strategic necessity rather than a voluntary choice.

By contrast, sectors such as ICT and transportation appear to perceive lower legitimacy threats or societal pressures regarding environmental impacts, which may explain their relatively low engagement in sustainable innovations. Inadequate sector-specific regulatory frameworks and muted stakeholder demands could be contributing to the ICT sector's drastic decline and transportation's stagnation. As Archel et al. (2009) note, the role of the state is crucial in shaping legitimacy-driven behaviours, suggesting that stronger governmental policies could incentivize broader sectoral engagement in sustainability practices.

The cross-sector variations are further illuminated through agency theory, which underscores the importance of aligning the interests of managers (agents) with those of shareholders and broader stakeholders (principals). In sectors like mining, where environmental risks and reputational stakes are high, enhanced sustainability disclosures serve to reduce information asymmetry and agency costs (Kuzey & Uyar, 2017). Mining firms are incentivized to proactively communicate their sustainability initiatives to assure stakeholders of responsible management, thereby safeguarding long-term shareholder value. Conversely, sectors with lower perceived environmental risks may not face the same pressure to disclose, leading to weaker sustainability commitments.

Applying the Resource-Based View (RBV), mining companies may also have developed internal capabilities that enhance their sustainability performance. Their operational scale and resource intensity necessitate the adoption of advanced environmental management systems and innovative technologies, which are valuable, rare, and difficult to imitate (Barney, 1991). These firms leverage sustainability not merely as a compliance exercise but as a competitive strategy that delivers operational efficiencies, access to environmentally conscious markets, and resilience against tightening regulations (Bhandari et al., 2022; Moussa et al., 2024). On the other hand, the ICT sector's sharp decline suggests a potential depletion of such internal capabilities or a strategic reprioritization of sustainable innovation in favour of short-term cost savings, risking long-term competitiveness.

Comparatively, in other African markets such as South Africa, Ghana, and Nigeria, there is evidence of broader sectoral participation in sustainability initiatives, driven largely by more stringent environmental regulations and active capital market requirements for ESG disclosures. For instance, South Africa's mandatory integrated reporting framework has elevated sustainability standards across sectors, including finance and ICT, leading to more consistent and robust ESG performance (Campbell, 2012; Eccles & Krzus, 2018; Mutisya & Yarime, 2014). This contrast highlights Botswana's regulatory gaps and the need for stronger policy interventions to stimulate sector-wide sustainability engagement.

7.4.1.2 Social Responsibility Initiatives

The findings reveal a heterogeneous commitment to social responsibility initiatives across sectors, with mining and agriculture demonstrating higher levels of engagement compared to sectors like tourism and property. The mining sector, in particular, leads with consistent and increasing community engagement, rising from 5.59% in 2022 to 6.02% in 2023, while agriculture maintains a steady 2.90% across both years. These trends can be effectively explained through the lens of Legitimacy Theory, which posits that organizations seek to align their activities with societal norms and expectations to maintain legitimacy (Archel et al., 2009; O'Donovan, 2002). The mining sector, often under intense public and regulatory scrutiny due to its environmental impact, appears to be strategically investing in social initiatives to secure its "social licence to operate." By demonstrating commitment to community welfare and local development, mining firms reinforce their legitimacy in the eyes of stakeholders, including regulators, communities, and investors. Agriculture, similarly, dependent on rural communities and natural resources, leverages social responsibility to align with societal expectations and sustain operational legitimacy. Conversely, the lagging performance in sectors such as tourism and property suggest a weaker alignment with legitimacy imperatives. The complete absence of recorded community investment in tourism in 2022 and marginal improvements in property highlight missed opportunities to strengthen stakeholder trust and societal acceptance. These sectors may underestimate the reputational risks of neglecting community engagement, especially as global sustainability standards intensify. Without meaningful investment in social responsibility, these sectors risk erosion of legitimacy, reputational damage, and diminished competitive positioning in both domestic and international markets.

From an Agency Theory perspective, these sectoral disparities also reflect differences in how firms manage principal-agent relationships, particularly regarding transparency and accountability. Mining and agriculture sectors, traditionally prone to agency problems due to complex stakeholder environments, appear to mitigate information asymmetry by visibly disclosing and committing to social initiatives. Sustainability reporting serves as a mechanism to reduce agency costs, aligning management actions with shareholder and stakeholder expectations (Jensen & Meckling, 1976; Kuzey & Uyar, 2017). These disclosures help address the inherent distrust between managers (agents) and shareholders or community stakeholders (principals), reinforcing the credibility of their sustainability claims. In contrast, sectors like tourism and property display limited engagement in sustainability reporting and investor dialogue, exacerbating information asymmetry and potential agency conflicts. The absence of investor dialogue across all sectors underscores a broader systemic gap in fostering transparent, trust-based relationships with key stakeholders. Without proactive communication, these sectors fail to signal their commitment to sustainable value creation, which can deter socially responsible investors and erode stakeholder confidence.

The Resource-Based View (RBV) further enriches the understanding of sector-specific dynamics by highlighting how internal capabilities drive sustainability performance. Mining firms, often endowed with significant financial and human resources, can invest strategically in sustainability initiatives that yield valuable, rare, inimitable, and non-substitutable (VRIN) capabilities (Barney, 1991). For example, investments in community development not only fulfil social expectations but also cultivate local partnerships and goodwill, creating barriers to entry for competitors and enhancing operational resilience. Similarly, agriculture's steady engagement in community welfare reflects its reliance on cultivating sustainable practices and community relationships as essential resources for long-term viability.

Sectors lagging in sustainability, such as tourism and property, may face internal resource constraints or lack strategic focus on integrating sustainability into their core competencies. This gap limits their ability to develop differentiated capabilities that align with evolving market demands for responsible business conduct. Comparative analysis with other African markets further illustrates this point. In countries like South Africa and Kenya, tourism sectors have increasingly prioritized community engagement and eco-tourism initiatives to enhance market appeal and secure competitive advantage. Botswana's tourism sector, by contrast, appears to underutilize such strategies, missing out on potential value creation through sustainability-led differentiation (Campbell, 2012; Mutisya & Yarime, 2014).

Moreover, the leadership of the mining sector in sustainability practices aligns with global and regional trends where extractive industries face heightened expectations for environmental stewardship and social accountability. Regulatory pressures, international reporting standards, and heightened activism necessitate robust sustainability frameworks in the mining industry. These external pressures incentivize mining firms to build comprehensive ESG strategies, aligning with RBV principles to transform compliance into a source of competitive advantage (Moussa et al., 2024).

These findings illustrate that mining and agriculture sectors strategically align their sustainability initiatives with societal expectations, effectively leveraging legitimacy, agency relationships, and resource-based advantages. However, the persistent underperformance in sectors such as tourism and property signals critical gaps in both strategic intent and operational capacity to embrace sustainability.

7.4.1.3 Governance and Ethical Business Conduct

The empirical results indicate a mixed and relatively inconsistent commitment to governance and ethical business conduct across sectors, particularly in ICT, agriculture, mining, tourism, retail and wholesale, and property from 2022 to 2023. A notable trend is the relatively stable but modest integration of social, financial, and environmental goals into annual reports, especially within the retail and wholesale sectors. The stagnation in governance practices, particularly the consistent but limited integration of sustainability reporting (e.g. retail and wholesale maintaining 2.37% from 2022 to 2023), suggests that while companies recognize the importance of governance, they may not be actively prioritizing or expanding these efforts at a significant pace.

The Legitimacy Theory provides valuable insight into these trends. According to this theory, organizations engage in activities that align their behaviour with societal expectations to gain legitimacy in the eyes of stakeholders (Archel et al., 2009; O'Donovan, 2002). In sectors like mining, where operations are heavily scrutinized due to environmental impacts, companies appear to be striving to align their practices with societal values, even though the slight decline in stakeholder engagement from 4.73% in 2022 to 4.62% in 2023 signals room for improvement. The stability in governance practices in sectors like retail and wholesale, and the relative stagnation across others, could be indicative of companies engaging in sustainability efforts not necessarily due to genuine concern but more to maintain or enhance their legitimacy within the market. In this respect, businesses may view governance and ethical conduct not only as a response to social and environmental issues but also as a tool for enhancing their reputations and reducing societal risks (Archel et al., 2009; Kuzey & Uyar, 2017).

The Agency Theory further illuminates these findings by explaining the role of managerial accountability in achieving sustainable business outcomes. Mining firms, despite a slight decline in stakeholder engagement, are still more active than other sectors in their governance efforts, which might reflect the ongoing tensions in the principal-agent relationship. According to agency theory, optimal outcomes for shareholders (principals) are achieved when managers (agents) are incentivized to align their goals with those of the stakeholders (Jensen & Meckling, 1976). The stagnation in sectors like property and retail, where governance practices and stakeholder engagement are minimal, may suggest a failure to address the agency problem effectively. This is especially evident in the lack of progress on critical governance dimensions such as supply chain management and management accountability, where businesses may not be disclosing or acting on sustainability issues that matter to shareholders or other stakeholders. This absence of effective agency mechanisms in these sectors increases information asymmetry and reduces transparency, which could hinder their ability to attract responsible investors and satisfy stakeholder expectations.

A comparison with other African markets, such as South Africa and Kenya, shows that mining and retail sectors are leading the way in sustainability reporting, whereas Botswana's retail and tourism sectors, for example, appear to lag behind in these practices. In South Africa, the mining sector has been more proactive in addressing governance issues, driven by global pressures and regulatory expectations (Campbell, 2012). In Kenya, retail businesses have made significant strides in integrating sustainability into their governance frameworks to meet the growing demand for responsible business practices from both local and international markets (Mutisya & Yarime, 2014). Botswana's sectors, however, appear less responsive to these trends, particularly in the case of tourism and retail.

Table 7.3 summarizes the frequency and percentage distribution of sustainability practices reported by the sampled firms. The analysis reveals that social sustainability practices are the primary drivers of performance, accounting for 52% of all sustainability disclosures in 2023 and 51% in 2022, making them the most frequently reported category among BSE-listed firms. This dominance is due to a higher number of disclosures and a more significant year-over-year increase in reporting, as shown in Table 7.3. In contrast, governance-related sustainability practices, focused on ethical business conduct, represented 30% of disclosures in 2023 and 33% in 2022. Environmental sustainability practices were the least reported, comprising 18% in 2023 and 16% in 2022. The significant representation of social sustainability practices suggests that companies strongly emphasize social responsibility initiatives, reflecting broader trends in corporate sustainability priorities.

Overall, the upward trend in sustainability practices disclosures, as shown in the Table 7.3, is in line with growing demands from investors and regulatory bodies for greater transparency (BSE and United Nations Development Program [UNDP], 2018). This increase in disclosure can be understood through the framework of stakeholder and legitimacy theories, which posits that firms increase their sustainability reporting to align their operations with societal values and norms, thereby gaining legitimacy and meeting stakeholders' expectations (Archel et al., 2009; O'Donovan, 2002). Sampled BSE-listed firms are likely enhancing their sustainability practices disclosures to demonstrate accountability and responsibility, aiming to bolster their legitimacy among stakeholders. Moreover, from an agency theory perspective, the shift towards more frequent sustainability practices disclosures, especially from the Mining and Agricultural sectors helps to mitigate agency costs by reducing information asymmetry, allowing firms to better satisfy shareholder demands for transparency and ethical behaviour (Kuzey & Uyar, 2017). This strategic approach not only satisfies immediate stakeholder demands but also positions the firms favourably for future investment and regulatory compliance.

In 2023, the predominant social sustainability practices disclosed by firms involved partnerships with communities and significant investments in community and employee welfare, with the latter showing a more substantial emphasis than in the previous year. In terms of governance, stakeholder engagement and sustainability reporting were the leading practices in both 2023 and 2022. For environmental impact mitigation, innovations and environmental management systems were the primary practices, with environmental management practices being more emphasized than innovations compared to 2022. The mining sector emerged as the top contributor to sustainability practices in their disclosures across all categories—environmental, social, and governance—in 2023, followed by the agriculture sector.

7.5 Conclusion, Recommendations, and Implications for the MENA Region

This study examined the adoption of sustainability practices among selected listed companies in Botswana during 2022 and 2023. It focused on six listed firms from various sectors, excluding banks and non-bank financial institutions, all of which adhered to integrated reporting guidelines. Data was gathered from the firms' annual integrated reports available on their websites and the BSE website. A quantitative content analysis was used to assess the frequency of sustainability practices reported by the sampled firms.

The findings reveal that social sustainability practices were the primary drivers of performance, accounting for 52% of all sustainability disclosures in 2023 and 51% in 2022, making them the most frequently reported category among BSE-listed firms. This prominence is attributed to a higher number of disclosures and a more significant year-over-year increase in reporting. In contrast, governance-related sustainability practices, focused on ethical business conduct, constituted 30% of disclosures in 2023 and 33% in 2022. Environmental sustainability practices were the least reported, comprising 18% in 2023 and 16% in 2022. The strong focus on social sustainability practices highlights companies' emphasis on social responsibility initiatives, reflecting broader trends in corporate sustainability priorities. Additionally, the study found that the mining sector was the leading contributor to sustainability practices in their disclosures across environmental, social, and governance categories in 2023, followed by the agriculture sector.

This research revealed crucial findings into the current state of sustainability practices, adoption, and integration across various sectors, with a particular focus on the African context. By addressing a significant gap in the existing literature, it sheds light on how different industries are engaging with sustainability, offering a comprehensive analysis that was previously lacking. The novelty of this study lies in its ability to provide actionable empirical evidence that can guide firms, industry leaders, regulators, and policymakers in making informed decisions about sustainable business practices.

Hence, the study provided the following specific recommendations for different key stakeholders. Firstly, firms should enhance their governance practices and integrate comprehensive sustainability reporting into their structures. This includes including social, environmental, and governance (ESG) metrics in annual reports, particularly in sectors like tourism, retail, and property. This will help companies align their practices with societal norms, enhancing legitimacy and reputation. Additionally, companies, especially in mining, should focus on strengthening stakeholder engagement beyond community investment, fostering partnerships with local governments, investors, and NGOs to create more inclusive sustainability strategies. By doing so, firms can improve transparency and reduce agency costs, creating more sustainable business practices.

In sectors like mining and agriculture, firms should prioritize long-term sustainability initiatives, focusing on improving supply chain management and reducing carbon footprints. By integrating sustainable technologies and practices into their core operations, these firms can gain a competitive advantage and improve resource efficiency. Innovation in these sectors will drive the development of sustainable solutions that can be scaled across the industry.

Secondly, regulators should establish comprehensive and enforceable ESG regulations across all sectors, particularly for those that are lagging behind in sustainability practices. These regulations should mandate sustainability reporting and align with global standards. Policymakers should also foster public-private partnerships to fund and scale sustainable innovation, particularly in resource-intensive sectors like mining. These partnerships will enable firms to access resources and knowledge that will enhance their sustainability efforts and reduce risks associated with environmental and societal expectations.

Also, industry associations and trade groups should work together to develop sector-specific sustainability benchmarks that help firms assess their performance and identify areas for improvement. These benchmarks will standardize sustainability practices across industries, promoting greater consistency. Additionally, associations should offer training programs on sustainability governance, ethical business conduct, and reporting to equip firms with the necessary skills to improve their sustainability practices and reduce agency costs.

Lastly, investors should prioritize companies that demonstrate strong sustainability practices, particularly those leading in governance and environmental stewardship. By incorporating ESG criteria into investment decisions, investors can encourage companies to adopt transparent and sustainable practices. This approach will reduce agency costs and align shareholder interests with broader societal and environmental goals, ultimately fostering long-term value creation.

Although this study focused on the Botswana market as a case study, which serves as a representative model for similar developing markets like Libya and Morocco in the MENA region, its findings offer valuable insights. These insights can drive sustainable business practices and policy development in MENA markets, with four key implications discussed below:

1. **Sustainability Framework Adoption**: Similar to Botswana, MENA markets could benefit from adopting integrated reporting frameworks that emphasize ESG factors. The Botswana case demonstrates that when companies align their practices with global sustainability standards, they improve their corporate reputation, reduce risks, and unlock new market opportunities. For MENA countries, the adoption of such frameworks would provide transparency, encourage investor confidence, and enhance global competitiveness.

2. **Regulatory Environment**: Botswana's proactive stance in sustainability through governmental policies and regulations provides a model for MENA countries to develop stronger sustainability guidelines. Countries such as Morocco and Libya, which are in varying stages of sustainability initiatives, can look to Botswana's regulatory approach as a reference for developing and implementing policies that promote long-term sustainable growth. Ensuring that regulations encourage transparency and sustainability reporting, while also balancing industry-specific needs, could drive more cohesive development in the MENA region.
3. **Sustainability Reporting and Governance Practices**: The relatively moderate engagement with governance and ethical business conduct in Botswana could highlight a potential gap in MENA markets as well. MENA companies could strengthen their sustainability efforts by enhancing governance frameworks, integrating environmental and social goals into their annual reports, and improving supply chain management and executive accountability. Such improvements would contribute to more transparent, ethical business practices, ultimately supporting long-term sustainability.
4. **Sample Size and Data Availability**: As seen in the Botswana context, the availability of reliable data is crucial for assessing sustainability efforts. For MENA markets, investing in reliable data collection, reporting mechanisms, and tools for sustainability monitoring would allow for more accurate assessments of corporate sustainability efforts. This would not only help businesses make informed decisions but also allow regulators and policymakers to craft more effective regulations.

While the study provided valuable insights and unique implications for the MENA regions, it faced a few limitations that should be considered when interpreting the findings. One major constraint was the relatively small sample size, which focused solely on companies listed on the Botswana Stock Exchange. This limited sample may not fully represent the broader business landscape in Botswana or the MENA region. Additionally, the study relied on data from only 2 years, primarily due to the fact that most companies began adopting the integrated reporting framework starting in 2022. As a result, the availability of relevant sustainability data from prior years was restricted. The study's narrative data analysis approach also imposed time constraints, limiting the depth of the investigation across various industries. These methodological constraints, particularly the small sample size and short time frame, may impact the generalizability of the findings to other regions or sectors. Future research could overcome these limitations by incorporating a larger and more diverse sample size, extending the data collection period, and exploring additional regions. Such improvements would strengthen the robustness of the findings and enhance their applicability across broader contexts. Despite these limitations, the study provides a solid foundation for further exploration into the adoption, integration, and impact of sustainability practices in African firms.

Appendix A: Description of Study Sample

Sector/Industry	Population (market capitalization)	Sampled/ Excluded	Sampled firms that met the inclusion and exclusion criteria
1. ICT	1. Botswana telecommunications corporation limited (P872M)	Sampled	1. Botswana telecommunications corporation limited
2. Security services	2. G4S Botswana limited (P196M)	Excluded	No integrated report available for the sampled period
3. Energy	3. Engen Botswana limited (P2.1B)	Excluded	No integrated report available for the sampled period
4. Agriculture	4. Seedco international limited (P1.38B)	Sampled	2. Seedco international limited
5. Mining	5. Minergy limited (P188M) 6. Botswana diamonds plc (P167M)	Sampled	3. Minergy limited
6. Tourism	7. Chobe holdings limited (P1.33B) 8. Cresta Marakanelo limited (P196M)	Sampled	4. Chobe holdings limited
7. Retail and wholesale	9. Choppies enterprises limited (P876M) 10. Sefalana holding company limited (P2.88B) 11. Sechaba brewery holdings limited (P2.54B) 12. Olympia Capital corporation (Botswana) limited (P16.7M)	Sampled	5. Sefalana holding company limited
8. Property	13. The far property company limited (P853M) 14. New African properties limited (P2.33B) 15. Turnstar holdings limited (P1.11B) 16. RDC properties limited (P1.82B) 17. PrimeTime property holdings limited (P439M) 18. Letlole La Rona limited (P759M)	Sampled	6. New African properties limited
9. Banks	19. Access Bank Botswana limited (P1.49B) 20. ABSA Bank of Botswana Limited (P5.44B) 21. First National Bank Botswana Limited (P11.1B) 22. Standard chartered Botswana limited (P1.61B)	Excluded	None
10. Financial services	23. Botswana insurance holdings limited (P5.73B) 24. Letshego holdings limited (P2.7B)	Excluded	None

Source: Botswana Stock Exchange

References

Abdelkafi, N., & Täuscher, K. (2016). Business models for sustainability from a system dynamics perspective. *Organization and Environment, 29*(1), 74–96. https://doi.org/10.1177/1086026615592930

Abdulaziz-al-Humaidan, A., Ahmad, N. H., & Islam, M. S. (2022). Investigating the mediating relationship between sustainability orientations and sustainable performance in the SME context of Tunisia. *Vision, 26*(3), 369–381. https://doi.org/10.1177/09722629211000481

Adams, C., Druckman, P., & Picot, R. (2020). Sustainable Development Goals Disclosure (SDGD) recommendations. ACCA, Chartered Accountants ANZ, ICAS, IFAC, IIRC and WBA.

Ahi, P., & Searcy, C. (2013). A comparative literature analysis of definitions for green and sustainable supply chain management. *Journal of Cleaner Production, 52*, 329–341. https://doi.org/10.1016/j.jclepro.2013.02.018

Ahmad, S., & Mokhchy, J. (2023). Corporate social responsibilities, sustainable investment, and the future of green bond market: Evidence from renewable energy projects in Morocco. *Environmental Science and Pollution Research, 30*(6), 15186–15197. https://doi.org/10.1007/s11356-022-23080-y

Ahmad, N., Mobarek, A., & Roni, N. N. (2021). Revisiting the impact of ESG on financial performance of FTSE350 UK firms: Static and dynamic panel data analysis. *Cogent Business and Management, 8*(1), 1900500. https://doi.org/10.1080/23311975.2021.1900500

Archel, P., Husillos, J., Larrinaga, C., & Spence, C. (2009). Social disclosure, legitimacy theory and the role of the state. *Accounting, Auditing & Accountability Journal, 22*(8), 1284–1307. https://doi.org/10.1108/09513570910999319

Barney, J. (1991). Firm resources and sustained competitive advantage. *Journal of Management, 17*(1), 99–120. https://doi.org/10.1177/014920639101700108

Bhandari, K. R., Ranta, M., & Salo, J. (2022). The resource-based view, stakeholder capitalism, ESG, and sustainable competitive advantage: The firm's embeddedness into ecology, society, and governance. *Business Strategy and the Environment, 31*(4), 1525–1537. https://doi.org/10.1002/bse.2967

Botswana Stock Exchange (BSE) & United Nations Development Programme (UNDP). (2018). *Profiling-Private-Sector-sustainability-practices-in-Botswana-_-BSE-listed-domestic--companies*. Retrieved from Committee of SADC Stock Exchange: https://www.cosse.africa/wp-content/uploads/2018/04/Profiling-Private-Sector-Sustainability-Practices-In-Botswana-_-BSE-Listed-Domestic-Companies.pdf

Buła, R., Foltyn-Zarychta, M., & Krawczyńska, D. (2024). Disentangling ESG: Environmental, social and governance ratings and financial performance of Polish listed companies. *Ruch Prawniczy, Ekonomiczny i Socjologiczny, 86*(1), 149–178. https://doi.org/10.14746/rpeis.2024.86.1.09

Business and Sustainable Development Commission. (2017). *Better business, better world.* Business and Sustainable Development Commission. Retrieved from https://unglobalcompact.org/library/5051

Campbell, B. (2012). Corporate social responsibility and development in Africa: Redefining the roles and responsibilities of public and private actors in the mining sector. *Resources Policy, 37*(2), 138–143. https://doi.org/10.1016/j.resourpol.2011.05.002

Chen, S., Song, Y., & Gao, P. (2023). Environmental, social, and governance (ESG) performance and financial outcomes: Analyzing the impact of ESG on financial performance. *Journal of Environmental Management, 345*, 118829.

Chtourou, H., & Triki, M. (2017). Commitment in corporate social responsibility and financial performance: A study in the Tunisian context. *Social Responsibility Journal, 13*(2), 370–389.

Clark, G. L., Feiner, A., & Viehs, M. (2015). *From the stockholder to the stakeholder: How sustainability can drive financial outperformance*. Available at SSRN 2508281.

Eccles, Robert G. & Krzus, Michael P. (2018). *Constructing ExxonMobil's first integrated report: An experiment*. Available at SSRN: https://ssrn.com/abstract=3145369. https://doi.org/10.2139/ssrn.3145369

Evans, S., Vladimirova, D., Holgado, M., Van Fossen, K., Yang, M., Silva, E. A., & Barlow, C. Y. (2017). Business model innovation for sustainability: Towards a unified perspective for creation of sustainable business models. *Business Strategy and the Environment, 26*(5), 597–608. https://doi.org/10.1002/bse.1939

Freeman, R. E. (1999). Divergent stakeholder theory. *Academy of Management Review, 24*(2), 233–236.

Garetti, M., & Taisch, M. (2012). Sustainable manufacturing: Trends and research challenges. *Production Planning & Control, 23*(2–3), 83–104. https://doi.org/10.1080/09537287.2011.591619

Geissdoerfer, M., Bocken, N. M., & Hultink, E. J. (2016). Design thinking to enhance the sustainable business modelling process—A workshop based on a value mapping process. *Journal of Cleaner Production, 135*, 1218–1232. https://doi.org/10.1016/j.jclepro.2016.07.020

Haji, A. A., Coram, P., & Troshani, I. (2022). Consequences of CSR reporting regulations worldwide: A review and research agenda. *Accounting, Auditing & Accountability Journal, 36*(1), 177–208.

Hu, A., Yuan, X., Fan, S., & Wang, S. (2023). The impact and mechanism of corporate ESG construction on the efficiency of regional green economy: An empirical analysis based on signal transmission theory and stakeholder theory. *Sustainability, 15*(17), 13236.

Hwara, T. D. (2023). *Sustainability-in-accounting-esg-reporting*. Retrieved November 2023, from bica.org: https://www.bica.org.bw/sustainability-in-accountingesg-reporting-19036.html

Hwara, T. D., Moyo, N. T., & Obadire, A. M. (2024). Exploring sustainability reporting and non-financial reporting in green finance: A literature review. In A. I. Hunjra & J. W. Goodell (Eds.), *The Palgrave handbook of green finance for sustainable development. Palgrave studies in impact finance*. Palgrave Macmillan. https://doi.org/10.1007/978-3-031-65756-6_8

Ibourk, A., & Amaghouss, J. (2016). Entrepreneurial activities and level of development in Morocco: Empirical investigation from global entrepreneurship monitor data. *Global Business Review, 17*(2), 241–256.

IFRS Foundation. (2022). *ISSB describes the concept of sustainability and its articulation with financial value creation*. Retrieved from www.ifrs.org: https://www.ifrs.org/news-and-events/news/2022/12/issb-describes-the-concept-of-sustainability/#:~:text=Sustainability%20is%20a%20condition%20for,regeneration%2C%20to%20achieve%20its%20goals

International Monetary Fund (IMF). (2023). *Country categorisation and classification*. Retrieved from https://www.imf.org/en/Videos/view?category=2. Accessed 20 Aug 2024

Jensen, M. C., & Meckling, W. H. (1976). Theory of the firm: Managerial behavior, agency costs and ownership structure. *Journal of Financial Economics, 3*(4), 305–360. https://doi.org/10.1016/0304-405X(76)90026-X

Khlif, H., Guidara, A., & Souissi, M. (2015). Corporate social and environmental disclosure and corporate performance: Evidence from South Africa and Morocco. *Journal of Accounting in Emerging Economies, 5*(1), 51–69.

Kozlowski, A., Searcy, C., & Bardecki, M. (2015). Corporate sustainability reporting in the apparel industry: An analysis of indicators disclosed. *International Journal of Productivity and Performance Management, 64*(3), 377–397.

Kumar, M., Raut, R. D., Mangla, S. K., Chowdhury, S., & Choubey, V. K. (2024). Moderating ESG compliance between industry 4.0 and green practices with green servitization: Examining its impact on green supply chain performance. *Technovation, 129*, 102898.

Kuzey, C., & Uyar, A. (2017). Determinants of sustainability reporting and its impact on firm value: Evidence from the emerging market of Turkey. *Journal of Cleaner Production, 143*, 27–39.

Mbekomize, C. J., & Wally-Dima, L. (2013). Social and environmental disclosure by Parastatals and companies listed on the Botswana stock exchange. *Journal of Mangement & Sustainability, 3*, 66.

Mohamed, A. M., Al-Habaibeh, A., Abdo, H., & Elabar, S. (2015). Towards exporting renewable energy from MENA region to Europe: An investigation into domestic energy use and householders' energy behaviour in Libya. *Applied Energy, 146*, 247–262.

Moldavska, A., & Welo, T. (2017). The concept of sustainable manufacturing and its definitions: A content-analysis based literature review. *Journal of Cleaner Production, 166*, 744–755.

Moussa, A. S., Elmarzouky, M., & Shohaieb, D. (2024). Green governance: How ESG initiatives drive financial performance in UK firms? *Sustainability, 16*(24), 10894.

Mutisya, E., & Yarime, M. (2014). Moving towards urban sustainability in Kenya: A framework for integration of environmental, economic, social and governance dimensions. *Sustainability Science, 9*, 205–215.

Noreen, E. (1988). The economics of ethics: A new perspective on agency theory. *Accounting, Organizations and Society, 13*(4), 359–369.

Nunes, B., & Bennett, D. (2010). Green operations initiatives in the automotive industry: An environmental reports analysis and benchmarking study. *Benchmarking: An International Journal, 17*(3), 396–420.

O'Donovan, G. (2002). Environmental disclosures in the annual report: Extending the applicability and predictive power of legitimacy theory. *Accounting, Auditing and Accountability Journal, 15*(3), 344–371.

Obadire, A. M., & Moyo, V. (2024). Exploring the dynamics of green banking and finance: A literature review. In A. I. Hunjra & J. W. Goodell (Eds.), *The Palgrave handbook of green finance for sustainable development. Palgrave studies in impact finance*. Palgrave Macmillan. https://doi.org/10.1007/978-3-031-65756-6_27

Obadire, A. M., Moyo, V., & Munzhelele, N. F. (2023). An empirical analysis of the dynamics influencing bank capital structure in Africa. *International Journal of Financial Studies, 11*(4), 127.

Obadire, A. M., Hwara, T. D., Magang, T., & Moyo, V. (2025). Beyond the green facade: Unravelling the nexus between sustainable business practices and firms' financial performance for Botswana's economic growth. *Cogent Business and Management, 12*(1), 2479182.

Patton, M. Q. (2007). Sampling, qualitative (purposeful). In *The Blackwell encyclopedia of sociology*. https://doi.org/10.1002/9781405165518.wbeoss012.pub2

Piketty, T., & Saez, E. (2014). Inequality in the long run. *Science, 344*(6186), 838–843. https://doi.org/10.1126/science.1251936

Radikoko, I., Mutobo, S. A., & Mphoeng, M. (2019). The impact of stock market development on economic growth: The case of Botswana. *International Journal of Economics and Finance, 11*(12), 149–159. https://doi.org/10.5539/ijef.v11n12p149

Rao, A., Dagar, V., Sohag, K., Dagher, L., & Tanin, T. I. (2023). Good for the planet, good for the wallet: The ESG impact on financial performance in India. *Finance Research Letters, 56*, 104093.

Ross, S. A. (1973). The economic theory of agency: The principal's problem. *The American Economic Review, 63*(2), 134–139.

Saygili, E., Arslan, S., & Birkan, A. O. (2022). ESG practices and corporate financial performance: Evidence from Borsa Istanbul. *Borsa Istanbul Review, 22*(3), 525–533.

Sharma, D., Bhattacharya, S., & Thukral, S. (2019). Resource-based view on corporate sustainable financial reporting and firm performance: Evidence from emerging Indian economy. *International Journal of Business Governance and Ethics, 13*(4), 323–344.

Sheffi, Y. (2018). *Balancing green: When to embrace sustainability in a business (and when not to)*. MIT Press.

Singh, A. P., & Rahman, Z. (2021). Integrating corporate sustainability and sustainable development goals: Towards a multi-stakeholder framework. *Cogent Business and Management, 8*(1), 1985686.

Sinha Ray, R., & Goel, S. (2023). Impact of ESG score on financial performance of Indian firms: Static and dynamic panel regression analyses. *Applied Economics, 55*(15), 1742–1755.

Strekalina, A., Zakirova, R., Shinkarenko, A., & Vatsaniuk, E. (2023). The impact of ESG ratings on financial performance of the companies: Evidence from BRICS countries. *Корпоративные финансы, 17*(4), 93–113.

Talan, G., Sharma, G. D., Pereira, V., & Muschert, G. W. (2024). From ESG to holistic value addition: Rethinking sustainable investment from the lens of stakeholder theory. *International Review of Economics & Finance, 96*, 103530.

The International Integrated Reporting Council. (2021). *International-ir-framework*. The International Integrated Reporting Council. Retrieved from https://integratedreporting.ifrs.org/the-iirc-2/council/

Tilt, C. A., Qian, W., Kuruppu, S., & Dissanayake, D. (2021). The state of business sustainability reporting in sub-Saharan Africa: An agenda for policy and practice. *Sustainability Accounting, Management and Policy Journal, 12*(2), 267–296.

United Nations. (2023). *Sustainability*. Retrieved January 18, 2023, from www.un.org: https://www.un.org/en/academic-impact/sustainability

Upward, A., & Jones, P. (2016). An ontology for strongly sustainable business models: Defining an enterprise framework compatible with natural and social science. *Organization and Environment, 29*(1), 97–123.

Wang, C. (2024). The relationship between ESG performance and corporate performance-based on stakeholder theory. In *SHS web of conferences* (Vol. 190, p. 03022). EDP Sciences.

Wells, E. C., Lehigh, G. R., & Vidmar, A. M. (2020). Stakeholder engagement for sustainable communities. In *The Palgrave handbook of global sustainability* (pp. 1–13).

Whelan, T., Atz, U., Van Holt, T., & Clark, C. (2021). *ESG and financial performance. Uncovering the relationship by aggregating evidence from 1,000 plus studies published between 2015-2020* (Vol. 1, pp. 2015–2020). NYU Stern Center for Sustainable Business and Rockefeller Asset Management.

Chapter 8
Sustainability for Competitive Advantage: Strategic Approaches and Firm Performance

Albana Berisha Qehaja

Abstract In today's world, sustainability is imperative for any business strategy that supports companies in improving performance and maintaining their competitive advantage. In this context, this chapter examines how businesses in the Middle East and North Africa (MENA) region are adopting sustainability practices to meet these goals. It begins with an analysis of current sustainability practices in the region, followed by a detailed exploration of strategic approaches to implementation. The practical impacts of these strategies on firm performance are illustrated through case studies of Masdar City in the UAE, Saudi Aramco in Saudi Arabia, and the Office Chérifien des Phosphates (OCP) Group in Morocco. These cases provide useful lessons for the successful implementation of sustainability in business. The chapter adds a new dimension to the literature by combining theory with practical policy relevance, specifically for the MENA region. This study also highlights the importance of embedding sustainability into the core business strategy because doing so has the potential to drive innovation, reduce costs, enhance stakeholder relations, and reduce environmental impacts. It ends with actionable suggestions to aid businesses in integrating sustainability into their strategic planning. Finally, it highlights how sustainability can drive competitive advantage for the company and the MENA region more broadly, such as achieving net-zero emissions and fostering sustainable growth in the MENA region.

Keywords Sustainability · Competitive advantage · Strategic approach · MENA · Firm performance · Sustainable growth

A. Berisha Qehaja (✉)
Faculty of Economics, University of Prishtina, Prishtina, Republic of Kosovo
e-mail: albana.berisha@uni-pr.edu

M. Elnahass et al. (eds.), *A Greenprint for Net Zero*, Gulf Studies 15,
https://doi.org/10.1007/978-981-92-1762-5_8

Abbreviations

AI	Artificial Intelligence
CCS	Carbon Capture and Storage
CEOs	Chief Executive Officers
EGA	Emirates Global Aluminium
ESG	Environmental, Social, and Governance
GCC	Gulf Cooperation Council
GRI	Global Reporting Initiative
HVAC	Heating, Ventilation, and Air Conditioning
IMF	International Monetary Fund
IoT	Internet of Things
ISO	International Organization for Standardization
KPIs	Key Performance Indicators
LCA	Life Cycle Assessment
MENA	Middle East and North Africa
NGOs	Non-Governmental Organizations
OCP	Office Chérifien des Phosphates
QNB	Qatar National Bank
R&D	Research and Development
RBV	Resource-Based View
SDGs	Sustainable Development Goals
SMEs	Small and Medium-sized Enterprises
TBL	Triple Bottom Line
UAE	United Arab Emirates
UN	United Nations
VNRs	Voluntary National Reviews

8.1 Introduction

In today's global business environment, sustainability has become a new imperative strategic path for reinforcing a firm's competitive advantage and performance (Gupta & Benson, 2011; Nassos & Avlonas, 2020; Shahbazpour & Seidel, 2006). Although the specific drivers may vary for each firm, there is a collective pressure on companies to adapt. Increasingly, businesses actively incorporate sustainability principles (Abdulla & Jawad, 2024; Adams et al., 2016; Alghafes et al., 2024; Ullah et al., 2022). York (2009) noted a worldwide rise in environmental awareness among major corporations, which are gaining measurable competitive advantages by addressing environmental issues, leading to cost savings and increased revenue.

The United Nations' (UN) 2030 Agenda for Sustainable Development, adopted in 2015, focuses on people, the planet, prosperity, peace, and partnership. It encompasses 17 Sustainable Development Goals (SDGs), with SDG 12 specifically targeting responsible consumption and production. Circular economy practices are integral to SDG 12, as they help manage toxic materials and promote environmental responsibility (UN, 2015). The strategic importance of the MENA region, particularly in terms of global energy demands and the extensive impacts of climate change, is evident. Addressing one of the twenty-first century's greatest challenges, global warming, many countries have made significant progress in implementing the roadmap to achieve the SDGs by 2030 (Ibrahim & Alola, 2020; Mertzanis et al., 2024).

Integrating the 2030 Agenda for Sustainable Development into national plans and policies in the MENA region is crucial for achieving the SDGs. Ben Hassen and El Bilali (2022) highlight the measures taken by countries such as Algeria, Qatar, Egypt, and Tunisia to incorporate the SDGs into their development strategies and examine the institutional frameworks established to support these goals. Using Voluntary National Reviews (VNRs) from 2016 to 2021, their findings show that these countries have utilized various policy instruments, such as national development plans, vision documents, and action plans, to implement the 2030 Agenda. Additionally, they have set up different institutional structures to ensure effective inter-ministerial cooperation, either through high-level coordination bodies or specific ministries.

The MENA region provides an intriguing context for studying the integration of sustainability into business strategies (Ahmed, 2008). The region's economic, social, and environmental challenges make it an ideal case study to explore the intersection of environmental and economic sustainability (Ibrahim, 2023). It uniquely combines theoretical perspectives with region-specific practices and case studies, providing a holistic view of sustainability as both a global necessity and a regional opportunity. The region's reliance on natural resources and rapid urbanization underscores the urgency of adopting sustainable practices for long-term economic and environmental stability (Buallay, 2022; Ibrahim & Alola, 2020).

The social systems in the region are fundamentally derived from the Quran, guiding individuals in their daily activities (Ahmed, 2008). The MENA region faces significant environmental challenges, including water scarcity, high energy consumption, and pollution (Ghaffour, 2009; Hoff et al., 2019). The region's economic dependence on fossil fuels or climate change's impacts worsen those issues (Aghahosseini et al., 2020; Razi & Dincer, 2022). Integrated management of natural resources is emphasized to enhance environmental, climate, human, and political security (Elnahass et al., 2022; Hoff et al., 2019).

However, with the challenges, there is growing recognition among businesses and policymakers for the need to transition to more sustainable practices, which can provide opportunities for economic diversification and resilience (Aker & Aghaei, 2019; Buallay et al., 2020; Ibrahim, 2023). The MENA region is seen as an emerging market with significant potential to balance economic growth and environmental sustainability, addressing the dual challenge of fostering economic development while ensuring environmental preservation.

This chapter reviews and synthesizes existing literature and practical cases to explore the relationship between sustainability and firm performance in the MENA region. It uniquely bridges theoretical insights and real-world practices tailored to regional needs. Leveraging regional examples and theoretical models provides a comprehensive understanding of sustainability's role in driving competitive advantage.

This chapter examines current sustainability practices and strategic approaches and their impact on firm performance to offer a comprehensive understanding of the role of sustainability in achieving business excellence. Additionally, it provides theoretical insights and practical recommendations supported by case studies from leading firms in the region.

Furthermore, this chapter emphasizes the practical challenges and solutions to implementing sustainability strategies. It focuses on identifying and analysing the strategic approaches businesses can adopt to integrate sustainability into their core operations. This involves exploring the role of leadership, organizational culture, and stakeholder engagement in driving sustainability initiatives (D'Amato & Roome, 2009; Hart & Milstein, 2003; Linnenluecke & Griffiths, 2010; Marie et al., 2024).

The chapter explores current sustainability practices in the MENA region, examines practical steps for businesses to integrate sustainability into their strategies, evaluates the impact of these practices on operational efficiency, innovation, and stakeholder relationships, and identifies common challenges and solutions in implementing sustainability strategies.

The chapter will address the following key questions and issues:

- What are the current sustainability practices in the MENA region?
- How can businesses incorporate sustainability into their core strategies?
- What are the practical steps for implementing sustainability strategies?
- How does sustainability impact firm performance in terms of operational efficiency, innovation, and stakeholder relationships?
- What are the common challenges and solutions in implementing sustainability strategies?

This chapter integrates theoretical perspectives with practical case studies to provide a thorough exploration of how sustainability can serve as a strategic tool for achieving a competitive advantage in the MENA region. The foremost goal is to encourage businesses to integrate sustainability into their strategic planning, thereby contributing to achieving net-zero emissions and sustainable growth in the region (Elkington, 1997; GRI, 2021; UN, 2015).

8.2 Literature Review

8.2.1 Theoretical Foundations of Sustainability and Competitive Advantage

The sustainability concept has gained significant recognition in business in recent years. The literature has begun to explore its strategic benefits, which can differentiate a firm through innovation, enhanced positioning, and strategic alignment (Gupta & Benson, 2011).

With the rise of pressing environmental issues worldwide, companies are shifting their focus towards embedding sustainability into their core operations. The focus is not solely a regulatory response but also reflects the growing consciousness among consumers and stakeholders about business operations' environmental and social consequences (Elkington, 1997; UN, 2015). The MENA region, with its unique economic, social, and environmental challenges, provides a compelling context for exploring the strategic integration of sustainability into business practices (Ahmed, 2008).

According to Ahmed (2008), defining its territorial boundaries is the initial challenge for most researchers examining issues within the MENA region. For this reason, we have relied on the International Monetary Fund (IMF, 2023). According to the IMF (2023, vi), the MENA region includes Algeria, Bahrain, Djibouti, Egypt, the Islamic Republic of Iran, Iraq, Jordan, Kuwait, Lebanon, Libya, Mauritania, Morocco, Oman, Qatar, Saudi Arabia, Somalia, Sudan, the Syrian Arab Republic, Tunisia, the United Arab Emirates (UAE), the West Bank and Gaza, and Yemen.

Sustainability, which refers to the ability to fulfil current requirements without threatening the capacity of future generations to meet their demands, includes environmental, social, and economic aspects. The aim is to strike a balance that supports the long-term sustainability and health of ecosystems, societies, and economies (Elkington, 1997; UN, 2015).

Since the 1960s, the concept of creating a competitive advantage has been explored through two main schools of thought. The external approach connects competitive advantage to how a firm responds to opportunities and threats in its competitive environment. In contrast, the RBV focuses on how a firm's unique internal resources and capabilities can be distinguished from those of other firms, leading to a competitive advantage (Barney, 1991).

Competitive advantage encompasses the qualities that enable an organization to surpass its competitors. This can be achieved through various means such as cost leadership strategy, differentiation strategy, cost focus strategy, and differentiation focus strategy (Porter, 1980). However, the first two strategies have received significant attention from researchers and academia, while the latter two are considered minor and rarely utilized (Ullah et al., 2022).

Sustainable competitive advantage refers to the capacity to preserve an advantage over competitors for an extended period, typically by leveraging distinctive resources and capabilities that are challenging for others to imitate (Khanra et al., 2022). According to Tu and Wu (2021), organizational learning is essential for integrating resources and generating new knowledge, which is crucial for executing business strategies and attaining sustainable competitive advantage.

Analysing sustainability through the lens of classic strategic theories reveals that firms recognized as sustainability leaders can leverage their initiatives as a source of competitive advantage. Integrating sustainability into a business strategy can enhance brand reputation, reduce costs, and drive innovation (Farida & Setiawan, 2022; Hermundsdottir & Aspelund, 2021; Tu & Wu, 2021; Zameer et al., 2022). At the very least, eco-efficiency enables firms to reduce costs over time, enhancing financial performance (Gupta & Benson, 2011).

This integration is supported by the *resource-based view (RBV)*, which posits that a firm's sustainable competitive advantage stems from its ability to manage and utilize resources efficiently (Barney, 1991). Barney (2018) argues that a firm's profit appropriation should consider the perspectives of social communities and the natural environment in addition to direct stakeholders like shareholders, suppliers, employees, and customers. In sustainability, resources such as green technologies, sustainable supply chains, and eco-friendly innovations can be crucial (Hermundsdottir & Aspelund, 2021; Ibrahim, 2023; Khanra et al., 2022; Tu & Wu, 2021).

The dynamic capabilities theory builds upon the RBV by highlighting the significance of a firm's capacity to adapt, integrate, and reshape internal and external competencies in response to rapidly evolving environments. Dynamic capabilities are crucial for firms to innovate and sustain competitive advantage in the face of environmental and market shifts (Teece et al., 1997).

According to **stakeholder theory**, organizations should generate value for all stakeholders, not solely for shareholders. By addressing the needs and concerns of different stakeholders, including employees, customers, suppliers, and communities, firms can build stronger relationships and enhance their competitive position (Donaldson & Preston, 1995; Freeman, 1984). Stakeholder theory fundamentally suggests that firms must effectively manage and nurture relationships with various stakeholders to ensure long-term survival and success (Freeman, 1994).

The triple bottom line (TBL) framework broadens the conventional reporting model by incorporating social and environmental performance alongside financial performance. It encourages businesses to focus on "people, planet, and profit," promoting a holistic approach to sustainability (Elkington, 1997; Gimenez et al., 2012).

Although not exclusively a sustainability model, *Porter's Five Forces framework* can be applied to analyse the competitive environment and identify opportunities for sustainable practices to enhance competitive advantage. For example, reducing dependency on non-renewable resources can mitigate the threat of supplier power (Porter, 1980; Porter & Kramer, 2011).

Recent advances in sustainability have brought forth new concepts and practices, reshaping the competitive landscape. The circular economy model, emphasizing the principles of reusing, recycling, and regenerating materials, is increasingly recognized as a viable and sustainable solution to replace the conventional linear economy approach (Geissdoerfer et al., 2017). Sustainable finance, including green bonds and impact investing, is also emerging as a vital driver of sustainability, providing the necessary capital for sustainable projects (Baker et al., 2018; Ullah et al., 2022). Additionally, the digital transformation of businesses through technologies such as artificial intelligence (AI), the Internet of Things (IoT), and blockchain is enabling more efficient and transparent sustainability practices (Bonilla et al., 2018).

Hermundsdottir and Aspelund (2021) conducted a literature review of 100 articles to explore the relationship between sustainability innovation and competitiveness. They found that most studies concluded that sustainability innovations positively impact firm competitiveness, with only a few reporting negative effects. The benefits of sustainability innovations include increased value creation, reduced costs, and non-financial assets. These findings suggest that sustainability innovations can ease the conflict between sustainability and economic results, supporting the view that the sustainability shift offers significant business opportunities that outweigh the costs.

Addressing sustainability challenges requires an interdisciplinary approach integrating environmental science, economics, and social sciences insights. Environmental science provides essential knowledge about ecological systems and examines the impact of human activities on the environment. Economics offers tools for evaluating sustainable practices' costs and benefits and designing policies that incentivize sustainable behaviour. Social sciences contribute an understanding of human behaviour, cultural norms, and social dynamics, which are crucial for driving the adoption of sustainable practices (Clark & Dickson, 2003).

Combining these diverse perspectives allows businesses to develop more holistic and effective sustainability strategies. For example, integrating environmental science with economics can help companies identify cost-effective ways to reduce their environmental footprint. Similarly, integrating social science perspectives can strengthen stakeholder engagement and cultivate a sustainability-focused culture within organizations (Ostrom, 2009).

8.2.2 Current Sustainability Practices in the MENA Region

Governments in the MENA region have increasingly recognized the importance of sustainability and have implemented various policies and regulations to promote sustainable development. For instance, the UAE's Vision 2021 and the UAE Green Agenda 2030 aim to transition the country towards a green economy (UAE Government, 2024a, 2024b). Saudi Arabia's Vision 2030 emphasizes sustainability, including goals for reducing dependence on oil and increasing renewable energy (Saudi Green Initiative, 2024). Egypt's Sustainable Development Strategy: Egypt Vision 2030 outlines the country's commitment to sustainable development (Egyptian Government, 2023).

Net-zero strategies are a key focus for businesses in the MENA region. They aim to achieve a balance where the greenhouse gases released into the atmosphere are equal to those removed. Net-zero emissions play a key role in combating climate change and can enhance a company's reputation while fostering stakeholder confidence (Buallay, 2022; World Economic Forum, 2023).

Various industries in the MENA region have implemented sustainability practices to comply with government policies and respond to the increasing demand for environmentally friendly products and services. The energy sector, especially, has experienced substantial investments in renewable energy initiatives, including projects like the UAE's Masdar City (Aghahosseini et al., 2020). The construction industry has embraced sustainability with projects like the King Abdullah Financial District in Saudi Arabia (Public Investment Fund, 2023). The agriculture sector in

Morocco promotes sustainable agriculture through the Green Morocco Plan (ADA, 2023).

Green finance encompasses investments in energy initiatives and efficiency improvements to advance sustainable development. It is essential for funding the transition to a low-carbon economy (Buallay et al., 2020; Nassos & Avlonas, 2020). The growth of green bonds and sustainable investment funds in the MENA region indicates a positive trend towards financing sustainability (Santamarta et al., 2022).

8.2.3 Review of Existing Studies on Sustainability Practices in the MENA Region

Research on sustainability practices in the MENA region reveals a diverse landscape shaped by varying levels of economic development, regulatory frameworks, and cultural attitudes towards sustainability (Ahmad et al., 2022; Buallay, 2022). Economic development is crucial for countries' overall sustainability and is essential for maintaining continuous competitiveness and engaging in market globalization (Ibrahim, 2023). While some countries have made significant progress, others face economic and political challenges (Aker & Aghaei, 2019).

For instance, the UAE and Saudi Arabia have launched ambitious sustainability initiatives, such as the UAE's Vision 2021 and Saudi Arabia's Vision 2030, which include investments in renewable energy and sustainable urban development (Saudi Green Initiative, 2024; UAE Government, 2024a). The UAE has also progressed in water desalination and waste management (Ghaffour, 2009). In contrast, other MENA countries face barriers such as limited financial resources and sociopolitical instability (Aghahosseini et al., 2020; Ahmed, 2008).

Climate risk is another critical factor influencing sustainability practices in the MENA region. Businesses face significant risks from climate change, including physical risks (e.g. extreme weather events) and transition risks (e.g. regulatory changes). Managing these risks ensures long-term business resilience (Hermundsdottir & Aspelund, 2021). Effective climate risk management can lead to cost savings and operational efficiencies (McKinsey & Company, 2022).

8.2.4 Relationship Between Sustainability and Firm Performance

Empirical studies generally support the idea that sustainable practices can enhance competitive advantage and firm performance (Shahbazpour & Seidel, 2006; Walsh & Dodds, 2017). For example, Tu and Wu (2021) found that green innovation positively impacts competitive advantage through organizational learning. Similarly, Zameer et al. (2022) demonstrated that green innovation mediates the relationship between business analytics and green competitive advantage. Thus, green innovation provides firms a competitive advantage while promoting sustainable development (Khanra et al., 2022).

Ullah et al. (2022) examined how small and medium-sized enterprises (SMEs) improve their green strategies. They found that financial resources do not directly influence green business strategies but are mediated by competitive business strategies like differentiation and cost leadership. The study suggests that SMEs should efficiently use financial resources to implement green practices. Gupta and Benson (2011) discuss how firms can leverage sustainability to gain a competitive advantage by applying classic corporate strategy principles. They explain how robust sustainability initiatives can lead to operational efficiencies, increased effectiveness, improved strategic positioning, innovation, commerce redesign, and enhanced value.

In the MENA region, research by Buallay et al. (2020) demonstrates that sustainability reporting positively impacts banks' performance. Similarly, Alghafes et al. (2024) highlight that key Environmental, Social, and Governance (ESG) factors significantly enhance the financial performance of Islamic banks in the Gulf Cooperation Council (GCC) countries. Their findings underscore the role of ESG integration in improving resilience and profitability in these institutions, particularly within the unique regulatory frameworks of Islamic finance. Another study by Ibrahim (2023) highlights the efficiency and productivity gains from integrating sustainability into business operations in MENA countries. Abdulla and Jawad (2024) further add that addressing climate risks through ESG practices positively impacts firm performance, providing MENA firms with a strategic advantage in adapting to environmental challenges.

Additionally, studies have shown that firms adopting sustainable practices can achieve cost savings, improve operational efficiency, and enhance brand reputation (Farida & Setiawan, 2022; Gupta & Benson, 2011; Walsh & Dodds, 2017). The study by Wollenberg et al. (2022) compares sustainable supply chains in the GCC, Europe, and Latin American countries, focusing on the GCC countries over 10 years. It evaluates performance based on the UN SDGs related to clean energy, innovation, sustainable cities, and climate action. The findings show that Kuwait, Qatar, and Saudi Arabia are inefficient due to high oil consumption and low investment in research and development (R&D) and clean energy. Western Europe leads in efficiency, but the UAE, Oman, Mexico, Panama, and Ecuador also perform well in sustainable supply chains. Mertzanis et al. (2024) contribute to this conversation by providing a holistic analysis of sustainable investment conditions in the MENA region. Their research emphasizes the significance of investments in environmental practices, showing clear links to improved firm performance across regional industries.

Furthermore, sustainability initiatives can drive innovation and open new market opportunities. For example, businesses that allocate resources towards renewable energy and environmentally friendly technologies can differentiate themselves from competitors and attract environmentally conscious consumers (Hoff et al., 2019; Wang & Juo, 2021). Leadership also plays a pivotal role in advancing sustainability. Marie et al. (2024) demonstrate that politically connected Chief Executive Officers

(CEOs) in MENA firms foster better ESG performance, which leads to enhanced financial outcomes. Elhoushy and Lanzini (2021) reviewed 71 articles published since 2000, challenging the assumption that developing countries exhibit similar behaviours to developed nations by examining sustainable consumer behaviours in the MENA region. Their findings indicate that MENA studies are predominantly based on rationalistic frameworks, with environmental values as key drivers, while habits and socio-demographic factors play minor roles.

Wiedmann and Minx (2008, p. 4) define the carbon footprint as "the exclusive total amount of carbon dioxide emissions that are directly and indirectly caused by an activity or is accumulated over the life stages of a product." Lowering the carbon footprint can result in cost savings and an enhanced brand reputation, thereby boosting overall firm performance. Elnahass et al. (2022) further note that sustainability practices enhance financial stability, even in adverse conditions such as those posed by external threats like terrorist attacks, specifically in MENA economies.

Despite the extensive research, there are still several gaps in the literature on sustainability and competitive advantage in the MENA region. First, more longitudinal studies are needed to assess the long-term impact of sustainability initiatives on firm performance. Second, research should explore the role of cultural and institutional factors in shaping sustainability practices in different MENA countries (Ahmad et al., 2022; Stephens & Moonesar, 2024).

Furthermore, research on the unique challenges and opportunities that SMEs in the region encounter concerning sustainability is limited. Future studies could also investigate the impact of emerging technologies, such as artificial intelligence (AI) and blockchain, on sustainable business practices (Guermit & Laouedj, 2023).

By addressing these gaps, future research can provide deeper insights into the strategic implementation of sustainability in the MENA region and its potential to drive competitive advantage and firm performance. This will help businesses and policymakers develop more effective strategies to promote sustainability and achieve long-term economic and environmental goals (Buallay, 2022; Ibrahim & Alola, 2020).

8.3 Strategic Approaches to Sustainability

8.3.1 Sustainability as a Core Business Strategy

Incorporating sustainability into the central business strategy is essential for long-term success and competitive advantage (Shahbazpour & Seidel, 2006). Sustainability is no longer a peripheral concern but a central component of strategic planning. Companies that embed sustainability into their core operations can achieve significant benefits, including enhanced brand reputation, increased operational efficiency, and improved stakeholder relationships (Eccles et al., 2014; Gupta & Benson, 2011; Walsh & Dodds, 2017).

Sustainable practices can help achieve cost efficiency by optimizing resource use and minimizing waste, unlocking new market potential by addressing the increasing demand for environmentally friendly products and services (Porter & Kramer, 2011). For instance, companies investing in renewable energy and green technologies can differentiate themselves from competitors and attract environmentally conscious consumers. Regions abundant in energy resources, such as the MENA, have developed strategies and policies to promote eco-innovation (Ibrahim, 2023).

However, integrating sustainability into core business strategies can also present challenges, such as the initial costs of implementing sustainable practices. Companies can tackle these challenges by prioritizing long-term cost savings and operational efficiencies, utilizing green financing opportunities, and cultivating a culture centred on innovation and ongoing improvement.

8.3.2 Role of Leadership, Organizational Culture, and Stakeholder Engagement

Leadership: Effective leadership is crucial for driving sustainability initiatives. Leaders set the vision and tone for sustainability, prioritizing it at all levels of the organization. They play a key role in fostering a culture of sustainability by promoting values such as environmental stewardship, social responsibility, and ethical behaviour (D'Amato & Roome, 2009). Specific leadership qualities that drive sustainability include:

Visionary thinking: Leaders must clearly understand how sustainability aligns with the company's long-term goals and communicate this vision effectively to all stakeholders (Maak & Pless, 2006).

Commitment: Leaders should exhibit an authentic dedication to sustainability by aligning their actions and decisions with sustainable principles, thereby inspiring others to follow suit.

Collaboration: Effective leaders engage with various stakeholders, including employees, customers, investors, and regulators, to build support for sustainability initiatives (Freeman, 1984).

Innovation: Leaders must encourage innovation and creativity in developing sustainable solutions, fostering a culture that values continuous improvement and adaptability (Ferdig, 2007).

For instance, Majid Al Futtaim, a leading retail and leisure pioneer in the Middle East, has committed to becoming net positive in water and carbon by 2040. Their leadership has been instrumental in setting ambitious sustainability goals and integrating them into their business strategy (Majid Al Futtaim, 2022). The research conducted by Çop et al. (2021) investigates how green transformational leadership influences the resilience of green teams, shedding light on its role in fostering adaptability and sustainability within organizations.

Organizational culture: A supportive culture is crucial for successfully implementing sustainability strategies. Companies must cultivate a culture that values sustainability, encourages innovation, and rewards sustainable practices. This involves integrating sustainability into the company's mission, values, and everyday operations (Linnenluecke & Griffiths, 2010).

For instance, Aramco, the Saudi Arabian oil giant, has embedded sustainability into its corporate culture by investing in renewable energy projects and reducing its carbon footprint.

Stakeholder engagement: Involving stakeholders is crucial for implementing and succeeding with sustainability strategies. Stakeholders, including employees, customers, suppliers, and communities, are vested in the company's sustainability performance. Engaging stakeholders in the sustainability journey enables companies to foster trust, strengthen their reputation, and generate shared value.

Etihad Airways, the UAE's national airline, has engaged with various stakeholders to achieve its sustainability goals. The airline has committed to achieving net-zero carbon emissions by 2050 and has implemented several initiatives to reduce its environmental impact (Etihad Airways, 2020).

8.4 Case Studies

8.4.1 Masdar City, UAE

Masdar City in Abu Dhabi is one of the most ambitious sustainability projects in the MENA region. Developed by Masdar, a subsidiary of Mubadala Investment Company, the city aims to be a model for sustainable urban development. It prioritizes renewable energy, water conservation, and sustainable transportation to achieve carbon neutrality and zero waste. The main characteristics feature widespread reliance on solar energy, energy-efficient infrastructure, and an integrated waste management system. The city also promotes sustainable transportation through electric vehicles and a personal rapid transit system. Despite challenges such as high initial costs and the need for continuous innovation, Masdar City has successfully reduced its carbon footprint and attracted international attention and investment, positioning the UAE as a leader in sustainable urban development (Masdar, 2019, 2021a).

Masdar City's initiatives align with several SDGs, highlighting its commitment to sustainability:

SDG 7: Affordable and Clean Energy: Masdar City relies on solar energy alongside other renewable resources to fulfil its energy requirements.

SDG 11: Sustainable Cities and Communities: The city promotes sustainable urban living through energy-efficient buildings and sustainable transportation.

SDG 12: Responsible Consumption and Production: Integrated waste management systems are implemented to reduce waste generation and encourage recycling practices (Masdar, 2019).

Strategies implemented: Masdar City employs a range of sustainability strategies, including solar power, energy-efficient buildings, and a comprehensive waste management system. The city also promotes sustainable transportation through electric vehicles and a personal rapid transit system (Masdar, 2019).

Challenges faced: One of the main challenges has been the high initial cost of developing sustainable infrastructure. Maintaining the city's sustainability goals while accommodating growth and development has also required continuous innovation and investment (Masdar, 2021a).

Outcomes achieved: Masdar City has reduced its carbon footprint and is a living laboratory for sustainable technologies. Waste management systems are strategically implemented to reduce waste production and foster recycling efforts (Masdar, 2021b).

8.4.2 Saudi Aramco, Saudi Arabia

Saudi Aramco, the world's leading oil producer, has made notable progress in integrating sustainability into its business practices. The company has launched various initiatives to reduce its environmental impact and promote sustainable practices.

Strategies implemented: Saudi Aramco's sustainability strategies include cutting greenhouse gas emissions, boosting energy efficiency, and investing in renewable energy projects. The company has also developed programs to enhance biodiversity and conserve water resources. Aramco's 2023 Sustainability Report outlines the company's commitment to achieving net-zero Scope 1 and Scope 2 greenhouse gas emissions by 2050. Additionally, Aramco aims to become one of the lowest upstream carbon intensity producers of hydrocarbon products.

Challenges faced: Balancing the core oil production business with sustainability goals has been a significant challenge. Additionally, transitioning to renewable energy sources while maintaining profitability has required substantial investment and strategic planning. The report also highlights the difficulty of integrating sustainability practices across its extensive operations while ensuring compliance with environmental regulations.

Outcomes achieved: Saudi Aramco has made notable progress in reducing its carbon emissions and improving energy efficiency. The company's investments in renewable energy and sustainability initiatives have enhanced its reputation and competitiveness in the global market. The 2023 report notes that Aramco has minimized environmental impact through innovative technologies and has created societal value by generating employment opportunities and supporting local communities (Saudi Aramco, 2023).

8.4.3 Office Chérifien des Phosphates (OCP) Group, Morocco

The OCP Group, a prominent phosphate mining and fertilizer company in Morocco, has adopted comprehensive sustainability initiatives to reduce its environmental footprint and support sustainable agriculture.

Strategies implemented: The OCP Group has adopted various sustainability strategies, including reducing water and energy consumption, minimizing waste, and promoting sustainable farming practices. The company has also heavily invested in R&D to enhance the efficiency and sustainability of its operations (OCP Group, 2022, 2023). For instance, the OCP Group has developed innovative technologies to optimize water usage in its mining processes and implemented energy-efficient practices. Additionally, the company has launched initiatives to support local farmers in adopting sustainable agricultural practices; in this way, it has contributed to the overall sustainability of the agricultural sector in Morocco (OCP Group, 2023).

Challenges faced: One of the primary challenges faced by the OCP Group is managing the environmental impact of its mining activities. The extraction and processing of phosphates can have significant environmental consequences, including habitat disruption and pollution. Additionally, ensuring the sustainability of water resources in a region prone to water scarcity has been a significant concern (Morocco World News, 2021). The OCP Group has had to navigate these challenges while maintaining its commitment to sustainability and operational efficiency.

Outcomes achieved: The OCP Group has effectively minimized its environmental impact and enhanced the sustainability of its operations. The company's initiatives have also contributed to advancing sustainable agriculture practices in Morocco and beyond (Atalayar, 2023). Through its sustainability efforts, the OCP Group has enhanced its corporate reputation and positioned itself as a leader in the global phosphate industry. The company's commitment to sustainability has attracted international attention and investment, further solidifying its role in promoting sustainable development in the region (OCP Group, 2022).

8.4.4 Role of Innovation and Technology in Advancing Sustainability

The role of innovation and technology is crucial in advancing sustainability. It enables businesses to develop innovative products, processes, and business models to reduce environmental harm and enhance resource utilization. Advancements in green technologies, including solar and wind energy, energy-efficient infrastructure, and electric vehicles, are revolutionizing industries and accelerating the shift towards a low-carbon economy (Wang & Juo, 2021).

In the MENA region, companies are increasingly investing in innovative solutions to address environmental challenges. As an illustration, the Mohammed bin Rashid Al Maktoum Solar Park stands as the largest single-site solar facility globally, with an anticipated production capacity of 5000 MW by 2030. It has positioned the UAE as a leader in renewable energy adoption (Dubai Electricity and Water Authority, 2023). Similarly, the development of smart cities and digital technologies to optimize resource use and reduce waste is gaining traction in the region (McKinsey & Company, 2018).

Technologies such as solar panels, wind turbines, energy-efficient heating, ventilation, and air conditioning (HVAC) systems, as well as electric vehicles, are being widely adopted. Additionally, digital technologies like IoT (Internet of Things) and AI (Artificial Intelligence) are being used to monitor and optimize resource use in real time (International Energy Agency, 2024).

It is important to note that technological improvements must be embedded in appropriate framework conditions, including policies, regulations, and monitoring mechanisms, to ensure the benefits of the approach materialize without causing negative environmental or socio-economic side effects in the MENA region (Hoff et al., 2019).

Leveraging regional examples and theoretical models provides a comprehensive understanding of sustainability's role in driving competitive advantage. This alignment with consumer preferences for sustainable products and services can increase customer loyalty and market share (Farida & Setiawan, 2022; Zameer et al., 2022). Moreover, adopting innovative technologies can result in cost savings, improved operational efficiency, and reduced environmental impact, further strengthening a company's position in the market.

8.5 Implementation of Sustainability Practices

8.5.1 Practical Steps for Implementing Sustainability Initiatives

Planning: The initial step in implementing sustainability strategies is creating a comprehensive plan. This involves conducting a sustainability assessment to identify key areas of impact and opportunities for improvement. Companies should set clear, measurable goals and objectives aligned with their business strategy. The planning process should also include identifying resources, assigning responsibilities, and establishing timelines for achieving sustainability targets (Epstein & Rejc Buhovac, 2014).

Execution: Successful execution of sustainability strategies requires integrating sustainability into all aspects of the business. This includes incorporating sustainability criteria into product design, supply chain management, and operational processes. Companies should implement energy efficiency programs, waste reduction efforts, and sustainable sourcing practices. It is also important to engage employees at all levels and provide them with the tools and resources needed to contribute to sustainability goals (Hart & Milstein, 2003).

Monitoring: Ongoing monitoring and evaluation are essential for determining the effectiveness of sustainability strategies. Companies should establish key performance indicators (KPIs) to track progress and measure outcomes. Regularly reporting and communicating sustainability performance to stakeholders is crucial for maintaining transparency and accountability. Monitoring also allows companies to identify areas for improvement and make necessary adjustments to their sustainability strategies (Alibašić, 2018; Hristov & Chirico, 2019).

8.5.2 *Challenges and Opportunities in the MENA Region*

Challenges: Implementing sustainability strategies can be challenging due to various factors, including financial constraints, resistance to change, and lack of expertise. Companies may face difficulties aligning sustainability goals with business objectives, securing necessary resources, and overcoming internal and external barriers (Santamarta et al., 2022).

Solutions: To overcome these challenges, companies can adopt several strategies:

Securing top management commitment and support is crucial for driving sustainability initiatives. This involves showcasing the business rationale for sustainability and highlighting the enduring advantages.

Investing in building internal capabilities through training and development programs.

Fostering a culture of innovation to identify creative solutions to sustainability challenges.

Collaborating with external partners, such as industry associations, Non-Governmental Organizations (NGOs), and academic institutions, to provide valuable insights and resources (McKinsey & Company, 2022).

8.5.3 *Tools and Frameworks for Implementation*

Several tools and frameworks can support the implementation of sustainability strategies. These include:

Global Reporting Initiative (GRI): The GRI framework offers guidelines for sustainability reporting, enabling companies to measure and communicate their sustainability performance. It covers different aspects of sustainability, including environmental, social, and economic impacts (GRI, 2021).

ISO 14001: ISO 14001 is an internationally acknowledged standard for environmental management systems. It offers a structured approach for organizations to manage their environmental responsibilities and enhance their environmental performance (ISO, 2015).

Sustainable Development Goals (SDGs): The United Nations' SDGs offer a comprehensive framework for addressing global sustainability challenges. Companies can align their sustainability strategies with the SDGs to contribute to broader societal goals and enhance their impact (UN, 2015).

Life Cycle Assessment (LCA): LCA is a tool for evaluating environmental impacts. It offers a holistic perspective on sustainability as both a global necessity and a measure of the ecological footprint of products and processes throughout their life cycle. It helps companies identify areas for improvement and make informed decisions to reduce their environmental impact (Baumann & Tillman, 2004).

Carbon Footprint Analysis: This tool quantifies the overall greenhouse gas emissions directly and indirectly generated by an organization. It helps companies identify emission sources and develop strategies to reduce carbon footprints (Wiedmann & Minx, 2008).

By leveraging these tools and frameworks, companies can effectively implement and manage their sustainability strategies, driving long-term success and competitive advantage.

8.6 Impact on Firm Performance

8.6.1 Analysis of How Sustainability Practices Affect Firm Performance

Over the past decade, most studies advocating for sustainability have emphasized its benefits for a firm's operational effectiveness, including resource efficiency, cost savings from waste reduction, and the advantages of sustainable supply chain management (Gupta & Benson, 2011). Empirical studies have consistently shown that sustainability practices can positively impact firm performance. Incorporating sustainability into business practices can result in cost reductions, increased operational efficiency, and a strengthened brand reputation. For instance, Tu and Wu (2021) found that green innovation positively impacts competitive advantage through organizational learning. Similarly, Zameer et al. (2022) demonstrated that green innovation mediates the relationship between business analytics and green competitive advantage. Alghafes et al. (2024) contribute to this body of research by emphasizing how key ESG factors enhance financial resilience and operational efficiency in Islamic banks in GCC countries, further supporting the positive impact of sustainability on firm performance.

In the MENA region, Buallay et al. (2020) found that ESG factors positively influence banks' performance. The study highlights that higher ESG scores correlate with better operational performance, financial performance, and market performance. This positive relationship is distinct from other regions where the impact of ESG on performance metrics can be inconsistent or even negative. Abdulla and Jawad (2024) add that MENA firms addressing climate risks through ESG practices gain operational and financial advantages, highlighting the unique benefits of sustainability practices tailored to the region's specific challenges. Another study by Ibrahim (2023) highlights the efficiency and productivity gains from integrating sustainability into business operations in MENA countries. Additionally, studies have shown that firms adopting sustainable practices can achieve cost savings, improve operational efficiency, and enhance brand reputation (Farida & Setiawan, 2022; Walsh & Dodds, 2017).

Sustainability initiatives can greatly improve operational efficiency by lowering costs and optimizing resource utilization. Companies that adopt sustainable practices often find that they can lower their operational expenses through energy efficiency, waste reduction, and improved resource management. For example, implementing energy-efficient technologies can reduce electricity consumption, leading to substantial cost savings. Similarly, waste reduction initiatives can minimize disposal costs and create opportunities for recycling and reusing materials, further reducing expenses (Epstein & Rejc Buhovac, 2014). Marie et al. (2024) emphasize the critical role of leadership in advancing ESG performance, showing how politically connected CEOs drive operational improvements and enhance resource utilization in MENA firms.

One notable example is the UAE-based company, Emirates Global Aluminium (EGA). EGA has implemented various sustainability initiatives to improve operational efficiency, such as using solar energy to power its smelting operations. This initiative reduces energy costs and minimizes the company's carbon footprint, contributing to significant cost savings and environmental benefits (EGA, 2021).

Another example is the Saudi Arabian company, Saudi Aramco, which has adopted advanced technologies to enhance operational efficiency and sustainability. Saudi Aramco's initiatives include using carbon capture and storage (CCS) technology to reduce greenhouse gas emissions and improve resource utilization. These efforts have resulted in substantial cost savings and improved environmental performance (Saudi Aramco, 2023). Elnahass et al. (2022) underscore how sustainability practices can enhance financial stability in MENA firms, even under adverse external conditions such as terrorist attacks, further highlighting the resilience built through sustainable initiatives.

8.6.2 Metrics and Indicators for Measuring Sustainability Impact

Companies must develop KPIs and metrics to assess how sustainability practices influence their performance effectively. These tools enable the monitoring of progress, identification of improvement areas, and demonstration of the value of sustainability efforts to stakeholders. Some frequently utilized metrics and indicators include:

Carbon footprint: Quantifies the overall greenhouse gas emissions generated both directly and indirectly by an organization. It helps companies identify emission sources and develop strategies to reduce carbon footprint (Wiedmann & Minx, 2008).

Energy efficiency: Tracks the amount of energy consumed relative to output. Improving energy efficiency can save costs and reduce environmental impact (ISO, 2015).

Water usage: Monitors the amount of water used in operations. Efficient water management can help companies reduce costs and mitigate the risk of water scarcity (U.S. Environmental Protection Agency, 2024).

Waste reduction: Assesses the volume of waste produced and the efficiency of waste management practices. Minimizing waste can result in financial savings and positive environmental outcomes (GRI, 2021).

Sustainable sourcing: Evaluates the degree to which a company procures materials and products from sustainable and ethical suppliers. Sustainable sourcing can enhance brand reputation and reduce supply chain risks (UN, 2015).

8.6.3 Case Studies and Empirical Evidence from the MENA Region

Several case studies and empirical evidence from the MENA region highlight the positive impact of sustainability practices on firm performance. For example:

Masdar City, UAE: Masdar City has successfully reduced its carbon footprint and serves as a living laboratory for sustainable technologies. The initiative has garnered global interest and secured investments from international stakeholders, positioning the UAE as a leader in sustainable urban development (Masdar, 2019).

Saudi Aramco, Saudi Arabia: Saudi Aramco has made notable progress in reducing its carbon emissions and improving energy efficiency. The company's investments in renewable energy and sustainability initiatives have also enhanced its reputation and competitiveness in the global market (Saudi Aramco, 2023).

OCP Group, Morocco: The OCP Group has effectively minimized its environmental footprint and bolstered the sustainability of its operations. The company's initiatives have also contributed to developing sustainable agriculture practices in Morocco and beyond (OCP Group, 2022).

Majid Al Futtaim, MENA Region: Majid Al Futtaim, a leading retail and leisure company in the MENA region, has successfully integrated sustainability into its business strategy, resulting in improved market position and brand reputation. Majid Al Futtaim's sustainability initiatives focus on reducing environmental impact, promoting social responsibility, and enhancing customer experience.

This commitment to sustainability has resonated with consumers, leading to increased brand loyalty and market share (Majid Al Futtaim, 2022).

Elsewedy Electric, Egypt: Elsewedy Electric has implemented various sustainability initiatives to engage with stakeholders and promote social and environmental responsibility. The company's efforts are centred around renewable energy projects, community development, and environmental conservation. By working closely with local communities, suppliers, and customers, Elsewedy Electric has built trust and strengthened relationships with its stakeholders (Elsewedy Electric, 2020).

Qatar National Bank (QNB), Qatar: QNB has integrated sustainability into its business operations and stakeholder engagement strategies. QNB's sustainability initiatives include green financing, environmental conservation projects, and community development programs. These efforts have enhanced QNB's reputation and strengthened its stakeholder relationships (Qatar National Bank, 2023).

These case studies demonstrate that sustainability practices can significantly improve firm performance, including cost savings, enhanced operational efficiency, and improved brand reputation. By adopting sustainable practices, companies in the MENA region can achieve long-term success and competitive advantage.

8.7 Conclusion

This chapter has explored how companies in the MENA region can leverage sustainability to enhance their competitive advantage and overall performance. The findings demonstrate that sustainability is not just an ethical imperative but a vital strategic asset for businesses in the region. By integrating sustainable practices, companies can address global challenges, such as climate change and region-specific issues, including resource scarcity and urbanization, while building resilience in competitive markets. Integrating sustainability into core business strategies has led to significant benefits, including enhanced brand reputation, increased operational efficiency, and improved stakeholder relationships (Eccles et al., 2014; Porter & Kramer, 2011; Walsh & Dodds, 2017).

Effective leadership, a supportive organizational culture, and active stakeholder engagement are crucial for driving sustainability initiatives (D'Amato & Roome, 2009; Linnenluecke & Griffiths, 2010; Marie et al., 2024). Strong leadership ensures that sustainability remains a priority, translating visionary goals into actionable strategies and embedding them within organizational operations. A supportive culture reinforces innovation, accountability, and collaboration, enabling firms to adapt to changing environmental and market demands. Stakeholder trust is particularly significant in the MENA region, where collaborative efforts among businesses, governments, and communities are key to achieving sustainable development goals.

Additionally, sustainability initiatives can reduce costs and optimize resource use, leading to substantial cost savings and environmental benefits (Alghafes et al., 2024; Epstein & Rejc Buhovac, 2014). Resource optimization is especially critical in regions where water scarcity and energy consumption demand efficient and innovative approaches. Firms that embrace green technologies and circular economy practices reduce expenses and strengthen their market positioning as environmentally responsible entities. Furthermore, sustainability can drive innovation, leading to new products, services, and business models that provide a competitive advantage (Hart & Milstein, 2003). Building trust and engagement with stakeholders is essential for the success of sustainability initiatives (Abdulla & Jawad, 2024; Freeman, 1984). Engaging stakeholders enhances transparency and shared value creation, strengthening business resilience and fostering long-term growth.

Implications for managers: Managers operating in the MENA region should prioritize embedding sustainability into their strategic planning and daily operations. This involves fostering a culture that values environmental stewardship and innovation while engaging stakeholders to build trust and accountability. Businesses should also tailor their sustainability strategies to regional challenges, such as water scarcity, dependence on fossil fuels, and urbanization, while leveraging opportunities in renewable energy and green technologies. By aligning sustainability with core business objectives, firms can drive operational efficiency, enhance brand reputation, and build resilience in competitive markets. Regularly revisiting and refining sustainability practices will ensure they remain effective and adaptive to changing market conditions.

Implications for policymakers: Governments across the MENA region have a critical role in creating a regulatory and economic environment conducive to sustainable development. Policymakers need to uphold environmental regulations while offering incentives like tax cuts and subsidies to motivate businesses to embrace renewable energy and resource-efficient technologies. Furthermore, promoting collaboration through public–private partnerships can tackle widespread environmental issues, such as climate change and resource shortages, while propelling the region towards achieving global sustainability objectives. Collaboration at both national and regional levels can promote knowledge-sharing and innovation to accelerate sustainable development.

Implications for researchers: Researchers have a valuable role in advancing the understanding of sustainability practices and their impact on firm performance. Conducting empirical studies can further explore the relationship between sustainability practices and firm performance in the MENA region. Identifying and documenting best practices in sustainability can provide valuable insights for businesses and policymakers. Additionally, investigating the role of innovation and technology in advancing sustainability and driving competitive advantage is crucial.

Key Takeaways:

- Integrating sustainability into strategic planning is essential for achieving long-term success and competitive advantage.
- Effective leadership and a supportive organizational culture are crucial for driving sustainability initiatives.
- Sustainability can act as a powerful driver of innovation, encouraging organizations to design new products, services, and business models that strengthen their competitive advantage and support long-term success.
- Sustainability can drive innovation, leading to new products, services, and business models that provide a competitive advantage.
- Building trust and engagement with stakeholders is essential for the success of sustainability initiatives.

8.8 Future Research

Future research and practice should focus on several key areas:

- Exploring the potential of the circular economy model in the MENA region and its impact on business performance and environmental sustainability.
- Investigating the growing trend of sustainable finance and its implications for businesses and investors in the MENA region.
- Evaluating the approaches that businesses can implement to enhance climate resilience and adapt to the challenges posed by climate change.
- Examining the role of emerging technologies, such as AI, blockchain, and IoT, in enhancing sustainability practices and driving innovation.
- Studying the shifting consumer preferences towards sustainable products and services, and how businesses can meet these demands.

Ultimately, sustainability is a valuable asset for securing a competitive advantage in the MENA region. By embedding sustainability into their strategic frameworks, businesses can achieve long-term success while supporting the overarching goals of net zero emissions and sustainable development in the region. Proactive collaboration among managers, policymakers, and researchers is essential to advancing sustainability and driving meaningful progress.

References

Abdulla, Y., & Jawad, A. (2024). Do MENA firms consider climate risks? Evidence from the relationship between ESG and firm performance. *Journal of Sustainable Finance and Investment, 14*(4), 870–888. https://doi.org/10.1080/20430795.2024.2334256

ADA. (2023). *Main achievements of the green Morocco Plan.* Retrieved from https://www.ada.gov.ma/en/main-achievements-green-morocco-plan

Adams, R., Jeanrenaud, S., Bessant, J., Denyer, D., & Overy, P. (2016). Sustainability-oriented innovation: A systematic review. *International Journal of Management Reviews, 18*(2), 180–205. https://doi.org/10.1111/ijmr.12068

Aghahosseini, A., Bogdanov, D., & Breyer, C. (2020). Towards sustainable development in the MENA region: Analysing the feasibility of a 100% renewable electricity system in 2030. *Energy Strategy Reviews, 28*, 100466. https://doi.org/10.1016/j.esr.2020.100466

Ahmad, N., Smajić, H., Palalic, R., & Al-Riyami, S. (2022). Sustainability management research in the Arab world: A citation-based systematic literature review. *Management and Sustainability: An Arab Review, 1*(1), 14–37. https://doi.org/10.1108/msar-01-2022-0004

Ahmed, A. (2008). Achieving business excellence and competitiveness in the MENA region. In *Business excellence and competitiveness in the Middle East and North Africa* (pp. 3–17).

Aker, Ş. L., & Aghaei, I. (2019). Comparison of business environments in oil-rich MENA countries: A clustering analysis of economic diversification and performance. *Emerging Markets Finance and Trade, 55*(12), 2871–2885. https://doi.org/10.1080/1540496x.2018.1537185

Alghafes, R., Karim, S., Aliani, K., Qureishi, N., & Alkayed, L. (2024). Influence of key ESG factors on Islamic banks' financial performance: Evidence from GCC countries. *International Review of Economics and Finance, 96*, 103629. https://doi.org/10.1016/j.iref.2024.103629

Alibašić, H. (2018). Measuring, tracking, monitoring, and reporting sustainability and resilience progress. In *Sustainability and resilience planning for local governments* (pp. 45–65). Springer. https://doi.org/10.1007/978-3-319-72568-0_4

Atalayar. (2023). *Morocco's OCP group emerges as a world leader in phosphates*. Retrieved from https://www.atalayar.com/en/articulo/economy-and-business/moroccos-ocp-group-emerges-world-leader-in-phosphates/20231209080000194484.html

Baker, M., Bergstresser, D., Serafeim, G., & Wurgler, J. (2018). *Financing the response to climate change: The pricing and ownership of U.S. green bonds*. National Bureau of Economic Research. https://doi.org/10.3386/w25194

Barney, J. (1991). Firm resources and sustained competitive advantage. *Journal of Management, 17*(1), 99–120. https://doi.org/10.1177/014920639101700108

Barney, J. B. (2018). Why resource-based theory's model of profit appropriation must incorporate a stakeholder perspective. *Strategic Management Journal, 39*(13), 3305–3325. https://doi.org/10.1002/smj.2949

Baumann, H., & Tillman, A. M. (2004). *The hitch hiker's guide to LCA: An orientation in life cycle assessment methodology and application* (Vol. 1). Professional Publishing House.

Ben Hassen, T., & El Bilali, H. (2022). Sustainable development goals in the Middle East and North Africa (MENA) region: Policy and governance. In *SDGs in Africa and the Middle East region* (pp. 1–16). Springer International Publishing. https://doi.org/10.1007/978-3-030-91260-4_20-1

Bonilla, S. H., Silva, H. R. O., da Silva, M. T., Gonçalves, R. F., & Sacomano, J. B. (2018). Industry 4.0 and sustainability implications: A scenario-based analysis of the impacts and challenges. *Sustainability, 10*(10), 3740. https://doi.org/10.3390/su10103740

Buallay, A. (2022). Toward sustainability reporting in the MENA region: The effects on sector's performance. *Managerial Finance, 48*(8), 1137-1155. https://doi.org/10.1108/mf-09-2021-0422

Buallay, A., Fadel, S. M., Al-Ajmi, J. Y., & Saudagaran, S. (2020). Sustainability reporting and performance of MENA banks: Is there a trade-off? *Measuring Business Excellence, 24*(2), 197–221. https://doi.org/10.1108/mbe-09-2018-0078

Clark, W. C., & Dickson, N. M. (2003). Sustainability science: The emerging research program. *Proceedings of the National Academy of Sciences, 100*(14), 8059–8061. https://doi.org/10.1073/pnas.1231333100

Çop, S., Olorunsola, V. O., & Alola, U. V. (2021). Achieving environmental sustainability through green transformational leadership policy: Can green team resilience help? *Business Strategy and the Environment, 30*(1), 671–682. https://doi.org/10.1002/bse.2646

D'Amato, A., & Roome, N. (2009). Toward an integrated model of leadership for corporate responsibility and sustainable development: A process model of corporate responsibility beyond management innovation. *Corporate Governance: The International Journal of Business in Society, 9*(4), 421–434. https://doi.org/10.1108/14720700910984972

Donaldson, T., & Preston, L. E. (1995). The stakeholder theory of the corporation: Concepts, evidence, and implications. *Academy of Management Review, 20*(1), 65–91. https://doi.org/10.5465/amr.1995.9503271992

Dubai Electricity and Water Authority. (2023). *Mohammed bin Rashid Al Maktoum Solar Park.* Retrieved from https://www.dewa.gov.ae/en/about-us/strategic-initiatives/mbr-solar-park

Eccles, R. G., Ioannou, I., & Serafeim, G. (2014). The impact of corporate sustainability on organizational processes and performance. *Management Science, 60*(11), 2835–2857. https://doi.org/10.1287/mnsc.2014.1984

Egyptian Government. (2023). *Egypt Vision 2030.* Retrieved from https://mped.gov.eg/DynamicPage?id=115&lang=en

Elhoushy, S., & Lanzini, P. (2021). Factors affecting sustainable consumer behavior in the MENA region: A systematic review. *Journal of International Consumer Marketing, 33*(3), 256–279. https://doi.org/10.1080/08961530.2020.1781735

Elkington, J. (1997). *Cannibals with forks: The triple bottom line of 21st century business.* Capstone.

Elnahass, M., Marie, M., & Elgammal, M. (2022). Terrorist attacks and bank financial stability: Evidence from MENA economies. *Review of Quantitative Finance and Accounting, 59*(1), 383–427. https://doi.org/10.1007/s11156-022-01043-1

Elsewedy Electric. (2020). *Sustainability strategy 2020–2025.* Retrieved from https://elsewedy-electric.com/pdf/PDFS/04191482/elsewedy-electric-sustainability-strategy-2020-2025.pdf

Emirates Global Aluminium (EGA). (2021). *EGA publishes fourth annual sustainability report.* Retrieved from https://media.ega.ae/ega-publishes-fourth-annual-sustainability-report/

Epstein, M. J., & Rejc Buhovac, A. (2014). *Making sustainability work: Best practices in managing and measuring corporate social, environmental, and economic impacts.* Greenleaf Publishing/Berrett-Koehler Publishers, Incorporated.

Etihad Airways. (2020). *Etihad targets zero net carbon emissions by 2050, in expanded commitment to environmental sustainability.* Retrieved from https://www.etihad.com/en/news/etihad-targets-zero-net-carbon-emissions-by-2050-in-expanded-commitment-to-environmental-sustainability

Farida, I., & Setiawan, D. (2022). Business strategies and competitive advantage: The role of performance and innovation. *Journal of Open Innovation: Technology, Market, and Complexity, 8*(3), 163. https://doi.org/10.3390/joitmc8030163

Ferdig, M. A. (2007). Sustainability leadership: Co-creating a sustainable future. *Journal of Change Management, 7*(1), 25–35. https://doi.org/10.1080/14697010701233809

Freeman, R. E. (1984). *Strategic management: A stakeholder approach.* Pitman. https://doi.org/10.1017/CBO9781139192675

Freeman, R. E. (1994). The politics of stakeholder theory: Some future directions. *Business Ethics Quarterly, 4*(4), 409–421. https://doi.org/10.2307/3857340

Geissdoerfer, M., Savaget, P., Bocken, N. M., & Hultink, E. J. (2017). The circular economy–A new sustainability paradigm? *Journal of Cleaner Production, 143*, 757–768. https://doi.org/10.1016/j.jclepro.2016.12.048

Ghaffour, N. (2009). The challenge of capacity-building strategies and perspectives for desalination for sustainable water use in MENA. *Desalination and Water Treatment, 5*(1–3), 48–53. https://doi.org/10.5004/dwt.2009.564

Gimenez, C., Sierra, V., & Rodon, J. (2012). Sustainable operations: Their impact on the triple bottom line. *International Journal of Production Economics, 140*(1), 149–159. https://doi.org/10.1016/j.ijpe.2012.01.035

Global Reporting Initiative (GRI). (2021). *GRI standards.* Retrieved from https://www.globalreporting.org/standards

Guermit, A., & Laouedj, Z. (2023). The duality of digital innovation and sustainable development in MENA countries. In *Competitive advantage, strategy and innovation in Africa* (pp. 125–151). Routledge. https://doi.org/10.4324/9781003306405-8

Gupta, N., & Benson, C. C. (2011, November). Sustainability and competitive advantage: An empirical study of value creation. *Competition Forum, 9*(1).

Hart, S. L., & Milstein, M. B. (2003). Creating sustainable value. *Academy of Management Perspectives, 17*(2), 56–67. https://doi.org/10.5465/ame.2003.10025194

Hermundsdottir, F., & Aspelund, A. (2021). Sustainability innovations and firm competitiveness: A review. *Journal of Cleaner Production, 280*, 124715. https://doi.org/10.1016/j.jclepro.2020.124715

Hoff, H., Alrahaife, S. A., El Hajj, R., Lohr, K., Mengoub, F. E., Farajalla, N., et al. (2019). A nexus approach for the MENA region—From concept to knowledge to action. *Frontiers in Environmental Science, 7*, 48. https://doi.org/10.3389/fenvs.2019.00048

Hristov, I., & Chirico, A. (2019). The role of sustainability key performance indicators (KPIs) in implementing sustainable strategies. *Sustainability, 11*(20), 5742. https://doi.org/10.3390/su11205742

Ibrahim, M. D. (2023). Efficiency and productivity analysis of innovation, human capital, environmental, and economic sustainability nexus: Case of MENA countries. *Environmental Science and Pollution Research, 30*(12), 34394–34405. https://doi.org/10.1007/s11356-022-24494-4

Ibrahim, M. D., & Alola, A. A. (2020). Integrated analysis of energy-economic development--environmental sustainability nexus: Case study of MENA countries. *Science of the Total Environment, 737*, 139768. https://doi.org/10.1016/j.scitotenv.2020.139768

International Energy Agency. (2024). *Digitalization and energy*. Retrieved from https://iea.blob.core.windows.net/assets/b1e6600c-4e40-4d9c-809d-1d1724c763d5/DigitalizationandEnergy3.pdf

International Monetary Fund (IMF). (2023). *Regional economic outlook: Middle East and Central Asia: Safeguarding macroeconomic stability amid continued uncertainty*. International Monetary Fund. https://doi.org/10.5089/9798400238239.086

International Organization for Standardization (ISO). (2015). *ISO 14001:2015 - Environmental management systems*. Retrieved from https://www.iso.org/iso-14001-environmental-management.html

Khanra, S., Kaur, P., Joseph, R. P., Malik, A., & Dhir, A. (2022). A resource-based view of green innovation as a strategic firm resource: Present status and future directions. *Business Strategy and the Environment, 31*(4), 1395–1413. https://doi.org/10.1002/bse.2961

Linnenluecke, M. K., & Griffiths, A. (2010). Corporate sustainability and organizational culture. *Journal of World Business, 45*(4), 357–366. https://doi.org/10.1016/j.jwb.2009.08.006

Maak, T., & Pless, N. M. (2006). Responsible leadership in a stakeholder society—A relational perspective. *Journal of Business Ethics, 66*(1), 99–115. https://doi.org/10.1007/s10551-006-9047-z

Majid Al Futtaim. (2022). *Sustainability report 2022*. Retrieved from https://www.majidalfuttaim.com/en/sustainability

Marie, M., Qi, B., Elgammal, M., & Elnahass, M. (2024). A more sustainable future: Can politically connected CEOs spur the nexus between ESG performance and firm financial performance? *Journal of International Financial Markets, Institutions and Money, 96*, 102056. https://doi.org/10.1016/j.intfin.2024.102056

Masdar. (2019). *Masdar City: A model for sustainable urban development*. Retrieved from https://masdar.ae/en/news/newsroom/masdar-city—a-template-for-sustainable-urban-development

Masdar. (2021a). *Sustainability at Masdar city*. Retrieved from https://masdarcity.ae/sustainable-urban-development/Sustainability-at-masdar-city

Masdar. (2021b). *Net Zero by 2050 | Net Zero targets*. Retrieved from https://masdarcity.ae/sustainable-urban-development/net-zero

McKinsey & Company. (2018). *Smart cities: Digital solutions for a more livable future*. McKinsey & Company. Retrieved from https://www.mckinsey.com/capabilities/operations/our-insights/-smart-cities-digital-solutions-for-a-more-livable-future

McKinsey & Company. (2022). *Organizing for sustainability success: Where, and how, leaders can start*. Retrieved from https://www.mckinsey.com/capabilities/sustainability/our-insights/-organizing-for-sustainability-success-where-and-how-leaders-can-start

Mertzanis, C., Marashdeh, H., Nasrallah, N., & Islam, A. R. M. (2024). Sustainable investment conditions and firm performance in the Middle East and North Africa: A holistic approach. *Journal of Sustainable Finance and Investment*, 1–38. https://doi.org/10.1080/20430795.2024.2303492

Morocco World News. (2021). *Water management drives OCP group's ambitious sustainability program*. Retrieved from https://www.moroccoworldnews.com/2021/04/338709/water--management-drives-ocp-groups-ambitious-sustainability-program

Nassos, G. P., & Avlonas, N. (2020). *Practical sustainability strategies: How to gain a competitive advantage*. John Wiley & Sons. https://doi.org/10.1002/9781119561132

OCP Group. (2022). *OCP Group launches a $13 billion Green Investment Strategy*. Retrieved from https://www.ocpgroup.ma/news-article/ocp-group-launches-its-new-green-investment-program-2023-2027

OCP Group. (2023). *Sustainability: Strategy and action*. Retrieved from https://www.ocpgroup.ma/sustainability

Ostrom, E. (2009). A general framework for analyzing sustainability of social-ecological systems. *Science, 325*(5939), 419–422. https://doi.org/10.1126/science.1172133

Porter, M. E. (1980). Industry structure and competitive strategy: Keys to profitability. *Financial Analysts Journal, 36*(4), 30–41. https://doi.org/10.2469/faj.v36.n4.30

Porter, M. K., & Kramer, E. J. (2011). Creating shared value. *Harvard Business Review, 89*(1/2), 62–77.

Public Investment Fund. (2023). *Saudi Arabia's KAFD sets new standards for smart cities*. Retrieved from https://www.pif.gov.sa/en/news-and-insights/global-insights/2023/saudi-arabias-kafd-sets-new-standards-for-smart-cities/

Qatar National Bank. (2023). *QNB sustainability framework and strategy*. Retrieved from https://www.qnb.com/sites/qnb/qnbqatar/page/en/ensustainabilityframeworkandstrategy.html

Razi, F., & Dincer, I. (2022). Renewable energy development and hydrogen economy in MENA region: A review. *Renewable and Sustainable Energy Reviews, 168*, 112763. https://doi.org/10.1016/j.rser.2022.112763

Santamarta, S., Seppä, T., Gruß, C., Bozic Mazzi, A., Cuellar, M., Catchlove, P., & Vikström, A. (2022, July). *The challenges of a sustainability transformation*. Boston Consulting Group. Retrieved from https://www.bcg.com/publications/2022/the-challenges-of-a-sustainability-transformation

Saudi Aramco. (2023). *Sustainability report 2023*. Retrieved from https://www.aramco.com/en/sustainability/sustainability-report

Saudi Green Initiative. (2024). *Saudi Green Initiative - Vision 2030*. Retrieved from https://www.vision2030.gov.sa/en/projects/saudi-green-initiative/

Shahbazpour, M., & Seidel, R. H. A. (2006). Using sustainability for competitive advantage. In *13th CIRP international conference on life cycle engineering*.

Stephens, M., & Moonesar, I. A. (2024). Developing cases for teaching social sustainability in business: Lessons from the MENA region. In *The Palgrave handbook of social sustainability in business education* (pp. 551–571). Springer Nature Switzerland. https://doi.org/10.1007/978-3-031-50168-5_28

Teece, D. J., Pisano, G., & Shuen, A. (1997). Dynamic capabilities and strategic management. *Strategic Management Journal, 18*(7), 509–533. https://doi.org/10.1002/(sici)1097-0266(199708)18:7<509::aid-smj882>3.0.co;2-z

Tu, Y., & Wu, W. (2021). How does green innovation improve enterprises' competitive advantage? The role of organizational learning. *Sustainable Production and Consumption, 26*, 504–516. https://doi.org/10.1016/j.spc.2020.12.031

U.S. Environmental Protection Agency. (2024). *Lean & water toolkit: Chapter 2*. Retrieved from https://www.epa.gov/sustainability/lean-water-toolkit-chapter-2

UAE Government. (2024a). *Vision 2021. The Official webpage of the UAE Government.* Retrieved from https://u.ae/en/about-the-uae/strategies-initiatives-and-awards/strategies-plans-and-visions/strategies-plans-and-visions-untill-2021/vision-2021

UAE Government. (2024b). *The UAE'S green agenda - 2030.* The Official webpage of the UAE Government. Retrieved from https://u.ae/en/about-the-uae/strategies-initiatives-and-awards/strategies-plans-and-visions/environment-and-energy/the-uaes-green-agenda-2030

Ullah, R., Ahmad, H., Rizwan, S., & Khattak, M. S. (2022). Financial resource and green business strategy: The mediating role of competitive business strategy. *Journal of Sustainable Finance & Investment, 14*(2), 410–429. https://doi.org/10.1080/20430795.2022.2031850

United Nations (UN). (2015). *Transforming our world: The 2030 Agenda for Sustainable Development.* Retrieved from https://sustainabledevelopment.un.org/post2015/transformingourworld

Walsh, P. R., & Dodds, R. (2017). Measuring the choice of environmental sustainability strategies in creating a competitive advantage. *Business Strategy and the Environment, 26*(5), 672–687. https://doi.org/10.1002/bse.1949

Wang, C. H., & Juo, W. J. (2021). An environmental policy of green intellectual capital: Green innovation strategy for performance sustainability. *Business Strategy and the Environment, 30*(7), 3241–3254. https://doi.org/10.1002/bse.2800

Wiedmann, T., & Minx, J. (2008). A definition of 'carbon footprint'. *Ecological Economics Research trends, 1*(2008), 1–11.

Wollenberg, A., Lazarini, J. G. O. C., Lazarini, J. J. C., Parra, L. F. O., & Kakade, A. S. (2022). A comparative analysis of green supply chains among countries in the Middle East, Latin America, and Europe. *Journal of Advanced Management Science, 10*(2). https://doi.org/10.18178/joams.10.2.58-64

World Economic Forum. (2023). *Closing the climate action gap: Accelerating decarbonization and the energy transition in MENA.* Retrieved from https://www3.weforum.org/docs/WEF_Closing_the_Climate_Action_Gap_2023.pdf

York, J. G. (2009). Pragmatic sustainability: Translating environmental ethics into competitive advantage. *Journal of Business Ethics, 85*, 97–109. https://doi.org/10.1007/s10551-008-9950-6

Zameer, H., Wang, Y., Yasmeen, H., & Mubarak, S. (2022). Green innovation as a mediator in the impact of business analytics and environmental orientation on green competitive advantage. *Management Decision, 60*(2), 488–507. https://doi.org/10.1108/md-01-2020-0065

Chapter 9
Determinants of Sustainability Disclosures in Moroccan Listed Companies: Insights from an Emerging MENA Market

Mohamed Anas Belidan and Halima Baghad

Abstract This study assesses the evolution of sustainability disclosure, identifies its determinants, and evaluates industry-level variations in sustainability reporting in the MENA region, with a particular focus on Morocco. A descriptive analysis utilizing 118 indicators to measure the Sustainability Disclosure Index (SDI) was conducted to examine trends and developments. This was followed by a random-effects linear and Feasible Generalized Least Squares (FGLS) panel regression to investigate the relationship between the SDI and its determinants. The findings reveal a significant improvement in the SDI over time, positively influenced by factors such as firm age, firm size, and indexation in the MASI ESG. However, despite mandatory ESG reporting in Morocco, the overall sustainability disclosure index remains mediocre, signaling the need for firms to enhance their practices. This study, while limited to listed firms due to the unavailability of data for non-listed entities, offers valuable insights for managers on the factors shaping the SDI and lays a foundation for further research in the region. To the best of our knowledge, this is the first empirical investigation in Morocco to analyze the determinants of sustainability reporting disclosure across all industries for the period 2014–2023.

Keywords Sustainability reporting · Sustainability disclosure · MENA · Emerging economy · Morocco

---

M. A. Belidan (✉) · H. Baghad
Department of Finance Research (LAREF), The Higher Institute of Commerce and Business Administration (Groupe ISCAE), Casablanca, Morocco

M. Elnahass et al. (eds.), *A Greenprint for Net Zero*, Gulf Studies 15,
https://doi.org/10.1007/978-981-92-1762-5_9

9.1 Introduction

Since the release of SDG17, numerous countries, including those in the MENA region, have initiated the implementation of strategic 2030 visions to achieve these goals (Morocco 2030, Qatar 2030, Saudi Arabia 2030). This development has implications for firms that must align their practices to contribute to the Sustainable Development Goals. Organizations can demonstrate their commitment through the disclosure of sustainability reporting practices, as such disclosure serves as a crucial means of communicating information to various stakeholders, including employees, clients, government entities, and regulatory agencies (Al Amosh et al., 2022; Joudeh, 2024).

Consequently, researchers must align their studies to examine these developments. Although there have been various studies in Morocco on how sustainability disclosures can be beneficial to firms, investors, and shareholders, as a means and incentive for them to make a more sustainable impact and to disclose it (Cherkaoui & Cherkaoui, 2021), limited studies in the MENA region, especially in Morocco, have investigated how to measure such disclosure, what constitutes good or less effective disclosure, and what factors determine those disclosures. While some international independent organizations provide frameworks to follow, it is the role of academics to investigate these aspects further (Machado et al., 2021).

In selecting the Morocco case, this study aimed to address some of the gaps in the literature. The Moroccan context presents a unique setting for the study of sustainability reporting practices, as there has been a mandatory requirement for listed firms since 2019, but no standardized framework to follow, making it more complex for companies to report sustainability disclosures. By examining an extensive sample of firms from all industries from 2014 to 2023, we aim to observe the evolution and determinants of sustainability disclosures in an emerging MENA country.

9.2 Literature Review, Theory, and Hypotheses

9.2.1 *Evolution of Sustainability Reporting in Morocco*

The earliest research to discuss the concept of corporate social responsibility (CSR) in Moroccan companies was conducted by El Malki (2010) as a thesis. This study examined the influence of the investment climate on firm performance using data collected between 1997 and 2005 and found that the investment climate significantly impacts firm performance, with specific institutional factors playing a crucial role. Additionally, the study revealed that Moroccan firms' commitment to CSR, particularly toward employees, is a significant aspect, while other dimensions appear to have less impact.

A subsequent thesis by Taoukif (2014) supported El Malki's (2010) findings by employing both qualitative and quantitative methods to ascertain that certain dimensions of corporate social responsibility (CSR) engagement, particularly in employee relations, exhibit positive outcomes. However, the overall societal engagement was found to be insufficient. In general, Research in Morocco has primarily focused on the CSR managerial dimension, rather than the disclosure and reporting dimension, as it was an emerging field and sustainability disclosure was voluntary (Zayer & Benabdelhadi, 2023) until 2019 when the AMMC, the highest regulatory body in Morocco for the financial market, made the disclosure of ESG information by listed companies mandatory. Although this provides general insights into disclosure, it does not offer a reference or standard to follow (Soussi, 2023).

Researchers in Morocco have examined the implications of such disclosures. A qualitative study of listed companies conducted by Cherkaoui and Cherkaoui (2021) concluded that there is a diverse landscape of extra-financial communication practices with varying engagement and transparency levels in reporting non-financial information. More recent quantitative studies, such as Farrat et al.'s (2024) findings on Moroccan-listed firms, identified a gap between GRI standards and the content of annual reports, although the study focused only on financial institutions and measured GRI standards through six principal dimensions: Accuracy, Balance, Clarity, Comparability, Reliability, and Timeliness. Another recent study conducted by Namoussi and Cherquaoui (2024) specifically examined industrial listed companies and evaluated environmental disclosure over 2 years: 2018 and 2020. Our study aims to fill the gaps left by these two studies by analyzing sustainability disclosure among Moroccan listed companies from all industries between 2014 and 2023 and assessing the reporting of sustainability disclosure based on 118 indicators based on the GRI 2016 and GRI 2021 guidelines.

9.2.2 Sustainability Disclosure Research in MENA Region

To present notable and pertinent regional studies pertaining to sustainability disclosure, we prepared Table 9.1. This table shows a prominent article and its respective research question and findings. Our study continues the progression of previous research on this topic.

9.2.3 Measurement of Sustainability Disclosure

Sustainability Reporting initially emerged within the critical accounting literature, characterized by the triple bottom line approach, which integrates three performance dimensions: social, environmental, and financial, encapsulating people, planet, and profit. The primary aim of Sustainability Reporting is to furnish stakeholders with insights into how companies conduct their operations in an environmentally and socially responsible manner, thereby enhancing transparency, accountability, and profitability (Kolk, 2008).

Table 9.1 Sustainability disclosure research in the MENA region

Article	Country	Research question	Findings
Al Amosh et al. (2022)	Jordan	What are the financial determinants that influence the level of integrated reporting disclosure among Jordanian companies?	The financial leverage, profitability, liquidity ratios, cash holdings, interest coverage ratios, company age, size, type of industry, and the level of internationalization have a significant impact on the level of integrated reporting disclosure by Jordanian companies
Aryan et al. (2022)	Saudi Arabia	What is the effectiveness of corporate governance on CSR performance and financial reporting quality in the manufacturing sector of Saudi Arabia?	Positive and significant relationship between corporate governance and CSR performance and financial reporting quality
Younis (2023)	Saudi Arabia	What is the impact of the disclosure of sustainability reports on the value of companies listed on the Saudi stock market?	Positive relationship between the level of sustainability report disclosure and firm value
Kouaib & Amara (2022)	Saudi Arabia	Does Corporate Social Responsibility (CSR) disclosure affect investment decisions in Saudi-indexed companies, and how does the quality of corporate governance moderate this relationship?	CSR reporting positively and significantly affects the investment decision at a 5% level CSR behavior improves the corporate investment strategies of Saudi companies
Joudeh (2024)	Jordan	Is there an impact of sustainability disclosure on value added from the perspective of Stakeholders in industrial companies listed on the Amman Stock Exchange?	Companies exhibited different levels of sustainability disclosure, with economic performance being disclosed more frequently compared to environmental and social performance There is a notable positive impact of sustainability disclosure on value added from the perspective of stakeholders, indicating that companies
Abdulla & Jawad (2024)	Bahrain, Egypt, Jordan, Kuwait, Morocco, Oman, Qatar, Saudi Arabia, Türkiye,, and the UAE	How do climate risks influence the relationship between Environmental, Social, and Governance (ESG) scores and firm performance in the MENA region?	ESG score has a negative impact on operational and financial performance among MENA firms. Conversely, the ESG score has a positive effect on market value In countries with high CO2 emissions, the ESG score has a more significant negative impact on firms' operational and financial performance compared to countries with low CO2 emissions

Source: Authors' own compilation

Subsequently, due to the lack of a unified international and national standard and framework for measuring sustainability reporting, numerous authors have defined this concept in alignment with the GRI-G4 framework (Appendix 9.1) (Kuzey & Uyar, 2017; Shad et al., 2019; Karaman et al., 2020; Giner & Luque-Vílchez, 2022; Al-Qudah & Houcine, 2023). These authors describe sustainability reporting as transparent business practices grounded in ethical values that respect the interests of various stakeholders, including shareholders, employees, communities, and the environment. They emphasize that sustainability reporting involves addressing the needs of direct and indirect stakeholders without compromising the ability to meet the needs of future stakeholders, which is essential for building trust and enhancing corporate reputation.

Despite some criticisms regarding the GRI framework, particularly concerning practical challenges in enhancing accountability for sustainability, these challenges include the complexity of the standards, the necessity for multi-stakeholder engagement, and the varying levels of corporate adoption (De Villiers et al., 2022). Research also indicates that while GRI standards are designed to assist organizations in integrating sustainability, they can still be perceived as demanding and difficult to implement, potentially leading to selective reporting (Hummel & Jobst, 2024). This selective reporting manifests in three main forms:

- Vague Disclosure: This involves providing general, incomplete, or ambiguous information that lacks clarity and fails to fully inform stakeholders about significant events.
- Avoidance: In this scenario, companies do not disclose any information about a specific negative event and effectively omit it from their reporting.
- Hypocrisy: This type refers to situations in which companies do not report negative performance related to a specific area but simultaneously present positive information about their sustainability initiatives in the same context.

Despite these criticisms, the GRI framework is regarded as one of the most reliable methods for assessing sustainability reporting quality, particularly since the establishment of the Corporate Sustainability Reporting Directive (CSRD) in the EU, which aligns more closely with the GRI requirements (Zimon et al., 2022).

9.2.4 Theories and Hypotheses

According to stakeholder theory, firms should consider the concerns and well-being of all parties that have the potential to influence an organization, going beyond shareholders (Meutia et al., 2022). This is especially relevant when it comes to sustainability reporting, as stated by ElAlfy et al. (2020), who argued that the successful integration of SDGs into the CSR practices of the private sector cannot be achieved without the active involvement of all stakeholders. This is because stakeholders can indirectly pressure firms to disclose more relevant and sustainable information (Torelli et al., 2020). This concept is related to legitimacy theory, which posits that companies can gain legitimacy not only from obligatory laws that require sustainability disclosures but also from other stakeholders, such as clients, suppliers, NGOs, and society (Boulhaga et al., 2023). With this type of legitimacy, firms can justify their financial and sustainability disclosures in relation to what stakeholders perceive as material information (Meutia et al., 2022).

Previous research on sustainability reporting (although there is no consensus on how to measure this variable) has demonstrated a positive relationship between the sustainability disclosure index (SDI) and firm age, as shown by Correa-Garcia et al. (2020). Although this study was conducted on Latin American firms, we expect that the variables will exhibit similar behavior in our study, conducted in an emerging economy context. Based on this, our first hypothesis is formulated as follows:

Hypothesis 1 Sustainability disclosure index is positively associated with the age of the firm

A company's size is often considered a key variable when evaluating the relationship between the SDI and other financial aspects, such as financial performance (Dase et al., 2024), governance (Aryan et al., 2022), and firm value (Gerged et al., 2023). A study conducted by Joudeh (2024) found a positive association between the SDI and the size of Jordanian listed companies. Given that the Moroccan economy shares similar attributes with Jordan's economy (Ezzahid & Maouhoub, 2020), we propose the following hypothesis:

Hypothesis 2 Sustainability disclosure index is positively associated with the size of the firm

Morocco has been developing a market index for Environmental, Social, and Governance (ESG) performance since 2018. This index, called ESG.10, was created by the AMMC, the highest regulatory body for the financial market in Morocco (Soussi, 2023). ESG.10 comprised the ten most relevant listed firms based on their ESG performance, and it was reevaluated annually. In 2022, the Casablanca Stock Exchange introduced the MASI.ESG index, which replaced ESG.10 with 15 firms. In the following year, the number of firms in the index increased to 20 (Namoussi and Cherqaoui, 2024). It is apparent that companies included in such indices are subject to regulatory controls to be included in the index or to remain in the index. Consequently, a high-quality SDI is required. Therefore, our hypothesis is as follows:

Hypothesis 3 Sustainability disclosure index is positively associated with the indexation of MASI.ESG.

Table 9.2 Sample selection

	Unique firms
Companies listed in Casablanca Stock Exchange between 2014 and 2023	83
Less: Companies that went out of business before 2023	4
Less: Companies that got introduced after 2014	10
Less: companies that are missing data	2
Total number of companies in the sample	67
	Firm years
Total number of firm-year observations in sample	670

Source: Authors

9.3 Research Design

9.3.1 Sample

Our study aims to address the gap in Moroccan empirical research by examining the sustainability reports of all firms continuously listed on the Casablanca Stock Exchange, regardless of their industry. To gather data, we manually collected information from annual reports, stand-alone sustainability reports, and equivalent reports (CSR reports, ESG reports, impact reports) from the AMMC website and the companies' websites. After excluding firms that were listed after 2014, those that went bankrupt before 2023, and those lacking data over the period of ten years from 2014 to 2023, we were left with a sample of 67 firms (Table 9.2). We collected data based on the indicators outlined in Appendix 9.1. The process involved checking the existence and disclosure of sustainability reports, as well as assessing whether the indicators were disclosed in integrated reports, annual reports, and sustainability/ESG/CSR/impact reports. If the indicators were disclosed, we assigned a value of 1; otherwise, we assigned a value of 0.

9.3.2 Dependent Variable

The dependent variable is the Sustainability Disclosure Index, which is based on the GRI G4 framework, as it is considered the most suitable framework for assessing sustainability disclosure (Sobhani et al., 2012; Utami, 2015; Gavana et al., 2017; Aifuwa, 2020; Gil-Marín et al., 2022; Abeysekera, 2022; Alodat et al., 2024). The index was calculated using the following formula (detailed in Appendix 9.1), where d is the indicator and n is the number of indicators present in the GRI guidelines:

$$SDI = \frac{\sum_{i=1}^{n} d_i}{n}$$

Table 9.3 Independent Variable definitions, calculations, and abbreviations

Variable	Description	Abbreviation
Firm age	Logarithm of the time interval between the date of creation and the date of observation	AGE
Firm size	Logarithm of total Assets	SIZE
Masi. ESG	A binary variable that assumes a value of 1 if the observation is indexed in Masi.ESG and 0 otherwise	MASI.ESG

Source: Compiled by the authors based on previous research (Abu Khalaf, 2024; Namoussi & Cherqaoui, 2024)

Table 9.4 Control Variable definitions, calculations, and abbreviations

Variable	Definition	Measurement	Sources
AQ	Audit quality	The quality of an audit is represented as a categorical variable, where a value of 1 is assigned if the company is audited by a Big Four firm, a value of 2 if audited by two Big Four firms, and a value of 0 if audited by non-big 4 firms	Orazalin and Mahmood (2019), Al Amosh et al. (2022; Marie et al. (2024)
SDG	Sustainable development goals	$SDGI = \frac{\sum_{i=1}^{n} d_i}{n}$ Where d = 1 if the information related to the corresponding SDG is disclosed, and 0 otherwise; n = 17, which is the maximum number of SDGs disclosed	Zampone et al., (2024)
SDG. IDX	Sustainable development goals' country index	The SDG Index is an annual score assessed by the United Nations, evaluating countries based on their fulfillment of the 17 Sustainable Development Goals	Diaz-Sarachaga et al. (2018)

Source: Compiled by the authors based on previous research

9.3.3 *Independent Variable*

To address our hypothesis and in line with previous research (Guix et al., 2018; Orazalin & Mahmood, 2019; Slacik & Greiling, 2019; Nandy et al., 2023; Aboud et al., 2024), our independent variables are those that, as mentioned earlier, are believed to have a positive impact on the sustainability disclosure index. These variables are firm size, firm age, and the indexation of the firm in MASI.ESG. Table 9.3 presents the independent variables, their descriptions, and their corresponding abbreviations.

9.3.4 Control Variables

To enhance our model, we incorporate the following control variables. Their inclusion allowed us to isolate the effect of the main independent variables and improve the model's overall validity. Table 9.4 provides a comprehensive overview of the measurement methods and data sources for each control variable.

9.3.5 Model

A panel linear regression model was selected to examine the research hypotheses, as our research sample incorporated firm and year variables, thus constituting longitudinal panel data. This methodology has been previously employed in sustainability reporting studies (Gerwanski et al., 2019; Velte, 2022; Meftah Gerged et al., 2023; Alodat et al., 2024), resulting in the following equation:

$$SDI_{it} = \beta_0 + \beta_1 AGE_{it} + \beta_2 SIZE_{it} + \beta_3 MASI.ESG_{it} + \beta_4 AQ + \beta_5 SDG + \beta_6 SDG.IDX + \sum Year + \sum Industry + \varepsilon_{it}$$

We conducted a Hausman test to determine whether to utilize the fixed effect (FE) or random effect (RE) models. The Hausman specification test indicates that (H0): "The unobserved individual-specific effects are uncorrelated with the independent variables, suggesting the f effects model is appropriate" (Hausman et al., 2005). We failed to reject the null hypothesis. This suggests that there is no systematic difference between the coefficients estimated by the FE and RE models. Consequently, the RE models are deemed most suitable for longitudinal data analyses such as ours. The estimated results from the FE models are presented in Table 9.9 in the subsequent section.

Random effects models typically assume homoskedasticity and the absence of serial correlation in the error terms. Violations of these assumptions, such as heteroskedasticity or autocorrelation, can lead to inefficient estimates and invalid inferences. To address these limitations, we employed Feasible Generalized Least Squares (FGLS) estimation. FGLS is particularly well suited for panel data, as it accounts for heteroskedasticity and serial correlation by estimating the error covariance structure and adjusting the estimation accordingly. This approach enhances the efficiency and reliability of the estimates, providing more robust inferences in the presence of complex error structures, as it is commonly used in sustainability reporting studies (Gallén & Peraita, 2017; Aliyu & Nguavese, 2024). This will result in a more robust analysis (Table 9.9).

Table 9.5 Descriptive statistics of SDI and its components 2014–2023

	SDI	*GD*	*S*	*G*	*SP*	*M*	*E*	*I*	*CE*
Mean	0.24,934	0.46,393	0.23,708	0.38,798	0.15,876	0.22,040	0.19,225	0.38,483	0.14,459
Median	0.16,949	0.66,667	0.15,385	0.31,579	0.05263	0.00000	0.03226	0.50,000	0.12,500
S. D	0.23,857	0.30,440	0.,25,878	0.32,318	0.23,195	0.39,006	0.25,498	0.28,983	0.18,381
S. Var	0.05692	0.09266	0.06697	0.10,445	0.05380	0.15,215	0.06502	0.08400	0.03379
Kurtosis	0.13,289	−1.21,115	−0.22,578	−0.97,633	3.27,942	−0.15,224	0.68,319	−1.17,201	9.78,366
Skewness	0.86,193	−0.79,517	0.87,702	0.36,800	1.97,676	1.30,070	1.24,263	−0.15,106	2.78,315
Min	0.00000	0.00000	0.00000	0.00000	0.00000	0.00000	0.00000	0.00000	0.00000
Max	0.99,153	1.00000	1.00000	1.21,053	1.00000	1.00000	1.00000	1.00000	1.00000
Sum	167.059	310.833	158.846	259.947	106.368	147.666	128.806	257.833	96.8750
Count	670	670	670	670	670	670	670	670	670

Source: Authors

9.4 Results

9.4.1 *Overview of Sustainability Disclosure Trends and Patterns*

The descriptive statistics presented in Table 9.5 offer a comprehensive overview of the variables under investigation, providing insights into their central tendencies, dispersion, and distributional characteristics. It is noted that components of the SDI are detailed in Appendix 9.1 and include the following: general disclosures: social, governance, stakeholder policy, materiality, environment, impact, and compliance and ethics. The mean values indicate a moderate level of engagement across most variables, with GD (General Disclosures) exhibiting the highest average value (0.46393), suggesting a relatively consistent emphasis on general disclosure among the sampled entities. Conversely, SP (Stakeholder Policy) demonstrates the lowest mean (0.15876), highlighting a potential area of underperformance or variability in stakeholder-related disclosure.

The variability across these variables was further elucidated by the standard deviations, with M (Materiality) and GD (General Disclosures) demonstrating considerable dispersion (0.39006 and 0.30440, respectively). This indicates that while some firms may excel in these areas, others significantly underperform, leading to a broad distribution of values. The presence of high kurtosis, particularly in CE (9.78366), indicates a leptokurtic distribution with heavy tails, implying the existence of outliers or extreme cases that could substantially influence the overall distribution.

Figure 9.1 shows the evolution of sustainability reporting over a decade, revealing a significant upward trend across all variables. The data highlight notable

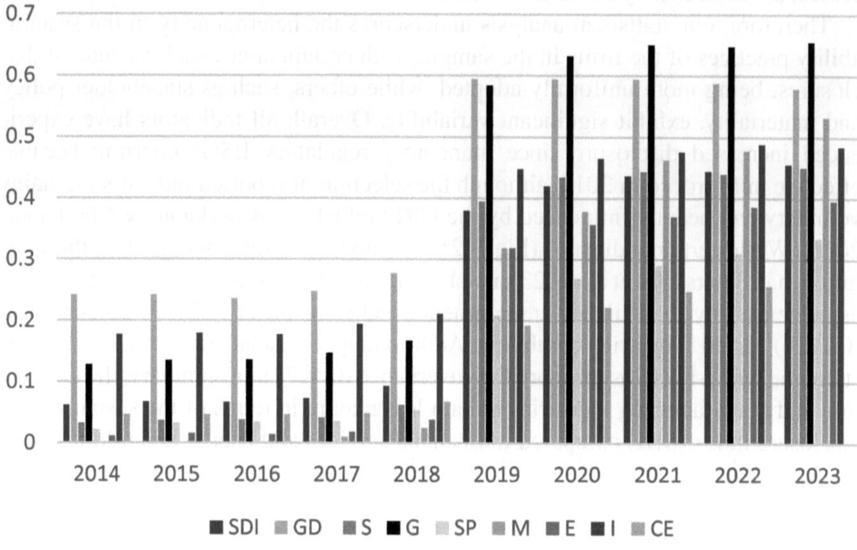

Fig. 9.1 Evolution of SDI and its components 2014–2023. Source: Authors

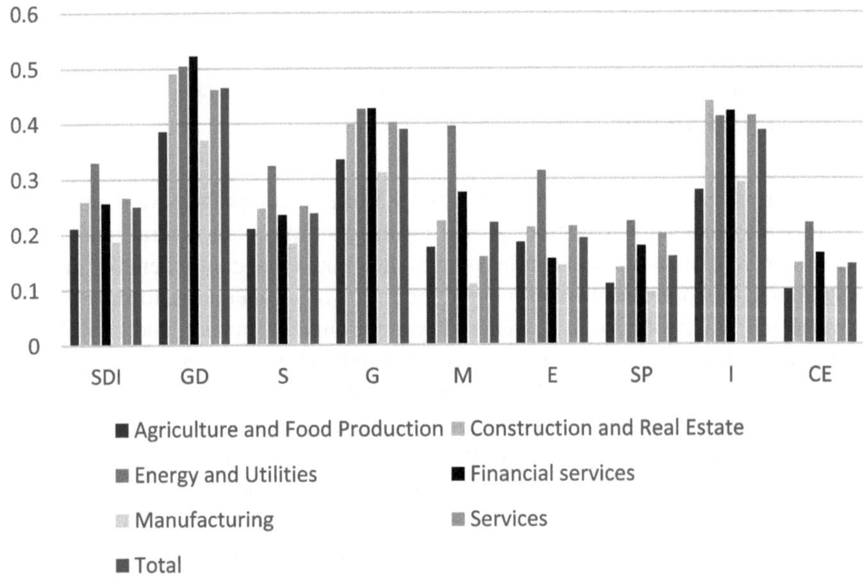

Fig. 9.2 Mean of SDI and its components by industry 2014–2023. Source: Authors

improvements in the scope and details of sustainability reporting over the past ten years. The acceleration observed after 2018 indicates a transition toward more comprehensive and integrated sustainability practices among organizations, which corroborates the findings of Soussi (2023). This trend is likely attributable to increasing external pressures and modifications in corporate governance, wherein sustainability has become a central component of strategic decision-making and reporting processes, as illustrated by Zhou et al. (2023).

Therefore, our statistical analysis underscores the heterogeneity in the sustainability practices of the firms in the sample, with certain areas, such as general disclosures, being more uniformly adopted, while others, such as stakeholder policy and materiality, exhibit significant variability. Overall, all indicators have experienced increased disclosure since mandatory regulatory ESG reporting became effective in Morocco in 2019, although the selection of reported indicators remains voluntary, yet heavily influenced by the GRI Guidelines (Cherkaoui & Cherkaoui, 2021). While certain industries (Fig. 9.2) exhibit greater concern regarding this evolution than others (Xie et al., 2023), notably the energy and utilities sector (Table 9.8), as evidenced by their higher environmental indicator mean (0.3143) compared to (0.1928) for all industries combined. Additionally, financial firms tend to demonstrate the highest disclosure score for governance (0.4274). In summary, the agriculture and manufacturing industries remain less mature in terms of the sustainability disclosure index (SDI) compared to financial services and energy and utilities.

9.4.2 Regression Analysis

This section presents the results of our panel data model testing, which jointly examines the influence of the firm's age (H1), size (H2), and indexation on MASI ESG (H3) on the sustainability disclosure index reporting in Moroccan listed firms throughout the period 2014–2023. We first present a discussion by analyzing the statistics of our variables. The mean SDI (Table 9.6) is particularly low (0.2493342), which can be attributed to our study including the period 2014–2018, during which such information was voluntarily disclosed. This also explains the significant standard deviation in the SDI (0.238574). During the period 2014–2016, only financial services by Moroccan listed firms published sustainability information, primarily focusing on governance and societal impacts (Menchi & Chemlal, 2017). Regarding the other variables, age and size are dependent on the type of industry. Industries such as financial services tend to have larger log assets (Janahi et al., 2023).

As shown in Table 9.7, there are no correlations between the independent variables that exceed 0.561. Although this indicates strong independence between our variables, it is insufficient to determine the degree of independence, as shown by Dong et al. (2023). Indeed, a variance inflation factor (VIF) test is necessary to assess multicollinearity between variables. A VIF of 1 indicates no multicollinearity, whereas a tolerance measured by 1/VIF, when closer to 1, indicates a low multicollinearity. This is evident in our results, as shown in Table 10, where the

Table 9.6 Descriptive statistics on variables

Variable	Observations	Mean	S.D	Min	Max
SDI	N = 670	0.249342	0.238574	0	0.991525
AGE	N = 670	1.708026	0.2284384	1.079181	2.064458
SIZE	N = 670	9.518372	0.9481648	7.392038	11.6278
MASI.ESG	N = 670	0.1044776	0.3061077	0	1
SDG	N = 670	0.8	0.4002988	0	1
AQ	N = 670	1.828	0.7541159	0	2
SDG.IDX	N = 670	69.295	1.292695	67.47	70.85

Source: Authors

Table 9.7 Pearson correlations among variables

Variables	(1)	(2)	(3)	(4)	(5)	(6)	VIF	1/VIF
(1) AGE	1.000						1.06	0.9424
(2) SIZE	0.179***	1.000					1.34	0.7440
(3) MASI.ESG	0.091**	0.342***	1.000				1.41	0.7094
(4) SDG	0.088**	0.029	0.171***	1.000			1.46	0.6834
(5) AQ	−0.100	0.272***	0.135**	−0.074	1.000		1.16	0.8645
(6) SDG.IDX	0.117***	0.036	0.280***	0.561***	−0.181***	1.000	1.65	0.6069

Note: *, **, and *** indicate significance at 10%, 5% and 1% levels, respectively
Source: Authors' own work via Stata

Table 9.8 Panel regression analysis

SDI	Coefficient	Std. Err.	*t*	*P* > *t*	[95% CI]	
AGE	0.2085495	0.09819	2.12	0.034	0.0161006	0.4009985
SIZE	0.0826964	0.0154441	5.35	0.000	0.0524266	0.1129663
MASI.ESG	0.0408369	0.0326998	1.25	0.212	−0.0232536	0.1049274
SDG	0.3789111	0.0418573	9.05	0.000	0.2968724	0.4609499
AQ	0.04333	0.0228153	1.90	0.058	−0.0013872	0.0880472
SDG.IDX	−0.0175444	0.0035081	−5.00	0.000	−0.0244203	−0.0106686

Note: *, **, and *** indicate significance at 10%, 5% and 1% levels, respectively
Source: Authors' own work via Stata

Table 9.9 Feasible Generalized Least Squares (FGLS) estimation regression analysis

SDI	Coefficient	Std. Err.	*t*	*P* > *t*	[95% CI]	
AGE	0.222134	0.0364656	6.09	0.000	0.1506632	0.2936057
SIZE	0.072286	0.0055415	13.04	0.000	0.0614252	0.0831476
MASI.ESG	0.072626	0.0152118	4.77	0.000	0.0428116	0.102441
SDG	0.365508	0.0184362	19.83	0.000	0.3293744	0.4016431
AQ	0.051264	0.0069274	7.40	0.000	0.0376873	0.0648422
SDG_IDX	−0.01654	0.0013654	−12.12	0.000	−0.0192211	−0.0138688

Note: *, **, and *** indicate significance at 10%, 5% and 1% levels, respectively

no-tolerance level is below 0.6, and the mean VIF is 1.65, suggesting that our independent variables do not face issues resulting from multicollinearity.

The results of the random-effects panel linear regression model are presented in

Table 9.8. The R-squared value was 0.7809. This indicates that over 78% of the variation in the sustainability disclosure index of Moroccan listed firms, as measured by Formula (1), is explained by the set of independent variables.

H1 proposes that the sustainability disclosure index is positively associated with firm age, as measured by the logarithm of the difference between the year of creation and the year of the data. The regression analyses conducted in this study revealed a significant positive relationship between firm age and the Sustainability Disclosure Index (SDI). In the Random Effects (RE) model (Table 9.8), the coefficient for firm age is 0.2085 with a *P*-value of 0.034, indicating statistical significance at the 5% level. The Feasible Generalized Least Squares (FGLS) model (Table 9.9) further reinforces this finding, presenting a coefficient of 0.2221 with a highly significant *P*-value (<0.001). These results suggest that older firms are more inclined to disclose sustainability information than are younger firms. This tendency can be attributed to accumulated experience, established reporting practices, and greater institutional pressures faced by older firms. Additionally, mature firms often have more developed stakeholder relationships and are subject to increased scrutiny, compelling them to be more transparent in their operations. This finding aligns with previous research indicating that firm age positively influences corporate social responsibility disclosures, contradicting the findings from another MENA country, Jordan (Al Amosh et al., 2022), where it was observed that the age of the firm is negatively associated with integrated reporting disclosure.

H2 tests the influence of company size, measured by the logarithm of assets, on the sustainability disclosure index and company size. The analysis demonstrates a robust positive association between firm size and the Sustainability Disclosure Index (SDI). In the Random Effects model (Table 9.8), the coefficient for firm size is 0.0827 with a *P*-value less than 0.001, indicating strong statistical significance. The FGLS model (Table 9.9) corroborates this result, showing a coefficient of 0.0723, also significant at the 1% level. These findings imply that larger firms are more likely to engage in sustainability disclosures. The propensity of larger firms to disclose such information can be attributed to greater resource availability, heightened public visibility, and increased pressure from stakeholders and regulatory bodies. Larger firms often operate in more complex environments and are subject to more rigorous scrutiny, necessitating comprehensive reporting practices. This observation is consistent with stakeholder and legitimacy theories, which posit that larger organizations are more responsive to stakeholder demands and societal expectations. This outcome aligns with the research conducted by Joudeh (2024) on Jordanian listed companies during the 2017–2022 period.

H3 states that firms covered by the Masi ESG index requirements to provide ESG information are more likely to deliver higher-quality sustainability reporting disclosures. The relationship between the MASI.ESG indexation and the Sustainability Disclosure Index (SDI) presents mixed results across the two models. In the Random Effects model (Table 9.8), the coefficient is 0.0408 with a *P*-value of 0.212, indicating a lack of statistical significance. However, the FGLS model (Table 9.9) reveals a coefficient of 0.0726 with a *P*-value less than 0.001, suggesting a significant positive association. The discrepancy between the models may be due to the FGLS model's ability to account for heteroskedasticity and provide more efficient estimates in the presence of panel data complexities. The significant result in the FGLS model implies that firms included in the MASI.ESG index are more likely to disclose sustainability information, possibly because of the rigorous ESG criteria required for index inclusion. This finding aligns with the notion that indexation in ESG-focused indices serves as an incentive for firms to enhance transparency and reporting practices. While this finding is novel in the context of MENA region research, it aligns with the study by Dong et al. (2023) on Chinese listed firms between 2014 and 2019.

9.5 Conclusion

The study of sustainability reporting has been increasing worldwide following the publication of the SDG17, although it remains limited in emerging countries, including those in the MENA region (Abdulla & Jawad, 2024). Our study aims to confirm findings previously tested in developed markets, which are still novel to the MENA market, especially in a developing African country within MENA, such as Morocco (Soussi, 2023). Through a comprehensive examination of the existing literature in the MENA region and by analyzing a sample of 67 Moroccan listed firms over a 10-year period, we arrive at the following conclusions: there is increasing awareness among companies regarding the significance of enhancing their sustainability disclosure index while simultaneously grappling with the lack of a standardized framework to follow. This leads to some industries, such as financial services and the energy sector, disclosing sustainability metrics more extensively than others. Additionally, older and larger firms tend to disclose more information, as they possess sufficient resources to extend beyond financial reporting and performance. Firms indexed in MASI ESG exhibiting more comprehensive sustainable reporting suggest that institutional and market pressures can lead to an improved sustainability disclosure index (SDI).

From the policymakers' perspective, the findings of this study underscore the pressing need for Moroccan authorities—particularly the AMMC, Bank Al-Maghrib, and relevant ministries—to design and implement a standardized and enforceable sustainability reporting framework. Despite the introduction of mandatory ESG disclosure, our results reveal persistent disparities in reporting quality across sectors and firm sizes. To address this, a national framework should be adopted, drawing on international benchmarks, such as the Global Reporting Initiative (GRI). This framework should mandate sector-specific ESG indicators, clearly define reporting obligations, and set a timeline for full compliance, ideally by 2030 (in alignment with Morocco 2030 vision). Furthermore, policymakers should establish an independent oversight body to supervise the sustainability disclosure index, provide regular evaluations, and issue public sustainability disclosure ratings to encourage companies to think beyond near-term targets, which most documentation referring to standards and guidance directs users toward, as suggested by Christy et al. (2023). This institution could also lead capacity-building initiatives, offering training and technical support to help firms—particularly small and medium-sized enterprises—align with reporting requirements and elevate the credibility of their sustainability disclosures.

From a managerial perspective, these findings underscore the significance of sustainability reporting as a critical element that organizations should approach with due diligence. This should be regarded as a means to demonstrate the firm's impact. Although certain factors may influence the sustainability disclosure index, efforts should be made to enhance it. Even after the implementation of mandatory ESG reporting in Morocco and an increasing trend in the Sustainability Disclosure Index (SDI), the mean SDI remains relatively modest (0.45535) in 2023. This should prompt firms to take appropriate action.

The present study had certain limitations. The findings are related only to listed firms, and the lack of data on non-listed firms is the primary reason for this choice. This study focuses on the impact of the disclosure of the metrics and not on the impact of certain metrics being performant. Based on the findings of our study, future research should prioritize investigating how other factors affect SDI, including financial performance (ROE, ROA, Tobin's q), capital structure (debt to equity, debt ratio), firm value (capitalization), and industry-specific characteristics. Future studies should also conduct empirical research on this matter in other North African MENA countries, such as Algeria, Tunisia, Libya, and Egypt.

Declaration of generative AI and AI-assisted technologies in the writing process During the preparation of this work the authors used open AI in order to improve language and readability, with caution. After using this tool/service, the authors reviewed and edited the content as needed and take full responsibility for the content of the publication.

Appendix

Appendix 9.1 List of SDI disclosures with classifications

Indicator	Classification
Operations assessed for risks related to corruption	Compliance and ethics
Communication and training about anti-corruption policies and procedures	
Confirmed incidents of corruption and actions taken	
Legal actions for anti-competitive behavior, anti-trust, and monopoly practices	
Approach to tax	
Tax governance, control, and risk management	
Stakeholder engagement and management of concerns related to tax	
Country-by-country reporting	
Operational sites owned, leased, managed in, or adjacent to, protected areas and areas of high biodiversity value outside protected areas	Environmental
Significant impacts of activities, products and services on biodiversity	
Habitats protected or restored	
IUCN Red List species and national conservation list species with habitats in areas affected by operations	
Financial implications and other risks and opportunities due to climate change	
Materials used by weight or volume	
Recycled input materials used	
Reclaimed products and their packaging materials	
Energy consumption within the organization	
Energy consumption outside of the organization	
Energy intensity	
Reduction of energy consumption	
Reductions in energy requirements of products and services	
Interactions with water as a shared resource	
Management of water discharge-related impacts	
Water withdrawal	
Water discharge	
Water consumption	
Direct (Scope 1) GHG emissions	
Energy indirect (Scope 2) GHG emissions	
Other indirect (Scope 3) GHG emissions	
GHG emissions intensity	
Reduction of GHG emissions	
Emissions of ozone-depleting substances (ODS)	
Nitrogen oxides (NOx), sulfur oxides (SOx), and other significant air emissions	
Significant spills	
Waste generation and significant waste-related impacts	
Management of significant waste-related impacts	
Waste generated	
Waste diverted from disposal	
Waste directed to disposal	

(continued)

Appendix 9.1 (continued)

Indicator	Classification
Organizational details	General disclosure
Entities included in the organization's sustainability reporting	
Reporting period, frequency and contact point	
Restatements of information	
External assurance	
Governance structure and composition	governance
Nomination and selection of the highest governance body	
Chair of the highest governance body	
Role of the highest governance body in overseeing the management of impacts	
Delegation of responsibility for managing impacts	
Role of the highest governance body in sustainability reporting	
Conflicts of interest	
Communication of critical concerns	
Collective knowledge of the highest governance body	
Evaluation of the performance of the highest governance body	
Remuneration policies	
Process to determine remuneration	
Annual total compensation ratio	
Statement on sustainable development strategy	
Policy commitments	
Embedding policy commitments	
Processes to remediate negative impacts	
Mechanisms for seeking advice and raising concerns	
Compliance with laws and regulations	
Activities, value chain, and other business relationships	Impact
Direct economic value generated and distributed	
Infrastructure investments and services supported	
Significant indirect economic impacts	
Incidents of violations involving rights of indigenous peoples	
Operations with local community engagement, impact assessments, and development programs	
Operations with significant actual and potential negative impacts on local communities	
Process to determine material topics	Materiality
List of material topics	
Management of material topics	

(continued)

Appendix 9.1 (continued)

Indicator	Classification
Employees	Social
Workers who are not employees	
Defined benefit plan obligations and other retirement plans	
Ratios of standard entry-level wage by gender compared to local minimum wage	
Proportion of senior management hired from the local community	
New employee hires and employee turnover	
Benefits provided to full-time employees that are not provided to temporary or part-time employees	
Parental leave	
Minimum notice periods regarding operational changes	
Occupational health and safety management system	
Hazard identification, risk assessment, and incident investigation	
Occupational health services	
Worker participation, consultation, and communication on occupational health and safety	
Worker training on occupational health and safety	
Promotion of worker health	
Prevention and mitigation of occupational health and safety impacts directly linked by business relationships	
Workers covered by an occupational health and safety management system	
Work-related injuries	
Work-related ill health	
Average hours of training per year per employee	
Programs for upgrading employee skills and transition assistance programs	
Percentage of employees receiving regular performance and career development reviews	
Diversity of governance bodies and employees	
Ratio of basic salary and remuneration of women to men	
Incidents of discrimination and corrective actions taken	
Security personnel trained in human rights policies or procedures	

(continued)

Appendix 9.1 (continued)

Indicator	Classification
Membership associations	Stakeholder Policy
Approach to stakeholder engagement	
Collective bargaining agreements	
Financial assistance received from government	
Proportion of spending on local suppliers	
New suppliers that were screened using environmental criteria	
Negative environmental impacts in the supply chain and actions taken	
Operations and suppliers in which the right to freedom of association and collective bargaining may be at risk	
Operations and suppliers at significant risk for incidents of child labor	
Operations and suppliers at significant risk for incidents of forced or compulsory labor	
New suppliers that were screened using social criteria	
Negative social impacts in the supply chain and actions taken	
Political contributions	
Assessment of the health and safety impacts of product and service categories	
Incidents of non-compliance concerning the health and safety impacts of products and services	
Requirements for product and service information and labeling	
Incidents of non-compliance concerning product and service information and labeling	
Incidents of non-compliance concerning marketing communications	
Substantiated complaints concerning breaches of customer privacy and losses of customer data	

Source: GRI guidelines

References

Abdulla, Y., & Jawad, A. (2024). Do MENA firms consider climate risks? Evidence from the relationship between ESG and firm performance. *Journal of Sustainable Finance & Investment*, 1–19. https://doi.org/10.1080/20430795.2024.2334256

Abeysekera, I. (2022). A framework for sustainability reporting. *Sustainability Accounting, Management and Policy Journal, 13*(6), 1386–1409. https://doi.org/10.1108/SAMPJ-08-2021-0316

Abu Khalaf, B. (2024). Impact of board characteristics on the adoption of sustainable reporting practices. *Cogent Business & Management, 11*(1), 2391563. https://doi.org/10.1080/23311975.2024.2391563

Aboud, A., Saleh, A., & Eliwa, Y. (2024). Does mandating ESG reporting reduce ESG decoupling? Evidence from the European Union's Directive 2014/95. *Business Strategy and the Environment, 33*(2), 1305–1320. https://doi.org/10.1002/bse.3543

Aifuwa, H. O. (2020). Sustainability reporting and firm performance in developing climes: A review of literature. *Copernican Journal of Finance & Accounting, 9*(1), 9. https://doi.org/10.12775/CJFA.2020.001

Al Amosh, H., Khatib, S. F. A., & Hussainey, K. (2022). The financial determinants of integrated reporting disclosure by Jordanian companies. *Journal of Risk and Financial Management, 15*(9), 375. https://doi.org/10.3390/jrfm15090375

Aliyu, Y. M., & Nguavese, Y. R. (2024). Sustainability reporting and firm value of listed oil and gas firm in Nigeria: The Moderating Role of Board Independence. *International Journal of Accounting, Finance and Administrative Research, 1*(2), 64–80.

Alodat, A. Y., Al Amosh, H., Alorayni, O., & Khatib, S. F. A. (2024). Does corporate sustainability disclosure mitigate earnings management: Empirical evidence from Jordan. *International Journal of Disclosure and Governance, 21*(1), 165–174. https://doi.org/10.1057/s41310-023-00213-4

Al-Qudah, A. A., & Houcine, A. (2023). Firms' characteristics, corporate governance, and the adoption of sustainability reporting: Evidence from Gulf Cooperation Council countries. *Journal of Financial Reporting and Accounting, 22*(2), 392–415. https://doi.org/10.1108/JFRA-02-2023-0066

Aryan, L. A., Owais, W. O., Dahiyat, A., Rahamneh, A. A. A., Saraireh, S., Haija, A. A. A., & Al-Hawary, S. I. S. (2022). The effectiveness of corporate governance on corporate social responsibilities performance and financial reporting quality in Saudi Arabia's manufacturing sector. *Uncertain Supply Chain Management, 10*(4), 1141–1146. https://doi.org/10.5267/j.uscm.2022.8.013

Boulhaga, M., Elbardan, H., & Elmassri, M. (2023). The effect of internal control and corporate social responsibility on conditional accounting conservatism: Evidence from France. *Journal of Corporate Accounting & Finance, 34*(2), 228–241. https://doi.org/10.1002/jcaf.22605

Cherkaoui, A., & Cherkaoui, Z. (2021). Du reporting de la performance globale à la stratégie de reporting ESG: Pratiques des entreprises faisant appel public à l'épargne au Maroc. *Revue Congolaise de Gestion, 32*(2), 131–176. https://doi.org/10.3917/rcg.032.0131

Correa-Garcia, J. A., Garcia-Benau, M. A., & Garcia-Meca, E. (2020). Corporate governance and its implications for sustainability reporting quality in Latin American business groups. *Journal of Cleaner Production, 260*, 121142. https://doi.org/10.1016/j.jclepro.2020.121142

Christy, A., Elnahass, M., Amezaga, J., Browne, A., & Heidrich, O. (2023). A dynamic framework to align company climate reporting and action with global climate targets. *Business Strategy and the Environment, 33*(4). https://doi.org/10.1002/bse.3635

De Villiers, C., La Torre, M., & Molinari, M. (2022). The Global Reporting Initiative's (GRI) past, present and future: Critical reflections and a research agenda on sustainability reporting (standard-setting). *Pacific Accounting Review, 34*(5), 728–747. https://doi.org/10.1108/PAR-02-2022-0034

Diaz-Sarachaga, J. M., Jato-Espino, D., & Castro-Fresno, D. (2018). Is the Sustainable Development Goals (SDG) index an adequate framework to measure the progress of the 2030 Agenda? *Sustainable Development, 26*(6), 663–671. https://doi.org/10.1002/sd.1735

Dong, Y., Liang, C., & Wanyin, Z. (2023). Board diversity and firm performance: I*mpact of ESG activities in China*. Economic Research-Ekonomska Istraživanj

Dase, G. V., Rura, Y., & Pontoh, G. T. (2024). The Impact of Sustainability Report Disclosure on Financial Performance and Firm Value: *A Moderation Analysis with Good Corporate Governance*. 974–988. https://doi.org/10.2991/978-94-6463-400-6_75

Elalfy, A., & Darwish, K. (2020). Corporations and sustainable development goals communication on social media: Corporate social responsibility or just another buzzword? *Sustainable Development, 28*, 1418–1430. https://doi.org/10.1002/sd.2095

El Malki, T. (2010), Environnement des entreprises, responsabilité sociale et performance: analyse empirique dans le cas du Maroc, Thèse de doctorat, Université Aix Marseille 2.

Ezzahid, E., & Maouhoub, B. (2020). Capital Account Liberalization in Morocco: Is it Compatible with Fixed or Flexible Exchange Rate Regime? *Journal of Central Banking Theory and Practice, 9*(1), 185–218.

Gallén, M. L., & Peraita, C. (2017). The relationship between femininity and sustainability reporting. *Corporate Social Responsibility and Environmental Management, 24*(6), 496–508. https://doi.org/10.1002/csr.1423

Gavana, G., Gottardo, P., & Moisello, A. M. (2017). The effect of equity and bond issues on sustainability disclosure. Family vs non-family Italian firms. *Social Responsibility Journal, 13*(1), 126–142. https://doi.org/10.1108/SRJ-05-2016-0066

Gerwanski, J., Kordsachia, O., & Velte, P. (2019). Determinants of materiality disclosure quality in integrated reporting: Empirical evidence from an international setting. *Business Strategy and the Environment, 28*(5), 750–770. https://doi.org/10.1002/bse.2278

Gil-Marín, M., Vega-Muñoz, A., Contreras-Barraza, N., Salazar-Sepúlveda, G., Vera-Ruiz, S., & Losada, A. V. (2022). Sustainability accounting studies: A metasynthesis. *Sustainability, 14*(15), 9533. https://doi.org/10.3390/su14159533

Giner, B., & Luque-Vílchez, M. (2022). A commentary on the "new" institutional actors in sustainability reporting standard-setting: A European perspective. *Sustainability Accounting, Management and Policy Journal, 13*(6), 1284–1309. https://doi.org/10.1108/SAMPJ-06-2021-0222

GRI Standards (2016). Consolidated Set of GRI Sustainability Reporting Standards. Global Sustainability Standards Board. http://www.ekvilib.org/wp-content/uploads/2018/03/GRI-standardi-2016.pdf

Guix, M., Bonilla-Priego, M. J., & Font, X. (2018). The process of sustainability reporting in international hotel groups: An analysis of stakeholder inclusiveness, materiality and responsiveness. *Journal of Sustainable Tourism, 26*(7), 1063–1084. https://doi.org/10.1080/09669582.2017.1410164

Gerged, A. M., Salem, R., & Beddewela, E. (2023). How does transparency into global sustainability initiatives influence firm value? Insights from Anglo-American countries. *Business Strategy and the Environment, 32*(7), 4519–4547. https://doi.org/10.1002/bse.3379

Hausman, J., Stock, J. H., & Yogo, M. (2005). Asymptotic properties of the Hahn–Hausman test for weak-instruments. *Economics Letters, 89*(3), 333–342. https://doi.org/10.1016/j.econlet.2005.06.007

Hummel, K., & Jobst, D. (2024). An overview of corporate sustainability reporting legislation in the European Union. *Accounting in Europe, 21*(3), 320–355. https://doi.org/10.1080/17449480.2024.2312145

Janahi, M., Millo, Y., & Voulgaris, G. (2023). Age diversity and the monitoring role of corporate boards: Evidence from banks. *Human Relations, 76*(10), 1599–1633. https://doi.org/10.1177/00187267221108729

Joudeh, M. (2024). The impact of disclosing sustainability on the value added from the stakeholders' perspective of industrial companies listed on the amman stock exchange. *Quality-Access to Success, 25*(201). https://doi.org/10.47750/QAS/25.201.07

Karaman, A. S., Kilic, M., & Uyar, A. (2020). Green logistics performance and sustainability reporting practices of the logistics sector: The moderating effect of corporate governance. *Journal of Cleaner Production, 258*, 120718. https://doi.org/10.1016/j.jclepro.2020.120718

Kolk, A. (2008). Sustainability, accountability and corporate governance: Exploring multinationals' reporting practices. *Business Strategy and the Environment, 17*(1), 1–15. https://doi.org/10.1002/bse.511

Kouaib, A., & Amara, I. (2022). Corporate social responsibility disclosure and investment decisions: Evidence from Saudi indexed companies. *Journal of Risk and Financial Management, 15*(11), 495. https://doi.org/10.3390/jrfm15110495

Kuzey, C., & Uyar, A. (2017). Determinants of sustainability reporting and its impact on firm value: Evidence from the emerging market of Turkey. *Journal of Cleaner Production, 143*, 27–39. https://doi.org/10.1016/j.jclepro.2016.12.153

Menchi, M., & Chemlal, M. (2017). La perception de la responsabilité sociale de l'entreprise par les banques au Maroc: Quel impact sur la performance financière? *Revue du contrôle, de la comptabilité et de l'audit, 1*(3). https://www.revuecca.com/index.php/home/article/view/114

Marie, M., Qi, B., Elgammal, M., & Elnahass, M. (2024). A more sustainable future: Can politically connected CEOs spur the nexus between ESG performance and firm financial performance? *Journal of International Financial Markets. Institutions and Money, 96*, 102056. https://doi.org/10.1016/j.intfin.2024.102056

Meftah Gerged, A., Kuzey, C., Uyar, A., & Karaman, A. S. (2023). Does investment stimulate or inhibit CSR transparency? The moderating role of CSR committee, board monitoring and CEO duality. *Journal of Business Research, 159*, 113762. https://doi.org/10.1016/j.jbusres.2023.113762

Machado, B. A. A., Dias, L. C. P., & Fonseca, A. (2021). Transparency of materiality analysis in GRI -based sustainability reports. *Corporate Social Responsibility and Environmental Management, 28*(2), 570–580. https://doi.org/10.1002/csr.206

Meutia, I., Kartasari, S. F., & Yaacob, Z. (2022). Stakeholder or Legitimacy Theory? The Rationale behind a Company's Materiality Analysis: *Evidence from Indonesia. Sustainability, 14*(13), 7763. https://doi.org/10.3390/su14137763

Nandy, M., Kuzey, C., Uyar, A., Lodh, S., & Karaman, A. S. (2023). Can CSR mechanisms spur GRI adoption and restore its lost value relevance? *Journal of Applied Accounting Research, 24*(4), 609–634. https://doi.org/10.1108/JAAR-03-2022-0068

Namoussi, F. Z., & Cherqaoui, M. (2024). Legal obligation of ESG reporting and the quality of environmental disclosure practices: The case of Moroccan listed industrial companies. *International Journal of Law and Management, 67*(4), 445–458. https://doi.org/10.1108/IJLMA-12-2023-0269

Orazalin, N., & Mahmood, M. (2019). Determinants of GRI-based sustainability reporting: Evidence from an emerging economy. *Journal of Accounting in Emerging Economies, 10*(1), 140–164. https://doi.org/10.1108/JAEE-12-2018-0137

Shad, M. K., Lai, F.-W., Fatt, C. L., Klemeš, J. J., & Bokhari, A. (2019). Integrating sustainability reporting into enterprise risk management and its relationship with business performance: A conceptual framework. *Journal of Cleaner Production, 208*, 415–425. https://doi.org/10.1016/j.jclepro.2018.10.120

Slacik, J., & Greiling, D. (2019). Compliance with materiality in G4-sustainability reports by electric utilities. *International Journal of Energy Sector Management, 14*(3), 583–608. https://doi.org/10.1108/IJESM-03-2019-0010

Sobhani, F. A., Amran, A., & Zainuddin, Y. (2012). Sustainability disclosure in annual reports and websites: A study of the banking industry in Bangladesh. *Journal of Cleaner Production, 23*(1), 75–85. https://doi.org/10.1016/j.jclepro.2011.09.023

Soussi, L. H. L., & Outanna, K. (2023). Le reporting RSE: Quelles tendances de la recherche scientifique au Maroc ? *Revue Internationale des Sciences de Gestion, 6*(4). https://revue-isg.com/index.php/home/article/view/1412

Taoukif, F. E.-Z. (2014). Analyse perceptuelle des déterminants de l'engagement sociétal des entreprises marocaines labellisées RSE: de la performance au développement durable – cas du Maroc. Gestion et management. Université de Toulon; Université Moulay Ismaïl (Meknès, Maroc).

Torelli, R., Balluchi, F., & Furlotti, K. (2020). The materiality assessment and stakeholder engagement: A content analysis of sustainability reports. *Corporate Social Responsibility and Environmental Management, 27*(2), 470–484. https://doi.org/10.1002/csr.1813

Utami, W. (2015). Financial performance and the quality of sustainability disclosure based on global reporting initiative: Value relevances study in Indonesia Stock Exchange. *Mediterranean Journal of Social Sciences*. https://doi.org/10.5901/mjss.2015.v6n5s5p243

Velte, P. (2022). Does sustainable corporate governance have an impact on materiality disclosure quality in integrated reporting? International evidence. *Sustainable Development, 30*(6), 1655–1670. https://doi.org/10.1002/sd.2333

Younis, N. M. M. (2023). Sustainability reports and their impact on firm value: Evidence from Saudi Arabia. *International Journal of Management and Sustainability, 12*(2), 70–83. https://doi.org/10.18488/11.v12i2.3275

Zampone, G., Nicolò, G., Sannino, G., & De Iorio, S. (2024). Gender diversity and SDG disclosure: The mediating role of the sustainability committee. *Journal of Applied Accounting Research, 25*(1), 171–193. https://doi.org/10.1108/JAAR-06-2022-0151

Zimon, G., Arianpoor, A., & Salehi, M. (2022). Sustainability reporting and corporate reputation: The moderating effect of CEO opportunistic behavior. *Sustainability, 14*(3), 1257. https://doi.org/10.3390/su14031257

Xie, J., Tanaka, Y., Keeley, A. R., Fujii, H., & Managi, S. (2023). Do investors incorporate financial materiality? Remapping the environmental information in corporate sustainability reporting. Corporate Social Responsibility and Environmental Management. *Advance online publication*. https://doi.org/10.1002/csr.2524

Zayer, C., & Benabdelhadi, A. (2023). Qualitative research on the drivers and barriers of CSR in the Moroccan SME context. *Revue Management & Innovation, 7*(1), 76–104. https://doi.org/10.3917/rmi.207.0076

Zhou, Y., Lamberton, G., & Charles, M. B. (2023). An explanatory model of materiality in sustainability accounting: Integrating accountability and stakeholder heterogeneity. *Sustainability, 15*(3), 2700. https://doi.org/10.3390/su15032700

Chapter 10
From Barrels to Balance Sheets: The Role of Accounting in Saudi Aramco's Sustainability Strategy to Net-Zero

Hrishikesh Desai

Abstract This chapter provides a comprehensive analysis of Saudi Aramco's sustainability accounting practices and offers insights into how the world's largest integrated oil and gas company manages the challenges of environmental responsibility and energy transition. Through a multi-faceted examination of Aramco's strategies, the study explores the intersection of sustainability, technology, and corporate governance in the oil and gas sector. The research employs a mixed-method approach by combining qualitative analysis of corporate reports with quantitative text analysis using Natural Language Processing (NLP) techniques. It examines Aramco's carbon accounting framework, ESG integration, and technological innovations in sustainability reporting. The study also presents a comparative analysis with other major players like ADNOC and ExxonMobil to highlight industry trends and best practices. Theoretical perspectives, including legitimacy, stakeholder, and institutional theories, are applied to interpret Aramco's actions that reveal complex and sometimes contradictory forces shaping its sustainability strategy. Key findings indicate Aramco's significant strides in integrating advanced technologies such as AI, blockchain, and IIoT into its sustainability accounting, improving data accuracy and reporting reliability. The company's multi-tiered ESG governance structure demonstrates a commitment to embedding sustainability at all decision-making levels. However, the lack of comprehensive Scope 3 emissions reporting remains a notable gap. The chapter concludes by discussing implications for industry and policymakers.

Keywords Sustainability accounting · Carbon intensity · ESG integration · Saudi Aramco · Oil and gas industry

H. Desai (✉)
Department of Accounting, Neil Griffin College of Business, Arkansas State University, Jonesboro, Arkansas, USA
e-mail: hdesai@astate.edu

M. Elnahass et al. (eds.), *A Greenprint for Net Zero*, Gulf Studies 15,
https://doi.org/10.1007/978-981-92-1762-5_10

10.1 Introduction

While many companies in the MENA region and worldwide are making significant strides in sustainability, Saudi Aramco's unique blend of scale, technology integration, and international influence makes it an exemplary case study. Other companies might replicate certain aspects of Aramco's approach, but few can match its impact on both regional and global scales. In this chapter, I examine the role of accounting in Saudi Aramco's sustainability strategy as it aims to achieve net-zero emissions. As the world's largest integrated oil and gas company, Aramco's approach to sustainability accounting and reporting provides important insights into how major energy players are adapting to the global energy transition.

Section 10.2 of this chapter justifies the *Choice of Saudi Aramco* as the primary case study. It discusses Aramco's global significance, its unique position in the energy sector, and its recent sustainability initiatives. Section 10.3 of this chapter presents a *Theoretical Framework*, where relevant theories such as legitimacy theory, stakeholder theory, and institutional theory are discussed in relation to Aramco's sustainability practices. The section explains how these theories help interpret Aramco's sustainability actions and strategies. Section 10.4 of this chapter explains Aramco's *Carbon Accounting Framework and Practices*. This section provides a detailed analysis of Aramco's approach to carbon accounting. It examines the company's measurement, reporting, and verification processes for emissions, as well as the challenges it faces in these areas. Section 10.5 of this chapter is about *Integrating ESG into Corporate Strategy*. This part explores how Aramco incorporates Environmental, Social, and Governance (ESG) factors into its business operations. It looks at the company's governance structure, financial planning, and investment decisions related to sustainability.

Section 10.6 discusses *Technology and Innovation in Sustainability Accounting*. This section examines the technological solutions Aramco employs in its sustainability accounting practices. It discusses innovations such as Artificial Intelligence (AI) and Machine Learning (ML), Blockchain, Digital Twin Technology, Robotics and Drones, Industrial Internet of Things (IIoT), Virtual/Augmented Reality (VR/AR), 3D Printing, Cloud Computing, Big Data Analytics, and Robotic Process Automation (RPA) and their impact on the accuracy and efficiency of Aramco's sustainability reporting. In Sect. 10.7, a *Comparative Analysis of Sustainability Reporting* is done to compare Aramco's sustainability reporting with that of other major oil companies, specifically ADNOC and ExxonMobil. The section highlights similarities and differences in approaches, methodologies, and effectiveness of reporting. Section 10.8 on *Conclusion and Future Directions* is the final section that summarizes key findings, discusses implications for the industry and policymakers, and suggests areas for future research in sustainability accounting within the oil and gas sector.

The following key variables are under investigation in my study: *Carbon Intensity* (measure of GHG emissions per unit of economic output); *Scope 1 Emissions* (direct GHG emissions from owned or controlled sources); *Scope 2 Emissions* (indirect GHG emissions from purchased energy); *Scope 3 Emissions* (all indirect emissions in a company's value chain); *ESG Integration* (incorporation of environmental, social, and governance factors into business decisions); and *Sustainability Accounting* (practices for measuring, analyzing, and reporting a company's social and environmental impacts). This study aims to (1) analyze Aramco's carbon accounting framework and practices, (2) evaluate the integration of ESG factors into Aramco's corporate strategy, (3) assess the role of technology in Aramco's sustainability accounting, and (4) compare Aramco's sustainability reporting with key industry peers. My key research questions include: *How does Aramco's carbon accounting framework align with global standards? To what extent has Aramco integrated ESG considerations into its financial planning and risk management? What technological innovations is Aramco employing in its sustainability accounting practices? How does Aramco's sustainability reporting compare to that of the Abu Dhabi National Oil Company (ADNOC) and ExxonMobil?*

This study is motivated by the urgent need to understand how a major oil and gas company with a disproportionate impact on climate is responding to the global climate crisis. Aramco's size, influence, and recent sustainability commitments make it an important case study. This urgency is indicated by recent scholarship that highlights the key role of fossil fuel firms in global carbon emissions (Cardoni et al., 2019; Heede, 2023). Research indicates that national oil companies (NOCs) like Aramco face mounting pressure to align with climate accountability frameworks, particularly given their outsized contribution to cumulative emissions (Grasso, 2019, 2023). NOCs produce approximately half of the world's oil and gas and invest 40% of capital in the sector (Muttit & Heller, 2021). NOCs are also known to lag behind publicly listed companies in setting ambitious climate targets but are important players in achieving global methane reduction goal (Cahill & Swanson, 2023). Studies on corporate sustainability reporting in high-emission industries emphasize the legitimacy challenges these firms encounter, as stakeholders increasingly scrutinize their environmental disclosures (Haffar & Searcy, 2020; Nwagbara & Belal, 2019; Megura & Gunderson, 2022). Aramco's case is particularly salient given its state ownership, which introduces unique tensions between national economic imperatives and global sustainability expectations (Krane, 2021).

This research is important for policymakers developing regulations for the energy sector, investors assessing long-term viability of oil and gas companies, industry peers looking to benchmark their sustainability practices, and researchers studying the intersection of accounting and sustainability in high-impact industries. This study employs a mixed-method approach that combines qualitative analysis of corporate reports with quantitative text analysis using Natural Language Processing (NLP) techniques. This approach is used to conduct an in-depth review of Aramco's sustainability reports and related reports, annual reports, and public disclosures from 2021 to 2023. The year 2021 was the first period when Aramco issued a detailed sustainability report. It is also used to conduct a comparative analysis of sustainability reports from ADNOC and ExxonMobil for the same period. Finally, this approach is also used to conduct an assessment of alignment with global reporting standards (Global Reporting Initiative [GRI], Task Force on Climate-Related

Financial Disclosures [TCFD], and Sustainability Accounting Standards Board [SASB]).

This NLP-based methodology enables a comprehensive, data-driven analysis of how specific terms and concepts are presented and evolved in sustainability reports. It allows for a more sophisticated comparison across companies and time periods. This chapter aims to contribute to the literature by (a) providing an in-depth analysis of sustainability accounting practices in a major national oil company; (b) providing insights into the challenges and innovations in carbon accounting for complex, multinational operations; (c) examining the role of advanced technologies in improving sustainability reporting and performance; (d) comparing sustainability strategies across different ownership models in the oil and gas sector; and (e) evaluating the effectiveness of current sustainability reporting frameworks for capturing the full environmental impact of oil and gas operations.

This study advances legitimacy theory by examining how Aramco's sustainability reporting navigates tensions between state ownership and global climate expectations (Patten, 2020). As Aramco operates in a petrostate, it has to manage dual legitimacy pressures that involve maintaining legitimacy within a national context where oil revenues are central to economic stability while simultaneously seeking international legitimacy in a carbon-constrained world (potentially irreconcilable stakeholder expectations).

It also expands institutional theory by analyzing how Aramco conforms to (or resists) transnational sustainability norms while operating within a hydrocarbon-dependent national context (DiMaggio & Powell, 1983). The study identifies specific decoupling mechanisms and ceremonial adoption practices that may be employed when core business activities remain fundamentally misaligned with sustainability goals. This contributes to our understanding of organizational responses to institutional complexity.

It also refines stakeholder theory by revealing how Aramco prioritizes certain stakeholders (e.g., the Saudi government, investors) over others (e.g., environmental groups) in its ESG disclosures (Freeman et al., 2021). It demonstrates how stakeholder salience may differ significantly in state-controlled enterprises compared to traditional Western corporations, which challenges conventional stakeholder prioritization models.

This study also demonstrates the value of mixed-methods approaches in sustainability accounting research, combining qualitative analysis of corporate reports with NLP-based text analysis to track discursive shifts in Aramco's reporting. It also innovates in applying NLP techniques to compare sustainability narratives across peer firms (ADNOC, ExxonMobil) and thus provides a replicable framework for cross-industry benchmarking. This research develops and applies a novel analytical framework for comparing sustainability accounting practices across different ownership models (state-owned versus publicly traded companies) and geographical contexts (Middle Eastern versus Western companies). This structured comparative approach enables more systematic identification of patterns and divergences in sustainability reporting practices than has been possible in previous research.

For investors assessing the long-term viability of oil and gas companies, this study provides a systematic framework for evaluating the credibility and comprehensiveness of sustainability claims. It identifies key indicators that can help distinguish between substantial sustainability efforts and superficial reporting. This study also identifies emerging best practices and potential areas for industry-wide standardization. These findings can guide sustainability accounting practitioners in high-impact industries toward more effective measurement and reporting approaches. It also highlights specific gaps in current reporting frameworks that may need regulatory attention from policymakers. Finally, this study also provides a detailed examination of how advanced technologies can enhance sustainability accounting by providing a practical roadmap for companies seeking to improve the accuracy, efficiency, and credibility of their environmental reporting through technological innovation.

10.2 Why Saudi Aramco?

Saudi Aramco stands as a colossus in the global energy landscape, which makes it an essential case study for understanding the intersection of sustainability and the oil and gas industry, particularly within the Middle East and North Africa (MENA) region. As the world's largest integrated oil and gas company, Aramco's influence extends far beyond its operational footprint and shapes global energy markets, regional economic policies, and international sustainability practices. In this section, I explore the key reasons why Saudi Aramco serves as an ideal subject for examining sustainability accounting and reporting in the context of the global energy transition. The company has also been recently targeted and singled out by a group of United Nations (UN) Experts Working Group (UNWG) amongst all the other oil companies for its massive contribution to climate change and human rights (Reuters, 2023). The UNWG has also accused Aramco of greenwashing and using its renewable project investments to further its core oil and gas business (ClientEarth, 2023).

10.2.1 Dominant Industry Position

Saudi Aramco's sheer scale and market dominance make it a key player in the global energy sector. As of 2023, the company reported a net income of $121 billion that makes it the world's most profitable company (Saudi Aramco, 2023). Its production capacity is equally impressive, with a maximum sustainable capacity of 12 million barrels per day of crude oil (Saudi Aramco, 2023). This unparalleled scale means that Aramco's strategies and operations have profound implications for global energy policies and practices.

The company's influence is particularly pronounced in the MENA region, where it serves as the de facto leader for industry practices. Aramco's operations directly impact Saudi Arabia's economy and contribute significantly to the country's GDP and government revenues. In 2023, the company's payments to the Saudi and foreign governments, including dividends, income taxes, and royalties, amounted to $205 billion (Saudi Aramco, 2023). This economic significance extends Aramco's influence beyond mere market share and positions it as a key driver of regional economic and energy policies.

Moreover, Aramco's global reach through its various subsidiaries and joint ventures, including its ownership of Saudi Basic Industries Corporation or SABIC (one of the world's largest petrochemicals manufacturers), amplifies its impact on international energy markets and sustainability practices (Saudi Aramco, 2023). This expansive influence makes Aramco an ideal subject for studying how major energy players can balance operational demands with evolving sustainability goals in a rapidly changing global energy landscape.

10.2.2 Leadership in Sustainability Efforts

Despite its position as a hydrocarbon giant, Saudi Aramco has emerged as a surprising leader in sustainability efforts within the oil and gas industry. The company's commitment to achieving net-zero Scope 1 (direct emissions from owned or controlled sources) and Scope 2 (indirect emissions from the generation of purchased electricity, steam, heating, and cooling consumed) greenhouse gas (GHG) emissions across its wholly owned operated assets by 2050 sets a new standard for environmental responsibility in the sector (Saudi Aramco, 2023). This ambitious target aligns with broader national initiatives, including Saudi Arabia's Vision 2030 and the Saudi Green Initiative, which aim to diversify the economy and reduce the country's dependence on oil (Kingdom of Saudi Arabia, 2016).

Aramco's sustainability efforts are particularly noteworthy given the challenges inherent in decarbonizing oil and gas operations. The company has implemented a multi-faceted approach to reducing its environmental impact, including investments in carbon capture, utilization, and storage (CCUS) technologies, methane emissions reduction, and energy efficiency improvements (Saudi Aramco, 2022, 2023). For example, Aramco has set interim targets to reduce upstream carbon intensity by at least 15% and GHG emissions (Scopes 1 and 2) by 52 million tonnes of CO_2e annually by 2035 (Saudi Aramco, 2023). These initiatives position Aramco as a leader in sustainability efforts within a region historically dependent on fossil fuels. The company's approach is important to understand how major oil companies can integrate environmental goals with their economic objectives. It also provides a potential roadmap for other firms in the MENA region and beyond.

10.2.3 Implementation of Advanced Technologies

Aramco's embrace of cutting-edge technologies to improve its sustainability measures sets it apart in the industry. The company has made significant investments in artificial intelligence (AI), Internet of Things (IoT), and big data analytics to optimize its operations and improve its environmental performance (Saudi Aramco, 2023). These technological advancements enable more accurate data collection, better transparency, and improved accountability in environmental reporting.

For example, Aramco has deployed advanced leak detection and repair (LDAR) programs using optical gas imaging cameras and other sensors to identify and reduce methane emissions (Saudi Aramco, 2023). The company has also implemented a real-time monitoring system for flare management, allowing for quick intervention and minimization of unnecessary flaring (Saudi Aramco, 2023). Flare management in the oil and gas sector refers to the practices used to control, reduce, and ideally eliminate the burning of excess natural gas at oil and gas production sites that release GHG (primarily CO2 and Methane) into the atmosphere. These technological solutions not only improve operational efficiency but also enhance the accuracy and reliability of Aramco's sustainability reporting. The integration of these advanced technologies into Aramco's sustainability practices offers a compelling case study on how digital transformation can drive environmental performance in the oil and gas sector. It demonstrates how technology can bridge the gap between traditional energy production and sustainability goals.

10.2.4 Impact on Regional Sustainability Practices

The MENA region presents a unique context for sustainability accounting due to its economic dependence on hydrocarbon resources and distinctive governance structures. Ismaeel & Zakaria, (2019) note that sustainability is not perceived the same way by companies in the MENA region, and the difference in perception is dependent on whether the company has global or local operations. Prior empirical research by Uyar et al. (2019), Ahmed et al. (2021), and Aljanadi & Alazzani (2023) that analyzes sustainability reporting among listed companies in MENA countries finds significant variation in disclosure practices both across and within countries. Findings from Qasem et al. (2024) also suggest that Corporate Social Responsibility (CSR) and sustainability reporting practices are still developing in Saudi Arabia, which is representative of broader trends across the MENA region. They identify cultural, regulatory, and institutional factors as barriers to comprehensive disclosure practices.

Aramco's efforts to reduce its carbon footprint and manage environmental impacts set a precedent for other firms in the region, where such practices are still evolving. Aramco's influence is particularly notable in its role in advancing the concept of the Circular Carbon Economy (CCE), which aims to reduce, reuse, recycle, and remove carbon emissions (Saudi Aramco, 2023).

Aramco's leadership in sustainability has catalyzed broader regional initiatives. For example, the company's commitment to net-zero emissions by 2050 has been followed by similar pledges from other national oil companies in the region. For example, the Abu Dhabi National Oil Company or ADNOC in the UAE pushed back its target date of achieving net-zero emissions by five years to 2045 from 2040 (ADNOC, 2023, Ipieca Ltd., 2024). This demonstrates how Aramco's actions can drive a shift toward more sustainable practices across the MENA energy sector.

Moreover, Aramco's sustainability efforts extend beyond its own operations. The company has been actively involved in promoting sustainability across its value chain by encouraging suppliers and partners to adopt more environmentally friendly practices. For example, through its In-Kingdom Total Value Add (iktva) program, Aramco has been promoting localization of the supply chain while also emphasizing sustainability criteria in supplier selection (Saudi Aramco, 2021, 2022, 2023). This approach showcases how a major company can utilize its market position to drive sustainability improvements throughout its ecosystem.

10.2.5 Challenges and Innovations in Sustainability Accounting

Saudi Aramco faces unique challenges in its sustainability journey due to its massive scale of operations, the geopolitical context in which it operates, and the inherent tension between maintaining oil production and reducing carbon emissions. These challenges necessitate innovative approaches to sustainability accounting and reporting. One of the primary challenges is accurately measuring and reporting emissions across Aramco's vast and complex operations. The company has responded by implementing advanced measurement technologies and developing sophisticated accounting methodologies. For instance, Aramco has been working on improving the accuracy of its methane emissions measurements, a critical area given methane's potent greenhouse effect (Saudi Aramco, 2023).

Another significant challenge is balancing the company's core business of hydrocarbon production with its sustainability goals. Aramco has addressed this through a strategy of focusing on operational efficiency and developing lower-carbon technologies. The company's investments in carbon capture and storage, blue hydrogen production, and non-metallic materials demonstrate its efforts to find innovative solutions to reduce the carbon intensity of its products (Saudi Aramco, 2022, 2023).[1] Aramco's approach to these challenges suggests how large, complex organizations

[1] Carbon capture and storage means catching the carbon dioxide that's produced and storing it safely underground instead of letting it go into the air. Blue hydrogen production is a way to make hydrogen fuel that produces less carbon than traditional methods. Non-metallic materials are materials that can replace metals in some uses, and often have a lower carbon footprint.

can develop comprehensive sustainability accounting practices and also illustrates the importance of integrating sustainability considerations into core business strategies and the need for continuous innovation in measurement and reporting methodologies.

10.2.6 Positive and Negative Contributions to Global Standards and Practices

Aramco's sustainability reporting practices and its involvement in global initiatives contribute significantly to shaping international energy and sustainability policies. The company prepares its sustainability reports in accordance with the Global Reporting Initiative (GRI) Standards and aligns its climate-related disclosures with the recommendations of the Task Force on Climate-related Financial Disclosures (TCFD) (Saudi Aramco, 2023).

However, the UNWG that targeted Aramco have accused the company of failing to account for Scope 3 GHG emissions that form the lion's share of the company's climate impact, thus disregarding the GHG Protocol and the Task Force of Climate-related Financial Disclosures (TCFD) standards (Nordsip, 2023). Scope 3 emissions or value chain emissions are indirect greenhouse gas emissions that occur in a firm's value chain, both upstream and downstream, that are not directly owned or controlled by the firm itself. Aramco did not respond to several UNWG accusations, including those relating to misrepresentation of their carbon intensity and emissions, within the time provided to them. In 2021, the UK-based NGO ClientEarth also filed a legal complaint against Aramco accusing it of the largest climate-related breaches of international human rights law (ClientEarth, 2021). However, such greenwashing allegations are levelled against several major corporates operating in almost every sector.

Aramco's active participation in global industry collaborations, such as the Oil and Gas Climate Initiative (OGCI), demonstrates its role in driving industry-wide sustainability efforts. As a founding member of OGCI, Aramco has been involved in collective efforts to accelerate the industry's response to climate change (OGCI, 2023). Aramco's contributions extend to the development of new sustainability metrics and methodologies. For example, the company has been at the forefront of efforts to standardize the measurement and reporting of carbon intensity in the oil and gas industry. In 2023, Aramco launched a pilot project to deliver Arabian Light crude oil with a product-level carbon emissions intensity, using a novel cradle-to-gate partial life cycle assessment approach. This was verified by a third party, which contributes to the standardization efforts in product-specific emissions reporting. Aramco has also consistently increased the number of sustainability metrics it reports, from 36 in 2021 to 61 in 2023. Its work in this area provides a benchmark for other companies and contributes to the ongoing evolution of global sustainability reporting standards.

10.3 Theoretical Framework

Several organizational theories provide valuable lenses through which to analyze Saudi Aramco's approach to sustainability accounting and reporting. This section examines how legitimacy theory, stakeholder theory, and institutional theory apply to Aramco's sustainability practices.

10.3.1 Legitimacy Theory

Legitimacy theory posits that organizations continually seek to ensure that they operate within the bounds and norms of their respective societies (Patten, 2020). It suggests that organizations must align their actions with societal expectations to maintain their "social license to operate" (Suchman, 1995). Suchman (1995) distinguishes between *pragmatic legitimacy* (stakeholder self-interest), *moral legitimacy* (normative approval), and *cognitive legitimacy* (taken-for-granted acceptance).

In the context of Aramco, this theory helps explain the company's extensive sustainability reporting and net-zero commitments. The distinctions made in Suchman (1995) are particularly relevant here. The company seeks pragmatic legitimacy through its contributions to Saudi Arabia's economy and its shareholders' returns. It pursues moral legitimacy through its sustainability initiatives and net-zero commitments by attempting to demonstrate that it is taking responsible action on climate change. However, Aramco faces particular challenges in establishing cognitive legitimacy in a global context increasingly questioning the very existence of fossil fuel businesses, which explains its emphasis on portraying oil and gas as an essential part of the energy transition.

Deegan (2002) emphasizes that organizations employ various strategies to gain, maintain, or repair legitimacy, including changing their activities to conform to stakeholder expectations, which can be seen as an attempt to alter stakeholder perceptions without changing actual behavior. Aramco's sustainability disclosures may reflect such *strategic legitimacy management*, wherein the company emphasizes operational efficiencies (e.g., carbon capture) to counterbalance its hydrocarbon core (Deegan, 2002). As a major contributor to global carbon emissions, Aramco faces significant pressure to justify its operations and demonstrate its commitment to addressing climate change. However, as Deegan (2002) notes, merely increasing the volume of environmental disclosures does not necessarily enhance legitimacy if these disclosures are perceived as incomplete or misleading. This helps explain why criticisms from the UNWG regarding Aramco's failure to account for Scope 3 emissions (Nordsip, 2023) pose a significant threat to the company's legitimacy efforts. Aramco's focus on Scope 1 and 2 emissions reporting, while omitting Scope 3, aligns with Deegan's (2002) observation that firms often disclose "safer" environmental data to mitigate reputational risks without fundamentally altering high-impact activities. This selective transparency highlights the tension between *symbolic legitimacy* (appearing compliant) and *substantive legitimacy* (implementing transformative change).

Aramco's adoption of advanced technologies for emissions monitoring and reduction, such as the flare monitoring system and leak detection programs, can be seen as efforts to maintain legitimacy in an increasingly carbon-conscious global environment (Saudi Aramco, 2023, 2024). By showcasing these initiatives, Aramco attempts to align its operations with societal expectations regarding environmental responsibility. However, the criticisms from the UNWG regarding Aramco's failure to account for Scope 3 emissions (Nordsip, 2023) highlight the challenges in maintaining legitimacy. This suggests that mere disclosure may not be sufficient; the content and comprehensiveness of sustainability reporting are important for legitimacy.

10.3.2 Stakeholder Theory

Stakeholder theory argues that companies should create value for all stakeholders, not just shareholders (Freeman et al., 2020, 2021). This theory is particularly relevant to Aramco given its diverse stakeholder base, including the Saudi government, international investors, employees, and global consumers.

Aramco's multi-tiered ESG governance structure, which includes board-level oversight and operational integration (Saudi Aramco, 2023), suggests an attempt to address various stakeholder concerns. The company's investments in new energy technologies and the establishment of a $1.5 billion Sustainability Fund (Saudi Aramco, 2022) can be interpreted as responses to stakeholder pressures for climate action. However, the company's reluctance to fully disclose Scope 3 emissions suggests a potential prioritization of certain stakeholders (e.g., government and shareholders concerned about short-term profitability) over others (e.g., environmental groups and long-term investors concerned about climate risks).

10.3.3 Institutional Theory

Institutional theory focuses on how organizational structures and practices are shaped by institutional pressures, including regulatory, normative, and cultural-cognitive elements (DiMaggio & Powell, 1983; Galbreath, 2013). This theory is particularly relevant given Aramco's position as a state-owned enterprise operating in a global context. A central concept in institutional theory is *isomorphism*, the process by which organizations within a field become increasingly similar as they respond to common pressures. DiMaggio and Powell (1983) identify three mechanisms of institutional isomorphism: coercive, mimetic, and normative.

Coercive isomorphism stems from formal and informal pressures exerted on organizations by other organizations upon which they are dependent and by cultural expectations in the society within which they operate. For Aramco, coercive pressures include regulations related to emissions reporting, international climate agreements like the Paris Agreement, and the Saudi government's Vision 2030 initiative, which emphasizes economic diversification and sustainability (Kingdom of Saudi Arabia, 2016). As a state-owned enterprise, Aramco is particularly subject to coercive pressures from its primary shareholder, the Saudi government.

Mimetic isomorphism occurs when organizations model themselves after others that they perceive to be more legitimate or successful, particularly in contexts of uncertainty. Aramco's adoption of global reporting standards such as GRI and TCFD (Saudi Aramco, 2023) and its participation in industry collaborations like the Oil and Gas Climate Initiative (OGCI, 2023) indicate mimetic isomorphism. The company also mirrors ADNOC's net-zero targets (2045 vs. 2050), which suggests industry benchmarking. By following reporting practices established by industry peers and endorsed by international bodies, Aramco signals its alignment with global best practices in sustainability reporting.

Normative isomorphism stems from professionalization, i.e., the collective struggle of members of an occupation to define the conditions and methods of their work and to establish a cognitive base and legitimation for their occupational autonomy. Aramco's approach to sustainability accounting is influenced by normative pressures from professional associations and standards-setting bodies in the accounting and sustainability fields. The company's engagement with external assurance providers and its adoption of standardized methodologies for emissions calculation reflect these normative influences.

Aramco's alignment with global reporting standards such as GRI and TCFD (Saudi Aramco, 2023) can be seen as a response to normative pressures in the international business environment. The company's participation in industry collaborations like the Oil and Gas Climate Initiative (OGCI, 2023) further demonstrates conformity to institutional norms. However, the institutional context of Saudi Arabia, with its historical reliance on oil revenues, creates competing pressures. This tension is evident in Aramco's dual focus on maintaining oil production while investing in sustainability initiatives. Thus, Aramco resists full institutional conformity, particularly on Scope 3 emissions, which demonstrates how national institutional logics (e.g., Saudi Vision 2030's economic diversification) compete with global sustainability norms.

While no single theory fully explains Aramco's approach to sustainability, the combination of legitimacy, stakeholder, and institutional theories provides a comprehensive framework for understanding the company's actions in the context of global sustainability challenges and local economic realities.

To sum it up, the application of these theories to Aramco's sustainability practices reveals a couple of interesting insights: Aramco's extensive sustainability reporting and technological investments can be seen as legitimacy-seeking behaviors in response to global climate concerns. The company's sustainability strategy indicates an attempt to balance diverse stakeholder interests, though challenges remain in fully addressing all stakeholder concerns. While Aramco shows conformity to global sustainability norms, its practices also reflect the unique institutional pressures of its home country. The simultaneous application of these theories highlights the complex and sometimes contradictory forces shaping Aramco's sustainability practices.

10.4 Research Design and Methodology

This research follows an explanatory sequential mixed methods design (Creswell & Clark, 2017), conducted in two distinct phases. In the first phase, a quantitative content analysis using Natural Language Processing (NLP) techniques was performed to identify patterns and trends in Aramco's sustainability reporting. The second phase involved a qualitative analysis to interpret these patterns and explore the underlying rationales, contextual factors, and implications. This design was selected because it allows for initial identification of reporting patterns and subsequent in-depth exploration of the meanings and motivations behind these patterns (Molina-Azorin, 2016).

10.4.1 Data Collection

The research employs a purposive sampling strategy focusing on official corporate reports from Saudi Aramco and selected industry peers. The primary documents analyzed for Saudi Aramco included the following: Sustainability Reports (2021–2023), Annual Reports (2021–2023), Corporate governance reports and policy documents, and Investor presentations related to sustainability. The following comparative documents were also analyzed: ADNOC Sustainability Reports (2021–2023), ExxonMobil Sustainability Reports (2021–2023), and relevant regulatory frameworks and standards (GRI, TCFD, SASB). This sampling approach enables both longitudinal analysis (tracking changes over time) and cross-sectional analysis (comparing practices across companies).

10.4.2 Quantitative Analysis: NLP-Based Content Analysis

This research used the latest available versions of the large language models (LLMs) including ChatGPT (GPT-4), Claude 3 Opus, and Grok 2 (xAI) to analyze sustainability reports. These advanced AI models provide significant advantages over traditional NLP techniques for analyzing complex sustainability disclosures (Hutt et al., 2024; Roumengas et al., 2023).

Each of these LLMs was used for initial content extraction and semantic pattern identification, with particular attention to sustainability terminology and contextual meaning. Their advanced reasoning capabilities enabled identification of implicit sustainability claims and commitments in each report. They were also used to **analyze sustainability narratives** and detecting greenwashing patterns, legitimacy-building language, and how environmental initiatives are framed. They were also used to extract quantitative metrics by systematically identifying and comparing numerical data and targets across companies. This comprehensive approach, based on employing multiple LLMs for all analytical tasks, provided a robust framework for triangulation and validation of findings.

To ensure reliability, a systematic cross-validation approach was implemented. Each of the three AI models independently performed identical analytical tasks on the same reports and documents and created a matrix of parallel analyses. This method allowed for direct comparison of findings across models and strengthened the validity of results through convergent evidence. All the three models consistently identified similar patterns in how companies reported sustainability issues.

Where discrepancies occurred between model outputs, a consensus approach was employed. Findings were considered reliable when at least two of the three models produced substantially similar results. In cases where models produced divergent analyses, additional prompt engineering was employed to explore the sources of disagreement and refine the analytical approach.

All the LLMs demonstrated sophisticated understanding of the contexts in which sustainability terms appear and were able to distinguish between substantive commitments and superficial mentions in ways that traditional keyword counting or statistical approaches cannot match. Each model showed the ability to make connections across different sections of reports and between different companies' documents and identified inconsistencies, common industry narratives, and distinctive approaches that might be missed in isolated text analysis. The models demonstrated capabilities in interpreting both textual content and graphical representations (when described or converted to text) and provided a more comprehensive analysis of how the sample companies present their sustainability data.

10.4.3 Qualitative Analysis: Interpretive Content Analysis

The qualitative phase employed an interpretive content analysis approach guided by the results of the quantitative analysis. A systematic coding process was implemented following Gioia et al.'s (2013) methodology. First, open coding was performed to identify basic themes and concepts in the reports in order to stay close to the language used in the documents. Then, first-order codes were abstracted into theoretical categories to draw connections to concepts from legitimacy, stakeholder, and institutional theories (referred to as second-order coding). Finally, second-order themes were integrated into overarching theoretical dimensions that explain Aramco's sustainability accounting practices. This coding process was applied to sections of the sustainability reports identified as significant through the quantitative analysis, as well as to comparative documents from ADNOC and ExxonMobil.

10.5 Aramco's Carbon Accounting Framework and Practices

This section explores Aramco's comprehensive carbon accounting framework and practices, with a focus on their measurement, reporting, and verification processes for Scope 1 and Scope 2 emissions. I also assess the internal and external challenges faced by Aramco, including technological limitations and regulatory compliance, and examine the impact of these practices on the company's sustainability and financial reporting.

10.5.1 Carbon Reporting Framework

Aramco's carbon accounting framework is built on the foundations of the GHG Protocol, which is widely recognized as the global standard for corporate GHG emissions reporting. Aramco's Scope 1 emissions are direct emissions from sources owned or controlled by Aramco, including production facilities, refineries, and transportation fleets. The company employs a combination of continuous emission monitoring systems (CEMS) and calculation methodologies based on fuel consumption and emission factors. Their Scope 2 emissions are indirect emissions that are primarily from purchased electricity and steam.

The company uses both location-based and market-based methods for calculating Scope 2 emissions. The location-based method calculates Scope 2 emissions based on the average emissions intensity of grids on which energy consumption occurs. This method uses regional or national grid average emission factors to determine the GHG emissions associated with the electricity consumed. The factors are generally provided by government agencies or grid operators and reflect the typical energy generation mix of a region (e.g., coal, natural gas, renewables). This method does not require tracking specific contractual agreements for electricity purchases, which simplifies the reporting process. The two advantages of this method are (1) *Standardization* (it provides a consistent approach that can be applied across different regions, which makes it easier to compare emissions across locations); and (2) *Benchmarking* (it allows firms to benchmark their performance against regional or national averages).

The market-based method reflects the GHG emissions from the electricity that companies have purposefully chosen or contracted. This method considers whether the electricity comes from renewable sources, like solar or wind power, which are often verified through special certificates. These certificates, such as Renewable Energy Certificates (RECs) in the United States or Guarantees of Origin (GOs) in Europe, are like proof of purchase. They confirm that certain amounts of electricity were produced from renewable sources. When companies use or buy these certificates, they can claim and report lower greenhouse gas emissions. This method requires detailed tracking of energy purchase agreements and certificates, which can be administratively complex. The two advantages of this method are (a) ***Choice and Impact*** (highlighting the choices a firm is making regarding its electricity sourcing and the direct impact of its procurement decisions on its carbon footprint); and (b) ***Incentivization*** (incentivizing investments in cleaner energy technologies by allowing firms to report lower emissions through the purchase of low carbon/renewable energy).

As of 2023, Aramco has not yet reported comprehensive Scope 3 emissions. The company acknowledges the challenges in measuring these value chain emissions and is working on methodologies to address this gap.

10.5.2 Measurement Practices

Aramco employs a range of advanced technologies and methodologies to measure its GHG emissions. Its *Flare Monitoring System* provides real-time, company-wide monitoring of all flare sources at an asset level. Gas flaring is when excess gas is burned off, creating a visible flame. The monitoring system watches how much gas is being flared and how efficiently it's burning. Its *Leak Detection and Repair (LDAR) Program* utilizes optical gas imaging cameras and other sensors to identify and reduce methane emissions. This is a set of practices used to find and fix leaks in equipment that handles gases or liquids and involves regularly checking pipes, valves, and other components for leaks using special tools.

Aramco's *GHGSat Satellite Constellation*, deployed in 2023, are to identify and monitor methane emissions from in-Kingdom operations. These satellites from space use special sensors to spot methane and other gases that contribute to climate change and can pinpoint emissions from specific industrial sites, which helps to identify major sources of pollution. Its *Energy Intensity Tracking* measures the ratio of total net energy consumption to total production. It involves measuring how much energy is used to produce a certain amount of goods or services. For example, Aramco might track how many kilowatt-hours of electricity or how many cubic feet of natural gas are consumed in the process of extracting, processing, and refining one barrel of crude oil. By monitoring this over time, the company executives can assess whether they're becoming more or less energy-efficient in their operations.

10.5.3 Reporting Practices

Aramco's reporting practices include (1) *Annual Sustainability Report*, which provides detailed disclosure of emissions data and sustainability initiatives; (2) *Alignment with GRI Standards and TCFD Recommendations*, which ensures comparability and credibility of reported data (though the company has been accused of greenwashing these disclosures as per the UN Experts Working Group); (3) *Expanded Metrics*, which increased from 36 in 2021 to 61 in 2023, thus demonstrating a commitment to comprehensive reporting; (4) *Internal Validation*, which requires their data undergo reviews by reporting businesses and internal subject matter experts; (5) *Third-Party Assurance*, which requires key metrics, including emissions data, undergo external limited assurance; and (6) *Continuous Improvement*, which involves regular review and improvement of data collection and reporting processes.

10.5.4 Challenges and Innovations

Aramco faces several challenges in its journey to net-zero, such as technological limitations, regulatory compliance, and Scope 3 emissions.

Aramco is developing and scaling carbon capture and storage technologies (CCS). CCS technology is important for capturing CO2 emissions at the source and storing them underground to prevent them from entering the atmosphere. However, the technology is still not fully matured and is costly. Scaling these technologies from pilot projects to full operational levels across all of Aramco's operations is a significant challenge due to high capital costs and technological complexities. Integrating CCS into existing infrastructure requires substantial modifications and presents operational risks. Ensuring the reliability and efficiency of these technologies once integrated is also a complex endeavor.

Aramco is also trying to navigate evolving global and national emissions regulations. The regulatory landscape for emissions is complex and varies by region. Aramco must comply with international agreements such as the Paris Agreement, as well as local regulations in the countries it operates. These regulations are also continuously evolving, which requires Aramco to be agile and proactive in compliance management. Compliance with stringent regulations often requires significant financial investment in technology, process improvements, and sometimes operational shifts, which can impact profitability.

Aramco is also developing methodologies to accurately measure and report value chain emissions. Aramco's operations involve extensive and complex supply chains that make it challenging to track and manage indirect emissions associated with activities like transportation, product use, and the services of third-party vendors. There is still no universally accepted methodology for calculating Scope 3 emissions comprehensively and accurately, which complicates efforts to standardize reporting and reduction strategies across the industry. The company is focusing on collaborations with suppliers to ensure transparency and engagement in emissions reduction. Aramco is also exploring ways to influence emissions reductions downstream through product design and supporting the use of renewable energy by its customers.

To sum it up, Aramco's path to net-zero is fraught with significant challenges that span technological, regulatory, and methodological aspects.

Aramco's selective disclosure patterns in carbon accounting indicate strategic legitimacy management as conceptualized by Suchman (1995). The company's emphasis on Scope 1 and 2 emissions, while minimizing attention to Scope 3, suggests what Deegan (2002) terms "strategic disclosure," which focuses stakeholder attention on areas where the company performs well while deflecting from more problematic aspects. This selective approach allows Aramco to maintain pragmatic legitimacy with immediate stakeholders while avoiding threats to its cognitive legitimacy that might arise from comprehensive Scope 3 disclosures.

From an institutional theory perspective, Aramco's adoption of GRI and TCFD frameworks represents a response to normative isomorphic pressures (DiMaggio & Powell, 1983). However, the company's reluctance to fully implement these frameworks (particularly regarding Scope 3 emissions) demonstrates the complex interplay between competing institutional logics, i.e., international sustainability norms versus the institutional logic of petrostate economic development that prioritizes continued hydrocarbon production. This tension highlights what Greenwood et al. (2011) describe as organizational responses to institutional complexity, wherein organizations selectively comply with certain institutional demands while buffering themselves from others.

10.5.5 Impact on Financial Reporting

Aramco's carbon accounting practices are also increasingly integrated into its financial reporting. For example, Aramco's *climate-related risk disclosures* include climate-related risks and opportunities in financial reports. The company's *investment decisions* factor in consideration of carbon intensity in capital allocation decisions. They also have internal *carbon pricing* mechanisms to guide investment and operational decisions.

10.6 Integrating ESG into Corporate Strategy

In this section, I analyze how ESG considerations are embedded in Aramco's financial planning, investment decisions, and risk management frameworks, and evaluate the influence of ESG on Aramco's operational and financial performance.

10.6.1 ESG Governance Structure

Aramco has established a governance structure across three levels to oversee its ESG initiatives (Saudi Aramco, 2023), that includes board-level oversight, executive management, and operational integration.

The Board of Directors' Sustainability, Risk and Health, Safety, and Environment (HSE) Committee provides leadership and oversight on ESG matters. This committee meets quarterly to assess performance, including health and safety incidents, and endorses sustainability reports. The Sustainability Steering Committee (SSC), led by the Executive Vice President of Strategy and Corporate Development, reports to both the Group Strategy Committee and Group Executive Committee. The SSC reviews and guides decisions related to sustainability for alignment with long-term value creation. The Group Health, Safety, Security, and Environment Committee (GHSSE), led by the President and CEO, shapes Aramco's health, safety, security, and environmental policies.

Thus, the company has a multi-tiered approach indicating that ESG considerations are integrated at all levels of decision-making within the company.

10.6.2 ESG in Financial Planning and Investment Decisions

Aramco seems to have incorporated ESG factors into its financial planning and investment decision-making processes.

In 2022, Aramco established a $1.5 billion Sustainability Fund to invest in technologies needed to support its net-zero ambitions (Saudi Aramco, 2022). In 2023, Aramco allocated around 10% of its capital investments to its New Energies organization, which demonstrates a significant financial commitment to sustainable initiatives (Saudi Aramco, 2023). However, as of mid-2024, there is no evidence of Aramco or its subsidiary SABIC directly issuing green bonds.

Aramco has linked senior executive remuneration to sustainability performance. The variable component of executive pay is contingent on meeting predefined goals, including sustainability metrics such as GHG emissions intensity, flaring, and safety (Saudi Aramco, 2023). While it's challenging to isolate the direct financial impact of ESG initiatives, Aramco's strong financial performance alongside its sustainability efforts suggests that its ESG strategy is not coming at the expense of profitability. Many of Aramco's sustainability initiatives, such as energy efficiency projects and flare gas recovery, have led to operational cost savings that contribute to the company's financial performance (Saudi Aramco, 2023). However, Aramco faces several challenges in fully integrating ESG into its corporate strategy. As discussed earlier, the company has been criticized for not reporting Scope 3 emissions, which are likely to be substantial given the nature of its business (ClientEarth, 2023). There has been increasing pressure from investors, regulators, and civil society for greater ESG performance and transparency from Aramco (Reuters, 2023).

The company's board-level oversight of sustainability and executive-level implementation structures demonstrate an attempt to address diverse stakeholder interests. However, the findings suggest a hierarchical prioritization of stakeholders, with government and majority shareholders' interests (economic returns and national development) generally prevailing over broader stakeholder concerns (comprehensive climate action). This aligns with Mitchell et al.'s (1997) stakeholder salience model, wherein stakeholders with power, legitimacy, and urgency receive priority attention. The Saudi government, as majority shareholder with all three attributes, shapes ESG integration in ways that protect the core business.

Aramco's multi-tiered ESG governance also indicates institutional isomorphism (DiMaggio & Powell, 1983). The Board-level Sustainability Committee mimics structures at BP and Shell (mimetic pressure), while adherence to GRI/TCFD reflects normative pressure from financial markets. However, the absence of Scope 3 targets reveals coercive pressure resistance, which could be a divergence tied to Saudi Arabia's rentier-state logic.

10.7 Technology and Innovation in Sustainability Accounting

Aramco has leveraged cutting-edge technologies in its approach to sustainability accounting, enhancing the accuracy, efficiency, and transparency of its reporting. The company's technological innovations span across data collection, analysis, and reporting processes.

10.7.1 Artificial Intelligence and Machine Learning

The company extensively uses AI and ML algorithms to analyze vast amounts of sustainability data. These technologies are particularly evident in predictive analytics and anomaly detection (Saudi Aramco, 2023).

Aramco employs machine learning models to forecast potential environmental impacts and energy consumption patterns. These models, likely using techniques such as random forests or gradient boosting, analyze historical data to predict future trends, which allow proactive measures to be taken (Yoon, 2021). The company's in-house AI-powered i4Safety 2.0 Hazard & Incident Prediction tool uses historical data to predict potential on-site incidents and hazards (Saudi Aramco, 2024). Such tools can predict and prevent potential environmental incidents and thus reduce the risk of spills or emissions.

Advanced AI algorithms, possibly based on unsupervised learning techniques like autoencoders or isolation forests, are used to identify unusual patterns in emissions or energy usage data that might indicate equipment malfunction or inefficiencies (Sadaf and Sultana, 2020).

10.7.2 Blockchain for Enhanced Transparency

Aramco is exploring and implementing blockchain technology to enhance the traceability and verification of its sustainability data (Saudi Aramco, 2024). This could involve creating an immutable ledger of emissions data, energy consumption figures, ensuring data integrity and facilitating third-party verification (Desai, 2023a). It can also facilitate carbon credit trading and offset verification with tamper-proof transaction records.

When a project reduces emissions or removes carbon from the atmosphere, it can be issued carbon credits. Each carbon credit would be tokenized as a unique digital asset on the blockchain, which prevents double-counting and allows for real-time traceability from issuance to retirement. Once recorded on the blockchain, these transactions cannot be altered or deleted due to the cryptographic nature of blockchain. Once a credit is used or retired to offset emissions, it would be marked as such on the blockchain and cannot be reused. Auditors and verifiers can access the blockchain to review the entire history of a carbon credit. Blockchain technologies can also automate sustainability-related payments and contract executions through smart contracts by improving transparency in sustainability investments (Desai, 2023b). For example, smart contracts can release carbon credits only when predefined conditions (like third-party verification) are met.

10.7.3 Digital Twin Technology

Aramco has implemented Digital Twin technology in its Hasbah field and aims to scale up across its engineering and project management organizations (Saudi Aramco, 2024). Digital Twin technology creates a virtual replica of physical assets or processes using real-time data from sensors and IoT devices (Opoku et al., 2021; Tuhaise et al., 2023). This digital model integrates historical and current operational data for sophisticated simulations and analyses. In Aramco's case, the Digital Twin of the Hasbah field likely combines 3D visualizations with live data streams on production rates, energy consumption, and emissions. Advanced algorithms process this data to predict performance, optimize operations, and identify potential issues before they occur in the physical asset. In sustainability accounting, Digital Twins can (a) simulate and optimize environmental performance of assets before physical implementation; (b) provide real-time monitoring of energy consumption and emissions at a granular level; and (c) enable more accurate allocation of environmental impacts to specific processes or products.

10.7.4 Internet of Things (IoT) and Industrial Internet of Things (IIoT)

Aramco's Smart Gas Detector utilizes IIoT technology for high-performing self-diagnostics (Saudi Aramco, 2024). IoT typically refers to consumer-oriented connected devices and applications, like smart home systems or wearable technology while IIoT focuses on industrial applications, connecting machines, sensors, and systems in manufacturing, energy, and other industrial sectors (Malik et al., 2021). In Aramco's context, IIoT works by embedding sensors and connectivity into industrial equipment and processes (Mao et al., 2021; Munirathinam, 2020). These sensors continuously collect data on various parameters like gas levels, energy consumption, and equipment performance.

This data is transmitted in real time to central systems via wireless networks. Advanced analytics and AI process this data, provide insights, and automate responses. For example, the company's flare monitoring system employs advanced sensors to provide real-time data on flare volumes and composition for more accurate reporting and immediate intervention when necessary (Saudi Aramco, 2022, 2023, 2024). For sustainability accounting, IIoT can (a) provide real-time, accurate data on energy consumption, emissions, and resource use; (b) enable more precise tracking of environmental Key Performance Indicators or KPIs across the entire value chain; and (c) automate data collection for sustainability reporting to reduce manual errors and improve data quality.

10.7.5 Robotics and Drones

Aramco's Shallow Water Inspection and Monitoring Robot (SWIM-R) enables efficient pipeline inspections (Saudi Aramco, 2024). Robotic and drone technologies work by deploying autonomous or remotely operated devices equipped with various sensors and cameras (Awwad et al., 2023). These robots and drones can access hazardous or hard-to-reach areas, such as underwater pipelines or tall structures, that would be dangerous or impossible for human inspectors (Aydin and Temizel, 2022).

They collect high-resolution visual data, as well as other sensor readings like temperature, pressure, or chemical composition. This data is then transmitted in real time or stored for later analysis. Advanced software processes this information to detect anomalies, potential leaks, or environmental issues. In sustainability accounting, robotics and drones can (a) enhance environmental monitoring capabilities, especially in hard-to-reach or hazardous areas; (b) improve the accuracy and frequency of asset integrity checks and thus reduce the risk of leaks or spills; and (c) Automate the collection of environmental data for reporting purposes.

10.7.6 Virtual/Augmented Reality (VR/AR)

Aramco's Digital Helmets utilize AR features to connect field workers (Saudi Aramco, 2024). AR overlays digital information onto the real world and improves the user's perception of their actual environment (Cheng et al., 2022). Users can still see and interact with the physical world around them, but with additional computer-generated information superimposed on it. This could be in the form of images, text, or 3D models. AR typically uses devices like smartphones, tablets, or special glasses/headsets that allow the user to see through to the real world (Koteleva et al., 2020).

In Aramco's case, their Digital Helmets use AR to provide workers with real-time data and instructions while they're performing tasks in the field. VR, on the other hand, immerses the user in a completely artificial digital environment (LaValle, 2023; Wanasinghe et al., 2021). It replaces the real world with a simulated one, typically using a headset that blocks out the physical surroundings. In VR, users are transported to a different place entirely, whether that's a computer-generated environment or a 360-degree video of a real location (Wohlgenannt et al., 2020). VR provides a fully immersive experience where users can interact with and navigate through the virtual world. For sustainability accounting, VR/AR can (a) provide real-time visualization of sustainability data to field workers, enabling immediate action on environmental issues; (b) improve training on sustainability practices and environmental compliance; and (c) improve the accuracy of data collection by guiding workers through proper measurement procedures.

10.7.7 3D Printing (Additive Manufacturing)

Aramco supports Saudi Arabia's first additive manufacturing center (Saudi Aramco, 2024). 3D printing, or additive manufacturing, works by building objects layer by layer from digital 3D models (Jandyal et al., 2022). The process typically begins with a computer-aided design (CAD) file that's sliced into thin layers. The 3D printer then deposits material (such as plastic, metal, or resin) in these precise layers, building the object from the bottom up (Karakurt and Lin, 2020). Each layer is fused to the previous one, usually through heat or light, depending on the material and printing technology used. This additive process allows for complex geometries and internal structures that would be difficult or impossible to create with traditional manufacturing methods (Thakar et al., 2022).

In Aramco's context, this technology enables on-demand production of parts and reduces inventory and transportation needs, while also allowing for design optimizations that can improve efficiency and reduce material waste. In sustainability accounting, 3D printing can (a) reduce waste in manufacturing processes and lead to more accurate reporting of material consumption; (b) enable localized production of spare parts and thus reduce transportation emissions in the supply chain; and (c) facilitate the development of more efficient components and potentially lower overall energy consumption.

10.7.8 Cloud Computing

Aramco's cloud-based eMarketPlace platform optimizes supply chain processes (Saudi Aramco, 2024). Cloud computing in Aramco's context works by utilizing remote servers hosted on the Internet to store, manage, and process data, rather than using local servers or personal computers. The eMarketPlace platform likely operates on a cloud infrastructure, and allows multiple users and systems to access and interact with it simultaneously from various locations (Sandhu, 2021). This cloud-based system can rapidly scale its computing resources up or down, based on demand, and efficiently process large datasets. It enables real-time data updates and access and facilitates immediate collaboration and reporting. The cloud also allows for the integration of data from various sources, such as IoT devices, enterprise systems, and external databases, into a centralized platform for comprehensive sustainability analysis and reporting (Sadeeq et al., 2021).

For sustainability accounting, cloud computing can (a) provide scalable storage and processing power for large volumes of sustainability data; (b) enable real-time collaboration and data sharing across the organization and with external stakeholders; (c) facilitate the integration of sustainability data from various sources for comprehensive reporting; and (d) provide robust cybersecurity solutions, including advanced encryption, multi-factor authentication, and regular security updates, to protect sensitive sustainability data from unauthorized access or breaches (Raja, 2024).

10.7.9 Big Data Analytics

Aramco's use of AI and IoT implies the application of big data analytics, as is also discussed in their sustainability report (Saudi Aramco, 2023). Big Data Analytics in Aramco's context works by collecting and processing massive volumes of structured and unstructured data from various sources across their operations. This data is stored in distributed systems capable of handling its scale and complexity (Wang et al., 2022). Advanced analytics algorithms, often powered by machine learning and AI, are then applied to this data to uncover patterns, trends, and insights that might not be visible through traditional analysis methods (Nti et al., 2022).

These algorithms can process data in real time or in batches for both immediate insights and deep, historical analysis. The results are typically visualized through dashboards or reports to enable decision-makers to quickly understand complex sustainability metrics and make data-driven decisions to improve environmental performance. In sustainability accounting, this can (a) process and analyze vast amounts of sustainability data from diverse sources; (b) identify correlations between operational activities and environmental impacts; and (c) generate predictive models for future sustainability performance.

10.7.10 Robotic Process Automation (RPA)

Aramco is researching RPA to automate human tasks (Saudi Aramco, 2024). RPA works by using software "bots" to automate repetitive, rule-based tasks typically performed by humans (Ribeiro et al., 2021). In Aramco's context, these bots would be programmed to interact with digital systems just as a human would—inputting data, moving files, filling out forms, and extracting information from various sources (Zhang et al., 2023).

RPA bots can be configured to automatically log into multiple systems, extract relevant environmental data, perform data validation checks, aggregate information into standardized formats, and even generate preliminary reports (Harrast, 2020). The bots operate 24/7, following predefined rules and workflows, significantly reducing the time and potential for human error in data processing and reporting tasks. In sustainability accounting, RPA can (a) automate data entry and validation processes for sustainability reporting; (b) streamline the collection and aggregation of sustainability data from various systems; and (c) automate the generation of standardized sustainability reports.

10.7.11 Analyzing Aramco's Technological Innovations in Sustainability Through Theoretical Lens

Aramco's investments in advanced technologies like AI, blockchain, Digital Twins, etc. represent what Suchman (1995) describes as "pragmatic exchange legitimacy," which demonstrates valuable innovation to stakeholders while also serving organizational interests in operational efficiency. These technological investments also function as a form of "symbolic management" (Ashforth & Gibbs, 1990) by showcasing the company's commitment to innovative solutions without necessarily transforming the underlying business model.

From an institutional perspective, Aramco's technological leadership creates mimetic pressure (DiMaggio & Powell, 1983) on other regional oil companies, as evidenced by ADNOC's subsequent adoption of similar technological approaches (as discussed in the next section). This demonstrates how even adaptation strategies can become institutionalized across an organizational field and create new norms that may paradoxically reinforce rather than challenge the existing institutional order.

10.8 Comparative Analysis of Sustainability Reporting

10.8.1 Comparison to the Abu Dhabi National oil Company (ADNOC)

As the two largest national oil companies in the Middle East, Aramco and ADNOC set the tone for energy transition and sustainability efforts in a region heavily dependent on hydrocarbon exports. Both are among the world's largest oil producers, significantly influencing global energy markets and climate change mitigation efforts. They are central to the economic diversification plans of Saudi Arabia

(Vision 2030) and the UAE (Vision 2031), which makes their strategies critical for these nations' futures. Their approaches to sustainability and energy transition serve as models for other state-owned oil companies worldwide. As partially public entities, their sustainability strategies are important for investors assessing long-term viability in a carbon-constrained world. Their strategies also impact regional geopolitics and global energy security. Table 10.1 outlines how Saudi Aramco and ADNOC report their sustainability initiatives and shows both where their approaches converge and where they take distinct paths (ADNOC, 2023; Saudi Aramco, 2023).

The similarities between Aramco and ADNOC in their reporting approaches demonstrate institutional isomorphism within the regional context of Gulf national oil companies. Both companies face similar stakeholder pressures and institutional environments, which leads to comparable strategies that balance international legitimacy with national economic priorities. This supports DiMaggio and Powell's (1983) contention that organizations within the same field develop similar structures and practices.

10.8.2 Comparison to ExxonMobil

As the world's largest state-owned oil company (Aramco) and the largest Western publicly traded oil company (ExxonMobil), respectively, their strategies and performances significantly impact global energy markets and prices. Aramco's state ownership versus ExxonMobil's public ownership offers insights into how different governance structures approach sustainability and energy transition. Both are leaders in oil and gas technology; comparing their R&D focus and investments in low-carbon technologies provides a window into potential industry futures. Their differing approaches to climate change mitigation and adaptation indicate broader tensions between state-controlled and market-driven responses to global environmental challenges. Comparing their financial strategies, particularly in relation to sustainability investments, provides important insights for investors and policymakers. Aramco's ties to Saudi foreign policy versus Exxon's alignment with US interests showcase how geopolitics shapes corporate strategy in the energy sector.

Table 10.2 outlines how Saudi Aramco and ExxonMobil report their sustainability initiatives, showing both where their approaches converge and where they take distinct paths.

The differences between these national oil companies and ExxonMobil highlight the influence of varying stakeholder configurations and institutional contexts. ExxonMobil's more detailed scenario analysis and greater emphasis on TCFD alignment indicate its accountability to a broader shareholder base and more stringent regulatory environments. This supports the stakeholder theory proposition that organizations adapt their behaviors to the specific expectations of their most salient stakeholders (Mitchell et al., 1997).

Table 10.1 Comparison of Aramco and ADNOC's sustainability strategies and reporting

Parameters	Similarities	Differences
Carbon accounting framework and practices	Both companies suggest they are aiming for net-zero emissions (Aramco by 2050, ADNOC by 2045) and employ advanced technologies for monitoring and reducing emissions. They also have ambitious targets for reducing carbon intensity and engage in rigorous external assurance of their emissions data	Aramco focuses on leveraging internal operational efficiencies and broader technological initiatives, while ADNOC emphasizes the integration of renewable energy sources like solar and nuclear for operational electricity, aiming for a more aggressive reduction in carbon intensity by 2030
Carbon intensity metrics	Both companies measure and report on carbon intensity as part of their sustainability strategies	Aramco reports one of the lowest carbon intensities in the industry, leveraging its efficient production techniques. However, its reporting of these metrics has been deemed misleading by a UN experts working group. ADNOC, on the other hand, has set specific targets for reducing carbon intensity and incorporates the use of renewable energy to achieve these goals
Use of green financial instruments	Both companies demonstrate a commitment to financing sustainability through significant capital allocation	Aramco has not directly issued green bonds but focuses on funding technologies that support its net-zero ambitions through its $1.5 billion sustainability fund. ADNOC has allocated $23 billion to scale up practical and commercially viable low-carbon initiatives, although it does not specify green bonds directly
Integrating ESG into corporate strategy	Both companies integrate ESG into their governance structures, ensuring strategic alignment with long-term value creation. They also link sustainability performance with executive remuneration	Aramco's approach is highly structured with specific committees and sub-committees overseeing different aspects of ESG integration. ADNOC integrates ESG performance factors into supplier qualifications and auditing processes, reflecting a broader operational integration
Technology and innovation in sustainability accounting	Both companies are advancing in utilizing AI and digital technologies to enhance operational efficiencies and environmental monitoring	Aramco is exploring blockchain technology for carbon tracking and emphasizes innovation through partnerships with tech companies and startups. ADNOC is developing a data management platform to enhance emissions monitoring and management across its operations and is active in deploying technologies like unmanned aerial vehicles (UAVs) for GHG leak detection

Table 10.2 Comparison of Aramco and ExxonMobil's sustainability strategies and reporting

Parameters	Similarities	Differences
Carbon accounting framework and practices	Both companies indicate they are committed to achieving net-zero emissions and use advanced technologies to monitor and reduce emissions. They also provide detailed external assurance of their emissions data	Aramco aims for net-zero by 2050, with a strong focus on integrating new technologies across operations. ExxonMobil, while also targeting net-zero by 2050, places a heavier emphasis on sector-specific roadmaps (based on SASB guidance) and detailed scenario analysis (in line with the TCFD guidelines)
Carbon intensity metrics	Both companies measure and report on carbon intensity as part of their sustainability strategies	Aramco reports one of the lowest carbon intensities in the industry, leveraging its efficient production techniques. However, its reporting of these metrics has been deemed misleading by a UN experts working group. Exxon Mobil, on the other hand, reports higher carbon intensity but is focused on reducing these metrics through advanced technology and operational efficiencies
Use of green financial instruments	Both companies emphasize the importance of financing in achieving their sustainability goals	While Aramco has not directly issued green bonds, it has set up a $1.5 billion sustainability fund. Exxon Mobil does not detail specific green financial instruments but discusses broader financial strategies for supporting low-emission technology investments
Integrating ESG into corporate strategy	Integration of ESG factors into corporate governance, with structured oversight and strategic alignment toward sustainability	Aramco has a well-defined ESG governance framework with specific committees overseeing sustainability. Exxon Mobil also incorporates ESG into its governance but focuses more on aligning with external frameworks and responding to shareholder requests
Technology and innovation in sustainability accounting	Both companies are leveraging technology to enhance sustainability reporting and operational efficiencies	Aramco is exploring blockchain for emissions tracking and extensively uses AI for operational efficiencies. Exxon Mobil emphasizes the role of scenario analysis in shaping technology strategies and highlights specific innovations like CCS and hydrogen as key to its sustainability roadmap

10.9 Conclusion and Future Directions

This comprehensive analysis of Saudi Aramco's sustainability accounting practices reveals complex challenges and innovations in the oil and gas industry's transition toward a low-carbon future. Several key findings emerge from this study.

First, Aramco has made significant strides in integrating advanced technologies like AI, blockchain, and IIoT into its sustainability accounting practices. This technological adoption improves the accuracy and reliability of environmental data reporting and sets a new standard for the industry. Second, the company's multi-tiered ESG governance structure demonstrates a commitment to embedding sustainability considerations at all levels of decision-making. However, the effectiveness of this structure in driving meaningful change remains to be fully assessed. Third, despite progress in Scope 1 and 2 emissions reporting, Aramco's lack of comprehensive Scope 3 emissions disclosure represents a significant gap in its sustainability reporting. This omission limits stakeholders' ability to fully assess the company's climate impact.

However, Aramco's sustainability initiatives have catalyzed similar efforts among other national oil companies in the MENA region, which demonstrates the company's potential to drive regional sustainability trends. The application of legitimacy, stakeholder, and institutional theories to Aramco's practices indicates the complex and sometimes contradictory forces shaping the company's sustainability strategy.

The oil and gas sector must accelerate the adoption of advanced technologies for sustainability accounting to improve transparency and credibility. Companies should prioritize the development of methodologies for comprehensive Scope 3 emissions reporting to address a critical gap in current practices. There is a need for more stringent and standardized regulations regarding sustainability reporting in the oil and gas sector, particularly concerning Scope 3 emissions. Policies should also incentivize technological innovation in sustainability accounting and reporting practices.

Christy et al. (2024) propose a dynamic framework that aligns corporate climate reporting with global climate targets. Their framework identifies three key dimensions that should be addressed in effective climate reporting: temporal alignment (short, medium, and long-term horizons), spatial alignment (local to global impacts), and sectoral alignment (industry-specific considerations). When applied to Aramco's sustainability reporting, this framework reveals both strengths and limitations in the company's approach. Aramco demonstrates temporal alignment through its establishment of both 2050 net-zero targets and interim 2035 goals (Saudi Aramco, 2023). However, as discussed earlier, the company's reporting lacks comprehensive spatial alignment, particularly in its limited attention to global value chain emissions (Scope 3). According to Christy et al. (2024), such gaps can significantly hinder a company's ability to contribute meaningfully to global climate objectives, regardless of the strength of its direct emissions management.

Additionally, Chen et al.'s (2023) research demonstrates that while renewable energy sources generally produce lower carbon emissions than fossil fuels, a comprehensive environmental assessment must consider multiple impact categories, including air pollution, water consumption, and land use. This multi-dimensional approach to environmental impact assessment is relevant for evaluating Aramco's sustainability claims regarding carbon intensity and operational efficiency. Chen et al.'s (2023) findings suggest that Aramco's focus on reducing operational carbon intensity, while important, represents only one dimension of environmental impact. The company's emphasis on technological solutions like carbon capture and blue hydrogen production (Saudi Aramco, 2022, 2023) should be evaluated within a broader environmental impact framework that considers potential trade-offs, such as increased water consumption or land use changes.

From a practical standpoint, the analysis of Aramco's technological innovations provides a foundation for developing structured technology integration roadmaps for sustainability accounting. Future research should explore how companies can strategically implement technologies like blockchain, AI, and IoT in phases to enhance reporting accuracy and credibility. This could include developing maturity models for technology adoption in sustainability accounting and identifying key success factors for technological implementation. The tensions between Aramco's sustainability initiatives and its core business model highlight the need for regulatory frameworks specifically designed to guide transition strategies in carbon-intensive industries.

Policymakers should explore how policy mechanisms can effectively incentivize genuine transition while acknowledging the economic and social complexities involved, particularly in regions economically dependent on fossil fuels. The technological innovations in sustainability accounting observed at Aramco have potential applications across sectors. Practitioners should explore how these technologies can be adapted and transferred to other industries, particularly those facing similar environmental accounting challenges but with different operational contexts.

Future studies should evaluate the long-term environmental and financial impacts of Aramco's sustainability initiatives, particularly as the company progresses toward its 2050 net-zero target. More comprehensive comparative analyses between national oil companies and international oil companies could provide insights into the influence of ownership structures on sustainability practices. Research into how different stakeholder groups perceive and value the sustainability efforts of oil and gas companies could inform more effective communication and engagement strategies. In-depth studies on the effectiveness of specific technologies (e.g., blockchain, AI) in improving sustainability accounting accuracy and transparency are needed. An analysis of how evolving international and national policies affect sustainability accounting practices in the oil and gas sector could guide future regulatory frameworks. To sum it up, the evolution of sustainability accounting practices will play a key role in shaping the future of energy production and consumption.

References

ADNOC. (2023). *Sustainability report 2023*. https://www.adnoc.ae/en/sustainability-report

Ahmed, A. H., Elmaghrabi, M. E., Dunne, T., & Hussainey, K. (2021). Gaining momentum: Towards integrated reporting practices in gulf cooperation council countries. *Business Strategy & Development, 4*(2), 78–93. https://doi.org/10.1002/bsd2.130

Aljanadi, Y., & Alazzani, A. (2023). Sustainability reporting indicators used by oil and gas companies in GCC countries: IPIECA guidance approach. *Frontiers in Environmental Science, 11*, 1069152. https://doi.org/10.3389/fenvs.2023.1069152

Ashforth, B. E., & Gibbs, B. W. (1990). The double-edge of organizational legitimation. *Organization Science, 1*(2), 177–194. https://doi.org/10.1287/orsc.1.2.177

Awwad, A., Albasha, L., Mir, H. S., & Mortula, M. M. (2023). Employing robotics and deep learning in underground leak detection. *IEEE Sensors Journal, 23*(8), 8169–8177. https://doi.org/10.1109/JSEN.2023.3248142

Aydin, H., & Temizel, C. (2022, April). Latest applications of UAVs, drones and robotics in the oil and gas industry. In *Offshore technology conference* (p. D021S019R007). OTC. https://doi.org/10.4043/31735-MS

Cahill, B., & Swanson, K. (2023, December 13). *National oil Companies, climate commitments, and methane*. Center for Strategic & International Studies (CSIS). https://csis-website-prod.s3.amazonaws.com/s3fs-public/2023-12/231213_Cahill_Oil_Companies.pdf?VersionId=4Ad82MnTnBPHJjhmZSyFSDmLb5NkdroZ

Cardoni, A., Kiseleva, E., & Terzani, S. (2019). Evaluating the intra-industry comparability of sustainability reports: The case of the oil and gas industry. *Sustainability, 11*(4), 1093. https://doi.org/10.3390/su11041093

Chen, X. H., Tee, K., Elnahass, M., & Ahmed, R. (2023). Assessing the environmental impacts of renewable energy sources: A case study on air pollution and carbon emissions in China. *Journal of Environmental Management, 345*, 118525. https://doi.org/10.1016/j.jenvman.2023.118525

Cheng, H., Liu, Z., Cai, F., Zhang, L., Ma, X., Sun, X., et al. (2022). AR technology in the design of oil and gas field surface engineering: Applications and prospects. *Wearable Technology, 3*(2), 23–31. https://doi.org/10.54517/wt.v3i2.1641

Christy, A., Elnahass, M., Amezaga, J., Browne, A., & Heidrich, O. (2024). A dynamic framework to align company climate reporting and action with global climate targets. *Business Strategy and the Environment, 33*(4), 3103–3128. https://doi.org/10.1002/bse.3635

ClientEarth. (2021). *ClientEarth Complaint Concerning Saudi Arabian Oil Company (Saudi Aramco) and the Kingdom of Saudi Arabia and JP Morgan, Citi, HSBC, SMBC, Crédit Agricole, Morgan Stanley, BNP Paribas, Goldman Sachs, Mizuho, Société Générale, and EIG Global Energy Partners*. https://www.clientearth.org/media/l44by31b/clientearth-complaint-concerning-saudi-arabian-oil-company.pdf

ClientEarth. (2023, August 25). *UN takes action against Saudi oil firm and its banks over climate and human rights*. https://www.clientearth.org/latest/news/un-takes-action-against-saudi-oil-firm-and-its-banks-over-climate-and-human-rights/

Creswell, J. W., & Clark, V. L. P. (2017). *Designing and conducting mixed methods research* (3rd ed.). SAGE Publications.

Deegan, C. (2002). Introduction: The legitimising effect of social and environmental disclosures–a theoretical foundation. *Accounting, Auditing & Accountability Journal, 15*(3), 282–311. https://doi.org/10.1108/09513570210435852

Desai, H. (2023a). Infusing Blockchain in accounting curricula and practice: Expectations, challenges, and strategies. *The International Journal of Digital Accounting Research, 23*, 97–135. https://doi.org/10.4192/1577-8517-v23_5

Desai, H. (2023b). An auditor's perspective on smart contracts and DAOs. *The. CPA Journal, 93*(7/8), 60–63. https://doi.org/10.13140/RG.2.2.18319.56481/1

DiMaggio, P. J., & Powell, W. W. (1983). The iron cage revisited: Institutional isomorphism and collective rationality in organizational fields. *American Sociological Review, 48*(2), 147–160. https://doi.org/10.1515/9780691229270-005

Freeman, R. E., Dmytriyev, S. D., & Phillips, R. A. (2021). Stakeholder theory and the resource-based view of the firm. *Journal of Management, 47*(7), 1757–1770. https://doi.org/10.1177/0149206321993576

Freeman, R. E., Phillips, R., & Sisodia, R. (2020). Tensions in stakeholder theory. *Business & Society, 59*(2), 213–231. https://doi.org/10.1177/0007650318773750

Galbreath, J. (2013). ESG in focus: The Australian evidence. *Journal of Business Ethics, 118*, 529–541. https://doi.org/10.1007/s10551-012-1607-9

Gioia, D. A., Corley, K. G., & Hamilton, A. L. (2013). Seeking qualitative rigor in inductive research: Notes on the Gioia methodology. *Organizational Research Methods, 16*(1), 15–31. https://doi.org/10.1177/1094428112452151

Grasso, M. (2019). Oily politics: A critical assessment of the oil and gas industry's contribution to climate change. *Energy Research & Social Science, 50*, 106–115. https://doi.org/10.1016/j.erss.2018.11.017

Grasso, M. (2023). *Climate justice from theory to practice: The responsibility and duties of the oil industry*. Retrieved from Link.

Greenwood, R., Raynard, M., Kodeih, F., Micelotta, E. R., & Lounsbury, M. (2011). Institutional complexity and organizational responses. *Academy of Management Annals, 5*(1), 317–371. https://doi.org/10.5465/19416520.2011.590299

Haffar, M., & Searcy, C. (2020). Legitimizing potential "bad news": How companies disclose on their tension experiences in their sustainability reports. *Organization & Environment, 33*(4), 534–553. https://doi.org/10.1177/1086026620942968

Harrast, S. A. (2020). Robotic process automation in accounting systems. *Journal of Corporate Accounting & Finance, 31*(4), 209–213. https://doi.org/10.1002/jcaf.22457

Heede, R. (2023). *History of Eni emissions 1953–2022 and attributed climate reparations*. Retrieved from Link.

Hutt, S., DePiro, A., Wang, J., Rhodes, S., Baker, R. S., Hieb, G., Sethuraman, S., Ocumpaugh, J., & Mills, C. (2024). Feedback on feedback: Comparing classic natural language processing and generative AI to evaluate peer feedback. In *proceedings of the 14th learning analytics and knowledge conference* (pp. 55–65).

Ipieca Ltd. (2024). *ADNOC brings net-zero target forward to 2045*. https://www.ipieca.org/news/adnoc-brings-net-zero-target-forward

Ismaeel, M., & Zakaria, Z. (2019). Perception of preparers of sustainability reports in the Middle East: Contrasting between local and global. *Meditari Accountancy Research, 28*(1), 89–116. https://doi.org/10.1108/MEDAR-03-2019-0459

Jandyal, A., Chaturvedi, I., Wazir, I., Raina, A., & Haq, M. I. U. (2022). 3D printing–A review of processes, materials and applications in industry 4.0. *Sustainable Operations and Computers, 3*, 33–42. https://doi.org/10.1016/j.susoc.2021.09.004

Karakurt, I., & Lin, L. (2020). 3D printing technologies: Techniques, materials, and post-processing. *Current Opinion in Chemical Engineering, 28*, 134–143. https://doi.org/10.1016/j.coche.2020.04.001

Kingdom of Saudi Arabia. (2016). *Vision 2030*. https://www.vision2030.gov.sa/media/rc0b5oy1/saudi_vision203.pdf

Koteleva, N., Buslaev, G., Valnev, V., & Kunshin, A. (2020). Augmented reality system and maintenance of oil pumps. *International Journal of Engineering, 33*(8), 1620–1628. https://doi.org/10.5829/ije.2020.33.08b.20

Krane, J. (2021). The bottom of the barrel: Saudi Aramco and global climate action. Working paper, Baker Institute for Public Policy. Retrieved from Link.

S. M. LaValle (2023). Virtual reality. Cambridge university press. doi:https://doi.org/10.1017/9781108182874.

Malik, P. K., Sharma, R., Singh, R., Gehlot, A., Satapathy, S. C., Alnumay, W. S., et al. (2021). Industrial internet of things and its applications in industry 4.0: State of the art. *Computer Communications, 166*, 125–139. https://doi.org/10.1016/j.comcom.2020.11.016

Mao, W., Zhao, Z., Chang, Z., Min, G., & Gao, W. (2021). Energy-efficient industrial internet of things: Overview and open issues. *IEEE Transactions on Industrial Informatics, 17*(11), 7225–7237. https://doi.org/10.1109/TII.2021.3067026

Megura, M., & Gunderson, R. (2022). Better poison is the cure? Critically examining fossil fuel companies, climate change framing, and corporate sustainability reports. *Energy Research & Social Science, 85*, 102388. https://doi.org/10.1016/j.erss.2021.102388

Mitchell, R. K., Agle, B. R., & Wood, D. J. (1997). Toward a theory of stakeholder identification and salience: Defining the principle of who and what really counts. *Academy of Management Review, 22*(4), 853–886. https://doi.org/10.2307/259247

Molina-Azorin, J. F. (2016). Mixed methods research: An opportunity to improve our studies and our research skills. *European Journal of Management and Business Economics, 25*(2), 37–38. https://doi.org/10.1016/j.redeen.2016.05.001

S. Munirathinam (2020). Industry 4.0: Industrial internet of things (IIOT). *In* Advances in computers (Vol. 117, No. 1, pp. 129–164). Elsevier. doi:https://doi.org/10.1016/bs.adcom.2019.10.010.

Muttit, G., & Heller, P. (2021, May 04). National oil Companies and climate change: Economic challenges and potential responses. In *International Institute for Sustainable Development (IISD)*. Link.

Nordsip. (2023, August 30). *UN Names and Shames Aramco's Banks.* https://nordsipcom/2023/08/30/un-names-and-shames-aramcos-banks/

Nti, I. K., Quarcoo, J. A., Aning, J., & Fosu, G. K. (2022). A mini-review of machine learning in big data analytics: Applications, challenges, and prospects. *Big Data Mining and Analytics, 5*(2), 81–97. https://doi.org/10.26599/BDMA.2021.9020028

Nwagbara, U., & Belal, A. (2019). Persuasive language of responsible organisation? A critical discourse analysis of corporate social responsibility (CSR) reports of Nigerian oil companies. *Accounting, Auditing & Accountability Journal, 32*(8), 2395–2420. https://doi.org/10.1108/AAAJ-03-2016-2485

Oil and Gas Climate Initiative (OGCI). (2023). *Annual report 2023.* https://www.ogci.com/progress-report/building-towards-net-zero/download

Opoku, D. G. J., Perera, S., Osei-Kyei, R., & Rashidi, M. (2021). Digital twin application in the construction industry: A literature review. *Journal of Building Engineering, 40*, 102726. https://doi.org/10.1016/j.jobe.2021.102726

Patten, D. M. (2020). Seeking legitimacy. *Sustainability Accounting, Management and Policy Journal, 11*(6), 1009–1021. https://doi.org/10.1108/SAMPJ-12-2018-0332

Qasem, A., Badru, B. O., Ghaleb, B. A., & AL-Duais, S. D., & Al-Qadasi, A. A. (2024). Corporate social responsibility disclosure in Saudi companies: Analysing the impact of board independence in family and non-family companies. *Humanities and Social Sciences Communications, 11*(1), 1–18. https://doi.org/10.1057/s41599-024-03495-3

Raja, V. (2024). Exploring challenges and solutions in cloud computing: A review of data security and privacy concerns. *Journal of Artificial Intelligence General science (JAIGS), 4*, 121–144. https://doi.org/10.60087/jaigs.v4i1.86

Reuters .(2023, August 25). *UN rights experts raise climate change concerns with Saudi Aramco.* https://www.reuters.com/business/energy/un-rights-experts-raise-climate-change-concerns-with-saudi-aramco-2023-08-25/

Ribeiro, J., Lima, R., Eckhardt, T., & Paiva, S. (2021). Robotic process automation and artificial intelligence in industry 4.0–a literature review. *Procedia Computer Science, 181*, 51–58. https://doi.org/10.1016/j.procs.2021.01.104

Roumengas, R., Di Lorenzo, G., Salhi, A., de Buyer, P., Chaudhuri, A., Lareyre, F., & Raffort, J. (2023). Natural language processing for literature search in vascular surgery: A pilot study testing an artificial intelligence based application. In *EJVES vascular forum* (Vol. 60, pp. 48–52). Elsevier. https://doi.org/10.1016/j.ejvsvf.2023.09.004

Sadaf, K., & Sultana, J. (2020). Intrusion detection based on autoencoder and isolation forest in fog computing. *IEEE Access, 8*, 167059–167068. https://doi.org/10.1109/ACCESS.2020.3022855

Sadeeq, M. M., Abdulkareem, N. M., Zeebaree, S. R., Ahmed, D. M., Sami, A. S., & Zebari, R. R. (2021). IoT and cloud computing issues, challenges and opportunities: A review. *Qubahan Academic Journal, 1*(2), 1–7. https://doi.org/10.48161/qaj.v1n2a36

Sandhu, A. K. (2021). Big data with cloud computing: Discussions and challenges. *Big Data Mining and Analytics, 5*(1), 32–40. https://doi.org/10.26599/BDMA.2021.9020016

Saudi Aramco. (2021). *Saudi Aramco Sustainability Report 2021.* https://www.aramco.com/-/media/downloads/sustainability-report/saudi-aramco-sustainability-report-2021-en.pdf

Saudi Aramco. (2022). *Aramco Sustainability Report 2022.* https://www.aramco.com/-/media/downloads/sustainability-report/report-2022/2022-sustainability-report-en.pdf

Saudi Aramco. (2023). *Saudi Aramco Sustainability Report 2023.* https://www.aramco.com/-/media/publications/corporate-reports/sustainability-reports/report-2023/english/2023-saudi-aramco-sustainability-report-full-en.pdf

Saudi Aramco. (2024). *Digital Technologies.* https://www.aramco.com/en/what-we-do/energy-innovation/digitalization/digital-technologies

Suchman, M. C. (1995). Managing legitimacy: Strategic and institutional approaches. *The Academy of Management Review, 20*(3), 571–610. https://doi.org/10.2307/258788

Thakar, C. M., Parkhe, S. S., Jain, A., Phasinam, K., Murugesan, G., & Ventayen, R. J. M. (2022). 3d printing: Basic principles and applications. *Materials Today: Proceedings, 51*, 842–849. https://doi.org/10.1016/j.matpr.2021.06.272

Tuhaise, V. V., Tah, J. H. M., & Abanda, F. H. (2023). Technologies for digital twin applications in construction. *Automation in Construction, 152*, 104931. https://doi.org/10.1016/j.autcon.2023.104931

Uyar, A., Ramadan, A., & Nimer, K. (2019). A critical evaluation of sustainability reporting in the Gulf cooperation council region. *International Journal of Sustainable Development, 22*(3–4), 158–185. https://doi.org/10.1504/IJSD.2019.105327

Wanasinghe, T. R., Trinh, T., Nguyen, T., Gosine, R. G., James, L. A., & Warrian, P. J. (2021). Human centric digital transformation and operator 4.0 for the oil and gas industry. *IEEE Access, 9*, 113270–113291. https://doi.org/10.1109/ACCESS.2021.3103680

Wang, J., Xu, C., Zhang, J., & Zhong, R. (2022). Big data analytics for intelligent manufacturing systems: A review. *Journal of Manufacturing Systems, 62*, 738–752. https://doi.org/10.1016/j.jmsy.2021.03.005

Wohlgenannt, I., Simons, A., & Stieglitz, S. (2020). Virtual reality. *Business & Information Systems Engineering, 62*, 455–461. https://doi.org/10.1007/s12599-020-00658-9

Yoon, J. (2021). Forecasting of real GDP growth using machine learning models: Gradient boosting and random forest approach. *Computational Economics, 57*(1), 247–265. https://doi.org/10.1007/s10614-020-10054-w

Zhang, C., Issa, H., Rozario, A., & Soegaard, J. S. (2023). Robotic process automation (RPA) implementation case studies in accounting: A beginning to end perspective. *Accounting Horizons, 37*(1), 193–217. https://doi.org/10.2308/HORIZONS-2021-084

Chapter 11
Exploring the Materiality and Disclosure Practices of Climate Change Risks Within the GCC Region

Tia Rebecca Driver

Abstract The disclosure of climate change risks is of growing importance for both firms and their stakeholders, with both physical and transition risks posing potential adverse effects to the firm. Recent literature has focused on quantitatively exploring the relationships between climate change risk disclosures and financial performance and evaluating the materiality assessments of firms from a European perspective. This study complements those works by applying a qualitative approach, utilising directed qualitative analysis and thematic analysis of corporate public disclosures, to understand how firms within the Gulf Cooperation Council (GCC) region adopt the recently developed ESG taxonomy. The study finds that the disclosure of climate change risks is superficial and lacks substantial discussion about the material nature of those risks. These insights reveal the current reporting practices of GGC firms and underscore the urgent need for MENA firms to enhance transparency in their disclosures.

Keywords Climate change risks · Materiality · Legitimacy theory · Gulf Cooperation Council (GCC)

11.1 Introduction

The management of risks by corporations is not a new phenomenon. Climate change has introduced a critical focus on risk management, particularly regarding physical and transition risks. Changes in weather patterns and an increased potential for extreme weather events expose firms to various physical risks (Schiemann & Sakhel, 2019). For firms, the impacts of physical risks could be seen in supply chain disruptions (Ghadge et al., 2020) or damages to physical assets or labour productivity

T. R. Driver (✉)
School of Environment, Enterprise and Development, University of Waterloo, Waterloo, ON, Canada
e-mail: trdriver@uwaterloo.ca

M. Elnahass et al. (eds.), *A Greenprint for Net Zero*, Gulf Studies 15,
https://doi.org/10.1007/978-981-92-1762-5_11

(Campiglio et al., 2023; TCFD, 2017). Furthermore, climate change can lead to other impacts generated by the shift towards a low-carbon economy, referred to as transition risks (Chabot & Bertrand, 2023). This risk category comprises several issues, including changes in consumer behaviour or preferences, costs associated with new clean technology, and exposing firms to additional regulations or reporting requirements (Campiglio et al., 2023). Given the potential impacts of physical and transition risks, there are growing calls from investors, consumers, and regulators for firms to disclose climate risks (Ben-Amar et al., 2023), and increased attention regarding climate risk disclosures within academic research (Shui et al., 2023).

Prior examinations of climate change risk disclosures have focused on quantifying the relationship between disclosure and financial performance (Arian & Sands, 2024; Ben-Amar et al., 2023; Benz-Saliasi, 2020; Megeid, 2024). While those studies provide valuable insights into the possible financial implications for firms, the quantitative focus provides limited insights into how firms discuss or perceive climate change risks within the context of their disclosure. Furthermore, the material nature of climate change risks has been explored through the analysis of the firm's materiality matrices (Ferrero-Ferrero et al., 2021), and the content analysis of corporate disclosures (Beske et al., 2020), which also includes a focus on firms within the Gulf Cooperation Council (GCC) region (Farooq et al., 2021). Such studies provide insights into the material nature of climate risks more generally, but the materiality of physical risks and transition risks still warrants further investigation, given their impacts on firms.

From a regional perspective, there has been a strong focus on climate risk disclosure within the European context, given the recent adoption of the European Union's Corporate Sustainability Reporting Directive (CSRD) (Dragomir et al., 2024) and the existing work centred on the Non-financial Reporting Directive (NFRD) (Andersson & Arvidsson, 2023). However, there has been less focus on climate risk disclosures within the Middle East and North Africa (MENA) region. This is unfortunate given the efforts by parts of this region in developing sustainability reporting standards (GCC Exchanges Committee, 2023) and the increased attention to climate change impacts by several nations (Kingdom of Saudi Arabia, n.d.; State of Qatar, n.d.; United Arab Emirates, 2024).

This chapter aims to enhance existing research on climate change risks and materiality in the MENA region by using firms in the GCC region as a sample. Applying legitimacy theory and a qualitative research approach provides an avenue to explore the following research questions: (1) How are physical and transitional climate change risks discussed within corporate disclosures by GCC firms? and (2) To what extent are climate change risks material for the firm? In doing so, this study finds climate risk disclosure to be poorly understood and superficial, but is optimistic that MENA firms can use these insights to refine their disclosure practices further to better integrate climate change risks.

This chapter will provide an overview of why climate change risks are important for firms and how previous literature has examined this topic. It will also outline the theoretical underpinnings of legitimacy theory and the research methodologies employed. The final section will present the key findings while discussing their relationship to the literature and research questions.

11.2 The Materiality of Climate Change Risks and Disclosures

The significance of climate change risks to companies stems from the possible financial consequences, which have been explored in extant literature. Financial-related stakeholders, such as investors and financial regulators, value physical climate change information within corporate disclosures for their analysis or evaluation processes (Ben-Amar et al., 2023; Ilhan et al., 2023; Megeid, 2024), and, at times, markets can expect such disclosures (Matsumura et al., 2024; Vestrelli et al., 2024).

Low disclosure of climate risks could put companies at higher risk of receiving lower valuations in terms of price-to-earnings and price-to-book ratios (Benz-Saliasi, 2020). Moreover, firms with lower quality disclosures could face additional costs or need additional collateral when obtaining loans from financial institutions (Huang et al., 2022). Financial performance is a critical focus for any for-profit firm; yet, it should also consider its stakeholders' information needs or interests.

Within the arena of public disclosures, it is expected that firms would abide by the principle of materiality, which broadly argues that firms ought to disclose information that is critical for stakeholders to make an informed decision regarding the firm and its future financial prospects (Font et al., 2016; Jebe, 2019). This principle is the basis for several popular sustainability reporting standards or frameworks, including the Global Reporting Initiative (GRI) and the S1 and S2 standards from the International Financial Reporting Standards (IFRS) Foundation (Global Reporting Initiative, 2022; International Financial Reporting Standards Foundation, 2022).

Firms determine material topics through a materiality assessment process. This assessment process requires the firm to review its current prioritisation of internal and external risks and opportunities, including climate change, and engage with its stakeholders to understand their perspectives on those risks (Ferrero-Ferrero et al., 2021; Geldres-Weiss et al., 2021; Remmer & Gilbert, 2019). The results of the materiality assessment are typically included within a firm's public disclosures (Kouloukoui et al., 2019), be it a standalone sustainability report or an integrated annual report. An important caveat is that the assessment process can be subjective and dependent on what information stakeholders need (Cerbone & Maroun, 2020; Mio et al., 2020; Puroila & Mäkelä, 2019; Reimsbach et al., 2020), and vague statements about the material nature of topics could lead stakeholders to draw unjustified conclusions (Jørgensen et al., 2022). Additionally, depending on the industry, the application of a materiality assessment might not be a common practice (Karagiannis et al., 2019).

The disclosure of climate change risks is not limited to the outcomes of the materiality assessment. Firms also disclose detailed descriptions of those risks and their relevance to the firm in an attempt to reduce information asymmetry (Schiemann & Sakhel, 2019) and bolster a positive public opinion of the firm (Herzig & Schaltegger, 2006; Setia et al., 2022). These actions can be examples of strategic disclosures or greenwashing. These types of disclosures illustrate attempts by firms to highlight positive performances through overdisclosure of that information to direct attention away from less favourable performances (Callery, 2023; Zharfpeykan, 2021). In doing so, firms enact obfuscation through information asymmetry, impression management, or exaggeration of benefits and value (Callery, 2023). The ability to engage in greenwashing is enhanced by the discretionary nature of sustainability disclosures, where the firm decides the materiality of the information (Ferrero-Ferrero et al., 2021), resulting in the cherry-picking of information that meets the minimum level of materiality (Bingler et al., 2022). Therefore, the material nature of a sustainability topic might not be solely determined by the stakeholders' information needs.

Moreover, the quality of climate change risk disclosures can vary (Arian & Sands, 2024) both within industries and between firms (Andersson & Arvidsson, 2023). For example, Machado et al. (2021) found that firms using the GRI standards exhibited considerable differences in the quality of information disclosure and the percentage of topics not disclosed, even though those disclosures received assurance from GRI. A similar outcome can be seen in the work of Farooq et al. (2021), who noted a decrease in the number and quality of disclosures related to materiality assessment and stakeholder engagement within the GCC region. A decrease in the quality of sustainability information could be attributed to several factors, including a shift in the social importance of the sustainability issue, a focus on less disclosure in order to maintain the current state of legitimacy, or an attempt to distract from or obfuscate poor performance (de Villiers & van Staden, 2006). As such, variation in disclosure quality can pose challenges for compatibility between firms and hinder the understanding of how firms integrate climate change risks.

To conclude, climate change risks are material to firms due to their potential financial implications, and disclosure of this information is needed to ensure that stakeholders can make informed decisions about the firm. Nevertheless, the quality of this information is not guaranteed, and firms may have ulterior motives behind their disclosures. This study seeks to build on the above insights by providing nuanced details into the disclosure of firms within the GCC region.

11.3 Theoretical Background

This study relies on legitimacy theory, which has been applied in prior research concerning corporate communication and public disclosure (Arvidsson, 2023; De Bernardi et al., 2019; Kouloukoui et al., 2019). Legitimacy theory helps us understand how corporate communications, including public disclosures, promote transparency and accountability of the firm (Arvidsson, 2023; De Bernardi et al., 2019).

It also explores how stakeholder expectations shape these communications and, in turn, how they influence stakeholder support (Arvidsson, 2023).

This theory argues that a firm's existence depends on whether it operates within the established parameters set out by society, effectively generating a social licence or social contract that the firm must abide by to maintain its status in society (Deegan, 2002). To prove that they are acting within the social contract, firms communicate their efforts through mandatory or voluntary disclosures. However, the firm must reach the minimum level of acceptable disclosure to maintain its legitimacy within society (Beske et al., 2020). Further, public disclosures should cater to the needs of all stakeholders, not only the financial stakeholders (Beske et al., 2020); yet, companies recognise that legitimacy comes from the collective opinion of all stakeholders, not just a select few (Suchman, 1995).

To build or maintain legitimacy, firms are able to employ several approaches that appeal to the practical needs and moral desires of stakeholders. As noted by Suchman (1995), there are three primary classifications of legitimacy: pragmatic legitimacy, moral legitimacy, and cognitive legitimacy. Each of these legitimacy classifications also contains subtypes, which will be explored in detail. Pragmatic legitimacy can be achieved through exchange legitimacy, where firms concentrate on providing direct benefits to stakeholders, or through dispositional legitimacy, which utilises the positive opinions or characteristics assigned to the firm by stakeholders (e.g. stakeholders perceive the firm to be trustworthy and honest). Influence legitimacy is the final pragmatic approach, and it considers the interests of stakeholders. Instead of providing direct benefits, it aims to ensure alignment with core stakeholder values, beliefs, or principles (Suchman, 1995).

The second type of legitimacy, moral legitimacy, is based on whether the firm's actions or disclosures align with the supporting social welfare or other social causes (Suchman, 1995). Moral legitimacy can be achieved through consequential legitimacy, whereby the outcomes of the firm's actions or operations meet the desired outcomes ascribed by stakeholders. Another option is to obtain legitimacy by applying socially acceptable processes and procedures, referred to as procedural legitimacy, or by employing internal structures (e.g. checks and balances or quality control) under the structural legitimacy classification. Finally, moral legitimacy might also be obtained when there is social approval of the charisma or integrity of the organisational leader(s), known as personal legitimacy (Suchman, 1995).

The final category, cognitive legitimacy, can be more challenging for firms to achieve since it relies on transcending moral and practical value so that firms are perceived as invaluable and irreproachable. Within this type of legitimacy, society perceives firms are legitimate because there is a match with the collective ethics and integration into the daily lives of social actors (Suchman, 1995). Further, firms must also exhibit a degree of social importance, such that their operations or mere existence would be either indispensable or would cause major disruption if they ceased to exist (Suchman, 1995).

Legitimacy is critically important for firms to achieve since it provides them with the social acceptability to maintain their operations (Beske et al., 2020; Deegan, 2002; Suchman, 1995). However, the risk of failing to meet society's expectations has prompted some firms to engage in activities that provide a false or misleading view of the firm, effectively employing greenwashing (Zharfpeykan, 2021). Firms can adopt different strategies to deflect or mislead a reader's interpretation of the disclosure, and this is particularly relevant when presenting negative information (Bingler et al., 2022; Callery, 2023; Hahn & Lülfs, 2014). In the context of this study, legitimacy theory offers a vital framework for comprehending the motivations behind firms' participation in voluntary disclosures. It also provides insights into their choices regarding the application of strategic disclosure and greenwashing.

11.4 Methodology

This study employed a directed qualitative analysis (Assarroudi et al., 2018) and thematic content analysis (Clarke & Braun, 2017) to examine the public disclosures of GCC firms from Qatar, Saudi Arabia, and the United Arab Emirates (UAE). By conducting both types of qualitative analysis, this study gathers both manifest content (e.g. content that is explicitly stated, the surface-level meaning) and latent content (e.g. implied or implicit meaning not directly stated) (Braun & Clarke, 2006; Bryman, 2012), together these insights provide a more holistic understanding of how firms discuss climate change risks and determine their material nature. Both methods will be discussed in turn.

11.4.1 Directed Qualitative Analysis

A directed qualitative analysis explores whether an established set of keywords, themes, or criteria is present within the sampled documentation with manifest content (Assarroudi et al., 2018; Braun & Clarke, 2006; Bryman, 2012). A framework or checklist can be utilised to assess the sample for the sought-after criteria (Assarroudi et al., 2018).

To explore how GCC firms disclose climate change risks, a framework was developed based on insights from existing sets of climate risk and GCC metrics. The framework was used to direct the initial analysis of the dataset to help identify what information each disclosure contained. The thematic content analysis builds upon the initial analysis to provide more details about what the firms discuss in those topical areas. Both sets of metrics will now be discussed.

The first set of assessment metrics comes from Andersson and Arvidsson (2023, Appendix 1), and is outlined in Table 11.1. These metrics are designed to determine which aspects of climate change risks the firm uses, the sustainability reporting framework applied, and whether firms adopt a long or short-term perspective of climate risks (Andersson & Arvidsson, 2023). Although these metrics provide a strong foundation to build, given their comprehensiveness, little consideration has

Table 11.1 Climate risk assessment checklist (Adapted from Andersson and Arvidsson (2023, Appendix 1))

1.	Does the firm disclose climate-related risks?
2.	Is the method used to map disclosed?
3.	Does the firm disclose climate change risks in general?
4.	Does the firm disclose physical risks?
5.	Does the firm disclose transitional risks?
6.	Does the firm report that they have a system for mapping climate-related risks?
7.	Does the firm report that they have adopted TCFD?
8.	Does the firm report that they map short-term climate-related risks?
9.	Does the firm report that they map long-term climate-related risks?
10.	Does the firm present the results from scenario analysis?
11.	Does the firm report that they educate its employees on UN sustainability development goals (SDGs)?
12.	Does the firm report that managers are evaluated on non-monetary sustainability goals?

been given to materiality perspectives and the reporting needs or interests of the GCC region.

The second set of metrics comes from the GCC's Exchanges Committee, which established a unified set of voluntary reporting metrics in January 2023 for use by the region's many security exchanges (Abu Dhabi Securities Exchange, n.d.). These metrics align with the Sustainable Stock Exchange Initiative and connect with several components of the GRI standards (Abu Dhabi Securities Exchange, n.d.), as well as highlighting what ESG issues are of interest or importance within the region. Twenty-nine individual metrics develop the EGS taxonomy and are categorised based on their association with environmental, social, and governance topics. This study will focus on the climate-related metrics, as outlined in Table 11.2. Applying this taxonomy to the sample set would provide valuable insights into whether firms were already considering these climate-related topics, thereby acting as a benchmark for future assessments.

The GCC ESG taxonomy provides a unified set of reporting metrics, but there is a strong focus on the physical aspects of climate change, such as emissions and energy usage, and less focus on transition risks. As such, these metrics were combined with the assessment checklist from Andersson and Arvidsson (2023) to create a comprehensive assessment framework that considers physical and transition risks, and includes measures relevant to the GCC region. The integration of the ESG taxonomy and climate risk assessment checklist is presented in Table 11.3, alongside the search terms used for the directed qualitative analysis. The combined framework includes all of the climate-related metrics from the ESG taxonomy. Regarding the climate risk assessment checklist, all components were included except for question 11, "Does the firm report that they educate its employees on UN sustainability development goals (SDGs)?", and question 1, "Does the firm disclose climate-related risks?". Question 11 was removed since it did not align closely enough with the aims of this inquiry, and question 1 was removed because it was too similar to question 3 on the checklist. In terms of the GCC taxonomy, all items related to climate change risks and emissions (e.g. total emissions and intensities) were included.

Table 11.2 ESG Taxonomy (Adapted from GCC Exchanges Committee (2023))

ESG category	Metric	Calculation(s)	Description	Associated GRI standard(s)
Environment	Greenhouse gas (GHG) emissions	Total amount in CO2 equivalents, for scope 1 Total amount, in CO2 equivalents, for scope 2 (if applicable) Total amount, in CO2 equivalents, for scope 3 (if applicable)	The company's overall GHG emissions, and/or disclosure of scope 1, 2, or 3 emissions	GRI 305: Emissions 2016
	Emissions intensity	Total GHG Emissions per output Scaling factor Total non-GHG Emissions per output Scaling factor	The amount of GHG emissions per specific scaling factor or unit (e.g. employee headcount, size of physical floor space occupied, revenue, or sales)	GRI 305: Emissions 2016
	Energy usage	Total amount of Energy directly Consumed Total amount of Energy indirectly Consumed	The amount of energy, from all sources, by the company	GRI 302: Energy 2016
	Energy intensity	Total direct energy Usage per output Scaling factor	The amount of energy used per specific scaling factor or unit (e.g. employee headcount, size of physical floor space occupied, revenue, or sales)	GRI 302: Energy 2016
	Energy mix	Percentage energy usage by generation type	The amount (absolute value or percentage) of energy used per energy source directly used by the company	GRI 302: Energy 2016
	Climate risk mitigation	Total amount invested, annually, in climate-related Infrastructure, Resilience, and product Development	The amount of money invested or provided for climate change-related research, climate change mitigation, or related R&D expenditures	N/A

Table 11.3 Integrated climate risk assessment framework (Adapted from Andersson and Arvidsson, (2023, Appendix 1), GCC Exchanges Committee (2023) and includes the author's elaboration)

Assessment criteria	Search terms
1. Does the firm disclose climate change risks in general?	Climate change, climate change risks, and climate risks
2. Does the firm disclose physical risks?	Physical risks, natural hazards, and extreme weather
3. Does the firm disclose transitional risks?	Transition risks, transitional risks, and regulatory risks
4. Does the firm disclose the annual amount invested in climate change-related research, mitigation efforts, or related R&D expenditures?	Climate mitigation, climate change mitigation, climate funding, climate research, climate change investment, and climate investment
5. Is the method used to map climate risks disclosed?	Risk assessment, risk methodology, and assessment method
6. Does the firm report that they have adopted GRI, SASB, and/or TCFD frameworks?	GRI, SASB, and TCFD
7. Does the firm disclose total GHG emissions for scope 1, 2, and/or 3?	Total greenhouse gas emissions, total GHG emissions, emissions, greenhouse gas, scope 1, scope 2, and scope 3
8. Are the firm's emissions intensities reported?	Emissions intensity, emission intensity, and GHG intensity
9. Does the firm report that it maps short--term climate-related risks?.	Short-term risks, present risks, current risks, and near-term
10. Does the firm report that it maps long-term climate-related risks?.	Long-term risks and future risks
11. Does the firm present the results from scenario analysis?	Scenarios, outlooks, and scenario analysis
12. Does the disclosure contain materiality map or assessment?	Materiality assessment, materiality map, materiality mapping, and materiality matrix

It should be noted that both the climate risk checklist and ESG taxonomy overlook the impact that materiality could have on reporting climate change information. To address this oversight, questions six and seven from Table 11.1 were condensed and revised to specifically look for climate-related frameworks, such as the GRI, SASB, and TCFD, and became question 6 in Table 11.3. Each of those frameworks adopts a specific materiality perspective that could impact their reporting of climate risks (Adams et al., 2021; Afolabi et al., 2022) and act as a proxy to determine the firm's materiality approach. This metric aids in identifying disclosures using the specific framework and allows for further investigation of disclosures not using either of the identified frameworks during the thematic analysis phase. Further, question 12 was generated by the author to elicit specific information about how material issues are identified, categorised, or mapped by the firms. Together, these additional metrics will support our understanding of the material nature of climate change risks at the firm level.

11.4.2 Thematic Content Analysis

The second phase of the data analysis included the application of thematic content analysis, which seeks to understand how a particular theme, topic, or criteria are discussed within the sample without using a pre-established framework and through latent content (Clarke & Braun, 2017). This analysis typically provides insights into how individuals or firms perceive the topic of interest (Braun & Clarke, 2006; Kleinheksel et al., 2020), but it can also aid in identifying patterns in corporate sustainability disclosures (Boiral et al., 2024; Jain et al., 2022).

For this study, only documents containing elements from the directed qualitative analysis were examined to see whether the firm provided additional details about its approach to addressing or measuring the assessment criteria or to determine if any justification regarding underreporting or non-disclosure was provided. To conduct the analysis, the documents were carefully read to understand what the firms were discussing concerning a given assessment topic, and notes were made on the statements' details or need for clarity. From those notations, general themes or trends in the data were collated. It should be noted that relative percentages for those themes were not measured since the purpose of the themes was to support the directed qualitative analysis, more detail about this is provided below, and because relative percentages are not common practice within qualitative studies (Boiral et al., 2024).

Unlike Boiral et al. (2024), these documents were not categorised based on the major themes arising from the analysis. Instead, the insights were used to develop a four-point scale to determine how well the topics were detailed within the disclosures, effectively generating a scale measuring the quality of the reports. This inductive process to generate the quality scale was selected to provide a meaningful expansion of the binary assessment approach used by Andersson and Arvidsson (2023) and to gather thematic insights to share alongside the results of the direct qualitative analysis, thereby providing more nuance to the numerical results.

The items of the four-point quality scale were constructed as such. Firms that did not appear in the results for the given assessment criteria were categorised as "No Disclosure" since the criteria's search terms did not appear within the body of the firm's disclosure document (e.g. sustainability, annual, or integrated reports). The documents that contained the search terms but did not provide clear discussions of the assessment criteria, or only provided part of the disclosure items, were listed as "Vague Disclosure" since it would be difficult to fully understand the firm's process or integration of those criteria. Those documents that provided clear discussion or integration of the assessment criteria were categorised as "Met Disclosure Requirements", while those that provided additional insights or more in-depth discussion of the criteria were labelled as "Exceeded Disclosure" since that level of insight was not expected by either the GCC Taxonomy or the Climate Change Risk Assessment frameworks. This scale would provide a better understanding of disclosure quality since it tracks the level of detail the firm includes within the discussions or descriptions of the assessment criteria.

11.4.3 Sample Selection and Data Collection

The sample for this study was selected from publicly listed firms within Qatar, Saudi Arabia, and the UAE. Those three nations were selected because they have the largest stock exchanges within the GCC region when considering market capitalisation (Mansy, 2018; Muzoriwa, 2023). Further, the documents collected in the sample were chosen due to their accessibility, in terms of locating and downloading files, and their readability in the selected qualitative analysis program, NVivo. Additionally, this study only considered the 2023 versions of the disclosures since the GCC ESG taxonomy was only announced in January 2023 (Macbeth & Otubu, 2023), disclosures prior to that time may not include the taxonomy items so the decision was to use the most recent 2023 disclosures since firms might have begun to incorporate the taxonomy.

The documentation utilised in the data analysis consisted of the firm's official sustainability, ESG, or integrated reports. If those documents were unavailable, the firm's annual report or governance reports were used. A total of 117 firms were examined from across all three countries, with 30 firms from Qatar, 34 from Saudi Arabia, and 53 from the UAE.

11.5 Results

This study aimed to understand how climate change risks are considered within the public disclosures of GCC firms. To accomplish that goal, the following research questions were posed: (1) How are physical and transitional climate change risks discussed within corporate disclosures by GCC firms? and (2) To what extent are climate change risks material for the firm?

The directed qualitative analysis highlighted the extent to which climate change risks are included in those firms' public disclosures. When considering the collated results across all three countries, Fig. 11.1 demonstrates a stark lack of disclosure for nine of the twelve assessment criteria, meaning few firms provided content aligned with the integrated assessment framework. The least disclosure criteria included the firm's short-term and long-term climate risks, with 92% and 91% non-disclosure rates, respectively. Of the remaining criteria, disclosure of climate change risks, applied frameworks, and GHG emissions had a non-disclosure rate of less than 50% of all documents. Relevant insights for each of the 13 criteria will be explored in turn.

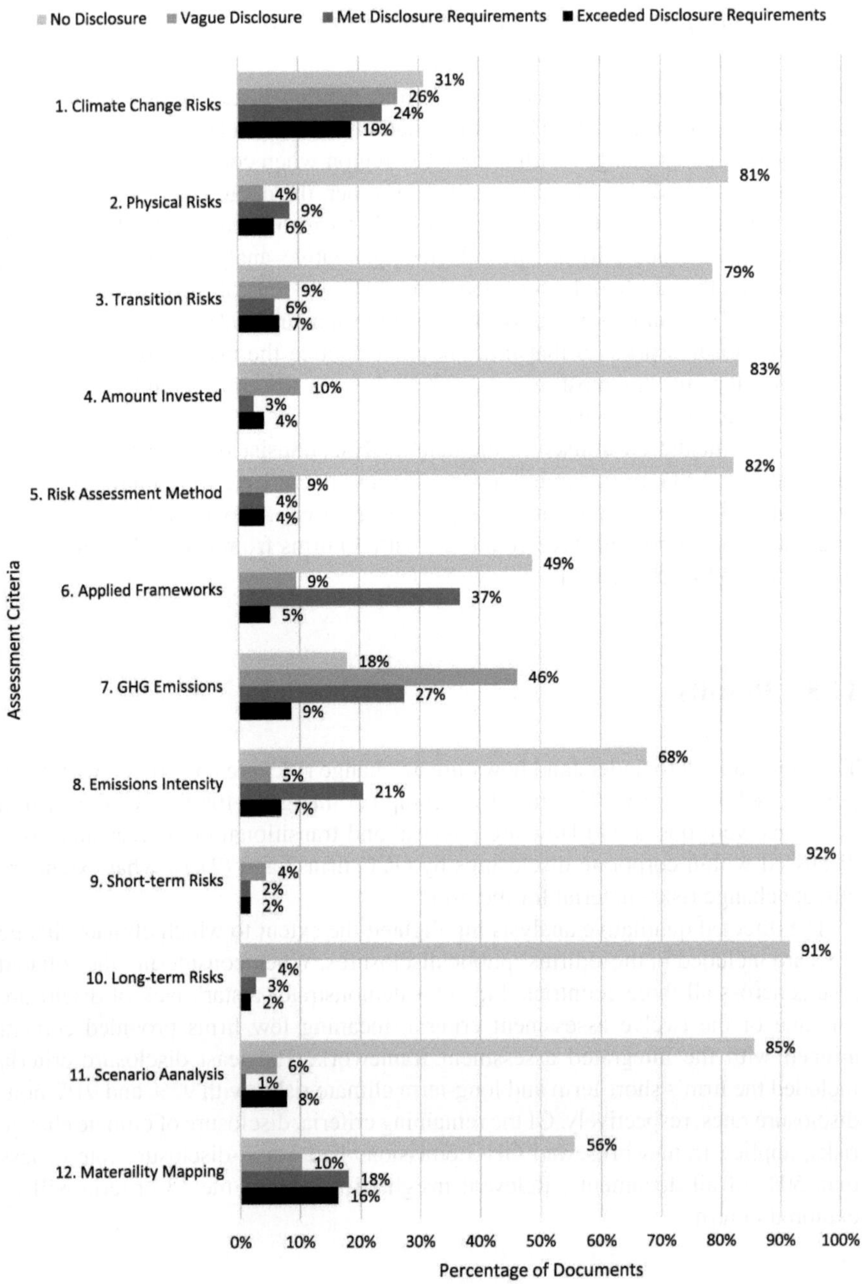

Fig. 11.1 Percentage of disclosure for all 13 criteria

11.5.1 Aggregated Results

For the climate change risk criteria, the difference between the categories of disclosure quality was not as drastic as it was for the applied frameworks and GHG emissions criteria. However, more firms provided vague disclosure of climate change risks by briefly mentioning the term and not providing context for why those risks are relevant or stating that they support the national climate efforts without discussing how the firm contributes to the national goal. The firms that met the disclosure requirements provided insights into the potential challenges they might face (e.g. increased costs or regulation) or explicitly acknowledged the overall risks of climate change on the industry and society. The firms that exceeded disclosure requirements did so by linking their discussion to Sustainable Development Goal (SDG) 13, explaining the connection between climate change and other aspects of their organisation (e.g. energy usage and waste generation), or providing details on their climate mitigation and adaptation strategies. Interestingly, firms that exceeded the disclosure requirements also tended to provide insights on physical and transition risks, with 73% and 77% disclosing some details, respectively. Only some firms that met disclosure requirements disclosed details for physical and transition risks, 14% for each risk type.

Regarding the quality of disclosures for physical and transition risks, firms met the disclosure requirements by presenting details about how those risks could manifest and impact their operations. For example, firms listed concerns with physical risks relating to the impacts of extreme weather events, droughts, and rising temperatures on the firm and its employees. On the other hand, transition risks included lower real estate values, lower asset performance, and the shifting supply and demand of markets. This level of detail regarding physical and transition risks was not present in the disclosures classified as "Vague Disclosure" since those firms typically only mentioned the terms physical or transition risk and did not articulate how those risks could impact the firm. Firms that exceeded disclosure expectations did so by highlighting the inclusion of physical or transition risks into their enterprise risk management framework or decision-making processes. Those firms also tended to provide more specific examples of the risks or explain the countermeasures they plan to adopt to address them.

When considering the amount of funds invested into climate change research and mitigation efforts, the firms that shared information about their investments focused on vague statements related to the intention to support these types of initiatives in the future or mentioned the various efforts they have undertaken but did not provide a specific monetary value attribute to those efforts. To meet the minimum disclosure requirements, firms needed to provide a monetary value for the initiative they supported, and 3% of firms disclosing this criterion were able to provide a value. Fortunately, 4% of disclosing firms exceeded expectations by specifically discussing what initiatives were supported, such as how the funds were used to develop a climate-related investment fund, to support enhancing the firm's infrastructure to make it more climate resilient or financing renewable energy projects.

When disclosing the risk assessment method used for climate change risks, firms tended to refer to their risk assessment process rather than a separate climate change risk process. Thus, firms that mentioned using a risk assessment or ESG assessment without clarifying the integration of climate change were considered "Vague Disclosures", resulting in 9% of firms disclosing risk assessments falling into this category. To be considered as meeting the disclosure requirements, the disclosures needed to contain additional insights into the role of climate change risks in the risk assessment of the firm or its portfolios. Firms within the "Exceeded Disclosure Requirements" typically have more information regarding the role of climate change risks as a specialised phase of the overall assessment process or as an integrated part of the entire process. Overall, evaluating the disclosures based on this criterion was difficult due to the varying approaches to describing the risk assessment process.

Alternatively, the applied framework criteria had the most firms meeting the disclosure requirements at 37% through explaining the use of GRI, SASB, or the TCFD frameworks, resulting in an easier understanding of how firms approached the structuring and development of their disclosures. Few firms, six in total, discussed the application of all three frameworks. This resulted in firms exceeding the disclosure requirements since using more than one framework was not required within the GCC taxonomy or the integrated assessment criteria. Much of that discussion related to the use of the frameworks in developing the reports, but there was a tendency for firms not to provide a finalised GRI, SASB, or TCFD summary or index. The lack of such indices resulted in the disclosures being labelled as "Vague Disclosure" since it would be difficult for a stakeholder to understand how the disclosures align with the content of those frameworks.

The GHG emissions criteria had 47% of firms vaguely reporting their emissions, resulting in either only discussing total emissions, reporting emission reduction percentage without any context for the total emissions for the current or previous year, or reporting one of Scopes 1, 2, and 3. For firms meeting the disclosure requirements, there was considerable discussion related to all three scopes. Further, of the 9% of firms that exceeded GHG emission disclosure, 40% provided information about all three scopes and their financed emissions, a topic not included within either of the assessment frameworks.

Although firms disclosed their GHG emissions, information related to their emissions intensity was not clearly discussed, with 68% not disclosing this information. Of the firms that did include emissions intensity, 21% provided enough information to meet the disclosure requirements. This information included a specific value for their emissions intensity along with a clear scaling factor. However, the "Vague Disclosures" did not present specific emissions intensities since they partially measured their intensity by including upstream emissions and discussed their intentions to reduce intensities or percentage of reduction in their intensities without presenting a clear intensity value. Firms that exceeded disclosure requirements provided insights into the intensities of their financed emissions, electricity generation, desalination processes, or provided separate intensities based on their end-users. These were topics not expected to be covered by the integrated assessment framework. However, the scaling factors in the emissions intensities were highly variable, which could make compatibility between disclosures challenging.

The discussions on short-term and long-term risks were not as prevalent as other topics in the assessment framework, as noted by the 92% and 91% non-disclosure percentages. If a firm did mention either of these time horizons, very little information was provided regarding the potential short- or long-term implications of those risks and the disclosures were listed as "Vague Disclosures". Some firms shared a list of potential risks for either time horizons or explained the impacts of short-term or long-term risks to the firm without providing a specific list of risks. Such discussions did result in those disclosures meeting the minimum disclosure requirements. Yet, some firms exceeded the requirements by clearly explaining how the short-term or long-term risks were identified and presenting a descriptive summary of those risks.

The application of scenario analysis to climate change risks was variable. Firms either vaguely discussed the outcomes or exceeded expectations by providing detailed explanations of the process. Application of the Network for Greening Financial System (NGFS) scenario analysis, climate stress testing, and Representative Concentration Pathways were methods used among the firms that exceeded the disclosure requirements. These processes allowed firms to articulate how evaluations of climate risks were conducted and highlight the most relevant issues the firm could face in the coming years. Regarding the firms that met the disclosure requirements, there was a tendency to briefly mention the results from their scenario analysis, with some description of the assessment process, but did not present as much detail as the disclosures labelled as "Exceeded Disclosure Requirements".

For the materiality mapping criterion, 56% of the sample did not discuss their materiality process. Of the disclosures that contained relevant material information, 18% met the disclosure requirements, and 16% exceeded the requirements. Those who met the requirements presented a list or summary of their material topics and a detailed assessment process description. Disclosures that did not describe the materiality assessment process were labelled vague since it was difficult to understand how their material topics were evaluated. Some disclosures included additional insights into how stakeholder engagement was integrated into their materiality process, exceeding the expected disclosure requirements. However, after completing the analysis of this criterion, the role of climate change risks within the material topics was unclear. Therefore, an additional analysis was conducted to understand whether the materiality maps or lists specifically mentioned climate change or climate change risks.

As presented in Fig. 11.2, 75% of disclosures that contained references to materiality mapping or assessments made no direct reference to climate change or climate change risks in their material topics. Some firms included general environmental topics in their materiality lists, but there was insufficient information to determine if climate change risks were addressed within that category. In other disclosures, climate change was addressed as a material topic; however, minimal information was provided regarding the significance or relevance of climate change risks. While

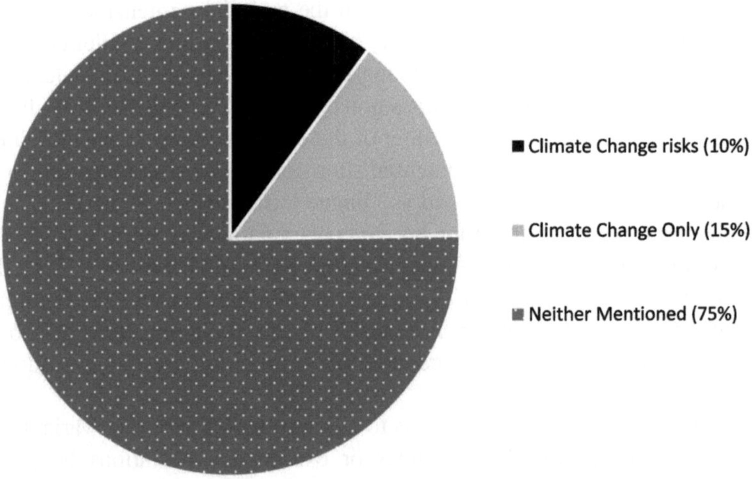

Fig. 11.2 Percentage of firms including climate change risks or climate change in materiality assessments

only 10% of firms with materiality assessment specifically mentioned climate change risks as a material topic.

To summarise the results thus far, the collated results from the directed qualitative analysis and thematic content analysis demonstrate that 9 of the 12 assessment criteria had non-disclosure percentages of over 50%. The one topic with the lowest percentage of non-disclosure is the GHG Emissions criterion, with 46% of the disclosures meeting the anticipated level of disclosure. When considering the disclosures mentioning the individual assessment criteria, there is variability in the disclosure quality, with firms electing to provide minimal or vague disclosures for several topics. However, the results in this section are the combined insights from Qatar, Saudi Arabia, and the UAE. As such, it would be prudent to examine the data from each nation.

11.5.2 Country-Specific Insights

The results for Qatari firms followed a similar trend to the aggregated results, with 10 of the 12 criteria having non-disclosure rates of over 50%, as highlighted in Fig. 11.3. However, evidence suggests that Qatar-based firms focus on disclosing information on climate change risks, physical risks, frameworks, and emissions intensity since firms disclosing those five assessment criteria managed to meet the disclosure requirements. Interestingly, GHG emissions disclosures surpassed the non-disclosure rate, with 50% being vague due to the firms providing aspirational reduction targets or indicating their desire to address their emissions without presenting their current emissions values. Furthermore, the firms disclosing under the

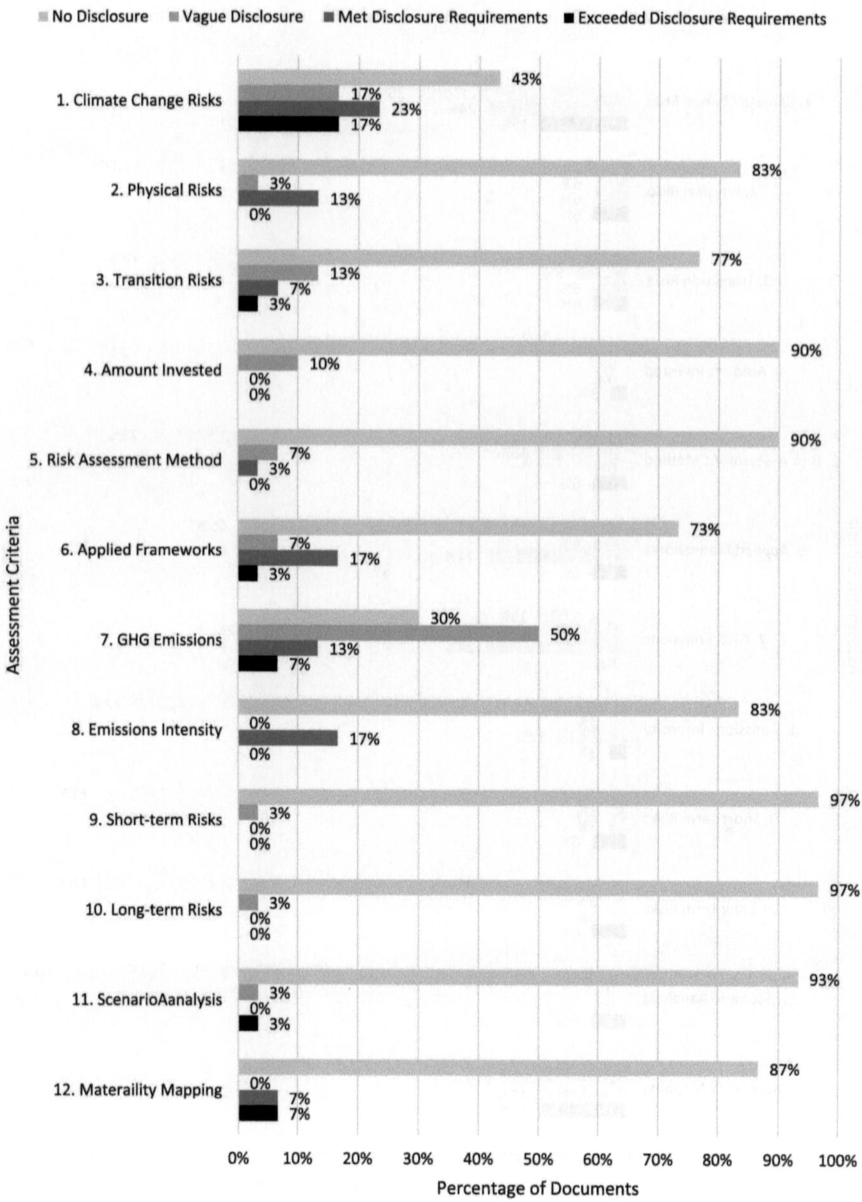

Fig. 11.3 Percentage of disclosure for Qatar

emissions intensity criterion did not provide vague or exceptional disclosures. Instead, firms adhered to the expected requirements by giving an exact value and the scaling factor.

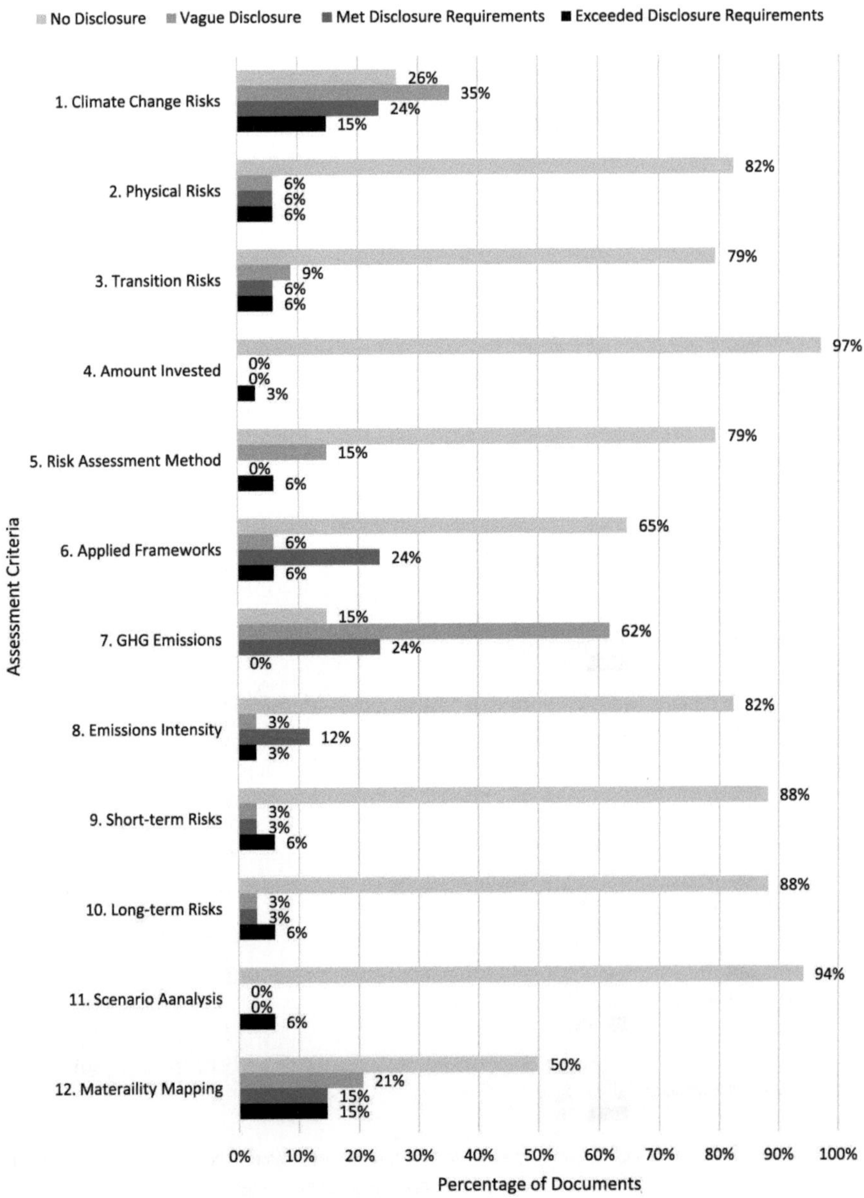

Fig. 11.4 Percentage of disclosure for Saudi Arabia

When examining the results for firms based in Saudi Arabia, there is a similar degree of non-disclosure rates as exhibited by Qatari firms; see Fig. 11.4. Fewer Saudi Arabian firms had "No Disclosure" for the climate change risks criterion than

Qatari firms, with 26% and 43%, respectively. However, Saudi Arabian firms tended to provide more vague climate change risk disclosures due to not clearly discussing the importance or relevance of climate change risks for their operations. When examining physical and transition risks, the disclosure percentages are close in value, with a nearly equal spread between firms providing vague disclosures, meeting the requirements, and exceeding disclosure expectations. Firms that met or exceeded the disclosure requirements for physical risks also tended to provide the same level of detail for transition risks, except for one firm that only provided a vague discussion of regulatory transition risks. Additionally, the GHG emissions criterion for Saudi Arabia did not exhibit any firms exceeding the disclosure requirement, which differed from the results of Qatar and the UAE. A review of the GHG emission disclosures for Saudi Arabian firms noted a lack of clarity on the exact emissions values of firms and a focus on future emissions reduction targets. This vague approach to discussing emissions is why only 24% of GHG emissions disclosures met the expected disclosure requirements.

For the UAE, seven of the 12 criteria exhibited non-disclosure percentages of over 50%, as outlined in Fig 11.5. For the climate change risks criterion, the non-disclosure rate was only 26%, and firms presented a range of insights that accounted for 25% of disclosures meeting the expected disclosure requirements. Firms that exceeded the disclosure requirements for this criterion did so by articulating the impacts of climate change on the firms throughout their disclosures or by connecting climate change risks to SDG 13. For physical and transition risks, more firms exceeded the disclosure requirements than those meeting the expected requirements, and firms did so by highlighting the most plausible transition risks, explaining the relevant acute and chronic physical risks, or providing detailed strategies to address those risks. Additionally, 57% of disclosures related to the applied frameworks criterion met the disclosure requirements, which was not true for Qatari and Saudi Arabian firms. Finally, the criterion with the lowest non-disclosure was GHG emissions at 13%. Still, the range in quality of those disclosures was limited, with 34% providing vague insights and only 38% meeting the expected disclosure requirements.

In summary, the analysis of Qatari, Saudi Arabian, and UAE firms reveals significant non-disclosure rates across various assessment criteria, particularly concerning physical and transition risks, short-term and long-term risks, and scenario analysis. The topics of GHG emissions and climate change risks saw a degree of non-disclosure, but firms within all three nations did provide enough disclosure to meet the minimum level of disclosure. To assert that one nation provided more or better disclosure than another would be impractical. A substantial number of firms from all three nations are not disclosing relevant information, and there is a considerable need to improve the quality of all disclosures across all 12 assessment criteria.

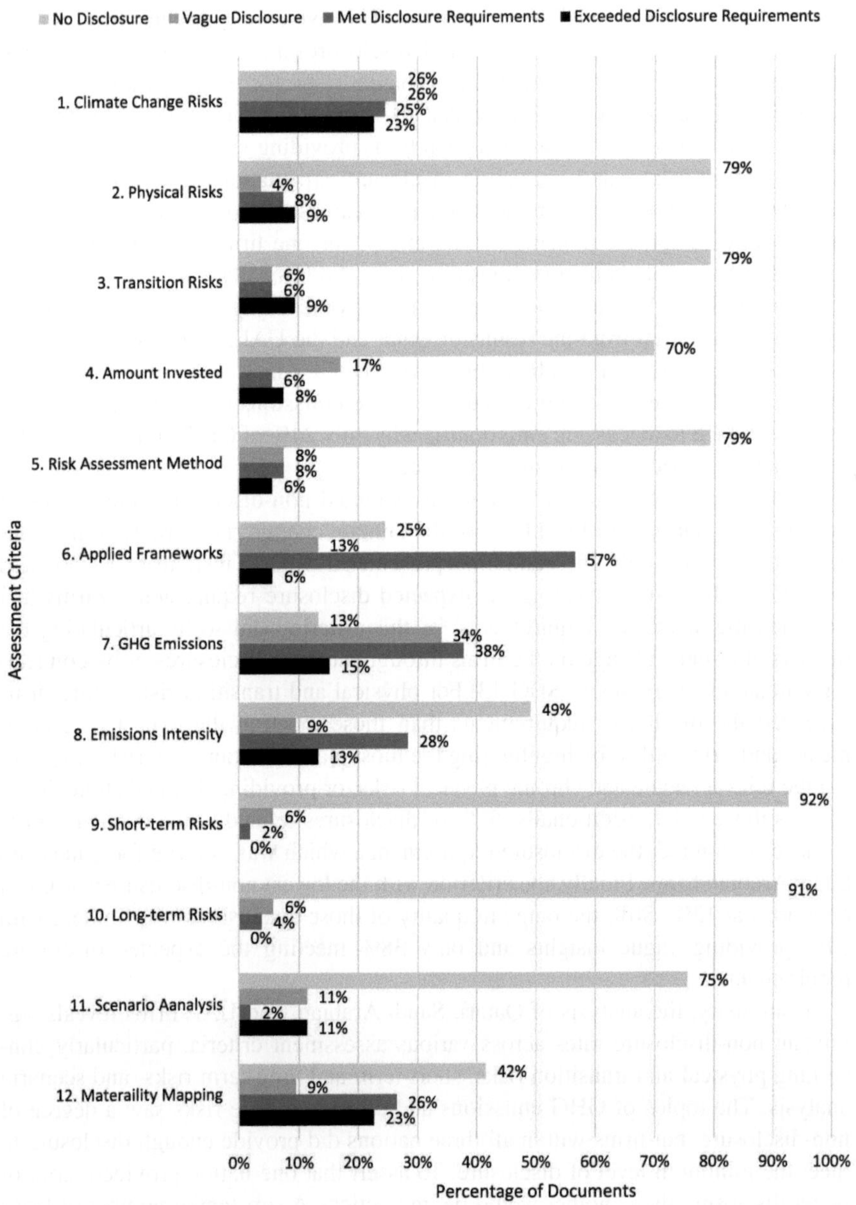

Fig. 11.5 Percentage of disclosure for the United Arab Emirates

11.6 Discussion

This chapter examined how climate change risks are addressed within public disclosures by exploring the following research questions: (1) How are physical and transitional climate change risks discussed within corporate disclosures by GCC firms?, and (2) To what extent are climate change risks material for the firm? The analysis of the disclosures of 117 firms from Qatar, Saudi Arabia, and the UAE has provided valuable insights into the firms' perspectives on climate change risks and the material nature of those risks. Based on those insights, the above research questions can be answered as follows.

11.6.1 Integration of Physical and Transition Risks

The sampled corporate disclosures offer a limited discussion of physical and transition risks, which are not well understood and are superficially integrated into the disclosures of the firms sampled, which aligns with the insights from Andersson and Arvidsson (2023) and Farooq et al. (2021). Although sampled firms met and exceeded the disclosure requirements for the climate change risk criterion, the insights provided only addressed the general impacts of climate change. The specific topics of physical and transition risks were underexplored by firms. Both Arian and Sands (2024) and Kouloukoui et al. (2019) noted limited disclosure of climate risks within their samples, suggesting that non-disclosure or vague disclosures highlighted in this study are not uncommon. By not providing clear descriptions of relevant physical and transition risks, it could be assumed that firms do not fully understand the importance of these risks in their operations (Arian & Sands, 2024).

Furthermore, the quality of the information shared exemplifies the superficial nature of these disclosures. The sampled firms tended to provide minimal information to meet the expected disclosure requirements, such as partially reporting emissions or providing elusive descriptions of the role of climate change in their risk assessment process. This use of vague disclosures is consistent with the strategic disclosure method of obfuscation as outlined by Callery (2023). Additionally, the lower degree of disclosure aligns with the findings from Karagiannis et al. (2019), who also showed a moderate degree of disclosure or completeness in their samples. Understandably, firms would provide enough information to meet stakeholders or the market's information needs if there is an expectation to do so (Matsumura et al., 2024) and providing too much information could harm a firm's value if the market intensifies its attention on a firm (Vestrelli et al., 2024). However, the vagueness of the disclosures for all 12 assessment criteria suggests an attempt to build legitimacy at the cost of fully understanding the ramifications of climate risks on the firm.

Legitimacy is a social construct that "reflects a congruence between the behaviours of the legitimate entity and the shared beliefs (or assumedly shared) beliefs of some social group…" (Suchman, 1995, p. 574). As such, firms seek the approval of society in order to maintain their operations (Deegan, 2002) and do so by voluntarily disclosing information to the public (Beske et al., 2020), which reduces the information asymmetry between themselves and their stakeholders (Schiemann & Sakhel, 2019). The frequent disclosure of specific topics within the sample, such as climate change risks, could be explained by legitimacy theory as the sampled firms attempt to meet the information needs of scrutinising stakeholders (Kouloukoui et al., 2019) and maintain their legitimacy. More specifically, GCC firms' disclosure of climate change risks could be facilitated by their interest in maintaining consequential legitimacy (Suchman, 1995) by demonstrating how the company implements its actions to address climate change, which is of considerable social concern in the region (Kingdom of Saudi Arabia, n.d.; State of Qatar, n.d.; United Arab Emirates, 2024).

Moreover, the less frequent disclosure of other topics in the sample, such as short-term risks and transition risks, could be indicators of strategic disclosure efforts employed by firms to safeguard their reputation from legitimacy risks arising from the detrimental effects of poor climate change performance (Beske et al., 2020; Callery, 2023; Hahn & Lülfs, 2014). As noted by de Villiers and van Standen (2006), lower levels of corporate disclosure regarding environmental issues may arise when a firm seeks to maintain its legitimacy by providing only the minimal information required or diverting attention from areas of poor performance. As such, the application of greenwashing within these disclosures cannot be discounted since many of the disclosures are incomplete, with several criteria having percentages of non-disclosure of over 50% (i.e. physical risks and transition risks) by firms from all three countries. Even when topics are disclosed, it is done so in a vague manner. For example, the aggregate percentage of vague disclosures for the GHG emissions criterion was 46%, with only 27% of firms meeting the disclosure requirements. These values demonstrate that firms perceive the disclosure of climate change risks as important for their legitimacy but are unwilling to provide the minimum information necessary to appeal to their stakeholders. Overall, the findings of this study align with legitimacy theory, indicating efforts to sustain current legitimacy and the potential for greenwashing through limited disclosures across various criteria.

11.6.2 The Material Nature of Climate Change Risks

Regarding the second research question, the materiality of climate change risks is prevalent for a small percentage of sampled firms, but it is not universally recognised. As noted in Fig. 11.2, only 10% of firms disclosing materiality assessment included climate change risks within their discussions. Many firms (i.e. 56%, as noted in Fig. 11.1) did not disclose any information related to materiality mapping. These results are not surprising since they align with the findings from Beske et al. (2020), who also noted missing or incomplete explanations for materiality disclosures. These insights echoed the climate change risk disclosure variability found in

Andersson and Arvidsson (2023). The above findings indicate that firms are willing to disclose general climate change risks, but most are not concerned about including them in their materiality assessments.

Fundamentally, materiality assessments intend to examine the shared concerns between the firm and its stakeholders (Geldres-Weiss et al., 2021). The materiality disclosures of firms that met disclosure requirements provided enough insights regarding their processes, while those that exceeded disclosure requirements highlighted their stakeholder engagement process, suggesting that firms are familiar with the process of materiality assessments. However, the level of detail provided concerning climate change and risks is limited, which creates challenges for stakeholders in understanding the connections between their climate-related concerns and the firm's perspectives (Jørgensen et al., 2022). This approach to disclosing the material nature of climate change risk could occur due to the innate subjectivity of materiality issues (Reimsbach et al., 2020) and the value-laden nature of corporate sustainability (Puroila & Mäkelä, 2019), which raises legitimacy concerns.

Through the lens of legitimacy theory, firms should ensure that their disclosures align with the present information needs since the minimum level of acceptable disclosure can change over time (de Villiers & van Staden, 2006), and disclosure should provide enough insights to support a wide range of stakeholders (Beske et al., 2020). However, firms are aware that legitimacy is built by the collective or aggregate acceptance (Deegan, 2002; Suchman, 1995). This allows them to prioritise the interests of influential stakeholders without worrying about the effects on their legitimacy, as minority stakeholder opinions may not significantly alter the overall acceptance level (Suchman, 1995). These power dynamics can impact the material nature of issues by directing the selection of topics based on the interests of influential stakeholders (Cerbone & Maroun, 2020; Remmer & Gilbert, 2019), leading to the strategic disclosure of information. This legitimacy-seeking approach may explain why GCC firms disclosed climate change risks but did not include or articulate them in their materiality mapping.

By incorporating this information into various sections of their reports instead of limiting it to the materiality mapping, GCC firms can offer valuable insights to specific stakeholders. This strategy helps fulfil the basic information requirements for dispositional legitimacy (Suchman, 1995) by showing that the firm is a good corporate citizen by sharing this information. On the other hand, firms ensured that the materiality assessments targeted the issues most pertinent to their key stakeholders. This approach helps maintain their influence legitimacy (Suchman, 1995), demonstrating that the company addresses the informational needs of investors and other important stakeholders. This practice is not surprising, as companies have demonstrated a tendency to favour the disclosure of positive climate change information (Bingler et al., 2022; Callery, 2023), use impression management to emphasise their strengths, and take advantage of the subjective nature of materiality assessments (Ferrero-Ferrero et al., 2021; Zharfpeykan, 2021).

The actions of the sampled disclosures raise questions as to the sincerity of their materiality assessments and whether greenwashing is intended or an accidental by-product of not being prepared to disclose. Given the recent establishment of the

GCC ESG taxonomy, the sampled firms might not have had enough time to embed it into their disclosures fully. Firms could alleviate some of this concern by including more detailed information about the role of climate change risks and the material nature of those risks. Firms within the GCC region should be mindful of how their disclosures align with stakeholder information needs and seek to establish legitimacy through transparency (Deegan, 2002; Machado et al., 2021).

11.6.3 Policy Implications

The findings of this study offer new insights into how companies in Qatar, Saudi Arabia, and the UAE disclose climate change risks, thereby enhancing the understanding of these risks within the GCC and broader MENA region. The results highlight a need for improved integration of climate change risks into corporate operations and public disclosures. There are financial risks that are linked to climate change risks (Benz-Saliasi, 2020; Huang et al., 2022) and providing vague or non-disclosure of those risks might lead firms to expose themselves to potential adverse financial impacts (Matsumura et al., 2024). MENA firms should consider expanding their understanding of climate change risks beyond the common concerns for temperature increases and extreme weather events. By providing more nuanced insights, firms would be better positioned to build legitimacy through insightful and reasonable strategies that stakeholders could support.

The GCC ESG taxonomy offers a chance to unify existing and emerging reporting standards while promoting greater adoption among companies not yet engaged in this practice. The GCC ESG taxonomy was developed based on GRI standards (Abu Dhabi Securities Exchange, n.d.), and the insights from this study illustrate how firms can disclose some information related to those standards. Given the recent developments of the IFRS's S1 and S2 standards (International Financial Reporting Standards Foundation, 2022), and the importance of climate change risk disclosures to financial stakeholders (Ben-Amar et al., 2023; Matsumura et al., 2024), it would not be unreasonable to assume that the taxonomy could be updated to include additional metrics or further evaluation details based on emerging reporting standards. Expanding the taxonomy might aid in addressing the declining quality of disclosures and use of international reporting standards (Farooq et al., 2021). The GCC ESG taxonomy was developed through the collective efforts of the stock exchanges from all three nations (Abu Dhabi Securities Exchange, n.d.), thereby suggesting that regional cooperation to develop unified climate change or net-zero standards is possible. However, the voluntary nature of this taxonomy should be reevaluated; otherwise, the issues of vague disclosures and potential greenwashing will persist (Callery, 2023; Zharfpeykan, 2021).

Regarding global climate commitments to net-zero and the IPCC efforts, expanding the GCC ESG taxonomy, or a similar taxonomy, to the MENA could further enhance the transition to a low-carbon economy. The country-level insights from this analysis highlight the inconsistencies with firms' disclosure in the GCC. For example, long-term and short-term risks are severely underreported by Qatari firms, while GHG emissions are disclosed less in the UAE when compared to the other two countries. The disparities between these three nations suggest national governments need stronger engagement to promote standardised disclosures. However, the framework promoted should ensure a problem-oriented approach (e.g. a net-zero, low-carbon economy focus) rather than a business-centred perspective. If a framework primarily prioritises the interests of businesses, then its disclosure requirements may not lead to significant progress towards collective goals. This is because the framework could emphasise company-level measures while neglecting the broader connections to higher level initiatives (e.g. the IPCC's efforts) (Christy et al., 2024; Mair et al., 2024).

By developing a deeper understanding of climate implications, businesses can enhance their legitimacy and build stakeholder trust while complying with evolving reporting standards. The potential expansion of the GCC ESG taxonomy could be crucial in standardising disclosures and promoting regional cooperation towards a unified approach to climate accountability. Ultimately, this collective effort is essential to support global climate commitments and facilitate the transition to a sustainable, low-carbon economy in the MENA region.

11.7 Conclusion and Future Research

This study explored how climate change risks were discussed in the public disclosures of 117 firms across Qatar, Saudi Arabia, and the UAE. The results of the directed qualitative analysis found that many firms do not disclose climate change-related topics within their disclosures. The thematic content analysis highlighted variability in the quality of those firms that did disclose relevant information. Examination of the materiality assessments highlighted little consideration for climate change risks. These insights highlight concerns about companies using disclosures to enhance legitimacy while omitting negative information about climate change (Bingler et al., 2022; Callery, 2023).

The disclosure of climate change risks is an essential task for firms since it provides stakeholders with the information needed to make informed decisions regarding the firm (Font et al., 2016; Ilhan et al., 2023; Jebe, 2019; Megeid, 2024). As well, these disclosures could assist firms in avoiding negative financial repercussions (Benz-Saliasi, 2020; Huang et al., 2022; Vestrelli et al., 2024) and provide an opportunity to maintain the legitimacy of the firm in society (Beske et al., 2020; Deegan, 2002). However, as this study and previous works have noted (Arian & Sands, 2024; Farooq et al., 2021; Kouloukoui et al., 2019; Machado et al., 2021), the current disclosure of climate change risks (e.g. physical and transition risks) is superficial, and the material relevance of these risks is not universally recognised.

Nevertheless, the recent development of the GCC ESG taxonomy (GCC Exchanges Committee, 2023), along with national commitments to climate change and net-zero (Kingdom of Saudi Arabia, n.d.; State of Qatar, n.d.; United Arab

Emirates, 2024), leave hope that firms within those nations will improve their disclosures in the coming years.

In terms of limitations and future research, this study could be improved in several ways. First, this study did not consider analysing the disclosure based on industry. Doing so might provide additional clarification regarding disclosure trends that would not be captured in the current approach. Second, applying an additional quantitative approach (e.g. regression analysis), perhaps alongside the industry analysis, could provide complementary insights into the relationships between the assessment criteria. Third, given the possibility of strategic disclosures, the analysis could be expanded to include an additional taxonomy to evaluate the degree to which firms adopt or enable greenwashing within their disclosures. Finally, this study could be expanded by including firms from additional GCC or MENA nations, which would support the understanding and development of climate change risk disclosures in the region.

References

Abu Dhabi Securities Exchange. (n.d.). *GCC ESG metrics*. https://www.adx.ae/english/pages/aboutadx/sustainability/unified-gcc-esg.aspx#:~:text=The%20metrics%20comprise%2029%20standards,data%20privacy%2C%20ethics%20and%20more

Adams, C., Alhamood, A., He, X., Tian, J., Wang, L., & Wang, Y. (2021). *The double materiality concept: Application and issues* [project report]. Global Reporting Initiative. https://www.globalreporting.org/media/jrbntbyv/griwhitepaper-publications.pdf

Afolabi, H., Ram, R., & Rimmel, G. (2022). Harmonization of sustainability reporting regulation: Analysis of a contested arena. *Sustainability, 14*(9), 5517. https://doi.org/10.3390/su14095517

Andersson, F., & Arvidsson, S. (2023). Understanding, mapping and reporting of climate-related risks among listed firms in Sweden. *Climate Policy, 23*(8), 945–958. https://doi.org/10.1080/14693062.2022.2116383

Arian, A., & Sands, J. (2024). Corporate climate risk disclosure: Assessing materiality and stakeholder expectations for sustainable value creation. *Sustainability Accounting, Management and Policy Journal, 15*(2), 457–481. https://doi.org/10.1108/SAMPJ-04-2023-0236

Arvidsson, S. (2023). CEO talk of sustainability in CEO letters: Towards the inclusion of a sustainability embeddedness and value-creation perspective. *Sustainability Accounting, Management and Policy Journal, 14*(7), 26–61. https://doi.org/10.1108/SAMPJ-07-2021-0260

Assarroudi, A., Heshmati Nabavi, F., Armat, M. R., Ebadi, A., & Vaismoradi, M. (2018). Directed qualitative content analysis: The description and elaboration of its underpinning methods and data analysis process. *Journal of Research in Nursing, 23*(1), 42–55. https://doi.org/10.1177/1744987117741667

Ben-Amar, W., Herrera, D. C., & Martinez, I. (2023). Do climate risk disclosures matter to financial analysts? *Journal of Business Finance & Accounting*, 1–26. https://doi.org/10.1111/jbfa.12778

Benz-Saliasi, E. (2020). Climate disclosure and climate risk for Asian companies. In J. Fu & A. W. Ng (Eds.), *Sustainable energy and green finance for a low-carbon economy: Perspectives from the Greater Bay Area of China* (pp. 25–45). Springer. https://doi.org/10.1007/978-3-030-35411-4_2

Beske, F., Haustein, E., & Lorson, P. C. (2020). Materiality analysis in sustainability and integrated reports. *Sustainability Accounting, Management and Policy Journal, 11*(1), 162–186. https://doi.org/10.1108/SAMPJ-12-2018-0343

Bingler, J. A., Kraus, M., Leippold, M., & Webersinke, N. (2022). Cheap talk and cherry-picking: What ClimateBert has to say on corporate climate risk disclosures. *Finance Research Letters, 47*, 102776. https://doi.org/10.1016/j.frl.2022.102776

Boiral, O., Brotherton, M.-C., & Talbot, D. (2024). Achieving corporate carbon neutrality: A multi-perspective framework. *Journal of Cleaner Production, 467*, 143040. https://doi.org/10.1016/j.jclepro.2024.143040

Braun, V., & Clarke, V. (2006). Using thematic analysis in psychology. *Qualitative Research in Psychology, 3*(2), 77–101. https://doi.org/10.1191/1478088706qp063oa

Bryman, A. (2012). *Social research methods* (4th ed.). Oxford University Press.

Callery, P. J. (2023). The influence of strategic disclosure on corporate climate performance ratings. *Business & Society, 62*(5), 950–988. https://doi.org/10.1177/00076503221115715

Campiglio, E., Daumas, L., Monnin, P., & von Jagow, A. (2023). Climate-related risks in financial assets. *Journal of Economic Surveys, 37*(3), 950–992. Scopus. https://doi.org/10.1111/joes.12525

Cerbone, D., & Maroun, W. (2020). Materiality in an integrated reporting setting: Insights using an institutional logics framework. *The British Accounting Review, 52*(3). https://doi.org/10.1016/j.bar.2019.100876

Chabot, M., & Bertrand, J. (2023). Climate risks and financial stability: Evidence from the European financial system. *Journal of Financial Stability, 69*, 101190. https://doi.org/10.1016/j.jfs.2023.101190

Christy, A., Elnahass, M., Amezaga, J., Browne, A., & Heidrich, O. (2024). A dynamic framework to align company climate reporting and action with global climate targets. *Business Strategy and the Environment, 33*(4), 3103–3128. https://doi.org/10.1002/bse.3635

Clarke, V., & Braun, V. (2017). Thematic analysis. *The Journal of Positive Psychology, 12*(3), 297–298. https://doi.org/10.1080/17439760.2016.1262613

De Bernardi, P., Venuti, F., & Bertello, A. (2019). The relevance of climate change related risks on corporate financial and non-financial disclosure in Italian listed companies. In P. De Vincentiis, F. Culasso, & S. A. Cerrato (Eds.), *The future of risk management, Volume I* (pp. 77–107). Springer. https://doi.org/10.1007/978-3-030-14548-4_4

de Villiers, C., & van Staden, C. J. (2006). Can less environmental disclosure have a legitimising effect? Evidence from Africa. *Accounting, Organizations and Society, 31*(8), 763–781. https://doi.org/10.1016/j.aos.2006.03.001

Deegan, C. (2002). Introduction: The legitimising effect of social and environmental disclosures—A theoretical foundation. *Accounting, Auditing & Accountability Journal, 15*(3), 282–311. https://doi.org/10.1108/09513570210435852

Dragomir, V. D., Dumitru, M., Chersan, I. C., Gorgan, C., & Păunescu, M. (2024). Double materiality disclosure as an emerging practice: The assessment process, impacts, risks, and opportunities. *Accounting in Europe*, 1–38. https://doi.org/10.1080/17449480.2024.2339264

Farooq, M., Zaman, R., Sarraj, D., & Khalid, F. (2021). Examining the extent of and drivers for materiality assessment disclosures in sustainability reports. *Sustainability Accounting Management and Policy Journal, 12*(5), 965–1002. https://doi.org/10.1108/SAMPJ-04-2020-0113

Ferrero-Ferrero, I., Leon, R., & Munoz-Torres, M. (2021). Sustainability materiality matrices in doubt: May prioritizations of aspects overestimate environmental performance? *Journal of Environmental Planning and Management, 64*(3), 432–463. https://doi.org/10.1080/09640568.2020.1766427

Font, X., Guix, M., & Bonilla-Priego, M. J. (2016). Corporate social responsibility in cruising: Using materiality analysis to create shared value. *Tourism Management, 53*, 175–186. https://doi.org/10.1016/j.tourman.2015.10.007

GCC Exchanges Committee. (2023, January 9). *ESG disclosure guidance for listed companies*. Gulf Cooperation Council. https://adxservices.adx.ae/WebServices/DataServices/contentDownload.aspx?doc=2693943

Geldres-Weiss, V. V., Gambetta, N., Massa, N. P., & Geldres-Weiss, S. L. (2021). Materiality matrix use in aligning and determining a firm's sustainable business model archetype and triple bottom line impact on stakeholders. *Sustainability (Switzerland), 13*(3), 1–23. https://doi.org/10.3390/su13031065

Ghadge, A., Wurtmann, H., & Seuring, S. (2020). Managing climate change risks in global supply chains: A review and research agenda. *International Journal of Production Research, 58*(1), 44–64. https://doi.org/10.1080/00207543.2019.1629670

Global Reporting Initiative. (2022). *GRI 3: Material topics 2021*. https://www.globalreporting.org/how-to-use-the-gri-standards/gri-standards-english-language/

Hahn, R., & Lülfs, R. (2014). Legitimizing negative aspects in GRI-oriented sustainability reporting: A qualitative analysis of corporate disclosure strategies. *Journal of Business Ethics, 123*(3), 401–420. https://doi.org/10.1007/s10551-013-1801-4

Herzig, C., & Schaltegger, S. (2006). Chapter 13: Corporate sustainability reporting—An overview. In M. Bennett, S. Schaltegger, & R. Burritt (Eds.), *Sustainability accounting and reporting* (pp. 301–324). Springer.

Huang, H., Kerstein, J., Wang, C., & Wu, F. (2022). Firm climate risk, risk management, and bank loan financing. *Strategic Management Journal, 43*(13), 2849–2880. https://doi.org/10.1002/smj.3437

Ilhan, E., Krueger, P., Sautner, Z., & Starks, L. T. (2023). Climate risk disclosure and institutional investors. *The Review of Financial Studies, 36*(7), 2617–2650. https://doi.org/10.1093/rfs/hhad002

International Financial Reporting Standards Foundation. (2022). *IFRS S1 general requirements for disclosure of sustainability-related financial information* [exposure draft]. https://www.ifrs.org/content/dam/ifrs/project/general-sustainability-related-disclosures/exposure-draft-ifrs-s1-general-requirements-for-disclosure-of-sustainability-related-financial-information.pdf

Jain, A., Islam, M. A., Keneley, M., & Kansal, M. (2022). Social contagion and the institutionalisation of GRI-based sustainability reporting practices. *Meditari Accountancy Research, 30*(5), 1291–1308. https://doi.org/10.1108/MEDAR-06-2020-0917

Jebe, R. (2019). The convergence of financial and ESG materiality: Taking sustainability mainstream. *American Business Law Journal, 56*(3), 645–702. https://doi.org/10.1111/ablj.12148

Jørgensen, S., Mjos, A., & Pedersen, L. (2022). Sustainability reporting and approaches to materiality: Tensions and potential resolutions. *Sustainability Accounting, Management and Policy Journal, 13*(2), 341–361. https://doi.org/10.1108/SAMPJ-01-2021-0009

Karagiannis, I., Vouros, P., Skouloudis, A., & Evangelinos, K. (2019). Sustainability reporting, materiality, and accountability assessment in the airport industry. *Business Strategy and the Environment, 28*(7), 1370–1405. https://doi.org/10.1002/bse.2321

Kingdom of Saudi Arabia. (n.d.). *Vision 2030*. https://www.vision2030.gov.sa/en/

Kleinheksel, A. J., Rockich-Winston, N., Tawfik, H., & Wyatt, T. R. (2020). Demystifying content analysis. *American Journal of Pharmaceutical Education, 84*(1), 7113. https://doi.org/10.5688/ajpe7113

Kouloukoui, D., Sant'Anna, A., Gomes, S., Marinho, M., de Jong, P., Kiperstok, A., & Torres, E. (2019). Factors influencing the level of environmental disclosures in sustainability reports: Case of climate risk disclosure by Brazilian companies. *Corporate Social Responsibility and Environmental Management, 26*(4), 791–804. https://doi.org/10.1002/csr.1721

Macbeth, C., & Otubu, M. (2023, February 24). *GCC launches unified sustainability disclosure metrics*. Cleary Gottlieb. https://www.clearygottlieb.com/news-and-insights/publication-listing/gcc-launches-unified-sustainability-disclosure-metrics.

Machado, B., Dias, L., & Fonseca, A. (2021). Transparency of materiality analysis in GRI-based sustainability reports. *Corporate Social Responsibility and Environmental Management, 28*(2), 570–580. https://doi.org/10.1002/csr.2066

Mair, L., Elnahass, M., Xiang, E., Hawkins, F., Siikamaki, J., Hillis, L., Barrie, S., & McGowan, P. J. K. (2024). Corporate disclosures need a biodiversity outcome focus and regulatory backing to deliver global conservation goals. *Conservation Letters, 17*(4), e13024. https://doi.org/10.1111/conl.13024

Mansy, O. (2018, February 1). *Top 5 Arab Stock Markets*. Forbes Middle East. https://www.forbesmiddleeast.com/industry/business/top-5-arab-stock-markets

Matsumura, E. M., Prakash, R., & Vera-Muñoz, S. C. (2024). Climate-risk materiality and firm risk. *Review of Accounting Studies, 29*(1), 33–74. https://doi.org/10.1007/s11142-022-09718-9

Megeid, N. S. A. (2024). The impact of climate risk disclosure on financial performance, financial reporting and risk management: Evidence from Egypt. *Future Business Journal, 10*(1), 21. https://doi.org/10.1186/s43093-024-00309-5

Mio, C., Fasan, M., & Costantini, A. (2020). Materiality in integrated and sustainability reporting: A paradigm shift? *Business Strategy and the Environment, 29*(1), 306–320. https://doi.org/10.1002/bse.2390

Muzoriwa, K. (2023, July 29). *GCC stock markets are hitting record highs. Here's why*. Gulf Business. https://gulfbusiness.com/gcc-stock-markets-are-having-a-field-day/

Puroila, J., & Mäkelä, H. (2019). Matter of opinion: Exploring the socio-political nature of materiality disclosures in sustainability reporting. *Accounting, Auditing & Accountability Journal, 32*(4), 1043–1072. https://doi.org/10.1108/AAAJ-11-2016-2788

Reimsbach, D., Schiemann, F., Hahn, R., & Schmiedchen, E. (2020). In the eyes of the beholder: Experimental evidence on the contested nature of materiality in sustainability reporting. *Organization & Environment, 33*(4), 624–651. https://doi.org/10.1177/1086026619875436

Remmer, S., & Gilbert, U. (2019). Applying materiality assessment in strategic management: The implicit coating of the materiality lens. In T. Wunder (Ed.), *Rethinking strategic management: Sustainability strategizing for positive impact* (pp. 267–291). Springer.

Schiemann, F., & Sakhel, A. (2019). Carbon disclosure, contextual factors, and information asymmetry: The case of physical risk reporting. *European Accounting Review, 28*(4), 791–818. https://doi.org/10.1080/09638180.2018.1534600

Setia, N., Abhayawansa, S., & Joshi, M. (2022). In search of a wider corporate reporting framework: A critical evaluation of the international integrated reporting framework. *Accounting in Europe*, 1–26. https://doi.org/10.1080/17449480.2022.2060752

Shui, X., Zhang, M., & Smart, P. (2023). Climate change disclosure and the promise of response-ability and transparency: A synthesizing framework and future research agenda. *European Management Review, 20*(1), 145–158. https://doi.org/10.1111/emre.12514

State of Qatar. (n.d.). *Qatar National Vision 2030*. Government communications The Office. Retrieved June 28, 2024, from https://www.gco.gov.qa/en/about-qatar/national-vision2030/

Suchman, M. C. (1995). Managing legitimacy: Strategic and institutional approaches. *The Academy of Management Review, 20*(3), 571. https://doi.org/10.2307/258788

TCFD. (2017). *Final Report: Recommendations of the Task Force on Climate-related Financial Disclosures*.https://assets.bbhub.io/company/sites/60/2021/10/FINAL-2017-TCFD-Report.pdf

United Arab Emirates. (2024, February 13). Vision 2021. United Arab Emirates Government Portal. https://u.ae/en/about-the-uae/strategies-initiatives-and-awards/strategies-plans-and-visions/strategies-plans-and-visions-untill-2021/vision-2021

Vestrelli, R., Fronzetti Colladon, A., & Pisello, A. L. (2024). When attention to climate change matters: The impact of climate risk disclosure on firm market value. *Energy Policy, 185*, 113938. https://doi.org/10.1016/j.enpol.2023.113938

Zharfpeykan, R. (2021). Representative account or greenwashing? Voluntary sustainability reports in Australia's mining/metals and financial services industries. *Business Strategy and the Environment, 30*(4), 2209–2223. https://doi.org/10.1002/bse.2744

Chapter 12
The Vulnerability Puzzle: Exploring Environmental and Financial Risks in the MENA Region

Naima Lassoued, Imen Khanchel, and Zahra Souguir

Abstract This study investigates the effects of environmental costs on corporate default risk using a panel of 326 firms across 10 Middle Eastern and North African (MENA) countries—Bahrain, Egypt, Jordan, Kuwait, Lebanon, Morocco, Oman, Qatar, Saudi Arabia, and the United Arab Emirates—analysed over the period from 2010 to 2022. Our findings show that high environmental costs increase corporate default risk, resulting in financial instability and reputational damage. This underlines challenges of MENA firms to maintain both legitimacy and resource adequacy, supporting the institutional theory and resource-based view perspectives. Specifically, five environmental sub-indices—natural resources, air pollution, land/water pollutants, greenhouse gas emissions, and water consumption—are associated with increased default risk, while waste management does not exhibit a significant impact.

Keywords Environmental costs · Default risk · MENA region · GCC countries · Non-GCC countries

N. Lassoued (✉)
University of Manouba Higher School of Commerce of Tunis, Manouba, Tunisia

QuAnLab, LR24ES21, ESCT, University of Manouba, Manouba, Tunisia
e-mail: naima.lassoued@esct.com.tn

I. Khanchel
University of Manouba Higher School of Commerce of Tunis, Manouba, Tunisia

LARIME LR11ES02, University of Tunis, Tunis, Tunisia
e-mail: imen.khanchel@esct.uma.tn

Z. Souguir
QuAnLab, LR24ES21, ESCT, University of Manouba, Manouba, Tunisia

University of Manouba Higher Institute of Accountancy and Business Administration, Manouba, Tunisia

M. Elnahass et al. (eds.), *A Greenprint for Net Zero*, Gulf Studies 15,
https://doi.org/10.1007/978-981-92-1762-5_12

12.1 Introduction

The Middle East and North Africa (MENA) region faces significant environmental challenges that often exceed global averages. These challenges include a 100% increase in greenhouse gas emissions from 1990 to 2020 (World Bank, 2023) and extreme air pollution, which causes approximately 270,000 deaths annually (World Bank, 2022). Severe water stress affects 60% of the population (World Bank, 2018), and desertification impacts 89% of dry lands (World Bank, 2019). These problems are compounded by a rapid population growth rate of 3% per year (World Bank, 2019). To address these issues, the region should invest at least 3% of its GDP in sustainable solutions by 2030, compared to less than 1% currently (World Bank, 2024). This data on climate change and environmental challenges in the MENA region highlights the urgency of assessing their economic impact. According to the World Economic Forum (2020), despite some government commitments, companies in the MENA region are significantly lagging in their environmental goals. Only 12% of large companies have a carbon neutrality target, and only 6% have formulated a roadmap to achieve it.

The reluctance of companies to commit to environmental initiatives remains a complex issue, as they may perceive that the costs of such activities represent a financial burden (GIZ, 2023) rather than a beneficial investment or competitive advantage (World Economic Forum, 2024; Zayer & Ben Abdelhadi, 2023). In this complex environmental landscape, these firms face increasing financial challenges. Specifically, corporate default risk, which represents the probability that a company will be unable to meet its financial obligations, is becoming a major concern. Environmental costs—whether related to regulatory compliance, investments in cleaner technologies, or direct consequences of climate change—can significantly impact companies' financial health. In this chapter, we examine the relationship between environmental costs and default risk among firms in the MENA region.

Previous studies have demonstrated the potential benefits of environmental commitment for corporate performance. For example, El Ghoul et al. (2011) found a negative correlation between corporate social responsibility, including environmental practices, and the cost of capital. Similarly, Oikonomou et al. (2012) documented an inverse relationship between environmental performance and systematic financial risk. Nevertheless, companies in the MENA region appear to lag considerably behind in their environmental commitments. This can be attributed to several specific regional factors, such as weak enforcement mechanisms for environmental regulations (Campbell, 2007; Chen et al., 2023; Cuesta et al., 2023; Jamali & Karam, 2018), a lack of public awareness concerning environmental issues (Darwish et al., 2018), and the situation in certain resource-dependent countries (Cuesta et al., 2023; Gorus & Aslan, 2019).

Furthermore, the commitment of MENA companies in environmental activities may arise from a perception that may be misguided and risky. For example, the claims brought against cement companies Lafarge Egypt and Suez Cement Group illustrate the legal and reputational risks to which companies that neglect their environmental responsibilities are exposed. Similarly, the waste crisis involving Sukleen in Lebanon in 2015 demonstrates how inadequate environmental management can rapidly escalate into a major crisis, resulting in the bankruptcy of the company and harm to public health and the environment.

These characteristics of the MENA region offer a convincing setting for conducting this study. The relationship between these environmental costs and corporate default risk in the MENA region is particularly crucial, given the region's unique challenges, including weak enforcement of environmental regulations, resource dependency in some countries, and the urgent need for sustainable investments. However, this relationship remains underexplored, despite its critical importance for the region's economic stability and sustainable development.

Consequently, the aim of this study is to test the impact of environmental costs on corporate default risk in the MENA region. By quantifying the direct economic consequences of climate change on corporate outcomes, we aim to provide insights to guide companies' adaptation and resilience strategies in the face of these growing challenges.

This chapter intends to contribute to the literature in several ways. *First*, previous research primarily focuses on an aggregate measure of CSR based on several outcomes such as community and employee relations, governance, environment, human rights, and diversity (e.g. Boubaker et al., 2020; Do, 2022; Shahrour et al., 2021; Sun & Cui, 2014). In this study, we investigate the outcomes of the environmental pillar as a driver of default risk—one of the more important dimensions of CSR—using a more precise proxy (i.e. the dollar value of environmental costs). *Second*, this study seeks to extend the literature by focusing on the effect of environmental costs on firm performance. Indeed, previous studies have examined the effect of these costs on various financial aspects, such as the cost of capital (Chava, 2014; El Ghoul et al., 2011), financial performance (Dallocchio et al., 2025; Feldman et al., 1997; Feng et al., 2024; Jo et al., 2015; Oikonomou et al., 2012), and firm value (Dixon-Fowler et al., 2013). Our study extends this field of research by specifically exploring the impact of environmental costs on corporate default risk. This sheds new light on the relationship between environmental responsibility and the long-term financial stability of companies. *Third*, this study contributes to the debate on the importance of corporate environmental commitment in the MENA region. Indeed, despite the urgency of the environmental challenges facing the region (World Bank, 2023) and the growing pressure for a transition to more sustainable practices (World Economic Forum, 2024), few studies have looked at the consequences of environmental commitment on companies themselves in this specific context. Some studies have examined corporate social responsibility practices in the region (AlHares et al., 2021; Jamali et al., 2020; Jamali & Karam, 2018), but analysis of the direct financial consequences of environmental commitment remains underdeveloped.

The remainder of this paper is structured as follows. Section 12.2 presents the context of the study. Section 12.3 outlines the theoretical background and develops the research hypotheses. Section 12.4 presents the methodology. Section 12.5 displays the empirical results. Section 12.6 concludes.

12.2 Context of the Study

The environmental landscape of the MENA region reveals a critical and complex ecological challenge, characterized by systematic underperformance across multiple environmental indicators. As shown in Table 12.1, critically high carbon emissions (5.03 tonnes per capita versus a global average of 4.29 tonnes) and an extremely low renewable energy consumption (2% compared to the global 19.77%) highlight the region's fundamental environmental vulnerabilities. This systemic environmental stress manifests through dramatically reduced forest cover (2.05% versus the global 31.18%), minimal protected areas (6.02% of territorial area), and unsustainable natural resource depletion (3.96% of GNI, significantly higher than the global 0.69%).

The waste management dimension further illustrates regional challenges, with municipal solid waste production slightly elevated at 0.81 kg/day compared to the global average of 0.74 kg/day. Moreover, the water management crisis emerges as a pivotal concern, with annual freshwater withdrawals reaching 133.11% of internal resources—a stark contrast to the global average of 9.11%—signalling extreme ecological fragility. These multifaceted environmental challenges transcend mere statistical indicators, showing deeper structural issues related to economic development, institutional capabilities, and policy priorities in the MENA region, requiring comprehensive and urgent ecological intervention strategies.

The institutional indicators in Table 12.2 reveal a nuanced landscape of environmental management across the MENA region. The Natural Environment Quality Index demonstrates significant variability, with Oman (6.01), Iran (5.27), Qatar (5.46), and United Arab Emirates (UAE) (5.46) emerging as top performers, while Lebanon (2.94) and Kuwait (3.04) lag. The environmental sustainability index further illuminates regional disparities, with Egypt (4.68), UAE (4.47), and Morocco (4.46) displaying the

Table 12.1 Key environmental indicators: MENA compared with global averages

	Indicators	MENA	Average	Year
Air pollution	CO2 emissions (metric tons per capita)	5.03	4.29	2020
Energy	Renewable energy consumption (% of total final energy consumption)	2.00	19.77	2020
Land use	Agricultural land (% of land area)	33.41	36.73	2020
	Forest area (% of land area)	2.05	31.18	2020
Natural resources	Natural resources depletion (% of GNI)	3.96	0.69	2020
Protected areas	Marine protected areas (% of territorial waters)	1.31	11.48	2018
	Terrestrial and marine protected areas (% of total territorial area)	6.02	14.34	2018
Waste management	Municipal solid waste generated (kg per capita per day)	0.81	0.74	2016
Water resources	Annual freshwater withdrawals, (% of internal resources)	133.11	9.11	2020

Note: Table 12.1 displays the key environmental indicators of the MENA region in comparison to global averages
Source: World Bank website

Table 12.2 Environment-related institutional indicators

Country*	Quality of the natural environment**	Environmental sustainability**	Enabling environment**	Enforcement of environmental regulations**	Environmental treaty ratification***
Algeria	3.38	3.53	4.64	2.83	19
Bahrain	4.23	4.14	5.45	4.21	17
Egypt	3.07	4.68	4.55	3.13	21
Iran	5.27	3.95	4.59	3.43	21
Jordan	4.23	4.32	4.93	4.12	24
Kuwait	2.94	4.00	5.18	3.66	20
Lebanon	3.04	4.08	4.47	2.69	21
Mauritania	3.73	3.91	3.66	1.74	23
Morocco	4.09	4.46	4.84	4.1	24
Oman	6.01	4.36	5.4	5.73	21
Qatar	5.46	4.37	5.58	5.18	20
Saudi Arabia	4.37	3.95	5.36	4.5	19
Tunisia	3.82	4.43	4.68	3.75	22
UAE	5.46	4.47	5.77	5.03	21
Yemen	3.63	3.38	3.23	1.96	19

Note: * We report all countries from the MENA region with available data, ** This score varies from 1 to 7 (best), *** The number of treaties ratified
Source: World Economic Forum, 2019

most robust environmental management capabilities, contrasting sharply with Yemen (3.38) and Algeria (3.53). The Enabling Environment sub-index highlights a critical correlation between economic development and regulatory frameworks, with UAE (5.77), Qatar (5.58), and Bahrain (5.45) leading, while Yemen (3.23) and Mauritania (3.66) struggle. The Environmental Enforcement Index reveals even more pronounced differences, with Oman (5.73), Qatar (5.18), and UAE (5.03) demonstrating strong implementation mechanisms, creating a stark 3.99-point gap that highlights significant institutional variations. Despite these disparities, the region shows a relatively consistent commitment to international environmental treaties, with Jordan and Morocco leading in treaty ratifications, suggesting a baseline of shared environmental diplomacy.

12.3 Theoretical Framework and Hypothesis Development

Institutional theory and the resource-based view provide complementary theoretical framework to understand how environmental costs increase corporate default risk.

Institutional theory (DiMaggio & Powell, 1983) demonstrates how environmental compliance has evolved from a peripheral concern to a fundamental determinant of corporate survival, where institutional pressures—regulatory, social, and market-based—create an environment where environmental non-compliance causes financial vulnerabilities. These institutional pressures manifest through increasingly stringent regulations, heightened stakeholder expectations, and market mechanisms that systematically penalize environmental underperformance, creating multiple pathways to financial distress.

According to the resource-based view (Barney, 1991; Wernerfelt, 1984), environmental capabilities are strategic resources that directly control financial risk. Environmentally harmful assets represent contingent liabilities that depreciate rapidly, increase compliance costs, and reduce organizational adaptability. These resources create a financial fragility where carbon-intensive and non-compliant assets become potential bankruptcy causes, limiting capital market access, reducing market valuation, and increasing conversion costs.

While these theories offer complementary insights, they also present tensions that influence corporate responses to environmental challenges. The resource-based view emphasizes internal capabilities and firm-specific advantages, suggesting heterogeneous approaches to environmental management based on unique resource configurations. In contrast, institutional theory highlights isomorphic pressures that drive firms towards standardized environmental practices, regardless of their internal resource endowments. This theoretical tension creates a dynamic where firms must simultaneously develop distinctive environmental capabilities (resource-based view) while conforming to institutionalized expectations (institutional theory).

This tension becomes particularly evident in how firms respond to environmental costs. From the resource-based view, firms with superior environmental capabilities can transform compliance costs into competitive advantages through innovation and efficiency gains (Hart, 1995). Conversely, institutional theory suggests that regardless of resource configurations, firms face legitimacy threats if they fail to meet minimum environmental standards, even when compliance imposes disproportionate resource burdens.

Under the institutional theory and the resource-based view, initial environmental non-compliance activates institutional penalties, reducing financial resources, which in turn limits environmental improvement capabilities, attracting additional institutional sanctions. This deterioration operates through multiple channels—operational, financial, market, and strategic—creating a threat to corporate financial sustainability. Environmental costs thus become systematic default risks, as they represent fundamental challenges to organizational legitimacy and resource adequacy.

Existing literature has limited investigation of the relationship between firms' environmental costs and default risk. Specifically, the financial implications of environmental costs present a multifaceted risk framework that can significantly compromise a company's financial stability and increase its vulnerability to bankruptcy. Environmental performance interacts with financial risk through several critical mechanisms that can create a compounding negative impact on corporate health. In an international context, Li et al. (2023) point out that a higher commitment to environmental and social practices leads to improved financial stability among banks, specifically during the COVID-19 pandemic.

First, environmental non-compliance introduces direct financial penalties that serve as exogenous shocks to corporate cash flows. Karpoff et al. (2005) demonstrate that these penalties extend beyond immediate fines, reducing profitability and compelling companies to accumulate short-term debt, thereby affecting liquidity and operational flexibility.

Second, evolving environmental standards can increase asset obsolescence and accelerated depreciation. Matsumura et al. (2014) show that carbon-intensive assets are particularly susceptible to this risk. When regulatory changes make assets non-compliant, their market value diminishes, causing debt covenant violations and restricting borrowing capacities, further increasing default probabilities. Dallocchio et al. (2025) highlight that environmentally prosperous firms benefit from increased financial profitability.

Third, environmental management influences investor confidence and capital accessibility. Chava (2014) and El Ghoul et al. (2018) show that environmental responsibility correlates with reduced equity capital costs. Companies demonstrating strong environmental risk management are perceived as less risky, consequently attracting more favourable investment terms, and enhancing financial resilience (Bargaoui et al., 2024). However, Feng et al. (2024) document that firms exposed to climate change face reduced operating cash flows and higher financial constraints, heightening their bankruptcy risk.

Fourth, corporate environmental performance impacts organizational legitimacy and market reputation. Venkatachary et al. (2017) and Khanifah et al. (2020) show that compliance with societal norms is critical for organizational survival. Flammer (2013) highlights how eco-initiatives can influence firm valuation. Based on a country-level sample, Kabir et al. (2021) document that carbon emissions increase firms' default risk, particularly under environmental controversies. Conversely, Park et al. (2023) indicate that the carbon intensity of South Korean firms negatively affects default risk, amplified by higher foreign ownership.

Goss and Roberts (2011) show how perceived social irresponsibility leads to less favourable financial conditions, potentially increasing bankruptcy risk.

Consequently, we hypothesize that environmental costs represent a significant and systemic threat to corporate financial stability:

H1: Environmental costs increase the risk of corporate default.

12.4 Methodology

12.4.1 Sample and Data Source

Our study examines 326 firms operating in 10 MENA countries over the period 2010–2022. *First*, we selected a sample comprising all the countries in the MENA region. As the data for this analysis comes from the Trucost and Refinitiv databases, we excluded certain countries for which data was missing during the study period to ensure the reliability and completeness of our analyses. Table 12.3 reports the sample distribution.

Table 12.3 Distribution of the sample

Countries	Percent
Bahrain	2.88
Egypt	17.83
Jordan	5.92
Kuwait	8.30
Lebanon	1.45
Morocco	9.15
Oman	4.81
Qatar	12.25
Saudi Arabia	19.98
UAE	17.42
Total	100.00

Note: Table 12.3 presents the distribution of the sample comprising 326 firms across 10 MENA countries

12.4.2 Variables

12.4.2.1 Default Risk

Building on previous research (Bharath & Shumway, 2008; Merton, 1974), we use the distance-to-default (DD) metric to evaluate a company's default risk. This metric measures the gap between the value of its assets and the extent of its debt:

$$DD_{it} = \log\left(\frac{Equity_{it} + Debt_{it}}{Debt_{it}}\right) + \left(r_{it-1} - \frac{\sigma^2 V_{it}}{2}\right) \times T_{it}\sigma V_{it}\sqrt{T_{it}} \tag{12.1}$$

Where:
$Equity_{it}$: The equity value of firm i at time t.
$Debt_{it}$: The debt value of firm i at time t.
r_{it-1} The return of firm i at time t-1.
σ: The standard deviation of returns.
V_{it}: The market value of firm i at time t.
T_{it}: The time to maturity of options for firm i at time t.

The DD metric measures the probability of a company defaulting by comparing the value of its assets with the amount of its debt. It quantifies the safety margin a company has before reaching a point where the risk of default becomes significant. A greater distance-to-default generally indicates a larger safety margin, suggesting a lower risk of default. Conversely, a smaller distance-to-default indicates a smaller safety margin and, therefore, a potentially higher risk of default.

12.4.2.2 Environmental Cost

The environmental cost variable (ENV_COST) is defined as the ratio of total external environmental costs to a company's total assets. External environmental costs are the direct costs associated with the environmental impacts generated by the company's activities, calculated in financial terms for each of the following categories (Khanchel et al., 2023: Lassoued & Khanchel, 2023):

- Direct costs of greenhouse gases (GHG): Costs of GHG emissions due to the combustion of fossil fuels and production processes, adjusted by their global warming potential (GWP) to obtain a carbon dioxide equivalence
- Direct water costs (WAT): Costs associated with water extracted directly by the company from natural sources and used in its operations, calculated based on the quantity of water used
- Direct waste costs (WAST): Costs associated with hazardous and non-hazardous waste produced by the company, considering the type of waste and its method of disposal
- Direct costs of terrestrial and aquatic pollutants (L_A_POL): Costs of pollutants released into water or soil, such as fertilizers, pesticides, and metal emissions
- Direct costs of air pollutants (AIR_POL): Costs of emissions to air, including acid rain precursors, ozone-depleting substances, and other pollutants
- Direct costs of natural resource use (NAT_RES): Costs associated with the direct extraction of minerals, metals, natural gas, oil, coal, and other resources

The ENV_COST variable is calculated by dividing these total external environmental costs by the company's total assets, thus providing an overview of the effectiveness of the company's environmental management.

12.4.3 Econometric Model

We use a regression analysis to examine the relationship between environmental costs and the default risk.

$$DD_{it+1} = \alpha_0 + \alpha_1 ENV\_COST_{it} + Control\,variables + Fixed\,effects + \varepsilon_{it} \quad (12.2)$$

where:

DD_{it} refers to default risk of firm i at year t and ENV_COST_{it} indicates environmental cost of firm i at year t.

We also introduce two categories of control variables that previous studies have associated with default risk (e.g. Abinzano et al., 2021; Atif & Ali, 2021; Cathcart et al., 2020; Chodnicka-Jaworska, 2022; Do & Vo, 2023; Nguyen & Nguyen, 2022; Sun & Cui, 2014).

The first category incorporates firm-specific characteristics, including (1) Profitability (return on assets—ROA) is calculated as net income divided by total assets. More profitable companies exhibit decreased default risk (Sun & Cui, 2014). (2) Leverage (LEV), captured by the total debt-to-equity ratio. Leverage exacerbates firms' financial risk, leading to higher default risk (Cathcart et al., 2020; Schultz et al., 2017). (3) Liquidity (CASH) is computed as cash and cash equivalents divided by total assets; companies with higher liquidity exhibit less default risk (Atif & Ali, 2021). Moreover, (4) Capital expenditure (CAPEX) denotes investment in long-term assets relative to total assets. Higher CAPEX is associated with higher default risk (Abinzano et al., 2021). (5) Firm size (SIZE), measured by the natural logarithm of total assets. Larger firms are less exposed to default risk due to their more stable financial situation (Atif & Ali, 2021).

The second category consists of country-level control variables, such as (6) Institutional quality (INST_QUAL), which is a composite indicator measuring governance effectiveness and (7) Economic and Political Uncertainty (EPU) represents an index assessing macroeconomic uncertainty. Higher institutional quality and lower economic and political uncertainty reduce default risk (Do & Vo, 2023). (8) Inflation rate (INF_RT) reflects the annual change in the consumer price index. Higher INF_RT indicates lower default risk (Chodnicka-Jaworska, 2022). (9) GDP growth (GDP_GTH) the annual percentage growth rate of GDP. Nguyen and Nguyen (2022) claim that GDP growth is negatively linked to default risk.

Additionally, we incorporate fixed effects for (i) countries to control for unobservable differences; (ii) years to eliminate temporal variations; and (iii) sectors to neutralize sector-specific influences on default risk.

To address potential heteroscedasticity issues and ensure more accurate and conservative estimates of standard deviations, we employ robust standard errors.

12.5 Results and Discussion

12.5.1 Descriptive Statistics

Table 12.4 reports environmental costs by country and industrial sector across the countries in our sample. At the country level, Oman stands out with the highest environmental cost (0.855), while Lebanon and the UAE have the lowest values (0.0026 and 0.055, respectively). In terms of industrial sectors, utilities exhibit a very high environmental cost of (1.7887), compared with the retail and wholesale

Table 12.4 Means of environmental costs

Country		Industry	
Bahrain	0.102	Agriculture and forestry	0.5340
Egypt	0.158	Mining	0.1003
Jordan	0.128	Utilities	1.7887
Kuwait	0.343	Construction	0.1926
Lebanon	0.0026	Manufacturing	0.0087
Morocco	0.204	Wholesale trade	0.0060
Oman	0.855	Retail trade	0.0116
Qatar	0.112	Services	0.0505
Saudi Arabia	0.550		
UAE	0.055		

sectors, which have low values (0.0116 and 0.0060, respectively). These data suggest significant disparities in environmental impact depending on geographical and economic contexts.

Table 12.5 (Panel A) presents descriptive statistics for several financial and economic variables. For the Distance to Default (DD), the mean of 0.488 suggests a moderate risk of default. In terms of other variables, the Return on Assets (ROA) has a low mean of 0.042, while Financial Leverage (LEV) has a mean of 2.579, indicating high dispersion. The mean of Cash Flow (CASH) is relatively modest at 0.067. Additionally, Company Size (SIZE), Institutional Quality (INST_QUAL), and Political Uncertainty (EPU) have respective averages of 8.177, 0.741, and 0.398. Macroeconomic variables reveal an average negative inflation rate of 0.127 and a GDP growth rate of 2.88%. Panel B of Table 12.5 presents the correlation matrix. The results reveal no excessively high correlations among the variables, indicating an absence of multicollinearity issues.

12.5.2 Main Findings

12.5.2.1 The Effect of Environmental Costs on Default Risk

Table 12.6 presents the results of a regression analysis examining the impact of ENV_COST on corporate default risk, measured by DD. In the first column, we test the effect of ENV_COST on DD only; in column 2, we include company control variables; and in the last column, we add country control variables. The main variable ENV_COST shows a negative and statistically significant coefficient (p value < 1%) in all three models, ranging from −0.0103 to −0.00857, confirming *hypothesis H1* that environmental costs increase the risk of corporate default. These findings are consistent with those of previous studies (Dallocchio et al., 2025; Kabir et al., 2021; Li et al., 2023; Marie et al., 2024), that highlight the complex relationship between environmental commitment and financial consequences. While proactive environmental management can foster long-term financial health and distance firms from default risk, our results specifically demonstrate that the immediate financial burden of environmental costs whether through compliance investments, regulatory penalties, or asset stranding weighs heavily on short-term financial stability. For example, Kabir et al. (2021) conclude that higher carbon-emitting companies are more vulnerable to default risk, leading to higher return volatility and cash-flow uncertainty, particularly when faced with stringent environmental regulations and reputational costs.

There are several reasons for this result. First, environmental fines and penalties, as well as environmental compliance constraints requiring unexpected investments, place a burden on liquidity and reduce the financial security margin. Moreover, the accelerated depreciation of non-compliant assets weakens the balance sheet and may initiate covenants on existing debt. Additionally, the cost of capital for polluting companies increases as investors perceive higher risk, thereby reducing access

Table 12.5 Summary statistics

Panel A: Descriptive statistics

Variables	Mean	Std. Dev.	Min	Max	p25	Median	p75
DD	0.488	0.259	0.067	0.973	0.259	0.491	0.696
ENV_COST	4.362	3.011	0.002	17.277	0.294	3.355	15.051
ROA	0.042	0.145	−2.446	0.529	0.01	0.028	0.076
LEV	2.579	7.883	0.039	16.13	0.441	1.066	3.086
CASH	0.067	0.086	−0.238	1.061	0.021	0.044	0.083
CAPEX	0.054	0.066	0	0.692	0.012	0.032	0.074
SIZE	8.177	2.169	−1.158	18.08	6.824	7.806	9.228
INST_QUAL	0.741	0.3	0	1	0.379	0.923	0.983
EPU	0.398	0.301	0.009	1.009	0.152	0.339	0.620
INF_RT	0.127	1.085	−4.0	12.561	−0.447	0.082	0.709
GDP_GTH	2.88	3.01	−6.592	11.075	1.135	3.01	5.111

Panel B: Correlation matrix

Variables	(1)	(2)	(3)	(4)	(5)	(6)	(7)	(8)	(9)	(10)	(11)
(1) DD	1.000										
(2) ENV_COST	−0.133*	1.000									
(3) ROA	0.008	0.005	1.000								
(4) LEV	−0.023*	0.014*	0.379*	1.000							
(5) CASH	−0.037*	−0.004	0.399*	0.172*	1.000						
(6) CAPEX	−0.039*	0.013*	0.276*	0.694*	0.097*	1.000					
(7) SIZE	−0.018*	0.130*	0.152*	0.209*	0.180*	0.187*	1.000				
(8) INST_QUAL	0.003	−0.048*	0.065*	0.072*	0.041*	0.085*	0.362*	1.000			
(9) EPU	−0.528*	0.022*	0.011*	0.022*	−0.006	0.036*	−0.009*	0.008	1.000		
(10) INF_RT	0.009*	−0.012*	−0.007	−0.009*	−0.005	−0.011*	0.010*	0.026*	0.006	1.000	
(11) GDP_GTH	0.008	−0.107*	0.013*	−0.022*	−0.021*	−0.004	−0.230*	−0.040*	0.017*	0.024*	1.000

Note: Panel A presents the descriptive statistics. Panel B reports the correlation matrix, * denotes $p < 0.05$

Table 12.6 Main findings

	DD	DD	DD
Variables	(1)	(2)	(3)
ENV_COST	−0.0103***	−0.00801***	−0.00857***
	(0.0009)	(0.0014)	(0.0012)
ROA		0.335***	0.340***
		(0.0904)	(0.0901)
LEV		−0.0257***	−0.0265***
		(0.0068)	(0.0068)
CASH		−0.0926	−0.105
		(0.0843)	(0.0892)
CAPEX		−0.295**	−0.275**
		(0.133)	(0.123)
SIZE		−0.111***	−0.110***
		(0.0360)	(0.0330)
INST_QUAL			0.102**
			(0.0439)
EPU			−0.101**
			(0.0408)
INF			0.0023
			(0.0020)
GDP_GTH			−0.0022**
			(0.0009)
Constant	0.425***	0.301	0.286
	(0.0006)	(0.299)	(0.277)
Country fixed effect	*Yes*	*Yes*	*Yes*
Year fixed effect	*Yes*	*Yes*	*Yes*
Industry fixed effect	*Yes*	*Yes*	*Yes*
Observations	2721	2721	2721
R-squared	0.087	0.522	0.542
N	326	326	326

Note: Table 12.6 displays regression results of the risk of default on environmental costs and control variables. ***significant at 1% level; ** significant at 5% level; * significant at 10% level

to financing and increasing the risk of default. Finally, environmental costs can be harmful; a loss of reputation due to poor environmental performance can lead to a decrease in market price on one hand, and less favourable lending conditions for companies perceived as environmentally irresponsible on the other hand, creating a financial vicious cycle.

Table 12.7 Additional evidence: The role of environmental cost sub-indices

	DD	DD	DD	DD	DD	DD
Variables	(1)	(2)	(3)	(4)	(5)	(6)
GHG	−0.0196***					
	(0.0055)					
WAT		−0.0596**				
		(0.0286)				
WAST			−0.0974			
			(0.1880)			
L_A_POL				−0.028***		
				(0.0101)		
AIR_POL					−0.0519**	
					(0.0081)	
NAT_RES						−0.118***
						(0.043)
ROA	0.292***	0.495***	0.501***	0.509***	0.484***	0.956***
	(0.107)	(0.0973)	(0.0952)	(0.0897)	(0.0937)	(0.184)
LEV	−0.0009***	−0.0013***	−0.0012***	−0.0011***	−0.0012***	−0.0034
	(0.0001)	(0.0002)	(0.0003)	(0.0002)	(0.0002)	(0.0026)
CASH	−0.0585	−0.141	−0.143	−0.150	−0.142	0.351**
	(0.108)	(0.0922)	(0.0922)	(0.0921)	(0.0926)	(0.141)
CAPEX	−0.443**	−0.332***	−0.322***	−0.294**	−0.323***	−0.429
	(0.197)	(0.122)	(0.122)	(0.113)	(0.118)	(0.271)
SIZE	−0.110***	−0.163***	−0.162***	−0.156***	−0.161***	−0.0295***
	(0.0186)	(0.0239)	(0.0235)	(0.0222)	(0.0239)	(0.0098)
INST_QUAL	0.0006	−0.0705*	−0.0716*	−0.0617	−0.0799*	−0.175
	(0.0646)	(0.0398)	(0.0420)	(0.0394)	(0.0446)	(0.123)

EPU	−0.0296	0.0638	0.0715	0.0634	0.0769	0.148
	(0.0502)	(0.0439)	(0.0467)	(0.0441)	(0.0491)	(0.0932)
INF	0.00237	0.00198	0.00173	0.00224	0.0019	0.0017
	(0.0025)	(0.0021)	(0.0022)	(0.0021)	(0.0021)	(0.0072)
GDP_GTH	−0.0007	−0.0013	−0.0015*	−0.0014*	−0.0015	−0.0063
	(0.0012)	(0.0008)	(0.0009)	(0.0008)	(0.0009)	(0.0046)
Constant	0.505***	0.876***	0.873***	0.810***	0.856***	0.702***
	(0.182)	(0.190)	(0.184)	(0.174)	(0.189)	(0.105)
Country fixed effect	Yes	Yes	Yes	Yes	Yes	Yes
Year fixed effect	Yes	Yes	Yes	Yes	Yes	Yes
Industry fixed effect	Yes	Yes	Yes	Yes	Yes	Yes
Observations	2721	2721	2721	2721	2721	2721
R-squared	0.207	0.447	0.443	0.466	0.451	0.356
N	326	326	326	326	326	326

Note: Table 12.7 reports regression results of default risk, environmental cost sub-indices, and control variables. ***significant at 1% level; ** significant at 5% level; * significant at 10% level

12.5.3 Additional Evidence

12.5.3.1 The Role of Environmental Cost Sub-Indices

To deepen our analysis, we examine the effects of environmental cost sub-indices on default risk. The results reported in Table 12.7 show negative and significant coefficients for all sub-indices, except for WAST, which is statistically insignificant. However, the effects vary in magnitude and importance across the analysed dimensions.

The NAT_RES variable exhibits the strongest effect (coefficient: −0.118, ($p < 0.01$)), suggesting that natural resource extraction generates substantial financial liabilities. This finding aligns with the high economic and ecological costs associated with resource diminution, such as land degradation, biodiversity loss, and long-term unsustainable resource use. The significance of this variable highlights the urgency of incorporating resource conservation measures into corporate environmental strategies to mitigate financial risks.

Air pollution (AIR_POL) and land/water pollutants (L_A_POL) also display significant negative effects (AIR_POL coefficient: −0.0505, ($p < 0.01$); L_A_POL coefficient: −0.0279, ($p < 0.01$)). These results emphasize the hidden costs of traditional industrial and agricultural practices, particularly concerning health impacts, remediation costs, and compliance with environmental regulations. Companies with higher pollutant emissions face increasing financial risks due to stricter environmental policies and rising societal expectations for sustainability.

Greenhouse gas emissions (GHG) show a modest, yet significant, effect (coefficient: −0.0196, ($p < 0.01$)), reflecting the financial implications of carbon-intensive operations.

This highlights the growing role of climate change-related risks in influencing firms' financial stability, particularly through mechanisms like carbon pricing, emission reduction mandates, and reputational risks. Water consumption (WAT) demonstrates a moderate negative impact (coefficient: −0.0596, ($p < 0.05$)), highlighting the financial consequences of unsustainable water use, including resource scarcity, higher extraction costs, and increased vulnerability to climate-induced water stress.

Interestingly, the waste management variable (WAST) does not show a statistically significant relationship with the DD metric (coefficient: −0.0974, $p = 0.1880$). This lack of significance likely stems from the diverse nature of waste streams and their varying management costs, which obscure a uniform financial effect, further compounded by several region-specific factors. *First*, waste management costs in MENA are often externalized to municipalities or informal sectors, reducing direct financial liabilities for firms (Negm & Shareef, 2020). *Second*, weak enforcement of waste-related regulations in many MENA countries limits penalties for non-compliance, diminishing its impact on default risk. *Third*, heterogeneous waste streams (e.g. industrial vs. municipal) and varying disposal methods (landfills vs. recycling) may obscure uniform financial effects. This finding highlights a critical gap in regional environmental governance: while waste management is a growing public health concern, its financial impact for firms remains muted due to institutional inefficiencies. Policymakers should prioritize standardizing waste regulations and incentivizing circular economy practices to internalize these costs, which may gain significance as global sustainability standards tighten.

12.5.3.2 GCC vs. Non-GCC Countries

Although the MENA region forms a cohesive bloc, Gulf Cooperation Council (GCC) countries have distinct characteristics that differentiate them from non-GCC nations.

Given the global calls for an energy transition, the demand for renewable energy is intensifying, placing increasing pressure on GCC states as they face a reduction in oil and gas revenues, which have long been the basis of their economic development (Shapland, 2023). Consequently, these countries have urged the need to follow the energy transition by reducing their dependence on hydrocarbons and fossil fuels. They are making efforts by prioritizing production efficiency, switching to sustainability of green energy, and mitigating the ecological harm caused by traditional energy sources (Rafindadi et al., 2018), thus taking a promising step towards lessening environmental costs (Al-Sarihi & Mansouri, 2022; Miniaoui, 2023). Furthermore, environmental regulations and standards differ significantly between these two groups of countries. Also, the financial capabilities and economic structures of companies vary between GCC (which are typically richer) and non-GCC countries (Alhmoud et al., 2023; Lassoued et al., 2024).

Based on these arguments, we deepen our analysis and re-estimate the model by distinguishing between GCC and non-GCC countries. The results reported in Table 12.8 indicate that the ENV_COST coefficient is negative and significant for both groups of countries. However, for GCC countries, the negative effect is relatively low (coefficient − 0.0104, p-value < 1%)), suggesting a greater ability to absorb environmental costs. In contrast, for non-GCC countries, the negative effect is approximately ten times greater (coefficient − 0.107, p-value < 5%), indicating increased vulnerability.

This unexpected result can be attributed to the greater financial resources of oil-producing countries, which are linked to oil revenues and allow for investment in environmental compliance technologies. Additionally, more developed financial markets in these countries facilitate access to capital.

12.5.4 Robustness Check: Alternative Estimation Methods

The model may present an endogeneity problem, likely arising from the simultaneity between default risk and environmental costs at the company level, which could bias our ordinary least squares estimates. To address this methodological challenge, we use an instrumental variable approach. Specifically, we instrument firm-specific environmental costs by their sector average, in line with the recommendations of previous studies (Alam, 2022; Kabir et al., 2021). This sector average, being

Table 12.8 GCC countries vs. non-GCC countries

	GCC countries	Non-GCC countries
Variables	(1)	(2)
ENV_COST	−0.0104***	−0.107**
	(0.000979)	(0.0385)
ROA	0.465***	0.175
	(0.0857)	(0.165)
LEV	−0.0270***	−0.0329***
	(0.00835)	(0.00882)
CASH	−0.0962	−0.0841
	(0.120)	(0.112)
CAPEX	−0.00434	−0.0892
	(0.184)	(0.133)
SIZE	−0.0751**	−0.197***
	(0.0338)	(0.0556)
INST_QUAL	0.110**	0.0352
	(0.0482)	(0.0651)
EPU	−0.153***	0.0363
	(0.0337)	(0.0715)
INF	0.00144	0.00276
	(0.00239)	(0.00411)
GDP_GTH	−0.00316**	−0.000122
	(0.00127)	(0.000967)
Constant	0.112	0.324
	(0.280)	(0.464)
Country fixed effect	Yes	Yes
Year fixed effect	Yes	Yes
Industry fixed effect	Yes	Yes
Observations	1731	990
R-squared	0.555	0.615
N	199	127

Note: Table 12.8 shows regression results of default risk, environmental cost, and control variables using two subsamples: GCC countries vs. Non-GCC countries. ***significant at 1% level; ** significant at 5% level; * significant at 10% level

correlated with the endogenous variable but exogenous to our dependent variable, satisfies both the relevance and exclusion criteria. We then estimate our model using two-stage regression (2SLS), obtaining consistent estimators in the presence of endogeneity. The findings, reported in column 1 of Table 12.9, are very similar to the main result.

We also use Generalized Method of Moments (GMM) estimation to account for potential endogeneity, using lags as instruments. This method also considers potential heteroscedasticity and autocorrelation, providing more efficient estimators in the presence of endogeneity. The results of this estimation (Table 12.9, column 2) confirm that environmental costs increase the risk of default.

Table 12.9 Alternative estimation methods

Variables	2SLS (1)	GMM (2)
ENV_COST	-0.0079**	−0.0415***
	(0.00380)	(0.0143)
ROA	0.349***	−0.0377
	(0.0942)	(0.101)
LEV	−0.0812***	−0.0336***
	(0.0036)	(0.0114)
CASH	0.291***	0.232**
	(0.0818)	(0.117)
CAPEX	−0.342**	−0.726***
	(0.138)	(0.151)
SIZE	−0.00460	−0.199***
	(0.00478)	(0.0308)
INST_QUAL	−0.170***	−0.0504
	(0.0546)	(0.0344)
EPU	0.133***	0.125***
	(0.0385)	(0.0275)
INF	0.0014	0.0007
	(0.0061)	(0.0014)
GDP_GTH	−0.0045**	−0.00127*
	(0.0017)	(0.00072)
L.DD		0.160*
		(0.0897)
Constant	0.659***	
	(0.0440)	
Country fixed effect	Yes	Yes
Year fixed effect	Yes	Yes
Industry fixed effect	Yes	Yes
Observations	2721	2405
R-squared	0.633	
N	326	326

Note: Table 12.9 presents 2SLS/GMM regression results of default risk, environmental cost, and control variables. ***significant at 1% level; ** significant at 5% level; * significant at 10% level

12.6 Conclusion

This chapter examines the impact of environmental costs on the corporate default risk of a sample of 326 MENA firms observed from 2010 to 2022. The MENA region has distinctive characteristics that make it a particularly interesting field of study: high dependence on natural resources, significant exposure to climate risks, and major challenges in terms of sustainable development. The results show that environmental costs increase the risk of corporate default. Furthermore, our estimation puts in evidence that all sub-indices of environmental cost—natural resources, air pollution, land/water pollutants, greenhouse gas emissions, and water consumption—negatively affect default risk, except for waste management. Additionally, the findings indicate that the impact of environmental costs on corporate default risk is more pronounced in firms from non-GCC countries compared to those in GCC countries, highlighting regional differences in vulnerability and response to environmental challenges.

This study has theoretical contributions. It enriches the literature on default risk by integrating the environmental dimension, which is particularly relevant in the context of MENA countries. It shows that the impact of environmental costs on default risk varies significantly according to the institutional context, especially between GCC and non-GCC countries. These results also reinforce institutional theory and resource-based view by showing how environmental costs affect default risk. In terms of implications for market finance research, the results suggest the need to integrate environmental costs into credit risk pricing models, while considering the institutional features of GCC/non-GCC countries. These differences call for the development of differentiated valuation models and further investigations into the impact of environmental costs on the cost of capital and financing structures.

Building upon these findings, practical implications for policymakers in non-GCC countries emerge with urgency, as environmental costs exert greater financial strain in these regions. Three actionable strategies are critical for these policymakers: strengthening regulatory frameworks by implementing enforceable standards for high-impact areas like natural resource extraction and air pollution, aligning with GCC nations' stricter protocols; offering financial incentives such as tax breaks or subsidies for firms adopting clean technologies, particularly in water-stressed sectors, to mitigate liquidity risks from environmental compliance; and creating regional collaboration through MENA-wide sustainability funds to support SMEs in transitioning to low-carbon practices, reducing default risks exacerbated by resource constraints.

Similarly, corporate managers, particularly in sectors with high environmental exposure such as utilities and mining, face strategic imperatives based on our findings. Prioritizing resource efficiency through investments in technologies that reduce natural resource extraction and greenhouse gas is crucial. Enhanced disclosure through transparent reporting on environmental costs, such as water withdrawal metrics, can improve investor confidence and reduce capital costs. Furthermore, capitalizing on partnerships with GCC firms provides opportunities to access advanced sustainability technologies and share best practices in regulatory compliance, creating a beneficial cycle of knowledge transfer across the MENA region.

The implications extend to financial institutions as well, where credit institutions and regulators must revise risk assessment models to integrate environmental impacts generated by the company's activities, for each category differentially. This requires weighting GCC versus non-GCC exposures by assigning higher risk premiums to firms in non-GCC countries with weak institutional quality and higher natural resource extraction and air pollution costs. Additionally, stress-testing climate scenarios by modelling default risks under varying carbon pricing and water scarcity scenarios is particularly important for agriculture and utilities sectors, enhancing the robustness of credit portfolios against environmental volatility.

On the broader regulatory front, this research highlights a pressing need to revise prudential frameworks to integrate environmental risks more systematically. This implies not only the adoption of stricter standards for assessing the exposure of companies to climate risks, but also the development of differentiated regulatory frameworks that take account of local institutional specificities. The harmonization of environmental standards between regions would reduce asymmetries and promote responsible financial practices on a global scale, fostering sustainable economic development throughout the MENA region.

This study has some limitations, which open promising avenues for future research. First, the study is conducted within a specific context (MENA region). It is important to note that environmental regulations, incentives, and enforcement may differ across countries, even within the MENA region. These factors can significantly shape the extent of environmental compliance among firms. Future studies could broaden our research question to an international context to further understand the dynamics of environmental costs and default risk. Beyond geographical expansion, several interesting research directions emerge from our findings. The waste management paradox deserves particular attention, as our analysis shows waste management lacks financial significance despite its ecological urgency. Future investigations could include case studies of firms facing reputational crises due to waste mismanagement or analysing sectors where waste costs are internalized, such as petrochemicals. Moreover, exploring how health or financial shocks, such as COVID-19 or oil price collapses, amplify the environmental cost–default risk link using longitudinal data would provide valuable insights into crisis interactions. Also, examining how institutional moderators such as civil society pressure or ESG investor activism modify the relationship in GCC versus non-GCC contexts would further enhance our understanding of the environmental costs–default risk relationship.

References

Abinzano, I., Corredor, P., & Martinez, B. (2021). Does family ownership always reduce default risk? *Accounting & Finance, 61*(3), 4025–4060.

Alam, M. S. (2022). Is trade, energy consumption and economic growth threat to environmental quality in Bahrain–evidence from VECM and ARDL bound test approach. *International Journal of Emergency Services, 11*(3), 396–408.

AlHares, A., Abu-Asi, T., Dominic, G., & Al Abed, R. (2021). The impact of corporate social responsibility on consumer's relationship intention: Evidence from MENA countries. *Journal of Governance and Regulation/Volume, 10*(4), 8.

Alhmoud, K. B. R., Ftomova, O. S., Wang, X., Koettl-Brodmann, J., Alrayess, D. J., Gomez Tamayo, S., Fostier de Moraes, G., Chartouni, C., Pankratova, E., Rivera Guivas, N., & Zeitoun, N. A. M. S. (2023). *Gulf economic update: Structural reforms and shifting social norms to increase women's labor force participation*. World Bank Group.

Al-Sarihi, A., & Mansouri, N. (2022). Renewable energy development in the gulf cooperation council countries: Status, barriers, and policy options. *Energies, 15*(5), 1923.

Atif, M., & Ali, S. (2021). Environmental, social and governance disclosure and default risk. *Business Strategy and the Environment, 30*(8), 3937–3959.

Bargaoui, I., Khanchel, I., & Lassoued, N. (2024). Are pollution control bonds the solution for energy firms under the magnifying glass? *Environment, Development and Sustainability, 27*, 1–22.

Barney, J. (1991). Firm resources and sustained competitive advantage. *Journal of Management, 17*(1), 99–120.

Bharath, S. T., & Shumway, T. (2008). Forecasting default with the Merton distance to default model. *Review of Financial Studies, 21*(3), 1339–1369.

Boubaker, S., Cellier, A., Manita, R., & Saeed, A. (2020). Does corporate social responsibility reduce financial distress risk? *Economic Modelling, 91*, 835–851.

Campbell, J. L. (2007). Why would corporations behave in socially responsible ways? An institutional theory of corporate social responsibility. *Academy of Management Review, 32*(3), 946–967.

Cathcart, L., Dufour, A., Rossi, L., & Varotto, S. (2020). The differential impact of leverage on the default risk of small and large firms. *Journal of Corporate Finance, 60*, 101541.

Chava, S. (2014). Environmental externalities and cost of capital. *Management Science, 60*(9), 2223–2247.

Chen, X. H., Tee, K., Elnahass, M., & Ahmed, R. (2023). Assessing the environmental impacts of renewable energy sources: A case study on air pollution and carbon emissions in China. *Journal of Environmental Management, 345*, 118525.

Chodnicka-Jaworska, P. (2022). Environmental, social, and governance impact on energy sector default risk—Long-term issuer credit ratings perspective. *Frontiers in Energy Research, 10*, 817679.

Cuesta, L., Alvarado, R., Ahmad, M., Murshed, M., Rehman, A., & Işık, C. (2023). Institutional quality, oil price, and environmental degradation in MENA countries moderated by economic complexity and shadow economy. *Environmental Science and Pollution Research, 30*(48), 105793–105807.

Dallocchio, M., Pistolesi, F., & Teti, E. (2025). Conciliating environmental and financial performance in emerging countries. *Finance Research Letters, 74*, 106781.

Darwish, S., Abdo, H., & AlShuwaiee, W. M. (2018). Opportunities, challenges and risks of transition into renewable energy: The case of the Arab Gulf Cooperation Council. *International Energy Journal, 18*(4), 391.

DiMaggio, P. J., & Powell, W. W. (1983). The iron cage revisited: Institutional isomorphism and collective rationality in organizational fields. *American Sociological Review, 48*(2), 147–160.

Dixon-Fowler, H. R., Slater, D. J., Johnson, J. L., Ellstrand, A. E., & Romi, A. M. (2013). Beyond "does it pay to be green?" A meta-analysis of moderators of the CEP–CFP relationship. *Journal of Business Ethics, 112*, 353–366.

Do, T. K. (2022). Corporate social responsibility and default risk: International evidence. *Finance Research Letters, 44*, 102063.

Do, T. K., & Vo, X. V. (2023). Is mandatory sustainability disclosure associated with default risk? Evidence from emerging markets. *Finance Research Letters, 55*, 103818.

El Ghoul, S., Guedhami, O., Kim, H., & Park, K. (2018). Corporate environmental responsibility and the cost of capital: International evidence. *Journal of Business Ethics, 149*, 335–361.

El Ghoul, S., Guedhami, O., Kwok, C. C., & Mishra, D. R. (2011). Does corporate social responsibility affect the cost of capital? *Journal of Banking & Finance, 35*(9), 2388–2406.

Feldman, S. J., Soyka, P. A., & Ameer, P. G. (1997). *Does improving a firm's environmental management system and environmental performance result in a higher stock price?* Environmental Group Study, ICF Kaiser International.

Feng, F., Han, L., Jin, J., & Li, Y. (2024). Climate change exposure and bankruptcy risk. *British Journal of Management, 35*(4), 1843–1866.

Flammer, C. (2013). Corporate social responsibility and shareholder reaction: The environmental awareness of investors. *Academy of Management Journal, 56*(3), 758–781.

GIZ. (2023). *The circular economy ecosystem in the MENA region*. Deutsche Gesellschaft für Internationale Zusammenarbeit (GIZ) GmbH. https://www.giz.de/en/worldwide/78869.html

Gorus, M. S., & Aslan, M. (2019). Impacts of economic indicators on environmental degradation: Evidence from MENA countries. *Renewable and Sustainable Energy Reviews, 103*, 259–268.

Goss, A., & Roberts, G. S. (2011). The impact of corporate social responsibility on the cost of bank loans. *Journal of Banking & Finance, 35*(7), 1794–1810.

Hart, S. L. (1995). A natural-resource-based view of the firm. *Academy of Management Review, 20*(4), 986–1014.

Jamali, D., Jain, T., Samara, G., & Zoghbi, E. (2020). How institutions affect CSR practices in the Middle East and North Africa: A critical review. *Journal of World Business, 55*(5), 101127.

Jamali, D., & Karam, C. (2018). Corporate social responsibility in developing countries as an emerging field of study. *International Journal of Management Reviews, 20*(1), 32–61.

Jo, H., Kim, H., & Park, K. (2015). Corporate environmental responsibility and firm performance in the financial services sector. *Journal of Business Ethics, 131*, 257–284.

Kabir, M. N., Rahman, S., Rahman, M. A., & Anwar, M. (2021). Carbon emissions and default risk: International evidence from firm-level data. *Economic Modelling, 103*, 105617.

Karpoff, J. M., Lott, J. R., Jr., & Wehrly, E. W. (2005). The reputational penalties for environmental violations: Empirical evidence. *The Journal of Law and Economics, 48*(2), 653–675.

Khanchel, I., Lassoued, N., & Bargaoui, I. (2023). Pollution control bonds and environmental performance in energy utility firms: is there an incantation effect? *International Journal of Energy Sector Management, 18*, 1066. https://doi.org/10.1108/IJESM-02-2023-0014

Khanifah, K., Udin, U., Hadi, N., & Alfiana, F. (2020). Environmental performance and firm value: Testing the role of firm reputation in emerging countries. *International Journal of Energy Economics and Policy, 10*(1), 96–103.

Lassoued, N., & Khanchel, I. (2023). Do green bonds alleviate environmental risk? In *Reference module in social sciences*. Elsevier. https://doi.org/10.1016/B978-0-44-313776-1.00133-1

Lassoued, N., Khanchel, I., Fakhfakh, I., & Etteyeb, M. (2024). Entrepreneur characteristics and financing patterns in SMES of MENA countries: Overcoming the burdens of liability of newness. *Journal of International Development, 36*(2), 1507–1535.

Li, T., Trinh, V. Q., & Elnahass, M. (2023). Drivers of global banking stability in times of crisis: The role of corporate social responsibility. *British Journal of Management, 34*(2), 595–622.

Marie, M., Qi, B., Elgammal, M., & Elnahass, M. (2024). A more sustainable future: Can politically connected CEOs spur the nexus between ESG performance and firm financial performance? *Journal of International Financial Markets Institutions and Money, 96*, 102056.

Matsumura, E. M., Prakash, R., & Vera-Muñoz, S. C. (2014). Firm-value effects of carbon emissions and carbon disclosures. *The Accounting Review, 89*(2), 695–724.

Merton, R. C. (1974). On the pricing of corporate debt: The risk structure of interest rates. *Journal of Finance, 29*(2), 449–470.

Miniaoui, H. (2023). *Climate change in the Middle East and North Africa: Between the repercussions of a lived reality and opportunities for a brighter future*. ERF Policy Brief No. 109.

Negm, A. M., & Shareef, N. (2020). Introduction to the "waste management in MENA regions". In A. Negm & N. Shareef (Eds.), *Waste management in MENA regions* (Springer water). Springer. https://doi.org/10.1007/978-3-030-18350-9_1

Nguyen, D. N., & Nguyen, C. P. (2022). Uncertainty and corporate default risk: Novel evidence from emerging markets. *Journal of International Financial Markets Institutions and Money, 78*, 101571.

Oikonomou, I., Brooks, C., & Pavelin, S. (2012). The impact of corporate social performance on financial risk and utility: A longitudinal analysis. *Financial Management, 41*(2), 483–515.

Park, D., Kang, Y. J., & Lee, J. (2023). Carbon intensity, default risk, and investors' attention to environment: Evidence from South Korea. *International Review of Economics and Finance, 88*, 1104–1121.

Rafindadi, A. A., Muye, I. M., & Kaita, R. A. (2018). The effects of FDI and energy consumption on environmental pollution in predominantly resource-based economies of the GCC. *Sustainable Energy Technologies and Assessments, 25*, 126–137.

Schultz, E. L., Tan, D. T., & Walsh, K. D. (2017). Corporate governance and the probability of default. *Accounting & Finance, 57*, 235–253.

Shahrour, M. H., Girerd-Potin, I., & Taramasco, O. (2021). Corporate social responsibility and firm default risk in the Eurozone: A market-based approach. *Managerial Finance, 47*(7), 975–997.

Shapland, G. (2023). Climate change and the GCC: Economic and environmental impact. In *GCC hydrocarbon economies and COVID: Old trends, new realities* (pp. 173–200). Springer Nature Singapore.

Sun, W., & Cui, K. (2014). Linking corporate social responsibility to firm default risk. *European Management Journal, 32*(2), 275–287.

Venkatachary, S. K., Prasad, J., & Samikannu, R. (2017). Challenges, opportunities and profitability in virtual power plant business models in Sub Saharan Africa-Botswana. *International Journal of Energy Economics and Policy, 7*(4), 48–58.

Wernerfelt, B. (1984). A resource-based view of the firm. *Strategic Management Journal, 5*(2), 171–180.

World Bank. (2018). *Beyond scarcity: Water security in the Middle East and North Africa*. MENA Development Report, The World Bank. https://doi.org/10.1596/978-1-4648-1144-9

World Bank. (2019). *Sustainable land management and restoration in the Middle East and North Africa Region: Issues, challenges, and recommendations (English)*. World Bank Group.

World Bank. (2023). *Climate watch historical GHG emissions (1990–2020)*. World Resources Institute.

World Bank Group. (2022). *Global economic prospects, January 2022*. World Bank Publications.

World Economic Forum. (2020). *How to cut direct emissions in MENA by 2030*. https://www.weforum.org/agenda/2023/12/cut-direct-emissions-mena-2030/

World Economic Forum. (2024, May 27). *Climate action: How MENA's biggest actors can help the region's suppliers and SMEs to decarbonize*. World Economic Forum. https://www.weforum.org/

Zayer, C., & Ben Abdelhadi, A. (2023). Qualitative research on the drivers and barriers of CSR in the Moroccan SME context. *Revue Management & Innovation, 2023/1*(7), 76–104.

Chapter 13
Impact of Belt and Road Initiative Investments on Sustainability in the MENA Region

Silvio Andrae

Abstract The Belt and Road Initiative (BRI) is playing an increasingly central role in the economic and energy transformation of the MENA region. This study examines the impact of BRI investments on sustainability in five selected countries—Morocco, Algeria, Tunisia, Libya, and Egypt. The initiative mobilizes much-needed funds to develop clean energy projects and helps increase the share of renewable energy in the energy mix. At the same time, energy is only the first step; in the long term, the BRI aims to provide Chinese companies with access to lucrative and emerging markets in the region. The analysis shows that the sustainability of BRI investments in the countries studied depends on several key factors. These include the political and regulatory environment, the economic interests of the recipient countries, and how China structures its investments. While the BRI offers significant economic opportunities, it poses challenges, particularly dependence on Chinese technologies and financing models, compliance with international environmental standards, and consideration of local socioeconomic conditions. This study contributes to understanding the complex dynamics between global investment flows and sustainable development in the MENA region. The results provide valuable insights for policymakers, international organizations, and academics by illustrating how different governance models and geopolitical strategies influence sustainability.

Keywords Environment and development · Environmental sustainability · International institutional arrangements · MENA · Sustainability

S. Andrae (✉)
Independent Researcher, Berlin, Germany
e-mail: andrae@zedat.fu-berlin.de

M. Elnahass et al. (eds.), *A Greenprint for Net Zero*, Gulf Studies 15,
https://doi.org/10.1007/978-981-92-1762-5_13

13.1 Introduction

Since its launch in 2013, the Belt and Road Initiative (BRI) has become one of the most ambitious international infrastructure and investment programs. Chinese investments under the BRI, with an estimated volume exceeding USD 1 trillion, have significantly reshaped the global infrastructure landscape in transportation, energy, and telecommunications (Liu et al., 2017). While the initiative initially focused on traditional infrastructure projects such as roads, ports, and railways, in recent years, it has been increasingly expanded to include sustainable investments, particularly in the energy sector. This shift reflects China's goal of making the BRI "greener" and the increasing need for partner countries to achieve sustainable development goals.

The BRI project, which spans more than 152 countries, including many low- and middle-income countries, holds the potential to significantly reduce global emissions. These countries currently contribute around 20% of total global greenhouse gas emissions (Hou et al., 2020), making their involvement crucial in the fight against climate change (Muhammad et al., 2020).

The MENA region is a unique region not only from a BRI perspective. It has been one of the fastest-growing regions in the last 10 years. At the same time, temperatures in the region have risen twice as fast as the global average. It poses several challenges to sustaining the livelihoods of 575 million people in the region. Climate shocks, such as rising temperatures and prolonged droughts, could severely impact agriculture and livelihoods and exacerbate the systemic problems of MENA countries (WEF, 2023). The WEF report (2023) notes that MENA countries need to catch up to comparable regions in terms of their progress toward sustainability. While local governments have pledged to bring 60% of emissions below the net zero target in the last 24 months, businesses have yet to close the gap with comparable global markets.

With its abundant natural resources, the MENA region has the potential to become a world leader in scaling new energy sources such as renewable energy and clean hydrogen. This potential holds geopolitical significance and inspires and motivates the region to take the lead in this crucial transition. Against this backdrop, this study examines the extent to which investments under the BRI influence the sustainability goals of the MENA region, particularly in the energy sector. It focuses on several key questions: What mechanisms enable Chinese foreign direct investment (FDI) to support the transition to renewable energies in the MENA region? Which political, economic, and institutional factors determine the success of sustainable energy projects under the BRI? And finally, what are the positive and negative externalities of the BRI for the environment and the region's socioeconomic development?

The study employs a combined methodological approach integrating qualitative and quantitative elements to answer these questions. Theoretically, the "FDI-SST" (Foreign Direct Investment-Supported Sustainable Transition) model is used to analyze the mechanisms by which foreign direct investment promotes the development of renewable energies in MENA countries. Empirically, the analysis is based on a comprehensive evaluation of secondary data from scientific studies, reports from international organizations such as the World Economic Forum (WEF) and the International Energy Agency (IEA), and official government documents from various MENA countries. In addition, comparative case studies from five countries—Morocco, Algeria, Tunisia, Libya, and Egypt—are used to examine differences in policy frameworks, regulatory challenges, and the chances of success for sustainable energy projects.

Several hypotheses are tested in the analysis. First, it is examined whether countries with well-defined and stable regulatory frameworks benefit more from sustainable BRI investments than countries with weak governance structures. Second, the study investigates how technological spillover effects can increase the efficiency of BRI investments in renewable energies. Third, the study addresses whether increasing economic dependence on China could constrain the autonomy of MENA countries in energy policy in the long term.

The background to this investigation is the changing dynamics of the BRI in the MENA region. While China focused heavily on fossil energy projects in the early phase of the initiative, there has been an increased focus on sustainable projects in recent years. This shift is influenced by global developments, regulatory adjustments, and growing environmental concerns. At the same time, many MENA countries face pressing environmental challenges, including water scarcity, rising temperatures, and a high dependence on fossil fuels. Integrating renewable energies into BRI investments presents both opportunities and risks. On the one hand, Chinese investments in solar, wind, and hydroelectric power plants enable a faster transition to sustainable energy systems. On the other hand, there are concerns about the long-term economic and geopolitical dependencies these investments could create.

This study aims to thoroughly understand the extent to which the BRI can act as a catalyst for a sustainable energy transition in the MENA region. The analysis, which is comprehensive in nature, will consider both the positive developments and the existing challenges in order to formulate recommendations for a more sustainable and inclusive design of the BRI. The results of this study will not only contribute to the scientific debate on the BRI, but also serve as a basis for policymakers, investors, and stakeholders in the energy sector. The comprehensive nature of the analysis provides a solid basis for a comprehensive understanding of the impact of BRI on sustainability in the MENA region, reassuring the audience about the thoroughness of the research. A valuable contribution to current research can be made through carefully selecting case studies and presenting the results.

13.2 Related Works

The theoretical link between FDI and sustainable transition can be analyzed from the perspective of economic modernization (i.e., neoclassical growth and endogenous growth) and dependency theory. According to Solow (1956), FDI is the most critical engine of economic growth. Neoclassical growth theory emphasizes that FDI not only promotes economic growth in host countries through capital accumulation and acquiring new inputs and foreign technologies (Raza et al., 2024), but also has the potential to foster sustainable economic growth in the long term. Foreign direct investment has become a key channel for knowledge and innovation (Awodumi, 2023), leading to technological progress in the field of renewable energies (Tiwari et al., 2022). The theory regards technological progress as exogenous and assumes that the marginal returns of capital decline in the long term, thus limiting its long-term growth effect. However, if FDI triggers significant technological progress, it can boost labor productivity and the overall efficiency of investment in host countries, promoting sustainable economic growth even in the long term (Barro & Sala-I-Martin, 1995). In the context of endogenous growth theory, FDI fosters rapid economic growth in host countries by increasing the stock of human capital and technological progress (Romer, 1994). The theory treats technological progress as endogenous and emphasizes its role in increasing the long-run rate of return to scale and the growth rate. Proponents of this theory argue that FDI can trigger a lasting increase in the growth rate of the host country through technology transfer, diffusion, and spillover effects (Borensztein et al., 1998). In this paradigm, foreign direct investment can expand knowledge, acquire new skills, and introduce modern management methods and organizational mechanisms (spillover effect) (Grossman & Helpman, 1991). Empirical evidence shows positive (e.g., Gui-Diby, 2014) and adverse growth effects due to foreign direct investment (e.g., Adams & Klobodu, 2017; Ofori et al., 2023).

Dependency theory, as formulated by Girling (1973), highlights the structural inequalities and dependency relationships that exist between industrialized countries (centers) and developing countries (peripheries). The analyses show a statistically significant correlation between foreign direct investment and income inequality. A central aspect of the theory is that the inequalities are not only due to economic factors but are deeply rooted in the social and political structures of the dependent countries. It is argued that the dependency of the periphery hinders social progress and limits the political autonomy of developing countries, leading to a vicious circle of poverty and underdevelopment. However, studies based on this show less the correlation mentioned above but rather capture geographical differences in inequality than the harmful influence of foreign direct investment (Ofori et al., 2023; Ravallion, 2018; Tsai, 1995).

The theoretical link between FDI and environmental sustainability is evidenced by the pollution halo (PH) and pollution haven hypotheses (PHH) (Ofori et al., 2023). The PH hypothesis posits that FDI can positively impact environmental conditions in host countries (Bakhsh et al., 2017). It is argued that multinational enterprises (MNEs) often bring more efficient and environmentally friendly technologies and stricter management and environmental standards. By transferring "clean" knowledge and technologies, MNEs can create a "halo effect" in developing countries, motivating local companies and industries to adopt similar ecological standards. In addition, FDI often brings international standards that go far beyond local requirements, as global compliance guidelines often bind multinational companies.

In the energy sector, foreign investment in renewable energy (RE) promotes the introduction of technologies such as wind, solar, or hydroelectric power. It reduces pollution and carbon emissions (Chen et al., 2023). PHH, on the other hand, assumes that companies could use FDI to relocate their environmentally intensive activities to countries with less stringent environmental regulations (Kastratović, 2019). This hypothesis argues that FDI flows to "pollution havens" countries because they have lower environmental standards and impose fewer environmental costs than industrialized countries. Companies that produce high emissions could relocate their production to countries with laxer regulations to save costs. It often leads to a deterioration in local environmental quality and increases the risk of ecological damage, particularly in sectors such as mining, chemicals, or fossil fuels (Garetto et al., 2025). This hypothesis could imply that investments are increasingly flowing into the energy sector but more into fossil fuels than RE if the countries do not have strict climate policies. The empirical evidence is mixed. Some studies show that FDI leads to improved environmental standards, especially when the investments come from companies in countries with high environmental standards (Chen et al., 2021). Other studies support the EHH, especially in resource-intensive industries, where companies use lax regulations (Opoku & Boachie, 2020).

The Low Carbon Transition Curve (LCTC) hypothesis assumes that the transition to a low-carbon economy behaves similarly to the Environmental Kuznets Curve (EKC), which describes the relationship between economic development and environmental pollution in an inverted U-curve. Initially, the consumption of fossil fuels dominated, but with increasing prosperity and technological development, it decreased while the use of RE increased. This approach thus posits that higher per capita income could correlate with higher consumption of RE.

Schulhof et al. (2022) examine a sample of 171 peer-reviewed academic articles to understand how sustainability is conceptualized and represented in the BRI. The authors conclude that despite the large amount of research, there still need to be more gaps between how the BRI is conceived and how China engages in practice. These gaps have significant implications for understanding the impact of the BRI on global sustainability. More research is needed to bridge these gaps and provide a comprehensive understanding of the BRI's impact on global sustainability. There is a risk that vague or incomplete representations obscure how the BRI undermines environmental goals. Latief et al. (2024) examine the relationship between sustainable tourism, renewable energy, technological innovation, and environmental quality. The study uses panel data from BRI countries without focusing on a specific region, such as MENA. The data covers the period from 1995 to 2020. The analyses show a significant negative correlation between sustainable tourism, technological innovation, and greenhouse gas emissions in the long and short term. The authors recommend the use of energy-efficient technologies, the construction of environmentally friendly buildings, and the introduction of sustainable, energy-efficient transportation systems. These initiatives are central to achieving the goals of environmental sustainability. In addition, comprehensive environmental management plans and tourism strategies should be developed to take advantage of regional technological advances. Such strategies attract investment in the technology and tourism sectors and mitigate the negative impacts of international tourism on the environment. Finally, the authors recommend actively promoting RE sources in all sectors. It can be achieved by setting quotas for RE in the overall energy mix, promoting the establishment of specialized RE agencies, and phasing out subsidies for conventional energy technologies.

Xiao and Qamruzzaman (2022) analyze the influence of green investments (proxied by the share of renewable energies), domestic investment (measured by gross fixed capital formation), and environmental sustainability (indicated by CO_2 emissions) on technological innovation in the countries of the Belt and Road Initiative (BRI) between 2000 and 2021. The results show that both green and domestic investment positively correlates with technological innovation performance. In contrast, a statistically significant negative relationship exists between environmental sustainability and technological innovation. The complementary asymmetric analysis also shows that all three influencing factors—green investment, environmental sustainability, and domestic investment—have asymmetric effects on technological innovation processes, with the directions and strengths of the effects differing significantly between positive and negative changes.

Ye et al. (2024) measured the green logistics efficiency of BRI countries using panel data from 2013 to 2019. They conclude that the green logistics efficiency of BRI countries is generally low. High-income countries have significantly higher efficiency than middle-income countries. However, the results show an upward trend, especially in middle-income countries.

Noureldin et al. (2024) analyze the relationship between intellectual property protection (IPP) and the influence of BRI on economic policies in MENA countries. The BRI framework emphasizes enhancing regional connectivity to achieve high standards of IP network for business model innovation and holds significant potential for positive change. The analysis of Alnafisah et al. (2024) shows a positive and significant impact of patents on CO2 emissions for selected MENA countries, confirming the impact of patents on environmental quality.

Hussain et al. (2020) provide insights into assessing Chinese foreign investment risk and natural resource potential in BRI countries. Their research forms the basis for Chinese companies to develop effective decision-making strategies for foreign investment in partner countries. By utilizing the Geographic Information System, potential risks and opportunities for Chinese companies investing in partner countries can be identified. The results highlight that most Central and Eastern European countries and other BRI countries, such as the United Arab Emirates, present favorable conditions for Chinese companies to invest overseas. These findings underscore the potential for Chinese companies to manage and execute BRI projects effectively, thereby minimizing potential risks and maximizing the benefits of their investments.

Mu et al. (2024) evaluated the role of BRI in improving export resistance for 145 products in 130 BRI countries. The results unequivocally show a significant positive impact of BRI on the export resistance of countries along the route. This underscores the BRI's effectiveness in improving export performance, risk resilience, and securing and stabilizing supply chains. The results also show that the impact of the

BRI is delayed by 3 years, consistent with its gradual implementation, and this demonstrates its effectiveness. The impact is significant in low-income countries.

Sun (2020) show that public–private partnership (PPP) projects under the BRI should focus on environmental protection and nature conservation. Given the high energy demand and concentrated energy consumption in most BRI countries, PPP projects in these areas face urgent ecological and environmental challenges. To ensure the safety of the ecological environment, the development of green energy technologies is necessary. In addition, notwithstanding the investment and trade activities, Chinese enterprises should fully consider the public interests and actively bear the cost of environmental pollution. The authors recommend setting up risk databases. Access to such databases containing contributions from all member countries would facilitate risk assessment and reduce information asymmetry between borrowers and lenders. Finally, China should learn from the rules for green finance, such as the "Equator Principles." These principles are a voluntary set of rules for banks to comply with environmental and social standards in project finance. Die Entwicklung grüner Finanzierungen ist jedoch von Natur aus langsam und erfordert mehr öffentlich-private Investitionen und eine umweltfreundliche öffentliche Beschaffung (Xu, 2023).

Xing et al. (2023) carry out an economic-historical analysis. Using Egypt as an example, they attempt to illustrate China's (re)emergence in the MENA region in the context of the BRI. The authors identify three key characteristics: connectivity, fluidity, and complementarity. They characterize the dynamics of historical Sino-Egyptian relations.

In their time series analysis, Yusoff et al. (2024) see a positive co-integration between RE deployment and gross domestic product (GDP) in six ASEAN countries. The positive relationship between GDP and RE use indicates that economic growth drives RE adoption. Carbon emissions, population density, and FDI hurt the demand for RE. Raihan et al. (2025) estimate that Egypt's increase in green energy consumption and foreign direct investment will lead to a long-term GDP increase. Carbon emissions that contribute to pollution do not motivate industry to invest in RE. This result calls into question the hypothesis of the Environmental Kuznets Curve (EKC). In this respect, promoting renewable energies depends heavily on political decisions, such as reforming fuel subsidies.

Awijen et al. (2022) analyze the factors driving RE use in MENA countries. They consider economic, financial, and political variables such as financial development, government effectiveness, and political stability.

13.3 FDI-Supported Sustainable Transition

Based on economic theory, the model examines how foreign direct investment (FDI) can support the transition to sustainable energy production (FDI-SST)—particularly in renewable energies—in the MENA region. For example, foreign direct investment enables technology spillovers, including various RE solutions based on cutting-edge clean and green technologies and their trading partners (Dussaux et al., 2022). It outlines the model structure and mechanisms of action, providing a practical framework for applying them to the case of Chinese FDI in selected MENA countries, making the research directly relevant to real-world scenarios.

13.3.1 The Model

The FDI-SST model is designed to illustrate how foreign direct investment inflows, particularly in the context of strategic partnerships such as the BRI, can help host countries transition away from fossil fuels and toward RE sources. The model includes the following core components:

- FDI capital inflow (C): capital from foreign direct investment specifically for RE infrastructure and technology transfer.
- Technology transfer (T) is the transfer of technology, knowledge, and know-how from investing countries, thereby improving local capacity for RE production.
- Domestic production capacity (D): The ability of host countries to utilize new technologies and infrastructure to increase their RE production.
- Environmental and economic outcomes (E): Measured by reducing carbon emissions and increasing energy independence from fossil fuels.

The equation can be summarized as:

$$D = f(C,T,P)$$

where:

C represents FDI inflow aimed at energy projects.

T represents technology transfer, a function of FDI moderated by factors such as local research and development (R&D) capacity.

P represents policy alignment and incentives, such as RE targets, subsidies, or regulatory support from the host government.

13.3.2 Mechanisms of Impact

Direct Capital Investment Chinese foreign direct investment in RE projects, particularly solar and wind farms, provides seed capital for developing these sectors, often requiring significant upfront investments. This capital injection alleviates budget constraints and reduces the initial financial risk for MENA countries with underdeveloped RE sectors.

Technology and Knowledge Transfer FDI brings advanced RE technologies to host countries, such as state-of-the-art solar panel manufacturing, energy storage solutions, or grid integration technology. This transfer enhances the host country's technological capabilities and empowers local workers and businesses, giving them the skills and technical expertise needed to produce renewable energy sustainably.

Political and Institutional Support Favorable policies in host countries that support sustainable energy enhance the effectiveness of foreign direct investment. For example, targeting BRI investments to MENA countries' RE targets can maximize impact by creating a regulatory environment encouraging RE adoption, including tax incentives, RE subsidies, and carbon reduction targets.

Supply Chain Development and Market Integration FDI often leads to local supply chain development, creating links supporting the RE industry, such as manufacturing components for solar panels or establishing local maintenance services. It promotes self-sufficiency, reduces dependence on fossil fuels, and holds the potential for significant economic growth in the RE sector.

Environmental and Economic Impact The model predicts that projects supported by foreign direct investment will, if successfully implemented, lead to a measurable reduction in carbon emissions and a transition to energy independence. It should also contribute to economic resilience by diversifying energy sources and reducing vulnerability to fossil fuel price fluctuations.

13.4 BRI and Sustainability: An Area of Tension

13.4.1 BRI as Global Governance

Below, we assess current developments within the BRI. In addition to governance and law issues, the main focus is on investments and their impact in the MENA region. The Belt and Road Initiative (BRI) is a development strategy adopted by the Chinese government in 2013. The major global project is based on the old Silk Road and offers a new concept of globalization and networking. BRI is to be interpreted as a complex, multi-layered, and adaptable project. BRI defies the usual taxonomies and typologies. Traditionally, BRI is seen as a plan to realize transport, road, infrastructure, communication, and other projects in different countries worldwide (Lu et al., 2024). In practice, BRI has proven to be a broad framework that engages partner countries in long-term, complex relationships with China (Ultimately, BRI was and is a framework to channel China's domestic and foreign policy needs) (Garlick, 2019). As such, the means and focus change frequently. Since 2013, the BRI has undoubtedly maximized China's influence and made the country an economic partner across the developing world.

By linking the concepts of norm localization, norm subsidiarity, and policy advice, Sun and Yu (2023) argue that China's efforts to transform the environment through the BRI initially began as a process of norm localization. In an open policy space created by the Chinese leadership, environmental standards developed that reflected China's overarching ambition to establish the BRI as a global governance model. After formulating a comprehensive norm for environmental responsibility, state bureaucracies identified opportunities to create procedural and operational regulations. It gave rise to a novel methodology for developing procedural rules that emphasizes inclusive dialogue with host countries. This process was significantly shaped by China's geopolitical ambitions for international leadership and the existing local South–South cooperation norms. At the operational level, a differentiated picture emerges: different actors within the BRI pursue different strategies, either adapting existing international standards to local conditions or developing entirely new norms. This dynamic process illustrates how norm localization functions as a mechanism of policy adaptation and contributes to the negotiation of new global environmental standards within the framework of the BRI.

Lorenzo (2021), on the other hand, is skeptical about the emergence of new environmental standards. From the perspective of international law and sustainable development, she identifies three central problematic features of the BRI. For example, the bilateral nature of the agreements between China and the participating states leads to fragmentation and significant discrepancies in the sustainability standards applied. These differences make implementing environmentally and socially sound measures coherently and effectively challenging. Furthermore, the informal nature of many agreements often prevents comprehensive parliamentary and public consultation in the host countries. This not only undermines democratic legitimacy but also weakens accountability mechanisms, which can jeopardize compliance with internationally recognized environmental and social standards. Finally, the decision-making structures of the BRI are largely state-dominated, excluding non-state actors—especially civil society organizations and local communities. This is particularly problematic because the voices of the communities directly affected by large-scale infrastructure projects are barely heard, and their interests are insufficiently considered in the planning and implementation processes.

At the core of the BRI lies a comprehensive, connectivity-driven development. China strategically utilizes the BRI to foster robust bilateral relations with third countries, with the aim of securing resource supplies, opening markets for its companies, and solidifying friendships that can be leveraged bilaterally and in multilateral forums. In the current geopolitical landscape, this all-encompassing approach, with its wide array of goals, has never been more crucial.

Furthermore, the BRI is set to serve as a platform for China's industrial and technology policy, particularly in times of crisis. As a result, cooperation in digital technology will be increasingly promoted to capitalize on the accelerated digitalization stemming from global lockdown measures. China views technological superiority as a prerequisite for economic and military strength, and thus, for its global political influence. The country is currently striving for independence from the USA's key digital technologies, with the next step being the global dissemination of its own technology. This strategy can only succeed by establishing a global technological sphere.

Projects such as the "Digital Silk Road," a component of the BRI that focuses on digital infrastructure and connectivity, will play a decisive role. BRI will serve both as a means to promote political communication and standardization and to facilitate the expansion of China's digital sector (and digital infrastructure). The Digital Silk Road will see China's digital champions take the lead rather than the megaprojects of state-owned enterprises. Much of this overlaps with the global rise of Huawei, Alibaba, Tencent, and other privately owned tech companies, which will also help to broaden the scope of players involved in the initiative.

China has adjusted the BRI strategy after more than 10 years of existence. The days of announcing a flood of megaprojects under the BRI umbrella are long gone. However, the initiative itself still needs to be remembered. Instead, it is becoming one of several instruments on the foreign policy agenda. The new strategy aims to include industries with higher added value, which China has consolidated over the last decade.

13.4.2 *BRI Investments: Greening the Future?*

According to the China Overseas Finance Inventory (WRI, 2023), in the first 10 years of the BRI, investments in fossil fuels dominated, with USD 52 billion invested in coal-fired power plants alone. In recent years, China has signaled that it is trying to address some of the leading causes of environmental damage associated with the BRI.

In 2015, China began to refer to the BRI as the Green Silk Road or Green BRI (CCICED, 2020). China made a landmark announcement at the 2021 UN General Assembly, a significant global platform. It declared its decision to halt the construction of new coal-fired power plants as part of the BRI. Instead, it pledged to channel more investments into renewable energy (RE) projects. This commitment was further solidified at the last BRI summit in October 2023, marking a clear shift toward a greener BRI. On March 28, 2023, four Chinese ministries published a crucial policy document, titled "Opinions on the Joint Implementation of Green Development in the Belt and Road Initiative." This proactive step is a significant part of China's commitment to a greener BRI. It reaffirms the commitment to refrain from building new coal-fired power plants and outlines strategies for retrofitting existing plants to make them more environmentally friendly and efficient. It also encourages the installation of technologies for capturing, using, and storing carbon dioxide, thereby addressing the environmental impact of existing infrastructure. In addition, "green" infrastructure projects in energy and transport, industry and manufacturing, environmentally friendly financing, and cooperation on standards will be initiated. The aim is to reduce and control environmental risks at the project level.

From 2013 to 2022, fossil fuel infrastructure—including coal-, gas-, and oil-fired power plants—accounted for more than two-thirds of Chinese investment in the energy sector in BRI countries (countries that have signed BRI agreements with China) (Jiang et al., 2025). Most of these investments have been made by the China

Development Bank, the Export-Import Bank of China, and state-owned commercial banks and companies that have historically favored fossil fuel investments. These projects are more significant and less risky than RE projects. Where BRI investments in renewables have taken place, they have primarily been made by profit-driven, privately owned Chinese companies that have been open to smaller projects and have been able to capitalize on comparative advantage in renewables.

In September 2021, a significant turning point was reached when China decided to cease financing coal-fired power plants abroad and promote greener financing. This shift, though gradual, is a major step toward a more sustainable BRI. The immediate halt and subsequent drop to zero in 2022 of investment in new coal-fired power plants abroad, coupled with the unchanged Chinese wind and solar energy investment, demonstrate a steadfast dedication to greener financing. China's leadership in the global wind turbine export market not only positions it well to spearhead wind energy cooperation among BRI countries but also impresses upon the audience the significant role China plays in promoting renewable energy.

One reason for this delay could be that China's announcement should have included concrete targets and incentives for investment in RE. It was followed by some guidelines for greening foreign investment and cooperation. Under these guidelines, Chinese contractors are encouraged to comply with international or Chinese environmental standards, especially in countries whose national environmental standards and policies do not meet international standards (Wong et al., 2021). This is a formalization of a series of previous guidelines, including the "Green Development Guidance for BRI Projects Baseline Study," a comprehensive study that provides a baseline for understanding the environmental impact of BRI projects, and the "Application Guide for Enterprises and Financial Institutions," a practical guide that helps these entities implement green practices in their BRI-related activities. Both were supported by various relevant Chinese ministries and published by the BRI Green Development Coalition (BRIGC) in December 2020 and October 2021, respectively. The BRIGC guidance is only advisory and cannot be enforced in court. However, they provide clear direction for the government agencies, companies, and financial institutions involved in the BRI. In this respect, all key players will read and implement the document. Further implementing measures and regulations are likely to follow.

According to the BRI Investment Report 2023 (Nedopil, 2024), the following vital data can be noted:

- Ten years after the BRI launch, the cumulative BRI commitment amounts to USD 1.053 trillion. Around USD 634 billion is attributable to construction contracts and USD 419 billion to non-financial investments.
- In recent years, Africa has become the largest recipient of BRI commitments (USD 21.7 billion), overtaking the Middle East countries (USD 15.8 billion). Nonetheless, Middle Eastern countries remained the top recipients of Chinese construction investment, receiving 36.7% of the total BRI construction investment volume in 2023.

- China's financing and investment were spread across 61 BRI countries in 2023 (up from 60 in 2022), with 37 countries receiving investment and 45 engaged in construction. Saudi Arabia had the highest construction volume in 2023, with around USD 5.6 billion (compared to USD 2.6 billion in 2022).
- In 2023, the technology (+1046%) and metals and mining (+158%) sectors grew the most compared to 2022. Chinese companies are investing heavily in metals and mining, which are particularly relevant for the green transition (e.g., lithium) and batteries for electric vehicles. Exposure related to batteries reached around USD 8 billion.
- Chinese BRI exposure focused on infrastructure, particularly in the energy (31%) and transportation (16%) sectors. Nevertheless, both sectors experienced declines compared to the previous year.

Viglioni et al. (2025) compare renewable energy policies between the BRI country groups using the "Regulatory Indicators for Sustainable Energy" (RISE) program. On average, from 2010 to 2021, all regions show an upward trend in their RE policies, except for Southeast Asian countries (members of ASEAN), which experienced a slight decline after 2018. Many of these countries still face barriers to RE investment, such as slow implementation of supportive policies, higher subsidies for fossil fuels, and a lack of affordability and accessibility when implementing RE projects (Erdiwansyah et al., 2019).

China's declaration of its intent to increase green investment through the BRI has the potential to significantly accelerate the global deployment of renewable energy. However, the successful execution of such a large-scale investment program will require concerted effort. The future of a green BRI is not only dependent on China's actions but also dependent on the global economic conditions and the investment environment in the BRI host countries (Zhang et al., 2023).

The demand for investment in RE in BRI host countries is high, and the potential for growth in the clean energy sector is significant. Zhou and Ma (2023) estimate that a six- to seven-fold increase in annual investment in clean energy is needed to achieve global climate targets. This growth potential is not just a possibility but a promising future, with investment expected to increase from 260 billion dollars in 2022 to 1.4 and 1.9 trillion USD by the early 2030s.

Over the past 10 years, China's foreign investment in solar and wind energy has been concentrated mainly in four Middle Eastern countries: the United Arab Emirates, Turkey, Saudi Arabia, and Oman. These countries were among the few where RE investments outside the advanced economies and China increased (IEA, 2024).

To spur demand for clean energy infrastructure, these governments have adopted aggressive RE targets and supportive measures, such as long-term power purchase agreements through government off-takers, to reduce risk and uncertainty for investors. In the United Arab Emirates, for example, the government is making land available for RE projects to reduce investor costs and risks. These measures and a competitive bidding process have helped reduce the cost of solar-generated electricity to record lows and accelerate the development of renewable capacity. However, these four countries are among the richer BRI host countries. While their strategies to attract funding for clean energy have been successful and can serve as a model for some countries, not all BRI countries will have sufficient resources and capacity to follow in their footsteps (Zhou & Ma, 2023).

Wang and Cheng (2024) show that BRI can significantly promote the green transformation of companies. BRI then promotes green transformation through three channels: promoting green technology spillovers, optimizing corporate resource allocation, and improving green governance capabilities. The study by Ullah et al. (2025) explores the influence of regional transformations on green development. It shows that regional transformation processes can significantly promote ecological sustainability and thus support climate policy objectives. However, within multilateral cooperation mechanisms, specific dimensions—trade linkages, investment flows, and institutional cooperation—tend to impact green development and climate resilience negatively. In contrast, monetary and fiscal policy aspects show only marginal adverse effects. There is a positive correlation between infrastructure development and connectivity and the promotion of green development. The mobility of labor and people has a positive but statistically insignificant impact. Importantly, the study highlights the role of the BRI, which contributes substantially to green development both before and after spillover effects, particularly in Asia and Europe. However, the results for Africa and the Middle East point to potential environmental impacts from large-scale infrastructure investments, which must be viewed critically. The study also underscores the complexity of the issue, as the effects of regional transformation processes vary significantly from region to region.

Finally, the conditions in China influence the appetite of BRI investors. On the supply side, it is the general economic conditions in China, how much capital is available to BRI investors, and what they will prioritize in their investment activities. Economic growth in China has slowed down. It could soon lead to less energy investment overseas, including in the clean energy sector. China's fiscal capacity is also limited. This makes it harder to offer incentives to encourage overseas investment, including clean energy and other infrastructure investments (Zhou & Ma, 2023).

One of the contradictions is that China has become a paradoxical double champion in polluting and greening. Conversely, 60% of the billion-strong nation's energy needs are still covered by coal. China still blows around eleven gigatons of CO2 into the atmosphere yearly, almost a third of all global emissions. On the other hand, the country is expanding renewable energies faster than any other country. China has the ambition to be the undisputed world champion in the expansion of green energy sources. According to the IEA (2023), 60% of the RE capacity added worldwide by 2028 will be generated in China alone. In 2023, as much photovoltaic capacity was connected to the grid in China as in the rest of the world combined. According to a recently published analysis by the China Renewable Energy Engineering Institute (CREEI), new systems with a capacity of 260 gigawatts are set to be installed in 2024 alone. China's total solar and wind power capacity could grow to more than 1310 gigawatts by the end of 2024. This means that the People's Republic will reach its 2030 expansion target for RE 6 years earlier than planned. Regardless of the problems with grid expansion, solar, wind, and hydropower have ensured that CO_2 emissions have not increased further despite rising electricity consumption. Thanks to cheap green technologies from China, countries in the global South can also afford the energy transition.

It is more critical to achieve the domestic growth target of around 5% than to implement the energy-saving targets. Due to their development- and policy-oriented lending practices, the foreign loans granted by the China Development Bank and the Export-Import Bank of China could be the most affected.

13.4.3 BRI and MENA: A Successful Cooperation?

With its strategic shipping lanes, abundant energy resources, diverse demographics, and robust financial capabilities, the MENA region is not just promising, but crucial for BRI players. Chinese companies have already made significant strides, constructing numerous railroads and other vital infrastructure projects in countries across the region, particularly Saudi Arabia and Iran. China's Maritime Silk Road's influence extends through the Arabian and Red Seas, facilitated by the Suez Canal, and into the Mediterranean, further underscoring the region's strategic importance.

Regarding the impact on the MENA region, countries have different institutional capacities, development financing, investment, and political needs. China is adapting or realigning strategies according to these dynamics. For example, China closed USD 5.5 billion worth of deals in Saudi Arabia as part of a rapid shift away from Russia and toward the Middle East regions, highlighting the suppleness of the BRI (Noureldin et al., 2024). Adaptation, therefore, also means focusing on specific markets.

Some states in the MENA region are more institutionally oriented and are actively working to enhance their fiscal capacity in infrastructure. Countries like Egypt and, to a certain extent, Iran have well-established state structures and institutions and are striving to accelerate fiscal transformation within their borders. This aligns with Chinese efforts to foster more investment, indicating a potential for fruitful collaboration.

However, the Gulf region presents a unique dynamic. Unlike other Arab states, it does not lack development finance and, therefore, does not require cheap long-term loans from China. This unique financial situation necessitates a more balanced approach from China in its relations with the MENA countries, with strategies adapted to these different dynamics. Paradoxically, this balanced approach would require' Beijing to take a more direct interest in the region's politics while still attempting to maintain its non-intervention strategy and avoiding military confrontation.

13.5 BRI and the Energy Sector

The BRI region's energy consumption growth is a significant global issue, accounting for 24%, 43%, and 62% of global household consumption, energy consumption, and population, respectively (Hafeez et al., 2019). The production methods associated with energy reinforce energy inequalities; the economies of the MENA region show a decreasing energy consumption pattern within the BRI region. At the same time, energy consumption in China continues to rise gradually. This mutual interdependency between China and the BRI countries puts China in a relatively passive position overall (Zhao et al., 2019).

Established interdependency relationships exist between China and the BRI countries regarding energy cooperation. Energy cooperation among the BRI countries could significantly impact the global energy market and the achievement of international climate and sustainability goals. Particularly noteworthy is how the BRI can promote or hinder the transition from fossil to RE sources.

Traditional and Renewable Energy Cooperation BRI countries are cooperating on fossil fuels and, increasingly, renewable energies. In the past, the focus was often on expanding traditional energy markets, such as oil and gas (Fang et al., 2021; Qi et al., 2019). However, the more recent focus is shifting to RE and the construction of green infrastructure. As a result, more and more collaborations are emerging to develop solar and wind energy projects and to promote hydrogen projects in the BRI countries (Lin et al., 2020; Schulhof et al., 2022). These collaborations support the diversification of energy sources and help make these countries' energy mix more sustainable.

Infrastructure Development for Renewable Energies The BRI encompasses extensive infrastructure projects that could improve access to energy sources and create new markets for RE. The construction of power grids and energy-efficient transportation systems enables more efficient integration of RE. This is crucial to ensure stable supply and the feed-in of variable energy sources such as wind and solar power (Bega & Lin, 2023; Wang & Lin, 2022). In countries with high RE potential, infrastructure development can attract investment and reduce dependence on fossil fuels.

Global Impact on Decarbonization Targets Many BRI countries are among the fastest-growing economies, so their energy consumption patterns will influence global carbon reduction efforts. There is a clear challenge as China, a leading BRI country, heavily relies on coal. Nevertheless, the BRI is expected to increasingly promote RE projects, which could support the global transition to a low-carbon economy (Rauf et al., 2020; Wang et al., 2023; Zhang et al., 2019).

China's energy-related exposure in 2023 was the greenest in absolute and relative terms since the BRI began. In 2023, China's engagement in green energy (solar and wind) amounted to about USD 7.9 billion, representing about 28% of its energy engagement, plus another 6% (USD 1.6 billion) for hydropower. Chinese engagement in the energy sector accounts for the largest share of China's BRI engagement. In 2023, total engagement in the energy sector reached just under USD 30 billion—the lowest BRI engagement in the energy sector, except in 2020. A particular development is the increased engagement in green energy (solar, wind, and biomass energy), which reached a record USD 7.9 billion; this figure does not include Chinese exports of solar installations. Exposure to distribution systems (e.g., substations and power lines) also accounted for more than 11% of China's BRI energy exposure (Nedopil, 2024).

On the one hand, the MENA countries are among the world's largest oil and gas exporters. Exports guarantee the region's economic growth and an improvement in living conditions. On the other hand, China is intensifying its involvement in the MENA region to respond to domestic energy problems and secure access to energy for its companies. Engagement across the energy supply chain has evolved in the 2022–2023 data frame. While energy production was most important for China's BRI engagement in both years, exploitation (particularly of oil and gas) has decreased significantly from USD 12.7 billion to USD 6.1 billion. By contrast, engagement in energy transmission, crucial for the green transformation, increased from USD 4.5 billion to over USD 7 billion. Similar to 2022, Saudi Arabia was also the country with the most significant energy commitment in 2023 (USD 3.7 billion). Saudi Arabia also received a significant share (19%) of the green energy exposure. Since the start of the BRI in' 2013, Saudi Arabia has been the third most crucial partner in China's BRI energy engagement (after Pakistan and Russia) (Nedopil, 2024).

The MENA region, known for its large oil and gas reserves, is now poised to become a significant player in the global energy transition. Half of the world's oil and a third of the world's gas reserves are located in the MENA countries. However, this statement conceals significant differences between the MENA countries. The energy transition is becoming an increasingly important political and economic issue in the MENA region. While in the past, most of them relied on their national oil and gas production to boost their economies, they now need to move away from the most polluting power generation systems. This transition, whether they are planning for a carbon-neutral economy or simply hoping to reduce their greenhouse gas emissions, presents a significant opportunity for the countries of the MENA region. It is a chance to reshape their economic landscape and create a more sustainable future.

Alongside the profound change and potential increase in energy dependency, the MENA region is also witnessing the emergence of new business opportunities. Their geographical location positions them as potential net exporters of green electricity—through solar energy or hydrogen. However, they all need more technology and companies to transition without foreign support. Here, competition between European and Chinese energy suppliers is in full swing. China wants to ensure global electricity interconnection with BRI and State Grid Corp (Marketos et al., 2022).

The energy sector is a crucial focus on China's path to a green and sustainable BRI, as this initiative is more about investing in energy than anything else. Thus, various policy documents issued by China have repeatedly emphasized the inevitability and importance of promoting energy cooperation and governance in this context. Given the situation in China and recent projects in North Africa, the pragmatic approach to energy transition clashes with the more environmentally oriented priorities of the United Nations 2030 Agenda. However, close cooperation between public and private actors, the mobilization of finance for RE and energy efficiency measures, and a solid legal framework to guide and coordinate the transition have proven effective in reducing greenhouse gas emissions in China (Zoppolato & Jiang, 2023).

In parallel with North Africa, other MENA countries are also increasingly at the center of China's renewable energy push. The acquisition of Alcazar, an energy company based in the United Arab Emirates with projects in Egypt and Jordan, by a Chinese consortium is worth highlighting. It is seen as an entry into the RE market. In this case, the priority is a shift away from securing energy sources from fossil fuels toward focusing on RE for the near future. Again, this is not only for energy security reasons but also because of the business opportunities that arise from cooperation between large state-owned companies and private companies (Zoppolato & Jiang, 2023).

13.6 Country Focus

The countries studied—Morocco, Algeria, Tunisia, Libya, and Egypt—are rich in natural resources. Their development strategy relies on the export of raw materials. A green energy transition in the region is seen as an opportunity to attract investment and create new jobs, particularly in the renewable energy sector (Table 13.1).

The five countries are analyzed in more detail below.

13.6.1 Morocco

Morocco has been focusing on renewable energies for quite some time. By 2020, almost half of the installed electricity capacity was to be covered by RE. This goal was narrowly missed. Nevertheless, the expansion of wind and solar energy has been massively promoted. Between 2004 and 2022, 20 billion US dollars in green foreign direct investment was attracted. Morocco has particular strengths in RE. The country has hardly any barriers to entry in developing RE projects. China is already working with Morocco on many RE projects, ranging from hydropower to solar and wind power.

The cooperation between the two countries is part of a broader Chinese initiative focusing on South–South technology transfer and China's increased attention to North Africa. Morocco's ambitious goal of achieving an 80% share of RE in its

Table 13.1 Influence of BRI on selected areas of the energy sector

Country	Investments in fossil fuels	Expansion of renewable energies	Energy efficiency	Regional energy cooperation
Morocco	• Expansion of oil and gas production	• Promotion of solar and wind energy	• Introduction of energy-efficient technologies in buildings and industry	• Strengthening cooperation with European partners on the energy transition
	• Modernization of refineries	• Development of green hydrogen projects		
Algeria	• Expansion of the gas infrastructure	• Diversification of the energy sector through the expansion of solar and wind energy	• Improving energy efficiency in the industry	• Strengthening cooperation with European partners on energy supply
	• Promotion of LNG exports			
Tunisia	• Expansion of oil and gas production	• Expansion of solar and wind energy	• Strengthening regional cooperation in the energy sector	
	• Extraction of natural gas as a transitional technology	• Promotion of energy efficiency measures for buildings		
Libya	• Reconstruction of the destroyed oil infrastructure	• Development of solar and wind energy projects, especially in remote areas	• Improving energy efficiency in industry and the building sector	• Strengthening regional cooperation in the energy sector
	• Promotion of oil exports			
Egypt	• Expansion of the gas infrastructure	• Expansion of solar and wind energy	• Improving energy efficiency in industry and the building sector	• Strengthening regional cooperation in the energy sector, especially with the Gulf states
	• Promotion of LNG exports	• Development of green hydrogen projects		

Source: Own representation

energy portfolio by 2050 requires increasing the deployment of RE projects in the country (Zoppolato & Jiang, 2023). Morocco is also a leading contender in the race to build low-carbon hydrogen production and export hubs. The country has entered into several partnerships (including one with the International Renewable Energy Agency) and bilateral agreements.

The necessary legal regulations to authorize the export of energy generated in Morocco are crucial. These requirements can lead to better integration into the regional energy market, a development that interests China. China also expects to gain access to the EU energy market through its partnerships in the MENA region. In this respect, the energy market in Morocco serves as the entry point for China to realize further interests and expand its influence in the global energy market, a strategic move that could have significant implications for the region's energy dynamics.

The Moroccan Agency for Sustainable Energy (MASEN) plays a pivotal role in the country's RE expansion. Initially founded in 2010 as the Moroccan Agency for Solar Development, it regained expanded responsibilities in 2016 and was renamed. As a public–private partnership, MASEN is at the forefront of integrating all major RE projects' development, financing, and management. This includes the flagship Noor Solar Plant in Ouarzazate, the world's largest concentrated solar power plant, which is a testament to Morocco's commitment to sustainable energy.

13.6.2 Algeria

Algeria's BRI activities will begin in 2019. An agreement between Algeria and China sets out the conditions for cooperation between the two countries. However, the agreement is relatively general. Energy and renewable energies do not play a role in it. Instead, it concerns infrastructure development and cooperation in the extractive industry, particularly mining.

Algeria has set ambitious targets for RE in recent years, with plans to meet 27% of its energy needs from renewable sources by 2030. These targets, along with significant legislative reforms, notably removing a restriction that limited foreign companies to hold no more than 49% of the shares in Algerian companies, have created a more attractive environment for foreign direct investment in RE. This policy change in 2020 has particularly attracted interest from China, which has extensive experience and capabilities in solar energy technology (Zoppolato & Jiang, 2023). The potential of Algeria's RE initiatives is a cause for hope in the region's energy future.

Since then, China has partnered with Algerian state-owned enterprises to establish joint ventures in solar energy. One notable example is the construction of a large solar power plant in the Tamanrasset region, which signals a mutual commitment to sustainable development and marks a shift in the BRI's energy focus from fossil fuels to renewable sources. These projects, including the Tamanrasset solar power plant, contribute to Algeria's energy security and aim to fulfill the country's commitments under international climate agreements, including the Paris Agreement.

China–Algeria cooperation under the BRI is a model for RE projects in other MENA countries. Given Algeria's strategic location and significant solar potential, these efforts could contribute to a more extensive RE network across North Africa. However, challenges still need to be addressed, such as financial hurdles, technical requirements, and regulatory frameworks that need continuous improvement to sustain foreign investment. Furthermore, geopolitical tensions and regional stability could affect the consistency and expansion of RE projects under the BRI.

China's engagement in renewable energy in Algeria through the BRI could have a significant ripple effect on neighboring MENA countries. By leveraging Algeria's experience, China could introduce similar renewable energy initiatives across North Africa. This potential regional influence is a cause for optimism about the broader impact of these projects on reducing carbon emissions, improving energy security, and achieving climate targets in the region.

China's significant shift toward renewable energy under the BRI in Algeria is a testament to its commitment to sustainability. This shift, from traditional fossil fuel partnerships to sustainable, long-term energy solutions, not only benefits both countries but also aligns with global sustainability goals. China's leadership in this shift sets a high standard for other countries and underscores its influence in the global energy landscape.

13.6.3 Tunisia

China's strategic alignment with Tunisia's RE sector under BRI is a testament to a broader vision for sustainable energy in North Africa. Since Tunisia's inclusion in the BRI in 2018, a memorandum of understanding has been signed to formalize the cooperation, with RE identified as a critical area. Tunisia's focus on solar energy, where China holds significant expertise, presents substantial opportunities for joint ventures and technology transfers, reinforcing the strategic alignment between the two nations (Trabelsi, 2024).

Tunisia has set ambitious targets for reducing its environmental impact. Specifically, the country aims to reduce emissions by 45% by 2030 and increase the share of RE in its energy mix to 35%. This is an essential step for Tunisia, a country that traditionally relies on imported fossil fuels, and underscores its strong commitment to transitioning to sustainable energy sources. To support these goals, Tunisia has been actively soliciting offers for the construction of solar power plants since 2017, with Chinese firms often leading or partnering on these projects due to their competitive prices, technology, and experience (Zoppolato & Jiang, 2023).

China's partnerships in Tunisia under the BRI extend beyond immediate infrastructure projects to focus on knowledge transfer and capacity building. By establishing partnerships with local companies, China is not only helping Tunisia develop its technical expertise in RE, particularly solar energy technologies but also creating jobs and fostering technical skills among the RE workforce. This cooperation accelerates the adoption of RE and strengthens local industry, creating a positive socioeconomic impact.

A significant increase in climate finance is essential for Tunisia to achieve its ambitious climate goals. China's involvement in BRI projects may facilitate access to climate finance through Chinese banks increasingly financing RE projects worldwide. This access could be crucial for Tunisia and help close the financing gap needed for the country's transition to a low-carbon economy. Furthermore, Tunisia's vision of a green economic model that relies on decarbonization, energy security, and technological innovation aligns well with the BRI's objectives of promoting green, sustainable growth, which aims to foster economic development while minimizing environmental impact.

Despite the clear advantages, Tunisia faces challenges in fully capitalizing on Chinese RE investments. These include regulatory hurdles, the need for improved infrastructure to support large-scale RE facilities, and the political will to prioritize green policies. However, Tunisia's growing cooperation with China is poised to serve as a model for other North African countries exploring RE partnerships under the BRI. With sustained engagement and political support, this partnership could be a beacon of hope in advancing RE adoption in the region and achieving shared climate goals.

13.6.4 Libya

Although relatively limited, China's involvement in RE projects in Libya is a significant development. Despite the difficult situation in Libya after the civil war, an emerging partnership with China has the potential to exert a significant influence on Libya's energy landscape. The country's goals of breaking away from oil dependency and expanding the RE sector are slowly being realized. The 2018–2030 Strategic Plan for Renewable Energy in Libya, which aims to achieve a 22% share of RE in the energy mix by expanding capacity by 500 MW, serves as the basis for these efforts. However, the development of RE in Libya has been delayed due to political instability, ongoing conflicts, and the need to rehabilitate infrastructure (Zoppolato & Jiang, 2023).

Chinese companies, known for their competitive solar technology and infrastructure expertise, are beginning to play a role in Libya's RE sector. Under the BRI, China has expressed interest in RE projects that contribute to Libya's economic reconstruction and views the energy sector as a gateway to improving Libya's economic stability and aligning it with the BRI's long-term goals of regional stability and sustainable development. Among the first project approvals are plans for large solar parks, some involving Chinese companies.

For Libya, cooperation with China on RE is not just a strategic move but a necessary one. It's a step toward reducing dependence on oil, increasing energy security, and building resilient infrastructure that aligns with international climate goals. China's technological and financial resources could be the catalyst that accelerates Libya's RE adoption trajectory. Given China's potential to provide capital and technology for large-scale solar energy projects, this cooperation is beneficial and urgent. Libya's RE transition also aligns with the broader objectives of China's BRI, which increasingly focuses on "green" investments in partner countries, particularly in Africa.

There are still significant obstacles, particularly the need for a stable political environment and formalized diplomatic agreements to ensure consistent project development and the security of foreign investments. Libya's unstable regulatory environment and ongoing risks from past conflicts challenge the creation of solid investment frameworks. Chinese firms approach Libya cautiously and require solid diplomatic agreements and possibly international guarantees for project financing and protection against security risks. These risks include potential disruptions due to political instability, regulatory changes, and security threats.

However, if these hurdles are overcome, China's engagement could be a game-changer for Libya. It could help the country achieve energy diversification, job creation, and environmental goals. China could stabilize Libya's energy sector and economy by leveraging the BRI's influence. Continued cooperation could also promote Libya's integration into broader North African RE networks, which would benefit the region and support global climate goals.

13.6.5 Egypt

China's strategic engagement with Egypt under the Belt and Road Initiative (BRI) is characterized by its comprehensive focus on sustainable energy projects, especially given Egypt's leadership in the North African RE sector. Since formalizing their partnership through a memorandum of understanding in 2016, Egypt and China have pursued ambitious joint ventures to align BRI objectives with Egypt's Vision 2030 for sustainable development. This partnership emphasizes the central importance of RE to Egypt's strategy for curbing carbon emissions and promoting energy security (Raihan et al., 2023).

One of the world's largest solar plants, Benban, not only exemplifies China's commitment to Egypt's RE strategy but also holds the potential to inspire similar projects across the MENA region. Chinese companies have been actively involved in developing the park's infrastructure, providing technology, and managing certain aspects of its operations. This partnership highlights China's commitment to clean energy in Egypt and positions the Benban Solar Park as a model for large-scale RE projects throughout the MENA region.

While the construction of solar manufacturing plants has been delayed, China and Egypt have discussed establishing a solar panel manufacturing plant to meet local and regional demand. This delay is due to specific reasons, but the project is still on track. Once operational, this plant will reduce Egypt's reliance on imported solar technology, strengthen local capacity, and create employment opportunities. Regarding solar energy, it is pointed out that deploying megaprojects is central to the national strategy. China's cooperation is not limited to installation but also focuses on manufacturing and managing existing plants by transferring Chinese expertise. However, a solar factory in Egypt has yet to be realized. Several cooperation, research, and development centers exist between China and Egypt (Zoppolato & Jiang, 2023).

China's investment in Egypt's wind energy sector has brought significant benefits. Projects such as the wind farms on the Gulf of Suez, which benefit from Chinese technology and financing have expanded Egypt's portfolio of RE and helped achieve the 2030 target of 42% RE in the energy mix. The benefits of Chinese technology and financing include demonstrating the value of China's involvement in Egypt's wind energy sector.

Joint research and development centers are designed to promote technological innovation and support Egyptian expertise in sustainable energy. China's contribution to knowledge and technology transfer is crucial to Egypt's ability to independently manage and maintain its RE infrastructure. Collaboration between Chinese institutions and Egyptian universities supports these goals with training programs tailored to energy management, solar technology, and environmental sustainability.

China has significantly financed RE infrastructure in Egypt through BRI funds and the Asian Infrastructure Investment Bank (AIIB), demonstrating the impact of international collaboration in supporting sustainable energy. These investments are helping to alleviate Egypt's financing problems and accelerate large-scale RE projects critical to achieving national and regional sustainability goals.

The synergy between Egypt's Vision 2030 and China's BRI enables Egypt to showcase its progress in RE and attract further investment in the MENA region. China's cooperation with Egypt is often cited as a model for other BRI countries because it balances economic growth with environmental stewardship. This dynamic partnership highlights a shared commitment to reducing fossil fuel dependence, improving energy security, and supporting global climate initiatives. As Egypt continues to draw on Chinese expertise and financing, it is advancing its energy transition and strengthening its role as a regional leader in sustainable energy (Xing et al., 2023).

13.6.6 Comparative Analysis

The following insights are also influential based on the FDI-SST model presented in Sect. 13.3:

China's significant FDI capital inflow (C) holds the potential to revolutionize the energy infrastructure in the MENA region. By directing substantial investments toward RE, such as establishing large solar farms in Egypt's Benban Solar Park, China can equip the MENA countries with the resources and incentives necessary to significantly enhance their RE capacity, sparking optimism for a greener future.

China has become a global leader in solar and wind technology. Through investments under the BRI, it can technology transfer (T) its knowledge to MENA countries and help local companies and technicians develop skills in manufacturing, installing, and maintaining RE infrastructure. This fosters self-sustaining RE sectors in the MENA region over the long term. Zhang et al. (2025) untersuchten BRI-Länder von 2002 bis 2020 und zeigten, dass der Transfer grüner Technologien die

Energieintensität verringern kann. Die Autoren argumentierten, dass grüne Patenttransfers in den Gastländern zu einem Energie-Rebound-Effekt führen können. Trotz der Steigerung der Energieeffizienz nimmt der Rebound-Effekt ab, wenn die Länder eine höhere technologische Reife im Bereich der grünen Technologien erreichen.

Policy alignment (P) plays a crucial role in fostering sustainable investments. The collaboration between China and MENA governments under the BRI agreements can lay the groundwork for common frameworks that support sustainable investments. For instance, the establishment of targets for RE production or the provision of tax incentives for imported green technologies can significantly enhance the effectiveness of FDI and ensure the stability of long-term renewable energy projects. Ein rechtlicher Rahmen ist für die BRI-Länder von entscheidender Bedeutung ist, um RE zu unterstützen (Hussain et al., 2021; Khan et al., 2021). However, the regions have weak network connections and use of carbon pricing and monitoring (i.e., the mechanism for pricing greenhouse gases and the monitoring, reporting, and verification of greenhouse gases). Although support from regional regulators is generally consistent, financial incentives for regulating renewable energies remain low. Carfora et al. (2018) explain that fiscal incentives (capital subsidies, grants, tax credits, sales, and energy rebates) and public financing, such as productive and competitive government grants, are critical to promoting the renewable energy sector.

Yasmeen et al. (2023) examined the role of the quality of governance in BRI economies from 1996 to 2018. They found that effective governance promotes the adoption of renewable energy and reduces the carbon footprint. This research shows that better governance promotes green environmental technology and clean energy consumption through environmental taxes, thus encouraging foreign companies to use carbon capture technologies.

For Viglioni et al. (2025), the effect of FDI on RE policy remains vague, given the prevalence of fossil fuels in many BRI countries. This finding highlights the challenges associated with the transition to clean energy despite political support and the development of low-emission alternatives. Importantly for the authors, the relationship between RE policies and FDI is critical for the transition to RE along the BRI route.

The model used here for selected BRI countries in the MENA region assumes several positive outcomes if FDI is managed effectively:

1. With sustained foreign direct investment in RE, the selected MENA countries are poised to witness a significant increase in the share of RE in their total energy production. This promising trend offers hope for the region's greener and more sustainable energy landscape.
2. As RE reduces dependence on imported fossil fuels, countries in the MENA region can look forward to a more stable energy supply. This reduced vulnerability to global oil price fluctuations and supply disruptions guarantees these countries' future energy security.

3. The transition to RE sources is expected to reduce greenhouse gas emissions and help the selected MENA countries meet their international climate commitments.
4. Economic diversification and resilience are key benefits of RE projects. These initiatives create jobs and stimulate related industries, contribute to economic resilience, and offer long-term economic benefits beyond reliance on fossil fuels.

With the goal of economic growth and securing global influence in the energy sector, countries should focus on technology-based solutions that reduce emissions under challenging sectors, optimize consumption, switch to RE, and implement carbon capture on a large scale. (WEF, 2023).

Based on the model presented and various studies, positive and negative effects for the analyzed MENA countries can be identified, which are summarized in Table 13.2.

The positive effects include transnational infrastructure development in the energy supply sector. For the recipient countries, the BRI has the advantage of financing megaprojects. Even if the projects (except Egypt) are still tiny, it is possible to create jobs. The most significant advantage, however, is the transfer of

Table 13.2 Impact of BRI on the energy sector

Country	Positive effects	Negative Impacts
Morocco	• Expansion of renewable energies	• Dependence on Chinese technologies and investments
	• Technology transfer	• Possible neglect of local companies
	• Creation of jobs	• Ecological crises in some projects
	• Strengthening regional cooperation	
Algeria	• Diversification of the energy sector and the economy	• Dependence on Chinese technologies and investments
	• Expansion of the infrastructure	• Environmental pollution from new projects
	• Promotion of LNG projects	• Risks for the security of energy supply
Tunisia	• Expansion of renewable energies	• Dependence on Chinese technologies and investments
	• Attracting foreign investment	• Indebtedness due to Chinese loans
	• Strengthening energy efficiency	• Social conflicts due to land use changes
Libya	• Reconstruction of the destroyed infrastructure	• Political instability
	• Promotion of oil exports	• Dependence on Chinese technologies
	• Expansion of renewable energies	• Ecological risks in oil production
Egypt	• Expansion of the gas infrastructure	• Dependence on Chinese technologies and investments
	• Promotion of LNG projects	• Water scarcity due to energy-intensive projects (coal-fired power plants) and dams
	• Development of green hydrogen projects	• Social conflicts due to land use changes

Source: Own representation

technology. Environmentally friendly technologies and know-how are introduced relatively quickly in host countries (Hafeez et al., 2019). These projects can contribute to capacity building. Finally, the large number of cooperation projects in the field of energy supply should be emphasized. Almost all countries have such agreements for the expansion of renewable energies and the promotion of energy-efficient technologies.

Tripartite cooperation with international financial and implementation partners can support BRI projects through better access to funding, risk sharing, and knowledge exchange. In particular, non-state enterprises, which often find it more difficult to access investment from large Chinese financial institutions, could benefit from broader access to finance. Chinese financial institutions could also reduce project financing risk by expanding their international cooperation (Nedopil, 2023). From a liberal perspective, political stability is therefore considered essential to ensure the viability of capital-intensive energy infrastructure and to reduce the risk of financing RE technologies (Elfarra et al., 2024).

These advantages are offset by disadvantages frequently cited by the host countries. Firstly, the environmental impact should be emphasized. Infrastructure projects, in particular (including those in energy supply), lead to increased emissions, deforestation, and water pollution. Technology transfer can also lead to dependence on China. There is a risk of economic dependence and even loss of national sovereignty. However, if a country cannot repay the development loans, the poorer countries fall into a debt trap. Competition from Chinese companies can even mean the loss of jobs and the displacement of local companies. Finally, the intensification of existing social inequalities and land grabbing are mentioned negatively.

The argument that China is turning away from the BRI is reinforced by the increase in BRI debt defaults and the decline in infrastructure investment by Chinese banks abroad. The country's economic slowdown means that large infrastructure projects can no longer be the basis for interaction with third countries. It explains the BRI's new focus on investing in "small and beautiful" projects, which are smaller in scale but more sustainable and environmentally friendly. They are easier to finance and implement (Nedopil, 2023). Scalable solar and wind energy investments are feasible, particularly in infrastructure and energy investments, provided local conditions provide the appropriate grids for renewable energy supply. As energy costs for renewables fall, we also see the opportunity to invest in the early phase-out of existing older coal projects, which would make both economic and environmental sense.

China can help BRI countries manage their debt bilaterally and multilaterally in repaying (sovereign) debt for already invested BRI projects, e.g., through debt-for-nature swaps, nature performance bonds, and nature-based solutions (Gebrael et al., 2024). These financial instruments involve exchanging a portion of a country's debt for a commitment to protect nature or achieve specific environmental performance targets. Debt is a significant problem for future growth in many BRI countries. Tackling the debt problem is crucial to provide BRI countries with the necessary fiscal space for future investment. While swapping debt for resources or debt for equity may be beneficial for China in the short term to reduce the debt burden in

BRI countries, these swaps undermine the future domestic growth opportunities of BRI countries. Instead, the relevant Chinese actors and international partners should support a green recovery through multilateral agreements by swapping part of the debt for nature and creating the necessary framework to increase transparency and accountability in using funds (Nedopil, 2023).

In addition to cooperation, expansionary fiscal policy can also help promote the development of RE; in contrast, contractionary monetary policy is associated with reducing the production of renewable energies in 71 BRI countries. Therefore, policymakers in BRI countries should consider applying fiscal policy instruments to enhance the impact of the clean energy transition. For example, the selected BRI countries should promote the creation of favorable financial resources and fiscal incentives to encourage private sector engagement in RE development. However, regarding wind energy, Bega and Lin (2025) emphasize that support initiatives must consider the importing countries' contextual framework. Thus, while more financing solutions should be developed for countries with limited financial resources, care should be taken to strike a balance with stricter lending conditions and economic assessments to mitigate potential risks.

Integrating fiscal policies into RE projects can promote green economic growth and reduce the impact of climate change by creating jobs, increasing energy security, promoting innovation, and supporting long-term prosperity. A stable policy regime effectively drives the expansion of renewable energies, whereas unstable policy frameworks are not conducive to the growth of renewable energies in BRI countries (Li et al., 2025).

Several cooperation mechanisms have been established between China and MENA countries. In addition to bilateral relations, China and MENA countries have actively worked to establish cooperation platforms at the regional level, such as the China-Arab States Cooperation Forum (CASCF) and the China-Gulf Cooperation Council Strategic Dialogue. Both sides recognize clean energy resources such as RE and hydrogen as key to the medium- and long-term development strategies and cooperation patterns. Prioritizing RE will have a decisive impact on deepening relations between the two parties. While China continues to import conventional oil and gas from the region, it is also looking for new markets to expand its economic influence. The need to diversify the economy and mobilize economic resources for economic and social development is particularly crucial in North Africa (Zoppolato & Jiang, 2023).

While the current practice of the BRI may need to be more consistent, there is significant potential for improvement. For instance, host countries that signed a BRI agreement could work closely with China to identify priorities and opportunities for energy access and a clean energy transition. By updating their national climate commitments (NDCs) and creating country-specific climate strategies, they can better align demand and supply, making it easier to link commitments to resources. This stresses the importance of strategic planning and instills a sense of hope and optimism for the future of the BRI.

Overall, China is spreading "green" narratives, which are stories or messages that promote environmental sustainability, and repeatedly declaring its departure from the previous development model, leading to environmental degradation.

However, this rhetoric has yet to improve the implementation of development projects financed by China. The Chinese authorities are still more interested in promoting international infrastructure projects than in making concrete efforts to create a genuinely green BRI. The legal and regulatory framework of the BRI must become multilateral, transparent, and inclusive to pave the way for sustainability and bring it closer to the UN agenda (Lorenzo, 2021). However, China is still heavily involved in financing new projects, some of which hurt the local environment. China needs to be more consistent in regulating overseas development projects. Responsibility for regulation needs to be assigned. Much of the responsibility for environmental protection lies with the companies that build the projects (IRI, 2024).

The WEF (2024) has proposed four country archetypes regarding the political sustainability landscape. The five MENA countries analyzed can be classified on this basis. It can be seen that two countries (Tunisia and Algeria) can be classified as "ambitious developers." These are countries with solid policies and ambitious targets that are actively developing their capabilities, financial instruments, and technological development. Egypt can be characterized as a "pioneering leader," i.e., it has comprehensive sustainability strategies, strong political ambitions, and a mature framework. Libya is considered an "emerging starter," i.e., it is still in the initial phase of developing national strategies and sectoral sustainability goals. Morocco is transitioning to becoming an "ambitious developer" (Fig. 13.1).

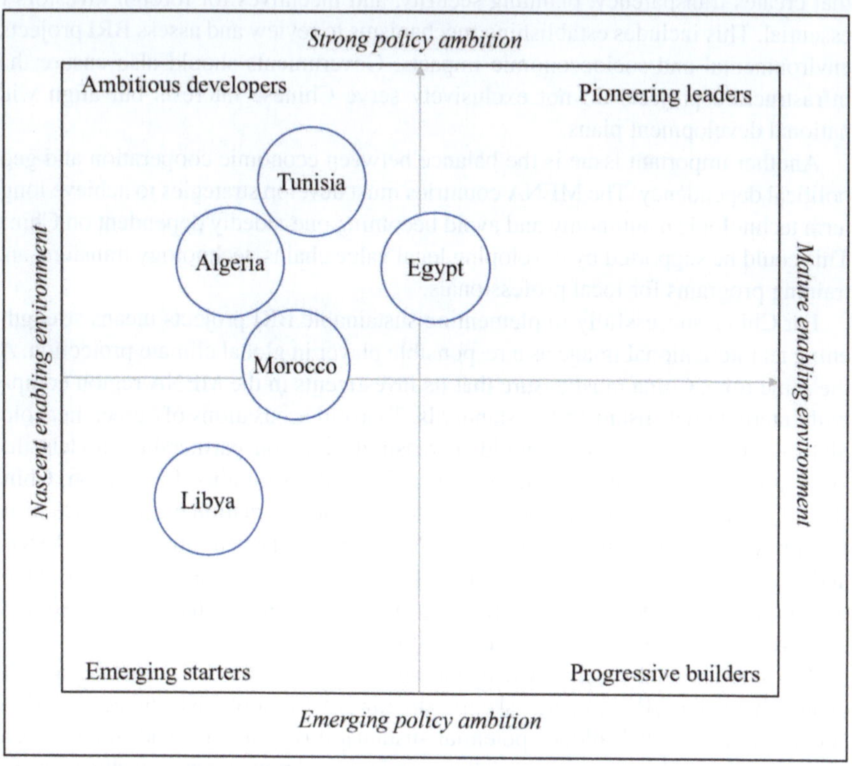

Fig. 13.1 MENA country archetypes about policy landscape parameters

New, more sustainable measures that have not been consistently implemented have slowed down the transition from the BRI to a more environmentally friendly model. However, there is significant potential for positive change. China could accelerate this transition by imposing stricter environmental regulations on companies, requiring them to comply with China's environmental laws when operating overseas. This stress on the potential for positive change should instill a sense of hope and optimism about the BRI's future.

The study provides some policy insights for promoting clean energy development in selected MENA countries of the BRI. It recommends increasing commitment to certain renewable energy policy framework aspects, independent of China's immediate engagement. Countries should move away from fossil fuel policies and toward the long-term benefits of mass renewable energy adoption. The MENA countries analyzed here need stronger renewable energy policies, including carbon pricing and monitoring systems, to reduce dependence on fossil fuels and subsidies. While direct Chinese investment in the analyzed countries is crucial for the start-up phase, developing institutional governance for renewable energies is necessary. In this respect, institution building is important in attracting more foreign direct investment in clean energy projects. MENA countries should jointly address the root causes of barriers to using clean energy to overcome institutional barriers and ensure a developed framework for renewable energies. A clear and consistent energy policy that creates transparency, planning security, and incentives for foreign investors is essential. This includes establishing mechanisms to review and assess BRI projects' environmental and socioeconomic impacts. Governments should also ensure that infrastructure projects do not exclusively serve Chinese interests but align with national development plans.

Another important issue is the balance between economic cooperation and geopolitical dependency. The MENA countries must develop strategies to achieve long-term technological autonomy and avoid becoming one-sidedly dependent on China. This could be supported by developing local value chains, technology transfers, and training programs for local professionals.

For China, successfully implementing sustainable BRI projects means strengthening its international image as a responsible player in global climate protection. At the same time, China must ensure that its investments in the MENA region comply with international sustainability standards. To avoid accusations of "green neocolonialism," Chinese investors should increasingly rely on partnership models that allow greater participation by local companies and institutions. In addition, China should promote multilateral cooperation to increase international acceptance of its BRI projects. Greater integration of the UN Sustainable Development Goals (SDGs) and closer coordination with international organizations such as the World Bank or the International Energy Agency (IEA) could help to ensure that BRI investments are perceived as more sustainable in the long term.

The international community, including the EU and the US, must balance its approach to the BRI. General skepticism toward Chinese investments could be counterproductive if it blocks potential sustainability gains. Instead, international actors should work with China to define standards and transparency mechanisms for sustainable investments. At the same time, alternative financing models for developing countries should be promoted so they do not have to rely exclusively on Chinese investments.

In summary, the increasing interconnection of energy infrastructure with neighboring BRI countries is crucial to leverage institutional funds, promote multilateral trade and investment, and prioritize projects such as wind and solar energy (Li et al., 2024). In this context, Bega and Lin (2025) found that China places a significant strategic emphasis on green energy investment, particularly on transferring cutting-edge wind energy technologies. Ultimately, foreign direct investment and support from Chinese policy banks play a crucial role in financing wind energy infrastructure projects along the BRI.

13.7 Conclusions

In the first decade of the BRI, investments in fossil fuels dominated. Critics argue that the BRI is selective and primarily concerned with China's energy security. China is a global manufacturing center and needs secure, cheap energy for its booming economy. However, in the second decade, there has been a significant shift to "greener financing," which refers to investment in environmentally friendly projects. This shift gives China the power to reshape the future of clean energy, especially in countries where it is needed most while maintaining the energy security strategy.

The FDI-SST model presented here provides a framework for understanding how Chinese FDI in the context of SD can promote sustainable energy development in MENA countries. The model describes the central mechanisms of sustainability in the BRI investment strategy and identifies several key factors:

- Governance structures of the recipient countries: A country's ability to enforce environmental and social standards varies considerably and influences the long-term success of sustainable investments.
- Type of Chinese investment: Direct investment in infrastructure projects with long-term financing models versus short-term capital flows in sectors with a high environmental impact.
- Regulatory and institutional adaptability: The capacity of a country to integrate external investments into existing development plans while safeguarding national interests.

The comparative country analysis empirically tests the model for Morocco, Algeria, Tunisia, Libya, and Egypt. By channeling Chinese FDI into RE projects and supporting them through policy and technology transfer, the selected MENA countries can improve energy security, reduce emissions, and achieve global climate targets, making FDI a catalyst for long-term sustainable development in the region. The results of the country analysis confirm the theoretical assumptions from Chap. 3: While well-established governance structures (e.g., in Morocco and

Tunisia) help to maximize sustainable investments, weak institutional frameworks (e.g., in Libya) have the opposite effect and lead to unsustainable development models. These findings are essential for policymakers, as they show that successful BRI implementation depends not only on financial factors but crucially on governance capacity. A key policy lever, therefore, is to strengthen the regulatory adaptability of MENA countries to optimize the positive effects of the BRI and minimize negative externalities.

The benefit of the BRI for MENA countries is the potential for job and wealth creation, allowing regional governments to manage their dynamic populations and steer their societies in a positive direction. An empirical correlation about the LCTC hypothesis cannot be established in the selected MENA countries. Nevertheless, in the BRI countries, the level of prosperity and energy intensity are the main factors determining energy consumption. The former contributes to the growth of energy consumption. It is relatively more important in BRI developing countries, while the latter has a negative impact, particularly in more developed BRI countries (Cao et al., 2024).

The analysis of BRI investments in sustainable energy projects in the MENA region shows that the initiative presents significant opportunities and challenges. On the one hand, Chinese FDI in RE offers a crucial opportunity to diversify the energy mix of many MENA countries and reduce their dependence on fossil fuels. On the other hand, structural and political obstacles impede these projects' successful and sustainable implementation.

Regarding political realities in the Middle East, some states in the MENA region (e.g., Egypt) are more institutionally oriented and interested in improving their infrastructure. The Gulf states are still in development funds but are seeking political benefits (stability and role as a regional player). Therefore, China's relations with MENA countries must be more balanced, and its strategies (re) must be adapted to these different dynamics, which include economic development, political stability, and regional influence. It would lead to Beijing becoming more actively involved in Middle Eastern politics, challenging the status quo of the established order led by the United States.

The abundance of oil and gas resources and low investment in the RE sector have slowed down the energy transition in the MENA region. However, China's expertise and model for RE industry development could facilitate and drive a dramatic increase in the share of RE in the MENA region, offering a promising future for clean energy.

This study highlights how China's approach to energy has proven successful for the country. China, now considered a leader in the RE sector, is gradually incorporating environmental concerns into its energy policy. China's energy sector aims to build infrastructure to ensure "cheap" energy access, a strategy implemented in the first phase of the BRI. Despite international resistance, the BRI is gradually becoming greener, with the pledge to ban coal and the drastic increase in funding for RE projects being two notable recent signs. This progress inspires hope for a cleaner, more sustainable future.

The analysis shows how the model used in China to develop the RE industry has been adopted in the MENA countries featured. However, implementing such RE sources requires a robust and efficient infrastructure that enables widespread adoption across all sectors of society (Yu et al., 2023), a significant financial burden for countries with more modest incomes (Qi et al., 2023). Large-scale projects and close cooperation between state and non-state actors are at the heart of changes to national legislation in the region to promote RE in the energy mix (Zoppolato & Jiang, 2023).

However, even if China's policy shift is a step in the right direction, realizing a green BRI will likely be easier said than done. Despite China's commitment to greening the BRI and phasing out overseas coal investments, the socio-environmental impacts of projects already planned, under construction, or in operation are likely to persist (Gu, 2024). Added to this are factors such as changing demand in the partner country and the uncertain economic situation in China. In some cases, financial headwinds in the partner countries will also determine whether China can realize a green BRI. Energy cooperation among BRI countries could have significant implications for the global energy market and the achievement of international climate and sustainability goals. Particularly noteworthy is how the BRI may encourage or hinder the transition from fossil to RE sources.

References

Adams, S., & Klobodu, E. K. M. (2017). Capital flows and the distribution of income in sub-Saharan Africa. *Economic Analysis and Policy, 55*, 169–178.

Alnafisah, H., Belaïd, F., Zaied, Y. B., Hussain, N., & Lahouel, B. B. (2024). Assessing the impacts of technological innovation on carbon emissions in MENA countries: Application of the innovation curve theory. *Energies, 4*(17), 904.

Awijen, H., Belaïd, F., Zaied, Y. B., Hussain, N., & Lahouel, B. B. (2022). Renewable energy deployment in the MENA region: Does innovation matter? *Technological Forecasting and Social Change, 179*, 121633.

Awodumi, O. B. (2023). Does foreign direct investment matter for environmental innovation in African economies? *Economic Change and Restructuring, 56*, 237–263.

Bakhsh, K., Rose, S., Ali, M. F., Ahmad, N., & Shahbaz, M. (2017). Economic growth, CO2 emissions, renewable waste and FDI relation in Pakistan: New evidences from 3SLS. *Journal of Environmental Management, 196*, 627–632.

Barro, R., & Sala-I-Martin, X. (1995). *Economic growth*. McGraw-Hill.

Bega, F., & Lin, B. (2023). China's belt & road initiative energy cooperation: International assessment of the power projects. *Energy, 270*, 126951.

Bega, F., & Lin, B. (2025). China's belt & road initiative wind energy cooperation: Wind of change for the Chinese energy finance? *Renewable and Sustainable Energy Reviews, 208*, 115069.

Borensztein, E., De Gregorio, J., & Lee, J.-W. (1998). How does foreign direct investment affect economic growth? *Journal of International Economics, 45*(1), 115–135.

Cao, X., Zhang, Z., Qian, Y., & Wen, Z. (2024). The spatial pattern, driving factors and evolutionary trend of energy cooperation and consumption in the 'belt and road initiative' countries. *Energy, 306*, 132416.

Carfora, A., Pansini, R. V., Romano, A. A., & Scandurra, G. (2018). Renewable energy development and green public policies complementarities: The case of developed and developing countries. *Renewable Energy, 115*, 741–749.

Chen, X. H., Tee, K., Elnahass, M., & Ahmed, R. (2023). Assessing the environmental impacts of renewable energy sources: A case study on air pollution and carbon emissions in China. *Journal of Environmental Management, 345*, 118525.

Chen, Z., Paudel, K. P., & Zheng, R. Z. (2021). Pollution halo or pollution haven: Assessing the role of foreign direct investment on energy conservation and emission reduction. *Journal of Environmental Planning and Management, 65*(2), 311–336.

China Council on Environmental Cooperation and Development (CCICED). (2020). *Green BRI and 2030 agenda for sustainable development*. September 2020. https://cciced.eco/wp-content/uploads/2020/09/SPS-4-1-Green-BRI-and-2020-Agenda-for-Sustainable-Development.pdf

Dussaux, D., Dechezleprêtre, A., & Glachant, M. (2022). The impact of intellectual property rights protection on low-carbon trade and foreign direct investments. *Energy Policy, 171*, 113269.

Elfarra, B., Yasmeen, R., & Shah, W. U. H. (2024). The impact of energy security, energy mix, technological advancement, trade openness, and political stability on energy efficiency: Evidence from Arab countries. *Energy, 295*, 130963.

Erdiwansyah, E., Mamat, R., Sani, M. S. M., & Sudhakar, K. (2019). Renewable energy in Southeast Asia: Policies and recommendations. *Science of the Total Environment, 670*, 1095–1102.

Fang, K., Wang, S., He, J., Song, J., Fang, C., & Jia, X. (2021). Mapping the environmental footprints of nations partnering the belt and road initiative. *Resources, Conservation and Recycling, 164*, 105068.

Garetto, S., Pavcnik, N., Ramondo, N., Alviarez, V., Fan, J., Zandalai-Nayar, N. et al. (2025, February 12). Foreign direct investment and development. *VoxDevLit, 13*(1).

Garlick, J. (2019). *The impact of China's belt and road initiative: From Asia to Europe*. Routledge.

Gebrael, K., Mitri, G., & Kalantzi, O.-I. (2024). Overview of nature-based solutions for climate resilience in the MENA region. *Nature-Based Solutions, 6*, 100159.

Girling, J. L. S. (1973). Dependency and persistent income inequality. In *Structures of dependency* (pp. 83–101).

Grossman, G. M., & Helpman, E. (1991). Trade, knowledge spillovers, and growth. *European Economic Review, 35*(2–3), 517–526.

Gu, B. (2024). Black gold and green BRI–A grounded analysis of Chinese investment in coal-fired power plants in Indonesia. *The Extractive Industries and Society, 17*, 101411.

Gui-Diby, S. L. (2014). Impact of foreign direct investments on economic growth in Africa: Evidence from three decades of panel data analyses. *Research in Economics, 68*(3), 248–256.

Hafeez, M., Yuan, C., Khelfaoui, I., Musaad, A. S., Akbar, M. W., & Jie, L. (2019). Evaluating the energy consumption inequalities in the one belt and one road region: Implications for the environment. *Energies, 12*, 1358.

Hou, J., Deng, X., Springer, C. H., & Teng, F. (2020). A global analysis of CO2 and non-CO2 GHG emissions embodied in trade with belt and road initiative countries. *Ecosystem Health and Sustainability, 6*(1), 1761888.

Hussain, J., Zhou, K., Guo, S., & Khan, A. (2020). Investment risk and natural resource potential in 'belt & road initiative' countries: A multi-criteria decision-making approach. *Science of the Total Environment, 723*, 137981.

Hussain, J., Zhou, K., Muhammad, F., Khan, D., Khan, A., Ali, N., & Akhtar, R. (2021). Renewable energy investment and governance in countries along the belt & road initiative: Does trade openness matter? *Renewable Energy, 180*, 1278–1289.

International Energy Agency (IEA). (2023). *Renewables 2023 analysis and forecast to 2028*. https://iea.blob.core.windows.net/assets/96d66a8b-d502-476b-ba94-54ffda84cf72/Renewables_2023.pdf

International Energy Agency (IEA). (2024). *World energy investment 2023*. https://iea.blob.core.windows.net/assets/8834d3af-af60-4df0-9643-72e2684f7221/WorldEnergyInvestment2023.pdf

International Republican Institute (IRI). (2024). *Environmental implications of the belt and road initiative*. April 2, 2024. https://www.iri.org/resources/environmental-implications-of-the-belt-and-road-initiative/

Jiang, G., Li, Y., Teng, X., Cheng, A., Mao, Y., Chang, T., et al. (2025). Performance improvement of energy extraction and energy utilization to reduce carbon emissions in resource-rich belt and road initiative countries: Applying a meta two-stage non-radial dynamic DEA model. *Energy, 328*, 136479.

Kastratović, R. (2019). Impact of foreign direct investment on greenhouse gas emissions in agriculture of developing countries. *Australian Journal of Agricultural and Resource Economics, 63*(3), 620–642.

Khan, A., Chenggang, Y., Hussain, J., & Kui, Z. (2021). Impact of technological innovation, financial development and foreign direct investment on renewable energy, nonrenewable energy and the environment in belt & road initiative countries. *Renewable Energy, 171*, 479–491.

Latief, R., Pei, Y., Javeed, S. A., & Sattar, U. (2024). Environmental sustainability in the belt and road initiative (BRI) countries: The role of sustainable tourism, renewable energy and technological innovation. *Ecological Indicators, 162*, 112011.

Li, P., He, Q., Zhang, J., & Xiong, J. (2024). Evaluating the impact and heterogeneity of China's OFDI on total-factor carbon emission performance in belt and road initiative countries. *Journal of Cleaner Production, 460*(1), 142607.

Li, Q., Khan, A., Xu, M., & Sun, C. (2025). The regulatory power of fiscal and monetary policies in driving renewable energy: The moderating influence of political stability in belt and road' initiative countries. *Environmental Impact Assessment Review, 110*, 107653.

Lin, B., Bae, N., & Bega, F. (2020). China's belt & road initiative nuclear export: Implications for energy cooperation. *Energy Policy, 142*, 111519.

Liu, H. Y., Kai, Y., Chen, X. L., & Poznanska, J. (2017). The determinants of Chinese outward FDI in countries along one belt one road. *Emerging Markets Finance and Trade, 53*(6), 1374–1387.

Lorenzo, J. A. P. (2021). A path toward sustainable development along the belt and road. *Journal of International Economic Law, 24*, 591–608.

Lu, S., Chen, N., Zhou, W., & Li, S. (2024). Impact of the belt and road initiative on trade status and FDI attraction: A local and global network perspective. *International Review of Economics and Finance, 89*(Part A), 1468–1495.

Marketos, T., Mazzucchi, N., & Alexopoulos, T. A. (2022). *Geostrategic alliances in the Eastern Mediterranean and MENA: A universal paradigm shift*. Springer.

Mu, X., Zhan, O., & Zeng, X. (2024). Can friend-shoring policies improve export resilience? A study based on the belt and road initiative. *International Review of Economics and Finance, 96*(Part A), 103535.

Muhammad, S., Long, X., Salman, M., & Dauda, L. (2020). Effect of urbanization and international trade on CO2 emissions across 65 belt and road initiative countries. *Energy, 196*, 117102.

Nedopil, C. (2023). *Ten years of China's belt and road initiative (BRI): Evolution and the road ahead*. Griffith Asia Institute and Green Finance and Development Center. https://greenfdc.org/wp-content/uploads/2023/10/Nedopil-2023_Ten-years-China-Belt-Road-Initiative-BRI.pdf

Nedopil, C. (2024). *China belt and road initiative (BRI) investment report 2023*. February 05, 2024. https://greenfdc.org/china-belt-and-road-initiative-bri-investment-report-2023/

Noureldin, A. A., Moussa, S., & Muhideen, S. (2024). Policy optimization of the belt and road and IP performance of Middle East and North Africa (MENA). *Research in Globalization, 8*, 100215.

Ofori, I. K., Figari, F., & Ojong, N. (2023). Towards sustainability: The relationship between foreign direct investment, economic freedom and inclusive green growth. *Journal of Cleaner Production, 406*, 137020.

Opoku, E. E. O., & Boachie, M. K. (2020). The environmental impact of industrialization and foreign direct investment. *Energy Policy, 137*, 111178.

Qi, G., Che, J., & Wang, Z. (2023). Differential effects of urbanization on air pollution: Evidences from six air pollutants in mainland China. *Ecological Indicators, 146*, 109924.

Qi, S.-Z., Peng, H.-R., & Zhang, Y.-J. (2019). Energy intensity convergence in belt and road initiative (BRI) countries: What role does China-BRI trade play? *Journal of Cleaner Production, 239*, 118022. https://doi.org/10.1016/j.jclepro.2019.118022

Raihan, A., Ibrahim, S., & Muhtasim, D. A. (2023). Dynamic impacts of economic growth, energy use, tourism, and agricultural productivity on carbon dioxide emissions in Egypt. *World Development Sustainability, 2*, 100059.

Raihan, A., Ibrahim, S., Ridwan, M., Rahman, M. S., Bari, A. B. M. M., & Atasoy, F. G. (2025). Role of renewable energy and foreign direct investment toward economic growth in Egypt. *Innovation and Green Development, 4*(1), 100185.

Rauf, A., Liu, X., Amin, W., Rehman, O. U., Li, J., Ahmad, F., et al. (2020). Does sustainable growth, energy consumption and environment challenges matter for belt and road initiative feat? A novel empirical investigation. *Journal of Cleaning Production, 262*, 121344.

Ravallion, M. (2018). Inequality and globalization: A review essay. *Journal of Economic Literature, 56*(2), 620–642.

Raza, M. S., Wang, Y., Rauf, A., Aziz, N., Khan, M. A., & Hussain, A. (2024). Unveiling the green horizon: A bibliometric analysis of global foreign direct investment research and its emphasis on climate change. *Journal of Cleaner Production, 446*(25), 141338.

Romer, P. M. (1994). The origins of endogenous growth. *Journal of Economic Perspectives, 8*(1), 3–22.

Schulhof, V., van Vuuren, D., & Kirchherr, J. (2022). The belt and road initiative (BRI): What will it look like in the future? *Technological Forecasting and Social Change, 175*, 121306.

Solow, R. M. (1956). A contribution to the theory of economic growth. *The Quarterly Journal of Economics, 70*(1), 65–94.

Sun, Y. (2020). Low-carbon financial risk factor correlation in the belt and road PPP project. *Finance Research Letters, 35*, 101491.

Sun, Y., & Yu, B. (2023). Greening China's belt and road initiative: From norm localization to norm subsidiarity? *Global Environmental Politics, 23*(1), 91–116.

Tiwari, A. K., Nasreen, S., & Anwar, M. A. (2022). Impact of equity market development on renewable energy consumption: Do the role of FDI, trade openness and economic growth matter in Asian economies? *Journal of Cleaner Production, 334*(1), 130244.

Trabelsi, E. (2024). Transition to sustainable environment and economic growth in Tunisia: An ARDL approach. *World Development Sustainability, 4*, 100165.

Tsai, P.-L. (1995). Foreign direct investment and income inequality: Further evidence. *World Development, 23*(3), 469–483.

Ullah, V., Nobanee, H., Ullah, S., & Khan, S. (2025). Addressing climate change and advancing green development: Regional transition mechanisms for a sustainable future in belt and road initiative nations. *Sustainable Development, 33*, 1–29.

Viglioni, M. T. D., Calegario, C. L. L., & Ferreira, M. P. (2025). Exploring the link between renewable energy policy and foreign direct investment in promoting renewable energy in belt and road initiative countries. *Renewable Energy, 246*, 122915122915.

Wang, B., & Lin, P. (2022). Whether China's overseas energy infrastructure projects dirtier or cleaner after the belt and road initiative? *Energy Policy, 166*, 113007.

Wang, B., Yang, H., Bi, C., & Feng, Y. (2023). Green infrastructure and natural resource utilization for green development in selected belt and road initiative countries. *Resources Policy, 85*, 103758.

Wang, L., & Cheng, Z. (2024). Impact of the belt and road initiative on enterprise green transformation. *Journal of Cleaner Production, 468*, 143043.

Wong, S. Y. S., Kong, J. C. S., Luo, G. W. T., & Lau, N. H. T. (2021). An integrated approach to sustainable infrastructure standards for the belt and road initiative. In F. M. Cheung & Y. Hong (Eds.), *Green finance, sustainable development and the belt and road initiative* (pp. 285–309). Routledge.

World Economic Forum (WEF). (2023). *Closing the climate action gap: Accelerating decarbonization and the energy transition in MENA*. https://www3.weforum.org/docs/WEF_Closing_the_Climate_Action_Gap_2023.pdf

World Economic Forum (WEF). (2024). *Prioritizing sustainability in MENA: Mapping critical environmental issues for regional businesses.* https://www3.weforum.org/docs/WEF_Prioritizing_Sustainability_in_MENA_2024.pdf

World Resources Institute (WRI). (2023). *China overseas finance inventory database.* https://doi.org/10.46830/writn.21.00003

Xiao, Z., & Qamruzzaman, M. (2022). Nexus between green investment and technological innovation in BRI nations: What is the role of environmental sustainability and domestic investment? *Frontiers in Environmental Science, 10*, 993264.

Xing, Y., Liub, Y., Cooper, S. C. L., & Vrontis, D. (2023). Reviving China's global footprint along the silk roads and the 'belt and road initiative': Chinese overseas industrial park in Egypt. *Business History, 67*, 318. https://doi.org/10.1080/00076791.2023.2233426

Xu, L. (2023). Role of belt and road initiative in the development of renewable energy: Mediating role of green public procurement. *Renewable Energy, 217*, 118963.

Yasmeen, R., Hao, G., Yusen, Y., Shah, W. U. H., & Kamal, M. A. (2023). The role of governance quality on mobilizing environmental technology and environmental taxations for renewable energy and ecological sustainability in belt and road economies: A methods of moment's quantile regression. *Energy Strategy Reviews, 50*, 101258.

Ye, C., Feng, H., & Cao, W. (2024). Does the belt and road initiative reduce the green logistics efficiency of partner countries? *Ecological Indicators, 163*, 112133.

Yu, J., Liu, P., Fu, D., & Shi, X. (2023). How do power shortages affect CO2 emission intensity? Firm-level evidence from China. *Energy, 282*, 128927.

Yusoff, N. Y. M., Aris, H., Nasiruddin, W. H. M., Rusli, A. K. C., Yurnaidi, Z., et al. (2024). The driving forces behind renewable energy consumption: A comprehensive analysis of key determinants and environmental concerns in ASEAN countries. *Environmental Science and Pollution Research, 31*, 48388–48405.

Zhang, B., Tian, X., He, B., Xin, Q., & Yang, Z. (2025). The impact of China's green technology transfer on energy intensity in countries along the belt and road. *Journal of Environmental Management, 373*, 123691.

Zhang, C., Fu, J., & Pu, Z. (2019). A study of the petroleum trade network of countries along 'the belt and road initiative'. *Journal of Cleaning Production, 222*, 593–605.

Zhang, N., Farooq, M. U., & Zhang, X. (2023). Studying green financing with economic development in BRI countries perspective: Does public–private investment matter? *Environmental Science and Pollution Research, 30*, 29336–29348.

Zhao, Y., Liu, X., Wang, S., & Ge, Y. (2019). Energy relations between China and the countries along the belt and road: An analysis of the distribution of energy resources and interdependence relationships. *Renewable and Sustainable Energy Reviews, 107*, 133–144.

Zhou, L., & Ma, Z. (2023). *After a decade of fossil fuel investing, can China fulfill its promise of a green belt and road initiative?* October 24, 2023. https://www.wri.org/insights/greening-chinas-belt-and-road-initiative

Zoppolato, D. G., & Jiang, S. (2023). China-MENA energy cooperation under the belt and road initiative: Megaprojects, economic planning, and a pragmatic approach to the 'green' transition. *Journal of World Energy Law and Business, 16*, 143–159.

Chapter 14
Board of Directors Compensation and Environmental Impact: A Study of Green and Brown MENA Firms

Alfred Wong and Alain Neher

Abstract This chapter explores the role of board of directors (BoD) compensation in managing the environmental impact experienced by firms in the Middle East and North Africa (MENA) region. We document the differences in environmental impact between Green and Brown firms and the extent to which Green firms are associated with lower environmental damage through higher BoD compensation relative to Brown firms. We find that for a 10% increase in BoD compensation by a Green firm, energy use decreases between 3.3% and 6.3% relative to Brown firms. Consistent with previous findings, our results suggest that BoD compensation plays a critical role in sustainable development in the MENA region. More specifically, Green firms are more likely to integrate compensation incentives as part of their sustainable strategies compared to Brown firms. Nevertheless, ethical thoughtfulness and reasonableness must be considered when creating BoD remuneration packages. Managers of MENA Brown firms should aim to adopt innovative approaches to tackle environmental challenges using existing or emerging technologies and processes. Collaboration and collective investments from the industry, as well as the private and public sectors, can help support these sustainability innovations in bringing the United Nations Sustainable Development Goals to fruition.

Keywords MENA · ESG · Environmental · Board of directors · Compensation · CO_2 emission · Green firm · Brown firm

14.1 Theoretical Background and Importance of the Study

The MENA region is a central global hub for exporting oil, natural gas, and solid minerals, containing some of the largest reserves in the world. It is home to nearly half of the world's proven natural gas reserves, the largest share of proven oil

---

A. Wong (✉) · A. Neher
School of Business, Charles Sturt University, Bathurst, NSW, Australia
e-mail: awong@csu.edu.au

M. Elnahass et al. (eds.), *A Greenprint for Net Zero*, Gulf Studies 15,
https://doi.org/10.1007/978-981-92-1762-5_14

reserves, and almost one-third of global mineral reserves (Olawuyi, 2021). This vast resource wealth has positioned the MENA region as a key player in global extractive industries, contributing a substantial share of the world's production and supply. As a result, the extractive sector in the MENA region holds significant potential to drive economic growth and social development, particularly in the region's low- to middle-income countries (Aoun, 2013; Luomi, 2011). The abundance of these natural resources provides a unique opportunity to transform the economic landscape, fostering sustainable development and improving living standards for millions of people across the MENA region and beyond.

By generating employment and stimulating economic opportunities, extractive industries have been instrumental in reducing poverty across MENA societies (Hertog, 2013; Kaya & Tsai, 2016). This contribution aligns with the United Nations Sustainable Development Goals (SDGs), particularly SDG 8, which focuses on promoting decent work and economic growth, and SDG 1, which aims at eradicating poverty (United Nations, 2015). The industry's impact on these goals underscores its importance in advancing sustainable development and improving the socio-economic conditions of the MENA region.

However, over the past decade, there has been a significant global shift away from carbon-intensive fossil fuels, which have historically been the economic foundation of many MENA countries. Numerous studies, including reports from the Intergovernmental Panel on Climate Change (IPCC), have identified fossil fuel combustion as the largest human source of greenhouse gas (GHG) emissions (IPCC, 2014). As a result, global energy policies and practices have increasingly prioritised reducing fossil fuel consumption, enhancing energy efficiency, and advancing the development of low-carbon technologies (Olawuyi, 2018). This transition reflects a broader commitment to mitigating climate change and reducing the environmental impact of energy production.

The global push towards a low-carbon energy system presents critical policy and practical challenges for resource-dependent MENA societies. Various studies have explored the potential threats that the global energy transition could pose to the long-term economic prosperity and development of MENA economies and societies (Olawuyi, 2021). These studies have identified that, without comprehensive economic diversification, the reduction of energy subsidies, and a move away from resource dependency, the global energy transition could lead to severe disruptions in MENA economies (e.g. El-Katiri & Fattouh, 2017; Tagliapietra, 2019). However, much of this discourse has largely overlooked the potential for MENA's extractive industries to play a pivotal role in facilitating the global energy transition. Hence, the implications of the global energy transition for MENA countries require a shift in focus. Instead of dwelling on potential economic disruptions, there is a need to embrace the global push for decarbonisation and the development of clean technologies (Olawuyi, 2021). This transition offers MENA countries a unique opportunity to adjust their energy systems and policies, positioning themselves as key contributors to the global energy transition.

MENA's extractive industries, particularly its companies, have a crucial role to play in this transition. With substantial reserves of natural gas—a resource that emits less GHG and is more environmentally preferable than other fossil fuels—these industries can significantly contribute to accelerating low-carbon energy systems worldwide (IEA, 2019). By leveraging their natural gas reserves and other less GHG-intensive resources, MENA economies can help drive the global transition towards a more sustainable energy future while also mitigating the economic risks associated with the decline of traditional fossil fuels (Olawuyi, 2021). This strategic shift towards promoting sustainability could ensure the region's continued relevance in global energy markets and open new avenues for economic growth and development.

Governance is pivotal in advancing sustainability in the MENA region, where effective policies and regulatory frameworks are essential for steering the transition towards sustainable development. Strong governance is crucial for integrating sustainability into national agendas, ensuring that development strategies align with global sustainability goals such as the SDGs (Eccles et al., 2014; Kilani & Babalola, 2020). Furthermore, governance plays an essential role not only at the national level but also within companies, particularly in driving corporate social and environmental sustainability (United Nations Global Compact, 2014), such as ESG (environmental, social, governance). Effective corporate governance structures contribute to embedding sustainability into companies' core business strategies. By establishing clear guidelines, accountability mechanisms, and ethical standards, governance frameworks within companies ensure that sustainability is given ample attention alongside profitability (Eccles et al., 2014). This includes setting sustainability goals, monitoring environmental and social impacts, and engaging stakeholders in decision-making. In the MENA region, where many businesses depend heavily on oil, gas and extractive industries, corporate governance can facilitate the transition towards more sustainable practices by encouraging investments in clean technologies and cleaner production methods, low-carbon energy infrastructure projects, resource efficiency, corporate social responsibility initiatives, and implementing low-carbon and renewable energy policies (Kilani & Babalola, 2020; Olawuyi, 2021). Moreover, governance within companies can promote ethics and fairness, drive transparency and accountability, and report on sustainability performance, enhancing corporate reputation and attracting investors, who increasingly prioritise ESG factors (Jamali et al., 2008). Hence, governance can be an enabler of corporate sustainability, aligning business strategy and operations with broader societal and environmental goals and contributing to the shift towards long-term perspectives, resilience, and success for companies in the MENA region.

Likewise, corporate governance is increasingly becoming more critical at the board level for companies. For example, the European Commission's 'Study on directors' duties and sustainable corporate governance' investigates how the incorporation of sustainability competencies into board operations might affect board composition and compensation (European Commission, 2020). Previous research on the compensation of board of director members has examined various factors, including effectiveness (Ang et al., 2002), enhancing corporate performance (e.g. Andreas et al., 2012; Lemma et al., 2020), firm valuations (Elnahass et al., 2022b), wealth maximisation (Brick et al., 2006), ESG disclosure (e.g. Suttipun, 2021), and

ESG ratings (Waterstraat et al., 2021). Recently, ESG metrics have been integrated into executive compensation (e.g. Cohen et al., 2023; Ratti et al., 2023). Only a few studies have further examined the relationship between ESG performance and board compensation (E-Vahdati et al., 2023; Wong & Neher, 2024). While E-Vahdati et al. (2023) investigated Japan and five member countries of the Association of Southeast Asian Nations (ASEAN), Wong and Neher (2024) focused their study on European firms. Both studies identified a clear relationship between Board of Directors (BoD) compensation and ESG performance. Some differences occurred among these regions. For example, while Japan demonstrated a strong connection to governance, the ASEAN countries were closely linked with the Social pillar. In Europe, the degree of connection between overall ESG and BoD compensation varied across the southern, eastern, northern, and western regions. Nevertheless, the link between board compensation and environmental impact remains under-researched. Consequently, this chapter seeks to fill this gap by investigating how BoD compensation influences the environmental impact of firms, with a particular focus on the MENA region.

Board compensation is pivotal in guiding companies' strategic direction and sustainability practices, with significant implications for corporate governance and long-term performance (Eccles et al., 2014; Wong & Neher, 2024), especially in Green and Brown firms. In Green firms, which put emphasis on environmental sustainability, compensation packages for board members are often incorporated, with incentives linked to ESG metrics (Eccles et al., 2014; Koors et al., 2019; Wong & Neher, 2024). Such incentives are intended to align the financial interests of BoD members with sustainability objectives, encouraging them to invest in long-term projects, such as renewable energy adoption, emissions reduction, and resource efficiency improvements (Eccles et al., 2014; Mahoney & Thorne, 2005). By integrating ESG criteria into BoD and executive compensation, Green firms signal their commitment to sustainable development and green innovations, which can enhance corporate reputation, stakeholder trust, and long-term financial performance (Flammer et al., 2019).

In contrast, Brown firms, characterised by higher environmental impacts due to their reliance on carbon-intensive operations, often structure board compensation around traditional financial performance metrics (Eccles et al., 2014), such as earnings per share and return on equity. Such financial incentives may lead to short-term profit maximisation at the expense of environmental sustainability (Eccles et al., 2014; Francoeur et al., 2017; Ikram et al., 2023). However, the increasing global pressure on sustainability and sustainability practices and the growing influence of investors advocating for responsible behaviour are also increasingly pressuring Brown firms to incorporate ESG factors into their compensation frameworks (Francoeur et al., 2017; Ioannou & Serafeim, 2019; Maas, 2018). This shift reflects a broader recognition that sustainable business practices are ethically vital for mitigating risks associated, for example, with regulatory changes and reputational damage (Ikram et al., 2023). As such, the evolution of board compensation in Brown firms towards incorporating ESG criteria is crucial for enabling these companies to transition to more sustainable operations and contribute positively to global efforts to combat climate change.

14.2 Research Questions

Our first research question concerns the extent of environmental impacts of MENA firms and how they may relate to BoD for Brown and Green firms. To address this question, first, we follow the approach of Hartzmark and Shue (2023) to summarise the selected MENA firms by ranking them using the emissions data collected for firms included in the sample. Firms are categorised into five different groups according to the size of their CO_2 emission. After adjusting for firm asset size, we examine the mean and median for BoD compensation and senior executive compensation to understand how they relate to CO_2 emissions. Our second research question attempts to answer, 'How does the level of board compensation impact the environmental performance of MENA firms?' Our analysis controls for a range of firm-specific factors, including financial and governance variables that may relate to the level of emissions recorded by each firm.

14.3 Methodological Approach

This section describes the data collection process, sampling and empirical research design. Following the methodology used in recent studies (Almaqtari et al., 2023; Hartzmark & Shue, 2023), we adopt a quantitative approach that includes descriptive and inferential statistics that align with the research aims of this study.

14.3.1 Data Collection, Sampling and Grouping

Secondary data sourced from LSEG (formerly known as Refinitiv) are used in this study. We obtained greenhouse gas emissions, energy consumption, governance and financial data from firms in various industries, as reported in the fiscal year 2023.[1] The sample of firms is headquartered in the MENA region, which consists of 22 countries according to the guidelines provided by the IMF (2023). This resulted in a total of 1655 publicly listed firms in the MENA region in our initial sample, which includes firms from Bahrain (n=39, 2.4%), Egypt (n=271, 16.4%), Iraq (n=108, 6.5%), Jordan (n=187, 11.3%), Kuwait (n=143, 8.6%), Lebanon (n=9, 0.5%),

[1] Nearly all firms recorded their fiscal year end on 31 December 2023, while a few firms end on 30 June 2024.

Morocco (n=75, 4.5%), Oman (n=115, 6.9%), Palestine (n=50, 3.0%), Qatar (n=54, 3.3%), Saudi Arabia (n=335, 20.2%), Sudan (n=2, 0.1%), Syria (n=28, 1.7%), Tunisia (n=83, 5.0%), and the United Arab Emirates (n=156, 9.4%). Seven other MENA countries (Algeria, Djibouti, the Islamic Republic of Iran, Libya, Mauritania, Somalia, and Yemen) were not represented as no data were available for these countries due to either a lack of reporting or failure to meet the data selection criteria.

To ensure representative sampling, we compared the total number of firms in our initial sample with the total number of listed domestic companies provided by the World Bank (2024). To ensure comparability, we only included firms located in the MENA region according to the definition of the IMF country groupings (IMF, 2023). Appendix 1 provides a list of countries included in MENA region. A total of 1882 listed domestic firms were identified in 2022[2] using the World Bank database, consistent with our initial sample. The Industry Classification Benchmark (IBC) is used to broadly classify firms into different industries for analysis (see FTSE Russell, 2024 for further details). Similar to Hartzmark and Shue (2023), the sample of firms is grouped into quintiles according to the size of CO_2 emissions per year. Firms in quintile 1 recorded the lowest emissions, representing the 'Green' firms, while those with the largest emissions in quintile 5 are referred to as 'Brown' firms. Firms that fall within quintiles 2–4 are referred to as 'Neutral' (see Results section).

14.3.2 Measure of Location

Total board compensation data for the MENA firms and the selected independent variables exhibit a skewed and kurtotic distribution. In most instances, the null hypothesis of the normal distribution is strongly rejected at the 1% level of statistical significance using the Kolmogorov-Smirnov test statistic. When the standard normality assumption for interval data is not met, the distribution-free approach (Keller, 2017) is the appropriate measure to determine if two population locations are the same. Accordingly, the nonparametric Mann–Whitney U test (Sekaran & Bougie, 2016; Zikmund, 2003) was used to test group differences between Brown and Green firms for the selected variables. To apply the statistical analysis, the two samples of firms are combined, and the entire dataset, including values from both Green and Brown firms, is ranked in ascending order. The ranks are summed for Green and Brown firms, and the U test statistic is calculated, respectively:

$$U_B = n_G n_B + \frac{n_B(n_B + 1)}{2} + T_B$$

[2] https://databank.worldbank.org/. See indicator name 'Listed domestic companies, total'. Data for 2023 were not available at the time of writing.

$$U_G = n_G n_B + \frac{n_G(n_G+1)}{2} + T_G$$

where U_B is the Mann–Whitney U statistic for Brown firms, while U_G is the corresponding statistic for Green firms, n_G and n_B represent the sample size for Green and Brown firms, respectively, and T_G and T_B are the sum of ranks for the Green and Brown firms, respectively. To test for any statistical difference between the two categories of firms, the lower of the two Mann–Whitney U statistics is reported. The computed U statistic is transformed into a normal z-score by subtracting its expected value of U and dividing by the standard error of U, which is corrected for rank ties within ranks.

The nonparametric statistical tests assess the null hypothesis that the sums of the rank data for Green and Brown firms are equal, as there should be no tendency for one group to exhibit a higher or lower rank than the other if the samples are from the same population. A statistically large value of U would provide evidence of inequality between Green and Brown firms. Furthermore, because the ranks are assigned in ascending order, the group of firms with a higher mean rank will have larger values. A statistical difference between the two groups can be concluded when the p-value associated with the z-score is <0.05 using a two-tailed test.

14.3.3 Regression Analysis

To better understand the environmental impact of Green or Brown firms, we also consider regression analysis to determine how much of the variation in environmental impact can be explained by the board of directors' compensation, $BoDComp_i$ for Green or Brown firms. The inclusion of $BoDComp_i$ is largely motivated by the findings of Wong and Neher (2024), which suggest a positive association between the size of BoD compensation and environmental performance. To this end, our empirical model also regresses the proxy for environmental impact on a range of firm-specific control variables, including financial and corporate governance variables. The main ordinary least squares (OLS) model takes the following form:

$$EnvImp_i = \alpha + \beta_1 \log(BoDComp_i) \times Green\ Firm_i + \beta_2 \log(BoDComp_i) + \sum_{l=1}^{m} \delta_l Control\,Var_{l,i} + \varepsilon_i$$

where the dependent variable, $EnvImp_i$ is a proxy of environmental impact for firm i and m=7 for the control variables. Following Boffo et al. (2020), the log of total energy used (in gigajoules) by individual firm i for the fiscal year 2023[3] is

[3] Firm-level environmental data in the MENA region is relatively sparse compared with Europe. For the former, data prior to 2023 are largely incomplete or not reported.

chosen for the dependent variable to capture the total direct and indirect energy use by Green or Brown firms on the environment. High energy consumption will result in negative impacts on the environment. The main independent variables are the terms involving $BoDComp_i \times Green\ Firm_i$ and $BoDComp_i$. Importantly, the key variable of interest is the dummy variable $Green_i$, which identifies whether the firm is Green or Brown. Because Green = 1 when a firm is in the first quintile (lowest CO_2 emission) and Brown = 0 when a firm is in the fifth quintile (highest CO_2 emission), the parameter β_1 measures the difference in energy consumption between Green and Brown firms when interacting with the variable $BoDComp_i$. Thus, the coefficient determines whether there is a positive environmental impact for Green firms relative to Brown firms when $BoDComp_i$ is varied. If β_1<0, then *ceteris paribus*, at a higher level of BoD compensation, Green firms will produce a lower environmental impact than Brown firms.

We control for the potential impact of other confounding factors by incorporating seven additional firm-specific variables into our model—first, the effect of firm size. Firm size is expected to be an important driver of energy use as larger firms tend to consume more energy than smaller firms. Therefore, the environmental impact arising from a larger firm will negatively affect the environment more than that of a smaller firm. Total revenue generated for the fiscal year is used as a proxy for firm size, following Drempetic et al. (2020). Additionally, any sustainability innovations are more likely to be achieved by large firms than their smaller counterparts (Hörisch et al., 2015). Second, market capitalisation is used as a measure of the aggregate value of the company, reflecting shareholder value. Malik (2014) argued that firms with good environmental and social practices tend to generate increased stock returns, leading to higher firm value. Third, a firm's profitability measured by return on assets (ROA) is known to be positively related to a firm's environmental score (Aydoğmuş et al., 2022), consistent with the stakeholder theory. Fourth, we used the cost of capital. Hartzmark and Shue (2023) asserted that sustainable investors tend to reward firms that generate positive environmental impacts with lower capital costs and punish firms with negative environmental impacts with higher capital costs. The cost of capital is proxied by the weighted average cost of capital, which comprises a mix of debt and equity. Fifth, governance control variables were utilised. Since sustainability strategies are designed to align with the BoD's vision (Welch & Yoon, 2023), the governance structure within a firm could drive the effective governance of environmental management. Accordingly, we included the size of the BoD at the end of the fiscal year; board independence, measured by the proportion of independent directors relative to the total number of directors (in percentage); and board gender diversity, measured by the percentage of BoD seats held by female directors. However, Arayssi and Jizi (2019) noted that the corporate governance system in the MENA region is less robust than that of its Western counterpart. Ownership is still concentrated compared to Western firms, and the BoD of MENA firms is mainly dominated by the state or family, inhibiting the role and effective functioning of the BoD. Log transformation is applied to the dependent variable and selected independent variables to alleviate the effect of skewness on the selected variables.

In addition, we addressed data-related issues, including the small sample size and robustness of our test study using the bootstrap quantile regression. Since a large sample size is necessary for obtaining an unbiased estimate of the true population characteristic (Keller, 2017), the estimated OLS standard errors may not accurately approximate the true sampling variation of an estimator due to the limited sample size. This is because the standard error varies inversely with the square root of the sample size. To address this concern, we also rely on a resampling method known as the nonparametric bootstrap (Greene, 2018) to validate our main results. More specifically, the bootstrap procedure can reduce the finite-sample bias induced by a small sample size (Horowitz, 2001). However, a possible issue with the OLS method using the bootstrap procedure is the high proportion of outliers in the resampled dataset (Salibian-Barrera & Zamar, 2002). Accordingly, the use of bootstrap quantile regression can help alleviate the methodological difficulties associated with small sample sizes and violations of the OLS assumptions (Nikitina et al., 2019). For a comprehensive description of all variables used in the nonparametric and regression analyses, see Appendix 2.

14.4 Empirical Analysis

In this section, we will first assess the selected MENA Green and Brown firms by industry category and region. This will be followed by an empirical examination of firm-specific characteristics that differentiate Green firms from Brown firms using the nonparametric Mann–Whitney U test statistic. Finally, we study the causal relationship between Green or Brown firms, BoD compensation, and the environmental impact using the proposed regression model. Hence, we intend to provide a comprehensive overview of Brown and Green firms that will inform market participants and policymakers about the environmental impacts of these firms in the MENA region.

14.4.1 Preliminary Analysis

Figure 14.1 provides a graphical display of MENA countries by industry using the IBC grouping constructed from a frequency table of the selected firms. Of the 11 key IBC industries, the 3 major industries are Financials (28.1%), followed by Industrials (17.5%) and Consumer Discretionary (12.0%). A similar pattern can be noted when the same analysis is separated into two main regions in Fig. 14.2, namely ME and NA. However, the ME region has a relatively higher proportion of Financials (30.8%) than NA, which has 9% fewer financial firms. These industries can be further divided into 45 ICB sectors. For example, in the MENA region, Financials includes mainly investment banking and brokerage services (34.7%), non-life insurance (28.6%), and banks (26.7%).

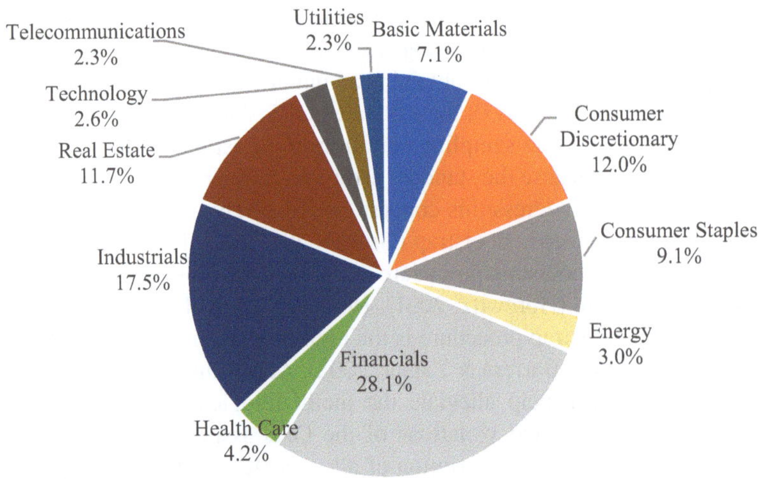

Fig. 14.1 Grouping of firms by industry: MENA region

Ranked by firm size (FY2023), Financials appears to be the largest industry in the MENA region, with an average asset size of USD33.9 billion. This is followed by the Energy sector, with an average asset size of USD32.8 billion, while Utilities has an average asset size of USD22.5 billion. Notably, the Energy sector appears to occupy only 3% of firms in the MENA region (n=43). However, these are very large firms with an average total asset size of more than USD30 billion and are involved in large-scale oil, gas, and coal operations in the MENA region. Overall, Brown firms are 2.43 times larger than Green firms measured by total assets.[4]

Figure 14.3 ranks the relative frequency of the selected sample grouped by Green and Brown firms. Evidently, Green firms consist mostly of Financials (71.6%), which sharply contrasts with the low proportion of Brown firms that operate in the financial industry (4.4%). Approximately one-third of Brown firms operate as Industrials (33.8%), followed by Basic Materials (22.1%) and Energy (11.8%). Figure 14.4 presents the same information, but further decomposes the firms by the ME and NA regions. As before, Green firms are mostly Financials located in the ME, while Brown firms mostly operate as Industrials (33.8%), Basic Materials (22.1%), and Energy (11.8%) industries in the ME countries.

To gain further insights into the environmental impacts resulting directly from sources that are controlled or owned by the selected firms, we employ Scope 1 greenhouse gas (GHG) emissions (WRI, 2004) data for the sample of MENA firms that reported a complete emissions record from 2019 to 2023. We ranked the firms by Scope 1 emissions (total estimated Scope 1 emissions in tonnes divided by net sales or revenues in million USD) and summarised the data using the ICB sector grouping to decompose the firms into industry subgroups. Figures 14.5 and 14.6 present the five lowest and highest emitters by industry sectors. As shown in

[4]The median total assets for Brown firms was USD6.74 billion, while Green firms recorded a median of USD2.77 billion in FY2023.

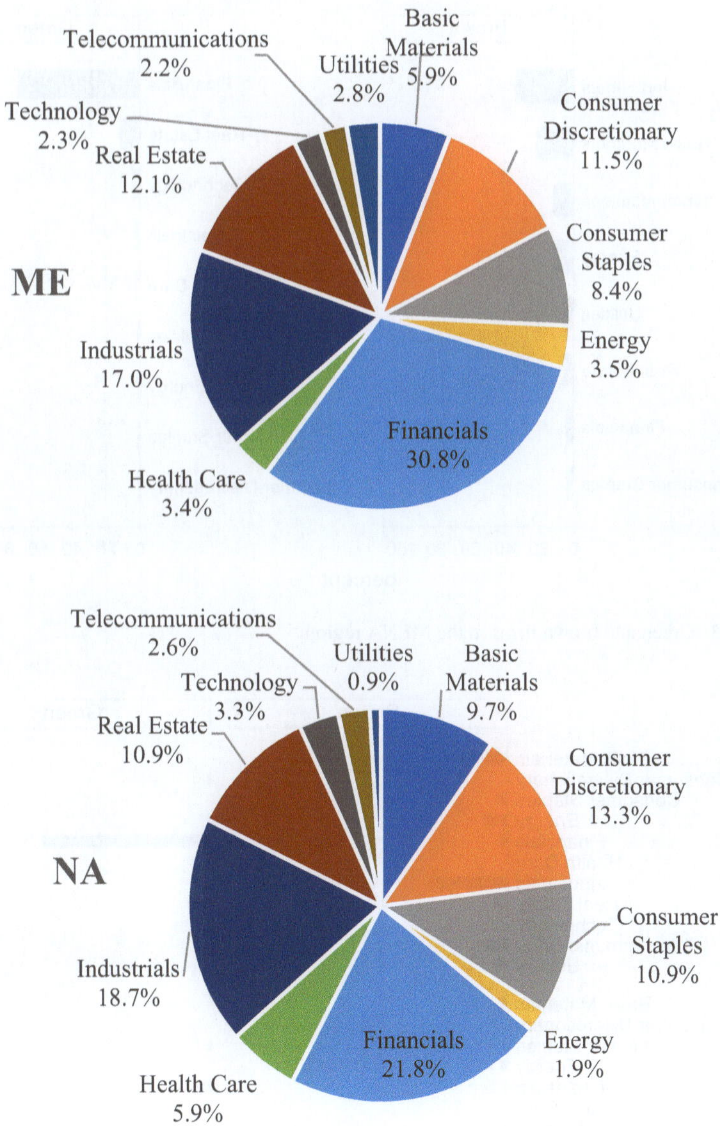

Fig. 14.2 Grouping of firms by industry: ME and NA regions

Fig. 14.5, the lowest emitters are Non-life Insurance; Banks; Media; Retailers; and General Industrials. Notably, General Industrials recorded the largest Scope 1 emissions with a steady downward trend over time. In Fig. 14.6, the five largest emitters are from sectors consisting of Electricity; Gas, Water, and Multi-utilities; Construction and Materials; Industrial Transportation; Oil, Gas and Coal. The yearly average Scope 1 emissions for the five largest emitters was USD3398.3 million tonnes per revenue. This contrasts distinctly with the average annual emission of 4.3 million tonnes per USD in revenue reported for the five lowest emitters. In

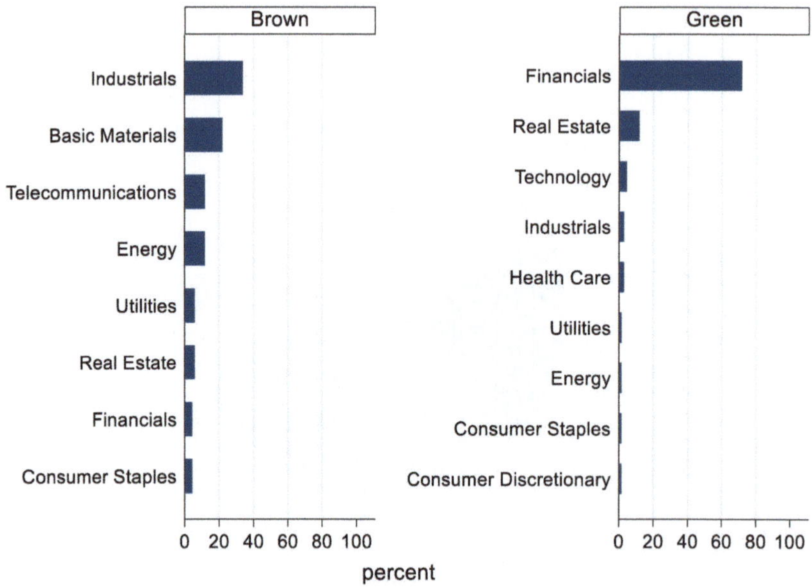

Fig. 14.3 Green and brown firms in the MENA region

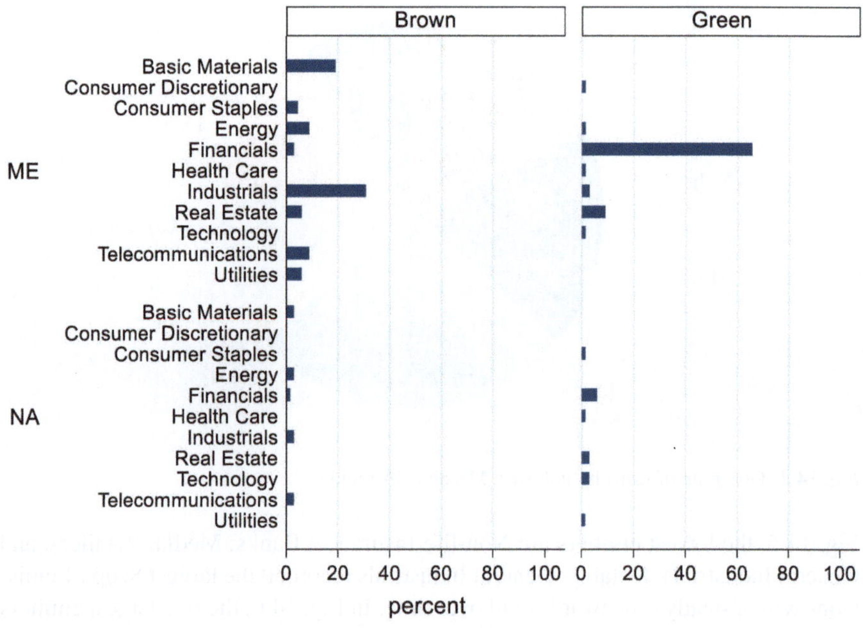

Fig. 14.4 Green and brown firms in the ME and NA regions

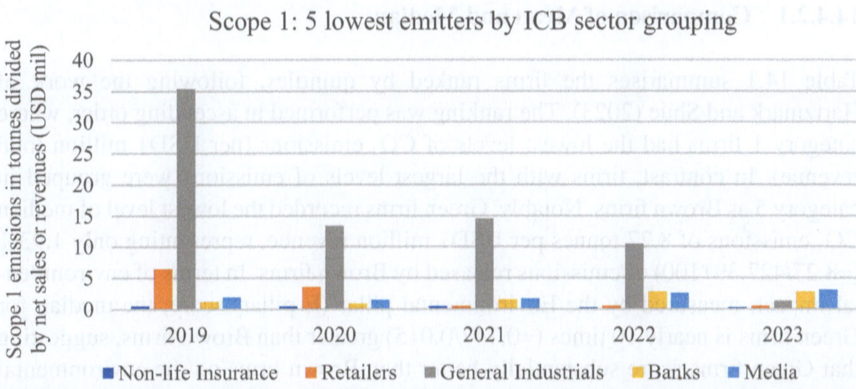

Fig. 14.5 Scope 1 low emitters by industry sector in the MENA region (Notes: Countries in the figures included Egypt, Kuwait, Morocco, Oman, Qatar, Saudi Arabia, and the United Arab Emirates)

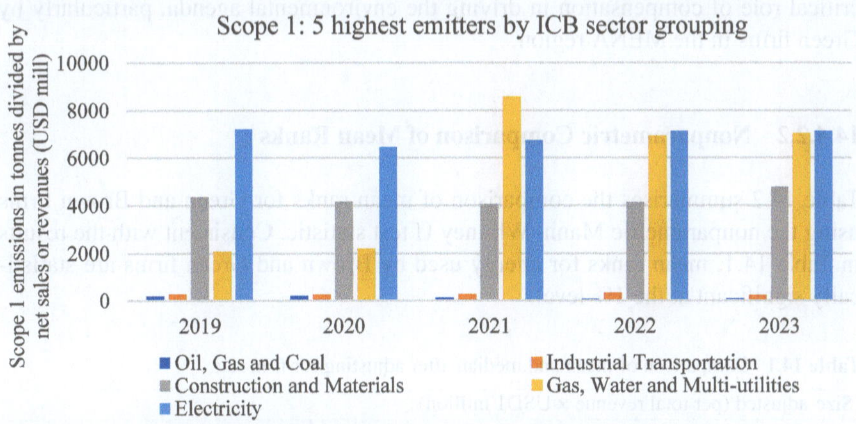

Fig. 14.6 Scope 1 high emitters by industry sector in the MENA region

summary, these figures indicate that Brown firms are mainly large emitters from the energy and construction industries, while Green firms are predominantly involved in the financial services industry.

14.4.2 Results

The following sections first provide a general comparison of selected variables that characterise Brown and Green firms using measures of central tendency. The non-parametric method is then applied to uncover any statistical differences between these variables of interest. Finally, regression analysis is used to better study the impact of the Green or Brown firm on the environment after controlling for cofounders previously discussed in Sect. 14.3.3.

14.4.2.1 Comparison of Mean and Median

Table 14.1 summarises the firms ranked by quintiles, following the work of Hartzmark and Shue (2023). The ranking was performed in ascending order, where category 1 firms had the lowest levels of CO_2 emissions (per USD1 million total revenue). In contrast, firms with the largest levels of emissions were grouped in category 5 as Brown firms. Notably, Green firms recorded the lowest level of median CO_2 emissions of 8.27 tonnes per USD1 million revenue, representing only 1.93% (=8.27/427.39*100) of emissions released by Brown firms. In terms of environmental impact, measured by the Environmental pillar (E pillar) score, the median for Green firms is nearly 10 times (=0.147/0.015) greater than Brown firms, suggesting that Green firms score substantially better than Brown firms on the environmental score. Regarding compensation to senior executives, Green firms tend to pay about 62 times more than Brown firms. Turning to the BoD compensation, Green firms compensated their board about 70 times more than Brown firms. These findings are broadly consistent with the findings of Wong and Neher (2024) and highlight the critical role of compensation in driving the environmental agenda, particularly by Green firms in the MENA region.

14.4.2.2 Nonparametric Comparison of Mean Ranks

Table 14.2 summarises the comparison of mean ranks for Green and Brown firms using the nonparametric Mann–Whitney U test statistic. Consistent with the results in Table 14.1, mean ranks for energy used by Brown and Green firms are statistically significant at the 1% level.

Table 14.1 Comparison of mean and median after adjusting for firm size

Size-adjusted (per total revenue × USD1 million)									
		CO_2 emissions (1)		Environmental pillar score (2)		Senior executive compensation (3)		BoD compensation (4)	
	N	Mean	Median	Mean	Median	Mean	Median	Mean	Median
1 (Green)	67	1.1	8.27	0.055	0.147	17,702.92	139,956.86	5333.89	33,434.37
2 (Neutral)	68	5.92	9.48	0.047	0.063	16,754.07	18,768.79	2288.09	3019.67
3 (Neutral)	68	20.67	30.4	0.04	0.043	15,442.95	16,863.18	2853.36	3337.88
4 (Neutral)	68	48.02	82.92	0.019	0.044	5994.94	7210.26	1349.10	1213.89
5 (Brown)	68	843.03	427.39	0.003	0.015	626.79	2275.68	122.13	475.23
Green/ Brown		0.13%	1.93%	19.79	9.86	28.24	61.5	43.67	70.35

Notes: Firms are ranked by quintiles following Hartzmark and Shue (2023). (1) CO_2 emissions are in tonnes; (2) the original scale for the Environmental pillar score ranges from 0 to 100; the raw values for senior executive and BoD compensation in (3) and (4) are in USD

Table 14.2 Nonparametric comparison of mean ranks by green and brown firms

Variables	N	Pair	Mean	Mean rank	MWU stats.	Z
Energy used	9	Green	4002.3[a]	5.10	45	−4.69***
	44	Brown	2.4×10[9a]	31.52		
Energy used per revenue	9	Green	0.16[b]	7.33	21	−4.15***
	42	Brown	20.93[b]	30.01		
Board comp. per revenue	13	Green	9.2[b]	47.77	94	−3.84***
	48	Brown	1.3[b]	26.46		
Senior-exec. comp./rev.	18	Green	82.56[b]	64.78	31	−5.98***
	57	Brown	5.03[b]	29.54		
Board size	66	Green	8[c]	59.53	1718	−2.37***
	68	Brown	9[c]	75.24		
Ind. board member	66	Green	39[d]	59.19	1695.5	−2.44***
	68	Brown	49[d]	75.57		
Female board member	66	Green	5[d]	59.63	1724.5	−2.57***
	68	Brown	8[d]	75.14		
Market value	67	Green	4.37[e]	40.01	403	−8.25***
	68	Brown	393.44[e]	95.57		
Total revenue	36	Green	1.35[e]	38.28	115	−7.55***
	67	Brown	114.34[e]	21.69		
WACC (%)	65	Green	8.53[d]	64.02	2016	−0.87
	68	Brown	8.57[d]	69.85		

Notes: 'MWU stats' denotes the Mann–Whitney U test statistics
***Significant at 1% level; **Significant at 5% level. The resulting p-values accounted for the effect of small samples (n≤40)
[a]in gigajoules; [b]in USD1000; [c]total headcount; [d]in percentage (%); [e]in USD100 million

More specifically, the energy used mean rank for Brown firms of 31.52 is statistically larger than the mean rank of 5.10 for Green firms. This suggests that, on average, the energy use values are much larger for Brown firms than those reported for Green firms. Analysis by GHG emissions Scopes 1, 2 and 3 shown in Table 14.3 further supports the view that Green firms are more likely to generate lower emissions according to the WRI (2004) international reporting standard. On the contrary, the mean rank for BoD compensation per revenue for Green firms is statistically larger (z-stat. = −4.15, p-value <0.01) than for Brown firms, and a similar result can be seen for senior executive compensation (z-stat. = −3.84, p-value <0.01). This result tends to support the recent evidence documented by Wong and Neher (2024).

A comparison of governance variables suggests that the size of BoD for Green firms is statistically smaller than that of Brown firms (z-stat. = −2.37, p-value <0.01) with fewer independent directors (z-stat. = −2.44, p-value <0.01). Since Brown firms are generally more than two times larger than Green firms (see Sect. 14.4.1), a smaller BoD size reported by Green firms is expected (see Table 14.2). Since a larger BoD size increases the likelihood of more independent board members (Almaqtari et al., 2023), this result highlights a potential governance issue inherent in the corporate structure of Green firms due to the firm size effect. Furthermore, the impact of ownership by the state or family, as suggested by Arayssi and Jizi (2019),

Table 14.3 Nonparametric comparison of mean ranks by GHG Protocol Corporate Standard GHG classification

GHG emissions	N	Pair	Mean[a]	Mean rank	MWU stats.	Z
Scope 1	67	Green	2.04	38.01	269	−8.968***
	68	Brown	1282.67	97.54		
Scope 2	67	Green	7.08	40.22	417	−8.307***
	68	Brown	144.08	95.37		
Scope 3	67	Green	33.15	41.67	514	−7.874***
	68	Brown	229.99	93.94		

Notes: 'MWU stats' denotes the Mann–Whitney U test statistics
***Significant at 1% level; **Significant at 5% level. The resulting p-values accounted for the effect of small samples (n≤40)
[a]Estimated total GHG emissions in tonnes to revenues (USD) in Million

may also play a role in driving the lower proportion of independent board members in Green firms.

In terms of gender diversity, interestingly, Green firms tend to report fewer females on the BoD (z-stat. = −2.57, p-value <0.01), which contradicts the findings of Wu et al. (2024) about the positive influence of female directors on the environmental performance of firms. Furthermore, Gallego-Álvarez and Rodriguez-Dominguez (2023) suggest that more diverse boards will promote environmental initiatives. However, recent evidence from Almaqtari et al. (2023) shows that for Asian companies (including firms from Saudi Arabia, the United Arab Emirates, Qatar, Kuwait and Egypt), board diversity tends to correlate negatively with environmentally friendly production through the minimisation of resource consumption. In contrast, board diversity for European companies correlates positively with environmentally friendly production.

For the financial variables, Green firms reported higher revenue, evidenced by a higher mean rank of 38.28, which is statistically larger than the corresponding mean rank of 21.69 for Brown firms (z-stat. = −7.55, p-value <0.01). However, by market value, Green firms are significantly lower than Brown firms in mean rank (z-stat. = −8.25, p-value <0.01) due mainly to the firm size effect as Green firms are overall smaller in asset size (see Sect. 14.4.1). This would suggest that Green firms are generally profitable, although they are smaller in size than Brown firms. Finally, a marginally lower weighted average cost of borrowing (WACC) is recorded for Green firms, but the difference in mean rank is not statistically significant. This result would suggest that while Green firms can fund their investment projects at a slightly more favourable rate, there is no evidence that debt and equity holders tend to penalise Brown firms for their higher levels of CO_2 emissions.

14.4.2.3 Regression Analysis

Table 14.4 summarises the results for the regression models estimated using the OLS approach. The dependent variable is the log of total energy used (in gigajoules), following Boffo et al. (2020). Robust standard errors in parentheses are

Table 14.4 Ordinary least square regression

Variables	(M1.1)	(M2.1)	(M3.1)	(M4.1)	(M5.1)	(M6.1)	(M7.1)
Green firm	−3.871***						
	(0.299)						
log(BoD compensation) × Green firm		−0.682***	−0.637***	−0.575***	−0.477***	−0.638***	−0.352**
		(0.053)	(0.078)	(0.088)	(0.037)	(0.088)	(0.113)
log(BoD compensation)			0.899***	0.428			
			(0.174)	(0.251)			
Log(Total revenue)				0.446**			
				(0.127)			
Log(Market capitalisation)					0.644**		0.735**
					(0.209)		(0.250)
Weighted average cost of capital					10.73		13.83
					(5.837)		(12.880)
Return on assets					−3.394		−3.913
					(2.663)		(2.093)
Board size						0.135	0.101
						(0.096)	(0.127)
Independent board member						−0.00166	0.0142
						(0.0053)	(0.0148)
Female board member						0.0138	−0.0235
						(0.0390)	(0.0409)
Constant	7.034***	7.009***	1.724	0.433	0.294	5.738***	−2.285
	(0.322)	(0.334)	(1.409)	(0.959)	(2.001)	(1.297)	(2.139)
Observations	53	53	42	36	26	53	26
Adjusted R−squared	0.641	0.612	0.713	0.652	0.420	0.649	0.493

Notes: The table shows coefficient estimates and statistical significance of the models. The dependent variable is the log of total energy used (in gigajoules) following Boffo et al. (2020). Robust standard errors in parentheses are clustered by country where the firms are headquartered. The VIFs of the independent variables (models M3.1 to M7.1) are all below 2.0, suggesting that multicollinearity is not a concern in the models

***p<0.01, **p<0.05, *p<0.1

clustered by country where the firms are headquartered. The variance inflation factors (VIFs) of the independent variables (models M3.1 to M7.1) are all below 2.0, suggesting that multicollinearity does not severely inflate the variance of the regression coefficients. Therefore, the models are expected to produce reliable and stable estimates of regression coefficients. The empirical models generally predict the dependent variable reasonably well, as evidenced by the adjusted R-squared, which explains between 42% and 71.3% of the variation in the dependent variable.

Given that the dependent variable is log transformed, the interpretation of the OLS estimates involves the exponentiated coefficients. In Table 14.4, the estimated coefficient on the Green firm dummy variable is negative and statistically significant at the 1% level. This suggests that energy use will be lower for Green firms than for Brown firms, as expected. More specifically, total direct and indirect energy use by all Green firms represents an annual reduction of 97.9% (=exp(−3.871)−1) relative to Brown firms. Economically, this is a non-trivial finding: the estimated total energy consumed (FY2023) in the MENA region is 1.06×10^{11} gigajoules for all Brown firms, and therefore, the expected energy saved in the MENA region is 1.04×10^{11}(=$1.06 \times 10^{11} \times -0.979$) gigajoules per year if all Brown firms are reclassified as Green firms.

A key finding in Table 14.4 highlights the role of Green firms when the BoD compensation is used as a strategic tool for driving positive environmental outcomes. Notably, when the Green firm dummy interacts with BoD compensation, the estimated coefficient for the interacting term is consistently negative and statistically significant for models M2.1 to M7.1. More specifically, the OLS estimated coefficient ranges from −0.352 (M7.1) to −0.682 (M2.1). This suggests that for a 10% increase in BoD compensation by a Green firm, energy use decreases annually between 3.3% (=$1-(1.1^{-0.352})$) and 6.3% (=$1-(1.1^{-0.682})$) relative to Brown firms. In M3.1, the estimated coefficient on BoD compensation is 0.899 (p-value <0.01), suggesting that BoD compensation is positively related to energy use. However, after adjusting for the size of total revenue, the result is no longer significant in M4.1. This result highlights the effect of size, which confounds the results reported in M3.1.

For the control variables, total revenue and market value for the selected firms are statistically significant at the 1% level. Therefore, firms with large business operations are expected to generate more environmental impact through higher levels of energy consumption. Similarly, the market value of the firm, which reflects the future potential to generate economic returns for investors, suggests that the cost of environmental impact is correlated with the firm's ability to generate increasing market value. The governance variables and WACC do not reveal any significant results, which is broadly consistent with the findings previously documented in Sect. 14.4.2.2.

A possible concern for the results presented in Table 14.4 relates to the challenge of low sample size, particularly in M5.1 and M7.1, which may undermine the statistical power of our analysis. In order to address this issue, we re-estimated the models using the bootstrap procedure with 1000 replications to alleviate finite-sample bias, thereby providing better estimates of standard errors for our models. Table 14.5

Table 14.5 Bootstrap quantile regression at median

Variables	(M1.2)	(M2.2)	(M3.2)	(M4.2)	(M5.2)	(M6.2)	(M7.2)
Green firm	−3.781***						
	(0.244)						
log(BoD compensation) × Green firm		−0.596***	−0.570***	−0.712***	−0.451***	−0.576***	−0.378***
		(0.064)	(0.052)	(0.100)	(0.075)	(0.069)	(0.201)
log(BoD compensation)			0.965***	0.918**			
			(0.334)	(0.349)			
Log(Total revenue)				−0.117			
				(0.370)			
Log(Market capitalisation)					0.369		0.485
					(0.456)		(0.752)
Weighted average cost of capital					20.37**		26.10*
					(8.678)		(15.021)
Return on assets					−2.602		−6.090
					(10.32)		(8.618)
Board size						0.137	−0.007
						(0.098)	(0.152)
Independent board member						−0.004	0.009
						(0.006)	(0.015)
Female board member						0.035	−0.026
						(0.028)	(0.041)
Constant	6.853***	6.828***	1.025	2.438	1.994	5.694***	0.432
	(0.117)	(0.116)	(2.005)	(2.990)	(4.231)	(0.998)	(8.009)
Observations	53	53	42	36	26	53	26
Pseudo R-squared	0.412	0.375	0.522	0.406	0.290	0.413	0.348

Notes: The table above shows coefficient estimates and the statistical significance of quantile regression models where the dependent variable is the log of total energy used (in gigajoules) following Boffo et al. (2020). Quantile regressions were performed for the 50th quantile. The bootstrap procedure was used to calculate the standard errors (in parentheses) by randomly resampling the data

***p<0.01, **p<0.05, *p<0.1

presents the bootstrap quantile regression at the median. Importantly, the re-estimated coefficients for the interaction term range from −0.378 to −0.712. This finding is similar to the main results previously documented in Table 14.4 and, therefore, provides further support for our main analysis using the standard OLS approach.

14.5 Conclusion, Discussion, Limitations and Future Research Avenues

14.5.1 Conclusion

This empirical research focuses on MENA firms and their integration of ESG considerations into BoD compensation. Specifically, this chapter explores the role of BoD compensation in managing the environmental impact using firm-level data from the Middle East and North Africa, a region of global economic importance, and yet it remains understudied. Our unique empirical study demonstrates the differences in environmental impact between Green and Brown MENA firms and the extent to which Green firms alleviate environmental damage through higher BoD compensation. Our main result suggests that BoD compensation is critical to sustainable development through energy use. Specifically, Green firms are more likely to integrate compensation incentives as part of their sustainable strategies than Brown firms. Furthermore, this paper contributes to the recent debate on the effects of board compensation on environmental performance, which may well be extended in future studies.

14.5.2 Discussion and Implications

Our study, employing Scope 1 GHG emissions of MENA firms, has found that the Green firms are predominantly involved in the financial services industry, including General Industrials, recording the largest emission; however, with a steady downward trend over time. Brown firms are mainly large emitters from the energy and construction industries, with distinctly higher annual emissions per revenue than Green firms. Given that environmental innovation is pivotal in advancing environmental sustainability, it can significantly contribute to reducing carbon emissions and enable firms to enhance their overall environmental performance (Többelmann & Wendler, 2020). Albitar et al. (2023) found that environmental innovation effectively reduces carbon emissions, including Scope 1 and Scope 2 emissions. Additionally, their results indicate a moderating role of environmental governance in strengthening the relationship between environmental innovation and CO_2 emissions reduction. Hence, it can be argued that integrating environmental considerations into governance structures, in combination with innovation, can drive carbon emissions reductions. We suggest that managers, guided and supported by BoDs, of

MENA Brown firms aim to adopt innovative approaches to tackle environmental issues using existing and emerging technologies and processes. Collaboration and collective investments within the industry and across the region may be considered to share financial investments. Additionally, policymakers need to put in place frameworks to support environmental innovation and, thus, the firms in the medium and long term. Ideally, policymakers also provide financial support to firms, for example, through grants and subsidies. Governments can further promote green innovation by implementing environmental policy tools, such as pollution abatement subsidies, environmental regulations, green credit initiatives, emission levies, and green procurement strategies (Li et al., 2020).

Implementing such policies significantly contributes to the United Nations SDGs, which aim to unite countries and regions around the globe in addressing international issues like environmental protection, while also considering socio-economic sustainability (Arora & Mishra, 2019; Henderson & Loreau, 2023). Henderson and Loreau's modelling confirms that innovation and technology must 'work in tandem' with environmental protection (p. 8). As the UN SDGs encourage the implementation of changes in specific regions and economic groups, the MENA region has the opportunity to cooperate and innovate, ultimately promoting environmental sustainability. Studies have shown (e.g. Neher et al., 2025; Pandey & Asif, 2022) that accelerating progress towards achieving the SDGs requires concentrated efforts, particularly related to governmental support, budget allocation, and strong policy formulation and implementation. Implementing sustainable development policies necessitates participation from all stakeholders, including scientists, environmentalists, economists, governments, NGOs, and indigenous communities, to conserve resources, biodiversity, and adopt sustainable technologies (Arora & Mishra, 2019). Therefore, by collaborating throughout the region, MENA countries are well positioned to take a leading role in sustainable development by harnessing innovation and regional cooperation in alignment with the UN SDGs. With focused policy backing and inclusive engagement from stakeholders, the region can promote environmental sustainability while fostering socio-economic advancement.

Regarding the BoD, this study highlights a correlation between board compensation and the environmental (E pillar score) impact of Green and Brown firms, suggesting that incentives like remuneration can play a pivotal role in driving the integration of ESG factors into firms' operations (Koors et al., 2019). The extent to which a firm prioritises ESG depends on how it aligns these factors with its strategic goals and integrates them into its business model. Tailoring this approach requires consideration of the firm's culture, values, resources, and capabilities. In the MENA region, Green firms, as demonstrated in this study, tend to achieve higher E scores than Brown firms, likely due to their industry characteristics. However, there may be other reasons. While not the focus of our study, the results show differences in the board structure (e.g. size, diversity, and independence) within the MENA region among Green and Brown firms and, to some extent, mixed findings compared with other geographical regions in the E context (e.g. Almaqtari et al., 2023), highlighting cultural differences. For instance, Green firms in the MENA region have both a higher financial performance (revenue) and environmental performance than Brown firms. This finding aligns with Almaqtari and colleagues, who found that elevated environmental performance and strong ESG scores contribute to increased adoption of environmentally friendly production practices. In contrast, gender diversity on boards is contentious due to cultural and societal differences between women and men (Elsheikh et al., 2022). Nevertheless, a higher proportion of gender diversity on the board can enhance a firm's responsiveness to social and environmental concerns (Birindelli et al., 2018). The smaller number of female board members in Green MENA firms appears to align with the findings of Almaqtari et al. (2023), indicating that board diversity correlates negatively in Asian firms. Consequently, by considering cultural and societal differences, policymakers may improve and enact corporate governance regulations to enhance various board characteristics, contributing to higher ESG levels, particularly environmental performance. For example, considering the lack of financial stability (Elnahass et al., 2022a) and the distinct political contexts in the MENA region, specific policy instruments might include tax incentives or regulatory mandates.

Besides the emerging scrutiny of board characteristics, firms' stakeholders increasingly assess the appropriateness of pay and compensation not only based on financial performance but also on non-financial factors, particularly ESG outcomes (Tharp & Freeman, 2021). Therefore, good practice remuneration programmes must consider these drivers, ensuring that they reflect both financial and ESG performance goals. Our research shows that the Green firm BoDs get paid higher compensation per revenue than the Brown firm BoDs. While this may appear appropriate as Green firm BoDs contribute to a greater extent to the reduction of global warming, at the same time, ethical thoughtfulness and reasonableness are key when creating BoDs' and executives' remuneration packages. For example, comparisons with similar-sized firms, industry standards, and performance metrics should be considered. Fair, appropriate and transparent board compensation demonstrates a commitment to accountability and fosters trust among stakeholders. It also serves as a strategic tool to signal corporate priorities, such as dedication to ESG principles and sustainable business practices, to investors, management, and employees (Koors et al., 2019). For example, misalignment, such as increasing board compensation during periods of decreasing firm performance, would undermine ESG values and erode a culture of fairness and ethical behaviour, potentially damaging organisational authenticity and trust (Neher et al., 2022; Tharp & Freeman, 2021). To uphold ethical standards and stakeholder confidence, BoD compensation must be carefully designed, aligned with corporate values, and transparently communicated as part of broader governance practices.

14.5.3 Limitations

There are limitations in this study that may restrict the scope of our present work. First, we find that firm-level environmental data in the MENA region is relatively sparse compared to that of European counterparts. Environmental data before 2023 are largely incomplete or not reported in the LSEG database, thus compromising the sample size available for analysis. Nevertheless, our study encompasses both non-parametric and bootstrap quantile regression to address data-related issues. In particular, our study uses the bootstrap quantile regression approach as a viable alternative to OLS for addressing the methodological challenges inherent in our small sample. A more comprehensive sample of MENA firms covering different governance variables (such as those considered in Arayssi & Jizi, 2019) can provide further insights into the relationship between corporate governance activities and environmental outcomes for MENA firms. Second, the data available for analysis are differentiated by various types of industries, and thus, there is potential for our results to be confounded by industry effects. A larger and more comprehensive sample that captures a broader definition of the MENA region grouping would help control for such effects in future studies. Third, related to the abovementioned issues, our cross-sectional analysis does not account for the fixed effects or random effects arising from country effects that are not captured by the independent variables. Future work would benefit from using panel regression methods to create richer models and estimation procedures. Lastly, while our study has demonstrated a relationship between BoD compensation and sustainability, very few studies to date have focused on sustainability performance related to board compensation (E-Vahdati et al., 2023). Hence, expanding this research focus would strengthen the evidence base and particularly enhance governance frameworks. Besides these methodological and conceptual aspects, the following section highlights additional avenues for future research based on our study's findings.

14.5.4 Future Research

Future research may examine the influence of ownership structures on environmental performance in the MENA region and, particularly, how these structures could affect our results. As mentioned, ownership is dominated by the state and family. Thus, in the BoD, this can inhibit the effective functioning of the board, potentially affecting the adoption and implementation of sustainability practices. As Arayssi and Jizi (2019) suggest, state and family ownership may also contribute to the lower proportion of independent board members in Green firms, as shown in our study. Hence, investigating how these ownership structures impact environmental outcomes can provide valuable insights into governance challenges and opportunities for fostering environmental sustainability. Such studies could also explore whether ownership and board composition reforms can enhance the effectiveness of green initiatives, offering practical recommendations for stakeholders in the MENA region.

Additionally, beyond CO_2 emissions, incorporating further E pillar measures, such as waste management, water consumption, water recycling, and water pollution, may provide a more comprehensive understanding of the environmental sustainability of Green and Brown firms in the MENA region. Such indicators are particularly relevant given the region's water scarcity and waste management issues

(Negm & Shareef, 2020). As data from providers like LSEG (Refinitiv) become increasingly complete and accessible over time, future research can leverage these metrics to provide a more granular and long-term understanding of environmental performance across industries in the MENA region. Extending our analysis across multiple years would further enhance the generalisability and veracity of our results. The increasing availability of firm-level data is promising, particularly with the introduction of International Standards for sustainability-related disclosures[5] and climate-related disclosures, which came into effect on 1 January 2024. This broader perspective will facilitate more targeted and effective strategies to address the region's unique environmental challenges, thereby allowing for a more comprehensive roadmap for further investigation. Additionally, by refining these tools and exploring innovative approaches, future studies can better inform policy decisions, guide sustainable practices, and track progress towards global environmental goals like the United Nations SDGs.

Also, more comprehensive data for the MENA region will enable comparative studies, such as those on BoD diversity, independence, and geographical contexts. For example, Almaqtari et al.'s (2023) cross-country investigation of Europe and Asia found that board size, independence, and industry expertise positively impact environmentally friendly production. However, as mentioned earlier, the authors observed that board diversity correlated differently with environmentally friendly production between European and Asian firms. Hence, further research could investigate how the MENA region compares with European and Asian firms, considering various BoD characteristics, compensation and ESG performance.

Finally, examining these empirical results through moral and ethical lenses is crucial for a better understanding of their implications. For instance, should enhanced environmental performance, or ESG performance in general, warrant higher compensation for board members, or is it an inherent expectation aligned with their fiduciary duties and responsibilities? This raises critical questions about the balance between incentivising desirable outcomes and upholding fundamental ethical standards. That is, to what extent is it acceptable for Green firms to pay more to the BoD? Is there a quantifiable threshold or a 'red line'? If yes, how would it be defined? Hence, a thorough discussion is needed to explore what these findings mean in practice, including the potential risks of unethical or immoral behaviour arising from poorly defined compensation structures. This debate ties back to the principles of reasonableness and fairness in designing board compensation packages (Wong & Neher, 2024), ensuring that they promote accountability and are aligned with the firm's long-term sustainability goals rather than encouraging short-term gains or exploitative practices. Addressing these issues, not only in the MENA region but also in general, is essential for fostering trust, enhancing good governance practices, and reinforcing the ethical underpinnings of board-level decision-making.

[5] See https://www.ifrs.org/issued-standards, accessed 10 April 2025.

Appendix 1

List of countries included in the MENA region

No.	MENA region (IMF, 2023)
1	Algeria
2	Bahrain
3	Djibouti
4	Egypt
5	Iran (the Islamic Republic of)
6	Iraq
7	Jordan
8	Kuwait
9	Lebanon
10	Libya
11	Mauritania
12	Morocco
13	Oman
14	Qatar
15	Saudi Arabia
16	Somalia
17	Sudan
18	The Syrian Arab Republic
19	Tunisia
20	The United Arab Emirates
21	West Bank and Gaza
22	Yemen

Appendix 2

Definition of variables

Variable	Definition
CO_2 emissions	The estimated total CO_2 and CO_2 equivalent emissions in tonnes
Environmental pillar score	The Environmental pillar measures a company's impact on living and non-living natural systems, including the air, land, and water, as well as complete ecosystems. It reflects how well a company uses best management practices to avoid environmental risks and capitalise on environmental opportunities to generate long-term shareholder value
Energy used	Total direct and indirect energy consumption (in gigajoules) that has been consumed within the boundaries of the company's operations. For utilities, transmission/grid loss as part of business activities is considered as total energy consumed. Raw materials such as coal, gas or nuclear materials used in the production of energy are not considered
BoD compensation	Total compensation of the board members in USD

(continued)

(continued)

Variable	Definition
Senior executive compensation	The total compensation paid to all senior executives in USD as reported by the company
Board size	The total number of board members at the end of the fiscal year
Independent board member	Percentage of independent board members as reported by the company
Female board member	Percentage of females on the board
Market value	The consolidated market value in USD of a company
Total revenue	Represents revenue in USD from all of a company's operating activities after deducting any sales adjustments and their equivalents
WACC (%)	Represents the Weighted Average Cost of Capital in percentage, as reported by the company
Scope 1	Total estimated scope 1 GHG emissions in tonnes divided by net sales or revenues in USD (million)
Scope 2	Total estimated scope 2 GHG emissions in tonnes divided by net sales or revenues in USD (million)
Scope 3	Total estimated scope 3 GHG emissions in tonnes divided by net sales or revenues in USD (million)

References

Albitar, K., Borgi, H., Khan, M., & Zahra, A. (2023). Business environmental innovation and CO2 emissions: The moderating role of environmental governance. *Business Strategy and the Environment, 32*(4), 1996–2007. https://doi.org/10.1002/BSE.3232

Almaqtari, F. A., Elsheikh, T., Al-Hattami, H. M., & Mishra, N. (2023). The impact of board characteristics on environmentally friendly production: A cross country study in Asia and Europe. *Journal of Cleaner Production, 392*, 136257. https://doi.org/10.1016/J.JCLEPRO.2023.136257

Andreas, J. M., Rapp, M. S., & Wolff, M. (2012). Determinants of director compensation in two-tier systems: Evidence from German panel data. *Review of Managerial Science, 6*(1), 33–79. https://doi.org/10.1007/s11846-010-0048-z

Ang, J., Lauterbach, B., & Schreiber, B. Z. (2002). Pay at the executive suite: How do US banks compensate their top management teams? *Journal of Banking & Finance, 26*(6), 1143–1163. https://doi.org/10.1016/S0378-4266(01)00165-0

Aoun, M. C. (2013). Oil and gas resources of the Middle East and North Africa: A curse or a blessing? In J.-M. Chevalier & P. Geoffron (Eds.), *The new energy crisis: Climate, economics and geopolitics* (2nd ed., pp. 133–160). Palgrave Macmillan. https://doi.org/10.1007/978-1-137-02118-2_5

Arayssi, M., & Jizi, M. I. (2019). Does corporate governance spillover firm performance? A study of valuation of MENA companies. *Social Responsibility Journal, 15*(5), 597–620. https://doi.org/10.1108/SRJ-06-2018-0157/FULL/PDF

Arora, N. K., & Mishra, I. (2019). United Nations Sustainable Development Goals 2030 and environmental sustainability: Race against time. *Environmental Sustainability, 2*, 339–342. https://doi.org/10.1007/s42398-019-00092-y

Aydoğmuş, M., Gülay, G., & Ergun, K. (2022). Impact of ESG performance on firm value and profitability. *Borsa Istanbul Review, 22*, S119–S127. https://doi.org/10.1016/J.BIR.2022.11.006

Birindelli, G., Dell'Atti, S., Iannuzzi, A. P., & Savioli, M. (2018). Composition and activity of the board of directors: Impact on ESG performance in the banking system. *Sustainability, 10*(12), 4699. https://doi.org/10.3390/SU10124699

Boffo, R., Marshall, C., & Patalano, R. (2020). *ESG investing: Environmental pillar scoring and reporting*. https://doi.org/10.1787/80680dc8-en

Brick, I. E., Palmon, O., & Wald, J. K. (2006). CEO compensation, director compensation, and firm performance: Evidence of cronyism? *Journal of Corporate Finance, 12*(3), 403–423. https://doi.org/10.1016/j.jcorpfin.2005.08.005

Cohen, S., Kadach, I., Ormazabal, G., & Reichelstein, S. (2023). Executive compensation tied to ESG performance: International evidence. *Journal of Accounting Research, 61*, 805–853. https://doi.org/10.1111/1475-679X.12481

Drempetic, S., Klein, C., & Zwergel, B. (2020). The influence of firm size on the ESG score: Corporate sustainability ratings under review. *Journal of Business Ethics, 167*(2), 333–360. https://doi.org/10.1007/s10551-019-04164-1

Eccles, R. G., Ioannou, I., & Serafeim, G. (2014). The impact of corporate sustainability on organizational processes and performance. *Management Science, 60*(11), 2835–2857. https://doi.org/10.1287/MNSC.2014.1984

El-Katiri, L., & Fattouh, B. (2017). A brief political economy of energy subsidies in the Middle East and North Africa. In G. Luciani (Ed.), *Combining economic and political development: The experience of MENA, International Development Policy series 7* (pp. 58–87). The Graduate Institute Geneva, Brill-Nijhoff. https://doi.org/10.4000/poldev.2267

Elnahass, M., Marie, M., & Elgammal, M. (2022a). Terrorist attacks and bank financial stability: Evidence from MENA economies. *Review of Quantitative Finance and Accounting, 59*(1), 383–427. https://doi.org/10.1007/s11156-022-01043-1

Elnahass, M., Salama, A., & Trinh, V. Q. (2022b). Firm valuations and board compensation: Evidence from alternative banking models. *Global Finance Journal, 51*, 1–19. https://doi.org/10.1016/j.gfj.2020.100553

Elsheikh, T., Hashim, H. A., Mohamad, N. R., Almaqtari, F. A., & Ettish, A. A. (2022). CEO facial masculinity, characteristics and earnings management. *Management and Accounting Review, 21*(3), 253–282. https://doi.org/10.24191/MAR.V21I03-11

European Commission. (2020). *Study on directors' duties and sustainable corporate governance: Final report*. Directorate-General for Justice and Consumers. https://doi.org/10.2838/472901

E-Vahdati, S., Wan-Hussin, W. H., & Ariffin, M. S. M. (2023). Sustainability performance and board compensation in Japan and ASEAN-5 countries. *Borsa Istanbul Review, 22*(2), S189–S199. https://doi.org/10.1016/j.bir.2022.12.004

Flammer, C., Hong, B., & Minor, D. (2019). Corporate governance and the rise of integrating corporate social responsibility criteria in executive compensation: Effectiveness and implications for firm outcomes. *Strategic Management Journal, 40*(7), 1097–1122. https://doi.org/10.1002/SMJ.3018

Francoeur, C., Melis, A., Gaia, S., & Aresu, S. (2017). Green or greed? An alternative look at CEO compensation and corporate environmental commitment. *Journal of Business Ethics, 140*(3), 439–453. https://doi.org/10.1007/S10551-015-2674-5

FTSE Russel. (2024). *FTSE Russel Industry Classification Benchmark (ICB)*. A&K Ventures OÜ and FTSE Russell. https://classification.codes/classifications/industry/icb

Gallego-Álvarez, I., & Rodriguez-Dominguez, L. (2023). Board of directors and environmental practices: The effect of board experience, culture, and tenure. *Environment, Development and Sustainability*, 1–26. https://doi.org/10.1007/S10668-023-03937-Z

Greene, W. (2018). *Econometric analysis* (8th ed.). Pearson Education.

Hartzmark, S. M., & Shue, K. (2023). Counterproductive sustainable investing: The impact elasticity of Brown and Green firms. *S&P Global*, 1–69. https://doi.org/10.2139/SSRN.4359282

Henderson, K., & Loreau, M. (2023). A model of sustainable development goals: Challenges and opportunities in promoting human well-being and environmental sustainability. *Ecological Modelling, 475*, 1–9. https://doi.org/10.1016/j.ecolmodel.2022.110164

Hertog, S. (2013). *The private sector and reform in the Gulf Cooperation Council*. https://eprints.lse.ac.uk/54398/1/Hertog_2013.pdf

Hörisch, J., Johnson, M. P., & Schaltegger, S. (2015). Implementation of sustainability management and company size: A knowledge-based view. *Business Strategy and the Environment, 24*(8), 765–779. https://doi.org/10.1002/BSE.1844

Horowitz, J. L. (2001). The bootstrap. In J. Heckman & E. Leamer (Eds.), *Handbook of econometrics* (Vol. 5, pp. 3159–3228). North-Holland.

IEA. (2019). *The role of gas in today's energy transitions: World energy outlook special report*. Paris. https://iea.blob.core.windows.net/assets/cc35f20f-7a94-44dc-a750-41c117517e93/TheRoleofGas.pdf

Ikram, A., Li, Z. (Frank), & Minor, D. (2023). CSR-contingent executive compensation contracts. *Journal of Banking & Finance, 151*, 105655. https://doi.org/10.1016/J.JBANKFIN.2019.105655

IMF. (2023). *Safeguarding macroeconomic stability amid continued uncertainty*. https://www.imf.org/en/Publications/REO/MECA/Issues/2023/04/13/regional-economic-outlook-mcd-april-2023

Ioannou, I., & Serafeim, G. (2019). *Corporate sustainability: A strategy?* (Working paper 19-065). https://www.hbs.edu/ris/Publication Files/WP19-065rev4-23-21_aefe2c98-5fa6-44c4-bbae-f68e6f283642.pdf

IPCC. (2014). Summary for Policymakers. In O. Edenhofer, R. Pichs-Madruga, Y. Sokona, K. Seyboth, P. Matschoss, S. Kadner, et al. (Eds.), *Climate Change 2014: Mitigation of climate change. Contribution of Working Group III to the Fifth Assessment Report of the Intergovernmental Panel on Climate Change* (p. 32). Cambridge University Press.

Jamali, D., Safieddine, A. M., & Rabbath, M. (2008). Corporate governance and corporate social responsibility synergies and interrelationships. *Corporate Governance: An International Review, 16*(5), 443–459. https://doi.org/10.1111/J.1467-8683.2008.00702.X

Kaya, A., & Tsai, I.-T. (2016). Inclusive economic institutions in the Gulf Cooperation Council States: Current status and theoretical implications. *Review of Middle East Economics and Finance, 12*(2), 139–173. https://doi.org/10.1515/RMEEF-2015-0015

Keller, G. (2017). *Statistics for management and economics* (11th ed.). Cengage Learning.

Kilani, J., & Babalola, A. (2020). Sustainable development goals, extractives industries and the energy nexus – insights in the MENA Region. *Journal of Sustainable Development Law and Policy, 11*(1), 185–203. https://doi.org/10.4314/JSDLP.V11I1.9

Koors, J. L., Pearl Meyer, & Partners LLC. (2019). *Executive compensation and ESG*. https://corpgov.law.harvard.edu/2019/09/10/executive-compensation-and-esg/

Lemma, T. T., Mlilo, M., & Gwatidzo, T. (2020). Board remuneration, directors' ownership and corporate performance: The South African evidence. *International Review of Applied Economics, 34*(4), 491–511. https://doi.org/10.1080/02692171.2020.1773654

Li, Z., Liao, G., & Albitar, K. (2020). Does corporate environmental responsibility engagement affect firm value? The mediating role of corporate innovation. *Business Strategy and the Environment, 29*(3), 1045–1055. https://doi.org/10.1002/BSE.2416

Luomi, M. (2011). Gulf of interest: Why oil still dominates Middle Eastern climate politics. *Journal of Arabian Studies, 1*(2), 249–266. https://doi.org/10.1080/21534764.2011.628499

Maas, K. (2018). Do corporate social performance targets in executive compensation contribute to corporate social performance? *Journal of Business Ethics, 148*(3), 573–585. https://doi.org/10.1007/S10551-015-2975-8

Mahoney, L. S., & Thorne, L. (2005). Corporate social responsibility and long-term compensation: Evidence from Canada. *Journal of Business Ethics, 57*(3), 241–253. https://doi.org/10.1007/S10551-004-5367-Z

Malik, M. (2014). Value-enhancing capabilities of CSR: A brief review of contemporary literature. *Journal of Business Ethics, 127*(2), 419–438. https://doi.org/10.1007/S10551-014-2051-9

Negm, A. M., & Shareef, N. (2020). *Waste management in MENA regions* (1st ed.). Springer Cham.

Neher, A., Wong, A., Adjei-Bamfo, P., Meinhold, R., Bawole, J. N., & Pawar, M. (2025). Social development funding desiderata: Key issues and financing gaps and options. *The International Journal of Community and Social Development*. https://doi.org/10.1177/25166026251345296

Neher, A., Wong, A., & Miles, M. P. (2022). The relationship between the level of a corporation's public disclosure of its managerial values and its level of organizational authenticity. *European Business Review, 34*(2), 277–296. https://doi.org/10.1108/EBR-06-2020-0150

Nikitina, L., Paidi, R., & Furuoka, F. (2019). Using bootstrapped quantile regression analysis for small sample research in applied linguistics: Some methodological considerations. *PLoS One, 14*(1), e0210668. https://doi.org/10.1371/journal.pone.0210668

Olawuyi, D. S. (2018). Advancing innovations in renewable energy technologies as alternatives to fossil fuel use in the Middle East. In D. Zillman, L. Godden, L. Paddock, & M. Roggenkamp (Eds.), *Innovation in energy law and technology: Dynamic solutions for energy transitions (Online Edition)* (Vol. 1). Oxford University Press. https://doi.org/10.1093/oso/9780198822080.003.0020

Olawuyi, D. S. (2021). Can MENA extractive industries support the global energy transition? Current opportunities and future directions. *The Extractive Industries and Society, 8*(2), 100685. https://doi.org/10.1016/J.EXIS.2020.02.003

Pandey, A., & Asif, M. (2022). Assessment of energy and environmental sustainability in South Asia in the perspective of the Sustainable Development Goals. *Renewable and Sustainable Energy Reviews, 165*, 1–14. https://doi.org/10.1016/j.rser.2022.112492

Ratti, S., Arena, M., Azzone, G., & Dell'Agostino, L. (2023). Environmental claims and executive compensation plans: Is there a link? An empirical investigation of Italian listed companies. *Journal of Cleaner Production, 422*, 1–12. https://doi.org/10.1016/j.jclepro.2023.138434

Salibian-Barrera, M., & Zamar, R. H. (2002). Bootstrapping robust estimates of regression. *The Annals of Statistics, 30*(2), 556–582. http://www.jstor.org/stable/2699968

Sekaran, U., & Bougie, R. (2016). *Research methods for business: A skill building approach* (7th ed.). Wiley.

Suttipun, M. (2021). The influence of board composition on environmental, social and governance (ESG) disclosure of Thai listed companies. *International Journal of Disclosure and Governance, 18*(4), 391–402. https://doi.org/10.1057/s41310-021-00120-6

Tagliapietra, S. (2019). The impact of the global energy transition on MENA oil and gas producers. *Energy Strategy Reviews, 26*, 1–6. https://doi.org/10.1016/J.ESR.2019.100397

Tharp, C., & Freeman, K. (2021). Executive compensation is a powerful communication to stakeholders. *People & Strategy, 44*(4), 36–41.

Többelmann, D., & Wendler, T. (2020). The impact of environmental innovation on carbon dioxide emissions. *Journal of Cleaner Production, 244*. https://doi.org/10.1016/j.jclepro.2019.118787

United Nations. (2015). *Transforming our world: The 2030 agenda for sustainable development*. https://documents.un.org/doc/undoc/gen/n15/291/89/pdf/n1529189.pdf

United Nations Global Compact. (2014). *Guide to corporate sustainability*. https://d306pr3pise04h.cloudfront.net/docs/publications%2FUN_Global_Compact_Guide_to_Corporate_Sustainability.pdf

Waterstraat, S., Kustner, C., & Koch, M. (2021). Does board composition taking account of sustainability expertise influence ESG ratings? An exploratory study of European banks. In A. Gerber & K. Hinkelmann (Eds.), *Society 5.0* (pp. 129–138). Springer International. https://doi.org/10.1007/978-3-030-86761-4_11

Welch, K., & Yoon, A. (2023). Do high-ability managers choose ESG projects that create shareholder value? Evidence from employee opinions. *Review of Accounting Studies, 28*(4), 2448–2475. https://doi.org/10.1007/S11142-022-09701-4

Wong, A., & Neher, A. (2024). ESG performance and board of directors compensation: A European perspective. In S. Boubaker & T.-H. Le (Eds.), *Handbook of environmental and green finance: Towards a sustainable future* (pp. 321–357). World Scientific. https://doi.org/10.1142/9781800614451_0011

World Bank Group. (2024). *Listed domestic companies, total – Middle East & North Africa*. World Federation of Exchanges database, World Bank Group. https://data.worldbank.org/indicator/CM.MKT.LDOM.NO?end=2022&locations=ZQ&start=2002&view=chart

WRI. (2004). *The greenhouse gas protocol: A corporate accounting and reporting standard* (revised ed.). World Resources Institute.

Wu, Q., Shahbaz, M., & Bigerna, S. (2024). Does board gender diversity benefit for corporate ESG performance: Evidence from Germany. *Journal of Environmental Management, 372*, 123304. https://doi.org/10.1016/J.JENVMAN.2024.123304

Zikmund, W. G. (2003). *Business research methods* (7th ed.). Thomson/South-Western.

Chapter 15
From Vulnerability to Opportunity: Policy, Market, and Institutional Lessons for MENA's Sustainability Future

Marwa Elnahass, Amr ElAlfy, and Toka S. Mohamed

Abstract The Middle East and North Africa (MENA) region stands at a critical crossroads in its transition towards sustainable development and net-zero emissions. This edited book brings together 13 chapters that collectively interrogate the interplay between sustainability governance, financial systems, corporate responsibility, and public policy frameworks. Drawing on diverse methodologies and disciplinary lenses, including economics, finance, environmental policy, and management, the book synthesizes empirical findings, theoretical perspectives, and practical insights to illuminate the institutional, structural, and behavioural dimensions of sustainability in MENA. The conclusion integrates contributions across four thematic domains. First, on climate governance and corporate performance, the chapters highlight how regulatory stringency, investor activism, and adaptive capacity shape corporate decarbonization and resilience. Second, on finance, markets, and sustainable development, the analysis underscores the double-edged role of financial systems: green finance and market innovation can catalyse sustainable growth; yet, weak standards risk greenwashing and instability. Third, on corporate governance and ESG integration, evidence demonstrates that board incentives, disclosure practices, and stakeholder engagement are decisive in embedding sustainability within firms. Fourth, on sectoral and regional transitions, the chapters reveal both the opportunities of diversification and foreign investment, and the challenges of dependency and fragmented governance. Collectively, the book advances scholarly debates and provides actionable guidance for policymakers, regulators, and business leaders. It argues that MENA's sustainability future hinges on integrative governance, harmonized finan-

M. Elnahass (✉)
King Fahd University of Petroleum and Minerals (KFUPM), Dhahran, Saudi Arabia
e-mail: Marwa.Elnahass@kfupm.edu.sa

A. ElAlfy
University of Waterloo, Waterloo, ON, Canada
e-mail: alelalfy@uwaterloo.ca

T. S. Mohamed
College of Business and Economics, Qatar University, Doha, Qatar
e-mail: tmohamed@qu.edu.qa

M. Elnahass et al. (eds.), *A Greenprint for Net Zero*, Gulf Studies 15,
https://doi.org/10.1007/978-981-92-1762-5_15

cial regulation, and inclusive institutional reform. The insights presented not only fill a critical gap in regional sustainability scholarship but also chart pathways for a just, resilient, and credible transition towards net-zero, with particular relevance for the upcoming UN COP 30 and beyond.

Keywords Climate risk · Corporate responsibility · Green investment · Sustainable finance · MENA · COP 30

15.1 Climate Governance and Corporate Performance

Several chapters in this book examine how regulatory frameworks, investor pressure, and institutional quality shape corporate climate outcomes. **Chapter 2** demonstrates that firms exposed to stricter environmental regulations and active investor scrutiny achieve superior emissions-reduction performance. This finding shows the importance of coercive institutional pressures, from both governments and financial markets, in steering companies towards beyond-compliance environmental practices. **Chapter 11** extends this analysis in the GCC context, showing that corporate climate disclosures often function as strategic reputation tools rather than substantive risk assessments. Selective reporting raises concerns about greenwashing, particularly where disclosures emphasize positive narratives while omitting material risks. Recent progress, such as the adoption of ESG taxonomies and net-zero pledges, signals a shift, but credibility will depend on standardized, enforced reporting norms and third-party verification mechanisms. At the macro level, **Chap. 12** demonstrates that climate vulnerability is mediated by adaptive capacity and governance quality, not simply exposure to physical risks. Stronger institutions, infrastructure investment, and social protection systems can offset vulnerabilities, while weak governance exacerbates them. The clear implication is that climate governance must integrate mitigation and adaptation, linking corporate targets with national strategies and ensuring resilience-building alongside emissions reduction.

15.2 Finance, Markets, and Sustainable Development

Chapters on finance emphasize the enabling but potentially destabilizing role of financial systems. **Chapter 4** shows that regulatory harmonization and financial innovation are key to mobilizing sustainable investment, with green bonds, Sharia-compliant instruments, and ESG funds emerging as tools to finance renewables and clean technology. **Chapter 5** highlights both the promise and pitfalls of green finance. While green bonds and funds are driving investment in renewables and sustainable agriculture, weak oversight risks speculative bubbles and greenwashing. The chapter's respective authors call for robust standards and oversight to ensure genuine impact. **Chapter 3** reveals how climate-induced disasters can trigger credit crunches in MENA's underdeveloped insurance and banking systems, exposing structural fragility. Similarly, **Chap. 6** highlights the challenge of "greenflation", which refers to short-term inflationary pressures from subsidy reform, carbon pricing, and large-scale energy transitions. Both contributions stress that sustainable finance requires not just capital mobilization but also financial resilience through climate risk stress-testing, insurance development, and phased fiscal reforms. Collectively, these chapters argue that financial systems must evolve into shock absorbers rather than amplifiers of climate risks, requiring coordination between regulators, central banks, and international institutions.

15.3 Corporate Governance, Strategy, and ESG Integration

Corporate governance emerges as a decisive factor for sustainability. **Chapter 7** highlights heterogeneity in ESG adoption, shaped by board commitment, stakeholder engagement, and resource availability. Firms with ESG committees, diverse boards, or proactive stakeholder dialogues outperform peers. **Chapter 8** reframes sustainability as a competitive advantage, aligning with Porter and Kramer's shared value framework. By embedding ESG into strategy, firms unlock innovation, cost savings, and reputational benefits, transforming sustainability from compliance into opportunity. **Chapter 9**, focusing on Morocco, demonstrates that ESG disclosure is influenced by firm size, foreign ownership, and regulatory mandates, with listed firms on ESG indices leading in transparency. Finally, **Chap. 14** shows how executive incentives tied to environmental KPIs can drive performance, provided that targets are ambitious, measurable, and independently verified. This reinforces the importance of aligning incentives with sustainability outcomes to embed ESG in corporate strategies. Across these contributions, the central message is that governance reforms, incentive structures, and stakeholder engagement are essential levers for mainstreaming ESG in MENA firms.

15.4 Sectoral and Regional Transitions

At the broader economic level, Chaps. 10 and 13 highlight the challenges and opportunities of sustainability transitions. **Chapter 10** shows that Gulf economies face structural constraints in shifting away from hydrocarbons: fiscal dependency, entrenched social contracts, and the political sensitivity of subsidy reforms. Diversification strategies, such as Saudi Vision 2030 and the UAE's green economy plans, are promising but require institutional coherence, human capital investment, and inclusive growth to avoid replicating rentier dynamics. **Chapter 13** situates foreign investment, particularly China's Belt and Road Initiative, as a double-edged sword. While it provides capital and technology transfer, it risks creating new dependencies unless MENA governments negotiate fair, transparent deals that include local capacity-building. Together, these chapters stress that MENA's sustainability future depends on coherent industrial policies, balanced foreign partnerships, and inclusive economic diversification.

15.5 Gaps and Directions for Future Research

While this volume provides a comprehensive overview of the sustainability landscape in the MENA region, the dynamic nature of the field presents numerous avenues for future scholarly inquiry. The following areas represent critical gaps in the literature and opportunities for deeper investigation:

1. **Longitudinal Dynamics of ESG Integration and Policy Efficacy:** Many of the analyses presented in this book, and indeed in the broader literature, are cross-sectional or based on short-term panel data. To truly understand the impact of ESG integration and the efficacy of sustainability policies, long-term longitudinal studies are imperative. Future research should track the evolution of corporate sustainability practices and performance over extended periods (e.g. 20–30 years), moving beyond the foundational work of scholars like Trinh et al. (2023), Marie et al. (2024), and Oyegunle et al. (2023). Such studies could address critical questions: Do early adopters of ESG maintain their competitive advantage? *Does sustained sustainability performance translate into long-term financial outperformance and reputational benefits? How do corporate carbon reduction commitments hold up against fluctuating external pressures?* Longitudinal analyses would also enable a more nuanced understanding of the real-world impact of major policy shifts, such as the implementation of national net-zero targets in the UAE (2050) and Saudi Arabia (2060), or the phased removal of fossil fuel subsidies.
2. **Beyond Carbon: Holistic Sustainability and the "S" and "G" in ESG:** The discourse on sustainability in the MENA region has been understandably dominated by climate change and the energy transition. However, a more holistic approach is needed, one that gives equal weight to the social (S) and governance (G) pillars of ESG. Future research should delve into the pressing issues of biodiversity loss, water security, and social equity. For instance, studies could examine how corporate strategies and public policies are addressing the degradation of marine ecosystems in the Gulf or the threats of desertification and deforestation, building on the work of Bayram and Öztürk (2021) and Qumsiyeh and Abusarhan (2021). The social dimension of sustainability is particularly critical in a region characterized by significant wealth disparities, high youth unemployment, and a large migrant workforce. Research is needed to explore how ESG frameworks are being adapted to address these unique social challenges. For example, *how do companies report on and manage issues related to workforce nationalization, labor rights for expatriate workers, and community impact?* Furthermore, the intersection of water security with both environmental

and social sustainability, as highlighted by Christy et al. (2024) and ElAlfy et al. (2025), remains a timely theme of research for investigation.

3. **Comparative Analyses and South–South Knowledge Exchange:** This book's regional focus is a key strength, but comparative studies that situate the MENA experience within a global context are essential for policy learning and knowledge transfer. Future research could compare the development of green finance in the MENA region with that of other emerging economies in Southeast Asia or Latin America. *What lessons can be learned from the European Union's advanced ESG regulatory framework, and how can they be adapted to the unique political and economic realities of the MENA region? Conversely, what insights can be gleaned from the state-led economic diversification strategies in the Gulf (Shehabi,* 2020*; Matallah,* 2022*) that could be relevant for other resource-dependent economies?* Such comparative work can foster valuable South–South knowledge exchange, for example, by sharing best practices in renewable energy deployment between the GCC and African nations (Chen et al., 2023).

4. **The Transformative and Disruptive Role of Emerging Technologies:** The rapid advancements in artificial intelligence (AI), big data, and blockchain technologies present both significant opportunities and profound challenges for sustainability governance (Zhang & Yang, 2024). Future research should explore how these technologies are being leveraged in the MENA region to enhance ESG monitoring, reporting, and performance. For example, *how are companies and regulators using AI-powered analytics to track and verify carbon emissions, or to assess climate-related financial risks? What is the potential for blockchain to enhance the transparency and credibility of carbon credit markets and green bond issuances?* At the same time, the governance of these technologies is a critical and under-researched area. What regulatory frameworks are needed to mitigate the risks of algorithmic bias, data privacy breaches, and "digital greenwashing"? Moreover, the implications of these technologies for cybersecurity and corporate governance, as explored by Elnahass et al. (2024) and Trinh et al. (2025), warrant further investigation.

5. **The Political Economy of Sustainability Transitions:** The transition to a sustainable, low-carbon economy is not merely a technical or financial challenge; it is a deeply political process. Future research should more explicitly address the role of political economy, power structures, and rent-seeking in shaping sustainability outcomes in the MENA region (ElAlfy et al., 2024). *How do different governance models, from the more centralized systems in the Gulf to the more pluralistic systems in North Africa, influence the design and implementation of climate policies? How do politically connected firms navigate the pressures and opportunities of the green transition (Marie et al.,* 2024*)? What is the role of civil society, the media, and youth movements in holding governments and corporations accountable for their sustainability commitments?* Answering these questions requires a multi-disciplinary approach that combines insights from political science, sociology, and economics.

15.6 Final Reflections and Recommendations

This edited book has traversed the complex and multifaceted landscape of sustainability in the Middle East and North Africa, revealing a region at an inflection point. The chapters collectively argue that the transition to a sustainable future is not a predetermined path but a contested journey, shaped by a confluence of global pressures, regional ambitions, and local realities. The MENA region is confronted with a dual imperative: to address its profound environmental vulnerabilities, from acute water scarcity to extreme climate impacts, while simultaneously transforming its economic foundations away from unsustainable, resource-dependent models. The insights gleaned from this book offer a roadmap for navigating this complex transition, with a particular sense of urgency and opportunity in the lead-up to the **UN Climate Change Conference (COP 30) in Brazil and beyond**.

A central theme that emerges is the **paramount importance of integrated and coherent governance** in advancing sustainability across the MENA region. Sustainability cannot be achieved in silos: climate policy must be embedded within national economic and industrial strategies; financial regulation must internalize environmental and social risks; and corporate governance frameworks must incentivize long-term value creation over short-term profit maximization. This alignment is particularly critical in hydrocarbon-dependent economies where diversification, resilience, and climate adaptation are existential priorities.

Qatar provides a prominent case study The **Qatar National Vision 2030** and the **National Environment and Climate Change Strategy (NECCS 2021–2030)** embed sustainability at the heart of economic and industrial transformation. These strategies are now being operationalized through sectoral decarbonization pathways in energy, water, waste, and biodiversity, supported by emerging sustainable finance mechanisms. The **Qatar Central Bank's (QCB) sustainable finance framework (2023–2024)** aligns with IFRS S1/S2 disclosure standards and explores integration with the Taskforce on Nature-related Financial Disclosures (TNFD), aiming to harmonize local reporting with global norms. Institutions such as **Earthna (Qatar Foundation)** are advancing capacity-building in carbon markets, biodiversity disclosures, and policy frameworks for net-zero transitions, while national dialogues increasingly emphasize the role of financial institutions in mobilizing private capital towards sustainability.

In the **GCC**, comparable integrated approaches are visible. The **Saudi Green Initiative (SGI)**, a pillar of Vision 2030, combines afforestation and renewable energy expansion with fiscal reforms, ensuring that natural capital is recognized within economic strategy. The **UAE's Green Economy for Sustainable Development initiative** and **Net Zero by 2050 strategic initiative** demonstrate how sustainability targets are institutionalized through regulatory reforms, green finance platforms, and corporate governance incentives. Both countries are also piloting alignment with international frameworks such as the EU's Corporate Sustainability Reporting Directive (CSRD), not only to strengthen domestic ESG standards but also to maintain global investment attractiveness. Oman and Sharjah have developed sustainable finance frameworks that underpin the growing issuance of green sukuk and bonds, reinforcing the integration of Islamic finance with sustainability goals.

Beyond the Gulf, **Egypt's Vision 2030** positions sustainability as a cornerstone of its economic modernization agenda. Egypt became the first MENA state to issue a sovereign green bond (2020), channelling proceeds into renewable energy, transportation, and sustainable infrastructure. Egypt is also progressively aligning disclosure practices with IFRS S1/S2 and exploring integration with CSRD-aligned requirements to enhance transparency and investor confidence. North African countries such as Morocco and Tunisia are advancing renewable energy transitions and embedding ESG criteria into national industrial strategies, thereby contributing to a wider regional momentum.

Taken together, these developments highlight the **growing convergence between regional strategies and global frameworks**. While the **IFRS S1/S2 standards** provide a universal baseline for climate and sustainability disclosures, the **TNFD** supports nature-positive reporting, and the **CSRD** reinforces comparability across markets. The real challenge, however, lies in moving from high-level commitments to implementation. This requires dismantling bureaucratic silos, fostering interministerial coordination, strengthening institutional capacity, and embedding sustainability into the practices of regulators, corporations, and financial markets. The MENA region is thus at a critical juncture: it has articulated ambitious strategies and is progressively aligning with global disclosure norms, but sustained progress will depend on governance innovation, regional cooperation, and the ability to translate policy design into measurable, on-the-ground outcomes.

Adaptability and resilience are also critical. The pace of global change—be it in the form of disruptive technologies, volatile energy markets, or escalating climate impacts—demands a shift away from rigid, top-down planning towards more agile and adaptive governance models. This means embracing a "learning-by-doing" approach, where pilot projects are scaled up, policies are regularly reviewed and adjusted, and innovation is encouraged. The MENA region, with its young and increasingly tech-savvy population, is well positioned to leverage emerging technologies like AI and big data to enhance its adaptive capacity. For instance, AI-powered analytics can be used to improve climate risk modelling, optimize water and energy use, and enhance the transparency of ESG reporting.

The imperative for a **just and inclusive transition** cannot be overstated. The shift to a low-carbon economy will inevitably create winners and losers. Without proactive policy interventions, there is a real risk that the costs of the transition will be borne by the most vulnerable segments of society, while the benefits will be captured by a select few. Therefore, social dialogue, involving labor unions, civil society organizations, and youth representatives, must be at the heart of the sustainability discourse. The public consultations that have accompanied the development of national vision documents in some MENA countries are a positive step in this direction, but more needs to be done to ensure that the voices of marginalized communities are heard and their concerns are addressed.

Looking ahead, the upcoming **UN COP 30 in Brazil** presents a critical opportunity for the MENA region to showcase its leadership and ambition on the global stage. Building on the momentum generated by COP 27 in Egypt and COP 28 in the UAE, MENA countries can play a pivotal role in shaping the future of the global climate regime. The focus of COP 30 on the Amazon and on the perspectives of developing nations provides a strategic platform for the MENA region to forge deeper South–South partnerships, amplify its collective voice, and advocate for a more equitable, inclusive, and effective global response to the climate crisis. The insights and analyses presented in this book are intended to serve as a timely and practical resource for policymakers, negotiators, and civil society representatives from the MENA region as they prepare for this critical summit.

In conclusion, the journey to a sustainable future in the MENA region will be long and arduous, but it is a journey that must be undertaken with a renewed sense of urgency, cautious optimism, and collective resolve. The challenges ahead are immense, ranging from water scarcity and energy transitions to social inclusion and institutional reform, but so too are the opportunities. By embracing integrated governance frameworks, fostering innovation and technological adaptability, and prioritizing a just and inclusive transition that leaves no community behind, the MENA region can not only secure its own long-term prosperity and well-being but also make a decisive and enduring contribution to the global effort to build a more sustainable and equitable world. We hope that this book will serve as both a catalyst for action and a guiding framework for reflection, dialogue, and transformation across the region and beyond.

References

Bayram, H., & Öztürk, A. B. (2021). Global climate change, desertification, and its consequences in Turkey and the Middle East. In K. E. Pinkerton & W. N. Rom (Eds.), *Climate change and global public health* (pp. 445–458). Springer International.

Chen, X. H., Tee, K., Elnahass, M., & Ahmed, R. (2023). Assessing the environmental impacts of renewable energy sources: A case study on air pollution and carbon emissions in China. *Journal of Environmental Management, 345*, 118525.

Christy, A., Elnahass, M., Amezaga, J., Browne, A., & Heidrich, O. (2024). A dynamic framework to align company climate reporting and action with global climate targets. *Business Strategy and the Environment, 33*(4), 3103–3128.

ElAlfy, A., Elgharbawy, A., Driver, T. R., & Ibrahim, A. J. (2025). Sustainability disclosure in the Gulf Cooperation Council (GCC) countries: Opportunities and challenges. *Green Finance, 7*(1), 40–82.

ElAlfy, A., El-Bassiouny, D., & Cochrane, L. (2024). Shifting geopolitical sands: COP 28 and the new BRICS+. *Management & Sustainability: An Arab Review, 3*(2), 197–206.

Elnahass, M., Ahmed, Y., & Trinh, V. Q. (2024). Empowering women to lead cybersecurity: The effect of female executives on disclosure sentiment. *International Journal of Finance & Economics*.

Marie, M., Qi, B., Elgammal, M., & Elnahass, M. (2024). A more sustainable future: Can politically connected CEOs spur the nexus between ESG performance and firm financial performance? *Journal of International Financial Markets, Institutions and Money, 96*, 102056.

Matallah, S. (2022). Economic diversification and governance challenges in MENA oil exporters: A comparative analysis. *The Journal of Economic Asymmetries, 26*, e00255.

Oyegunle, A., Weber, O., & Elalfy, A. (2023). Carbon costs and credit risk in a resource-based economy: Carbon cost impact on the Z-score of Canadian TSX 260 companies. *Journal of Management and Sustainability, 13*, 187.

Qumsiyeh, M. B., & Abusarhan, M. A. (2021). Biodiversity and environmental conservation in Palestine. In M. Öztürk, V. Altay, & R. Efe (Eds.), *Biodiversity, conservation and sustainability in Asia: Vol. 1: Prospects and challenges in West Asia and Caucasus* (pp. 1–22). Springer International.

Shehabi, M. (2020). Diversification effects of energy subsidy reform in oil exporters: Illustrations from Kuwait. *Energy Policy, 138*, 110966.

Trinh, V. Q., Cao, N. D., Li, T., & Elnahass, M. (2023). Social capital, trust, and bank tail risk: The value of ESG rating and the effects of crisis shocks. *Journal of International Financial Markets, Institutions and Money, 83*, 101740.

Trinh, V. Q., Elnahass, M., & Pasiouras, F. (2025). Personal traits of CEOs and cybersecurity-related disclosure. *Journal of International Accounting, Auditing and Taxation, 58*, 100680.

Zhang, C., & Yang, J. (2024). Artificial intelligence and corporate ESG performance. *International Review of Economics & Finance, 96*, 103713.

MIX
Papier aus verantwortungsvollen Quellen
Paper from responsible sources
FSC® C105338

If you have any concerns about our products,
you can contact us on
ProductSafety@springernature.com

In case Publisher is established outside the EU,
the EU authorized representative is:
Springer Nature Customer Service Center GmbH
Europaplatz 3, 69115 Heidelberg, Germany

Printed by Libri Plureos GmbH
in Hamburg, Germany